simple modern maths 2

F C Boyde
formerly Head of Mathematics
Tower Hamlets School, London.

R A Court
Head of Mathematics
Stationers Company School.

with

A M Court

and

J C Hawdon

Nelson

Contents

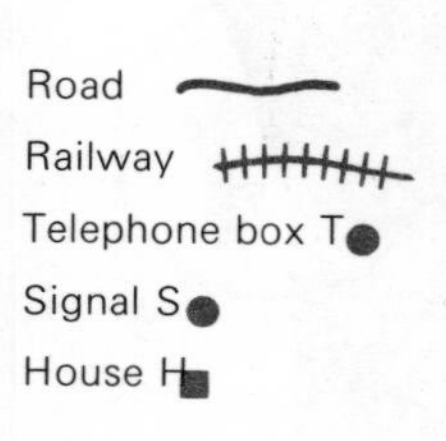

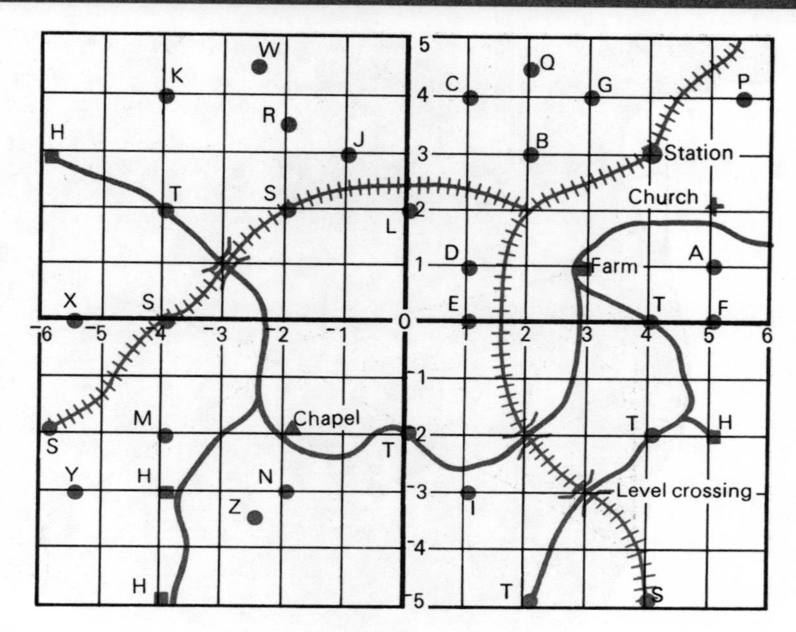

Exercise R 1

Write down the co-ordinates of

1 Station
2 Church
3 Farm
4 Railway junction
5 Point C
6 Point G
7 Point P
8 Point B
9 Point A
10 Point D
11 Point F
12 Point E
13 Point L
14 Point Q
15 Level crossing
16 Two bridges
17 Point I
18 Point K
19 Point J
20 Point W
21 Point R
22 Telephone boxes
23 Signal boxes
24 Houses
25 Point M
26 Point N
27 Point X
28 Point Y
29 Point Z
30 Chapel

On squared paper mark the horizontal axis from $^-10$ to 10, and the vertical axis from $^-10$ to 10. Mark in the following points

31 A(6,5)
32 B(2,8)
33 C(10,8)
34 D(10,2)
35 E(2,2)
36 F(6,0)
37 G(0,5)
38 H(2, $^-2$)
39 I(6, $^-5$)
40 J(10, $^-2$)
41 K(10, $^-8$)
42 L(2, $^-8$)
43 M($^-6$,5)
44 N($^-10$,2)
45 P($^-10$,8)
46 Q($^-2$,8)
47 R($^-2$, $^-2$)
48 S($^-6$,0)
49 T(0, $^-5$)
50 U($^-2$, $^-8$)
51 V($^-2$,2)
52 W($^-6$, $^-5$)
53 X($^-10$, $^-2$)
54 Y($^-10$, $^-8$)

unit 1 Finding directions

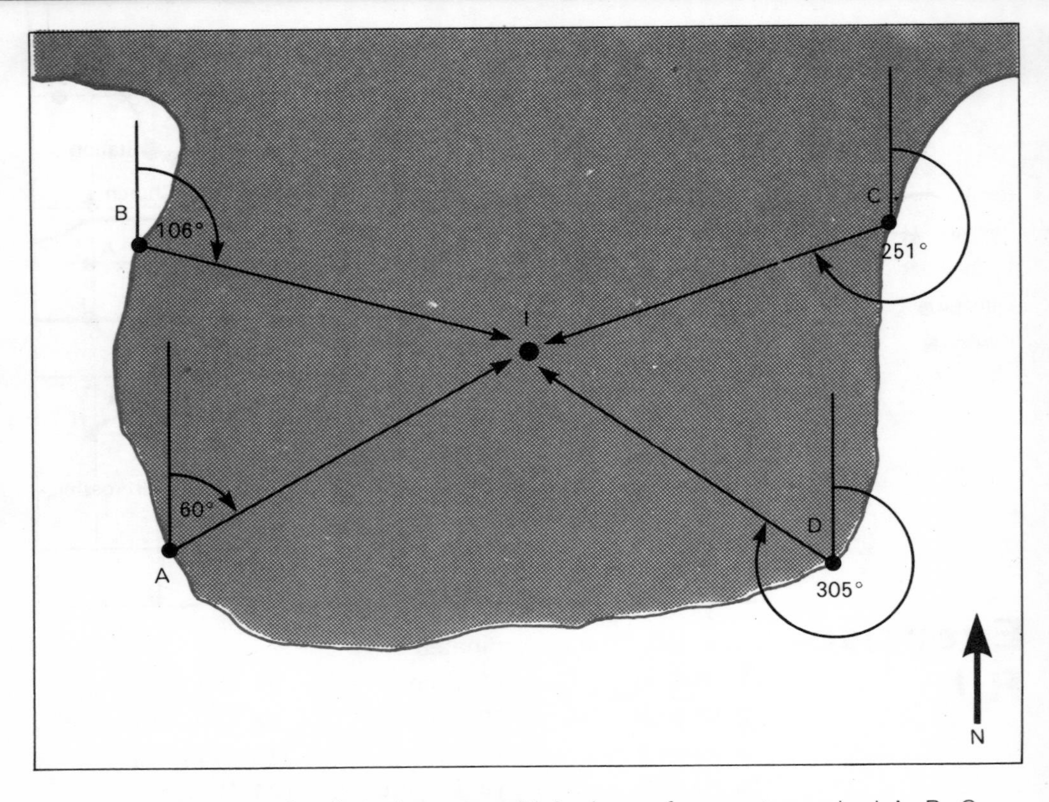

Look at the map which shows four ports marked A, B, C, and D, and an island marked I. If a sailor wanted to get to the island from A and it was a misty day how would he do it?

One way to do it would be to place a protractor on the map and measure the angle between North and the direction he wants to go. He would then use the compass on his boat to sail in the right direction.

The angle he measures is called the **bearing**. The bearing of I from A is 60°. This is often written as 060°.

The angle he sets on the compass is not quite the same as the bearing. This is because the magnetic North pole is not in the same place as the North pole. A book on map reading will tell you more about this.

Some other bearings are shown on the map.

The bearing of I from B is 106°
The bearing of I from C is 251°
The bearing of I from D is 305°

Remember bearings are always measured from North in a clockwise direction. Measure the four bearings on the map and check that you get the correct answers.

Exercise 1.1

Measure the bearings of each of these points from the dot in the centre of the diagram.

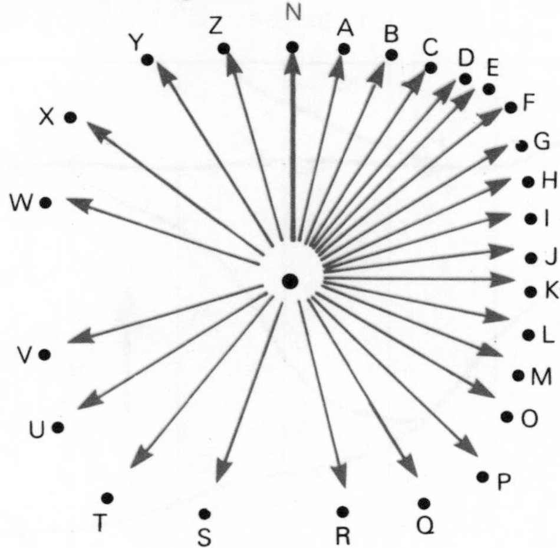

Measuring bearings and distances

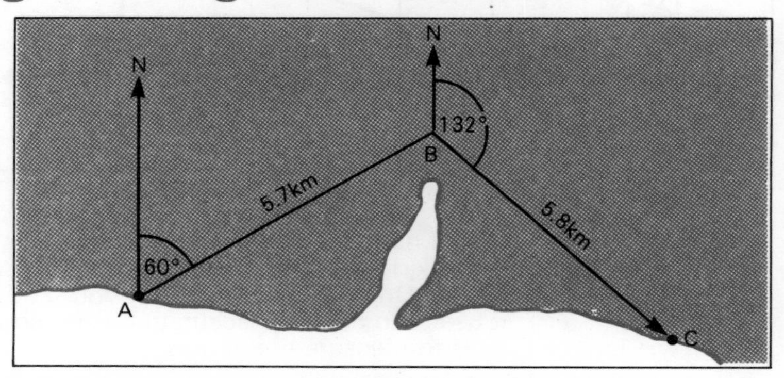

Example

Two boys are on a walking holiday along a part of the coast where there are no roads. They want to walk from A to C as quickly as possible, but not going too near the coast because of marsh land. They mark a point B on the map and decide to go from A to B, and then from B to C.

So we can say:

> They have to walk 5·7 km on a bearing of 060°,
> and then 5·8 km on a bearing of 132°.

They will use a compass to go in the right direction, but how do you think they will know when they have walked 5·7 km to B and have to change direction?

Exercise 1.2

The map shows towns marked with dots and letters, and several roads. One cm on the map represents one km. The vertical lines will help you to measure the bearings.

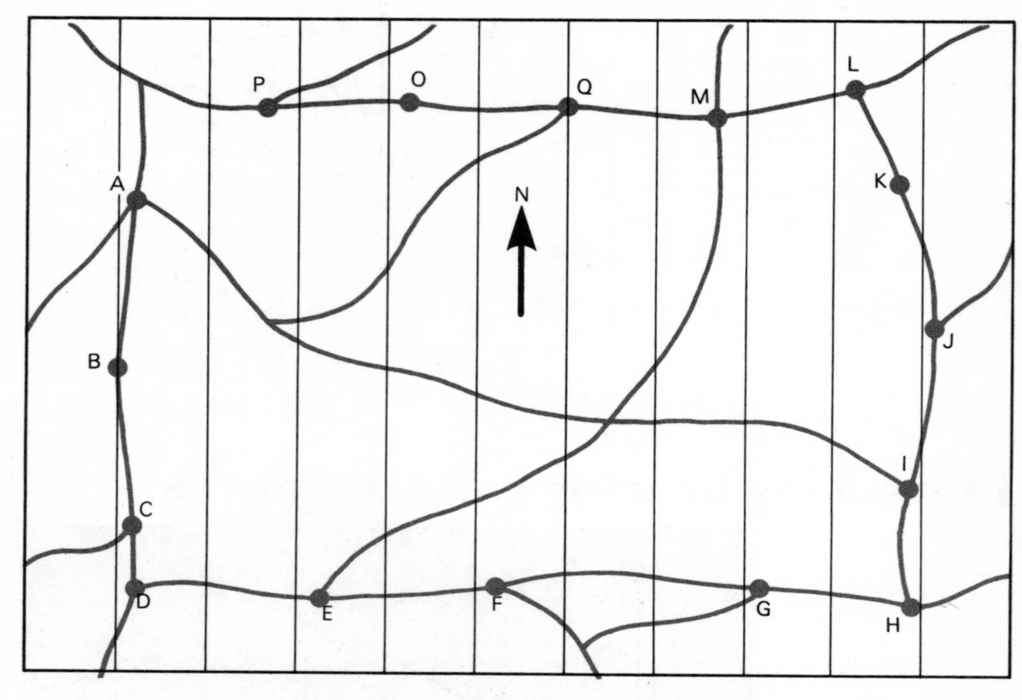

Use a ruler and protractor to measure the bearings you would have to follow, and the distances you would have to walk, to go directly between the following pairs of towns. One has been done for you. You will probably find it useful to make an accurate tracing so that you can draw lines to measure the angles.

Example

What is the distance and bearing of Q from A?
6 km at 078°. We may also write this as (6km,078°)

Write down the distance and bearing of

1 M from A
2 L from A
3 O from B
4 Q from B
5 M from B
6 P from C
7 O from C
8 Q from C
9 L from C
10 L from D

11 K from P
12 J from P
13 I from P
14 H from P
15 K from O
16 J from O
17 I from O
18 H from O
19 J from Q
20 I from L

21 G from K
22 F from K
23 E from K
24 D from K
25 F from J
26 E from J
27 D from J
28 F from I
29 E from I
30 D from I

31 D from G
32 C from G
33 B from G
34 A from G
35 C from F
36 B from F
37 A from F
38 B from E
39 A from E
40 B from D

Exercise 1.3

So far in this Unit you have had to measure angles and distances. The next examples will give you practice in drawing angles and lines.

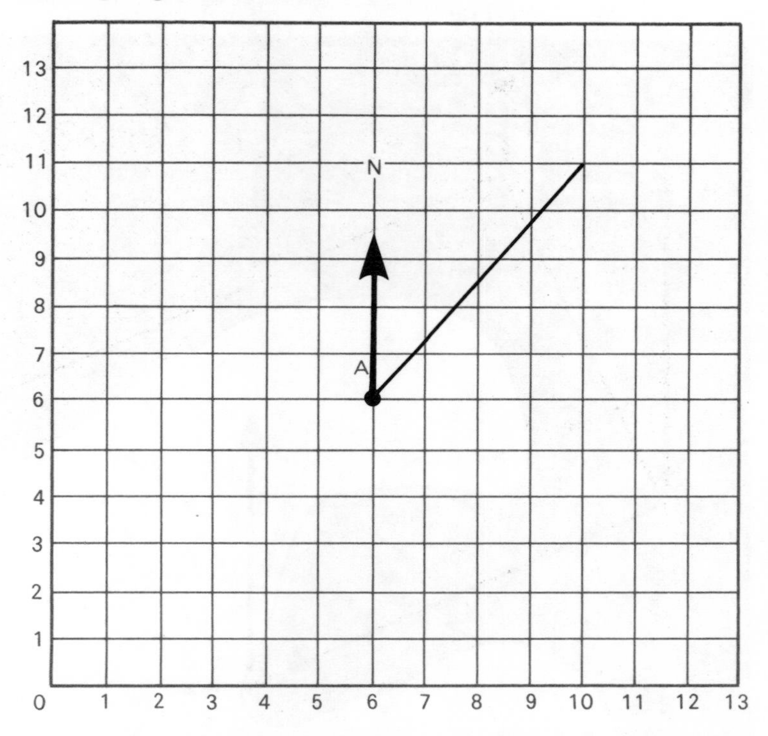

Make a copy of this on paper ruled with cm squares, and then do the following examples. The first one has been done for you.

Example

Starting from A draw a line 6·4 cm long on a bearing of 038° and write down the position of the other end.

This has been done and you will see that the other end is at (10,11)

Do the same with the following bearings and distances starting from A.

1 6·1 cm at 010°
2 5·8 cm at 031°
3 7·8 cm at 040°
4 7·1 cm at 045°
5 5·7 cm at 045°
6 6·4 cm at 051°
7 7·2 cm at 056°
8 5·8 cm at 059°
9 6·7 cm at 063°
10 7·1 cm at 082°
11 6·1 cm at 100°
12 5·4 cm at 112°
13 7·2 cm at 124°
14 6·4 cm at 142°
15 5·4 cm at 158°
16 5·9 cm at 211°
17 5·7 cm at 225°
18 6·4 cm at 232°
19 5·4 cm at 248°
20 5·1 cm at 281°
21 5·4 cm at 292°
22 7·3 cm at 304°
23 5·7 cm at 315°
24 6·7 cm at 334°
25 6·1 cm at 351°

Exercise 1.4

A boat has to be sailed from Breem to Frant. To avoid the underwater rocks which cannot be seen the boat goes from Breem to P, then from P to Q, and then from Q to Frant.

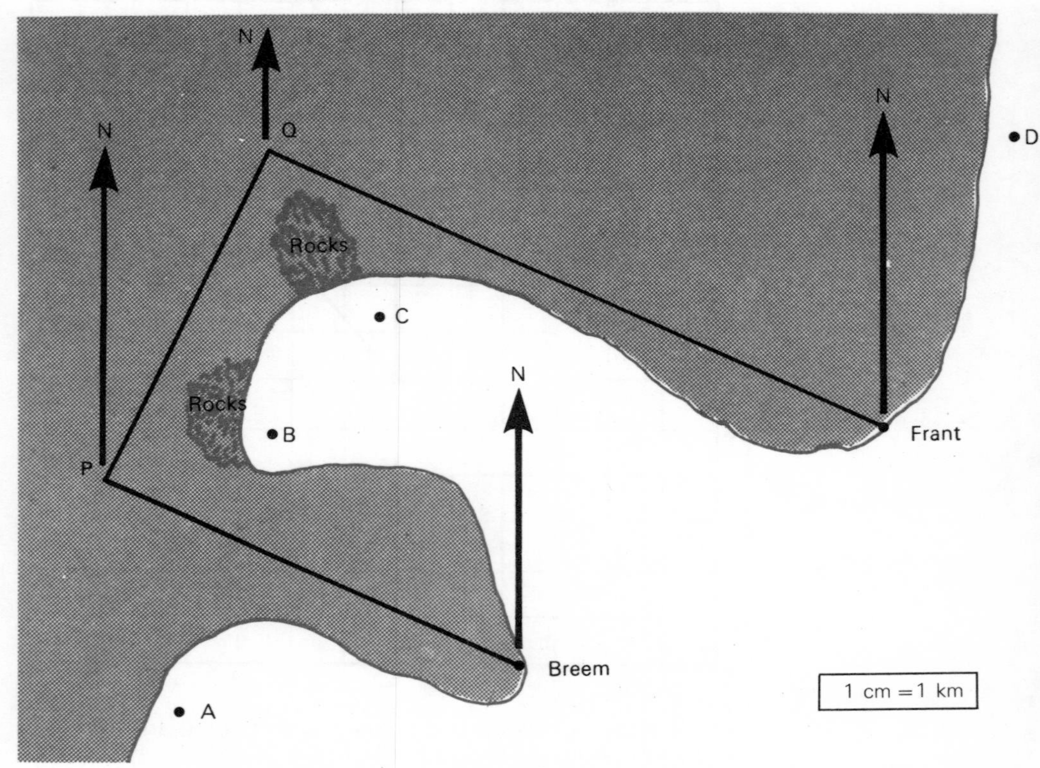

1 What is the bearing and distance of P from Breem?

2 What is the bearing and distance of Q from P.

3 What is the bearing and distance of Frant from Q?

The Captain checks when he is at P by taking the bearings of lighthouses at A and B.

4 What should be the bearing of B from P when he turns?

5 What should be the bearing of A from P when he turns?

He checks when he is at Q by taking the bearings of lighthouses at D and C.

6 What should be the bearing of C from Q when he turns?

7 What should be the bearing of D from Q when he turns?

8 Make a tracing of the map and mark on it the following course.

Breem to S. 5 km on a bearing of 289 km, then S to T, $5\frac{1}{2}$ km on a bearing of 3°, then from T to Frant.

a What is the distance and bearing of Frant from T?
b What can you say about the course you have marked?

unit 2 Movements, journeys, and vectors

If you wanted to ask someone to move from A to B you could ask them to go 5 km on a bearing of 037°
Another way of describing the movement from A to B is to say that we can go 3 along and then 4 up.

We can say: From A to B is 3 along, 4 up

It is usually written like this $\overrightarrow{AB} = \begin{pmatrix} 3 \\ 4 \end{pmatrix}$

A movement like this is often called a **vector**.

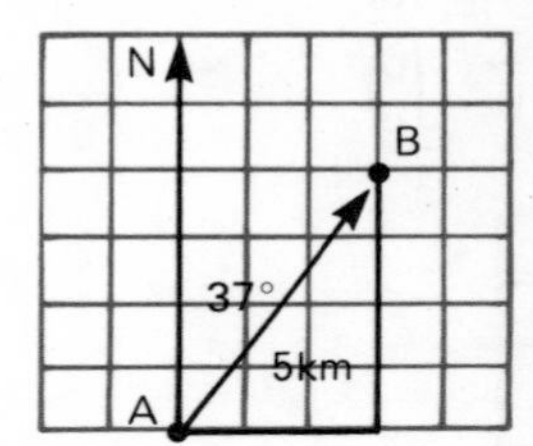

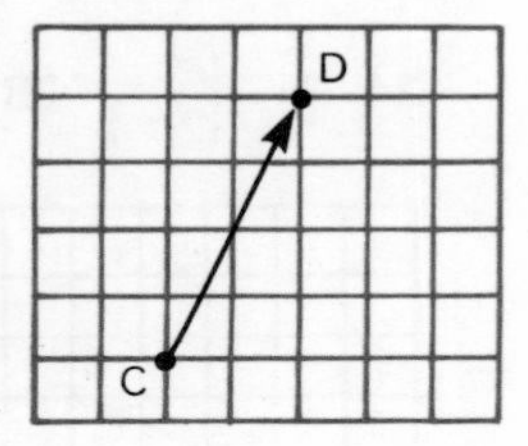

Example

Copy and complete

$\overrightarrow{CD} = \begin{pmatrix} \ \\ \ \end{pmatrix}$

From the diagram we see that from C to D is 2 along and 4 up, so we can write

$\overrightarrow{CD} = \begin{pmatrix} 2 \\ 4 \end{pmatrix}$

Exercise 2.1

Copy and complete these examples

1 $\overrightarrow{OK} = \begin{pmatrix} \ \\ \ \end{pmatrix}$	**6** $\overrightarrow{KG} = \begin{pmatrix} \ \\ \ \end{pmatrix}$	**11** $\overrightarrow{QH} = \begin{pmatrix} \ \\ \ \end{pmatrix}$	**16** $\overrightarrow{LN} = \begin{pmatrix} \ \\ \ \end{pmatrix}$
2 $\overrightarrow{OL} = \begin{pmatrix} \ \\ \ \end{pmatrix}$	**7** $\overrightarrow{KH} = \begin{pmatrix} \ \\ \ \end{pmatrix}$	**12** $\overrightarrow{QM} = \begin{pmatrix} \ \\ \ \end{pmatrix}$	**17** $\overrightarrow{RN} = \begin{pmatrix} \ \\ \ \end{pmatrix}$
3 $\overrightarrow{OF} = \begin{pmatrix} \ \\ \ \end{pmatrix}$	**8** $\overrightarrow{KI} = \begin{pmatrix} \ \\ \ \end{pmatrix}$	**13** $\overrightarrow{OP} = \begin{pmatrix} \ \\ \ \end{pmatrix}$	**18** $\overrightarrow{RI} = \begin{pmatrix} \ \\ \ \end{pmatrix}$
4 $\overrightarrow{OG} = \begin{pmatrix} \ \\ \ \end{pmatrix}$	**9** $\overrightarrow{PL} = \begin{pmatrix} \ \\ \ \end{pmatrix}$	**14** $\overrightarrow{KM} = \begin{pmatrix} \ \\ \ \end{pmatrix}$	**19** $\overrightarrow{MH} = \begin{pmatrix} \ \\ \ \end{pmatrix}$
5 $\overrightarrow{ON} = \begin{pmatrix} \ \\ \ \end{pmatrix}$	**10** $\overrightarrow{PG} = \begin{pmatrix} \ \\ \ \end{pmatrix}$	**15** $\overrightarrow{PF} = \begin{pmatrix} \ \\ \ \end{pmatrix}$	**20** $\overrightarrow{KK} = \begin{pmatrix} \ \\ \ \end{pmatrix}$

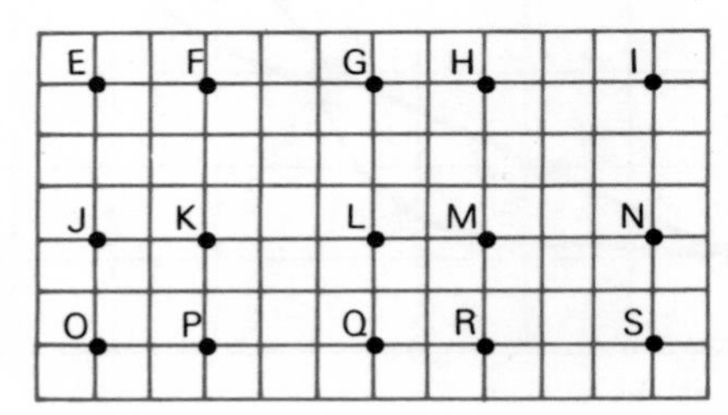

Mark a point I on some squared paper and draw these vectors. The first one has been done to show you what to do.

21 $\vec{IG}=\begin{pmatrix}4\\2\end{pmatrix}$ **25** $\vec{IK}=\begin{pmatrix}4\\0\end{pmatrix}$ **29** $\vec{IE}=\begin{pmatrix}0\\2\end{pmatrix}$

22 $\vec{IH}=\begin{pmatrix}6\\2\end{pmatrix}$ **26** $\vec{IA}=\begin{pmatrix}0\\5\end{pmatrix}$ **30** $\vec{IC}=\begin{pmatrix}4\\5\end{pmatrix}$

23 $\vec{IB}=\begin{pmatrix}2\\5\end{pmatrix}$ **27** $\vec{ID}=\begin{pmatrix}6\\5\end{pmatrix}$ **31** $\vec{IJ}=\begin{pmatrix}2\\0\end{pmatrix}$

24 $\vec{IF}=\begin{pmatrix}2\\2\end{pmatrix}$ **28** $\vec{IL}=\begin{pmatrix}6\\0\end{pmatrix}$

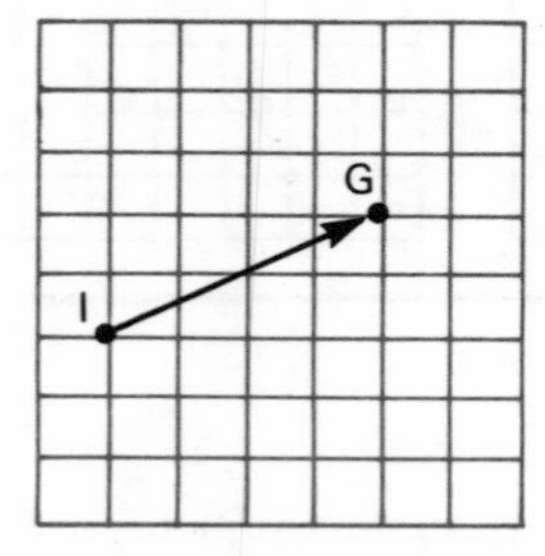

Adding vectors together

A ship goes from A to B, and then from B to C, and then from C to D.
How can we write down this journey using vectors?
One way of doing it is like this

$$\begin{pmatrix}4\\2\end{pmatrix}+\begin{pmatrix}2\\3\end{pmatrix}+\begin{pmatrix}3\\1\end{pmatrix}$$

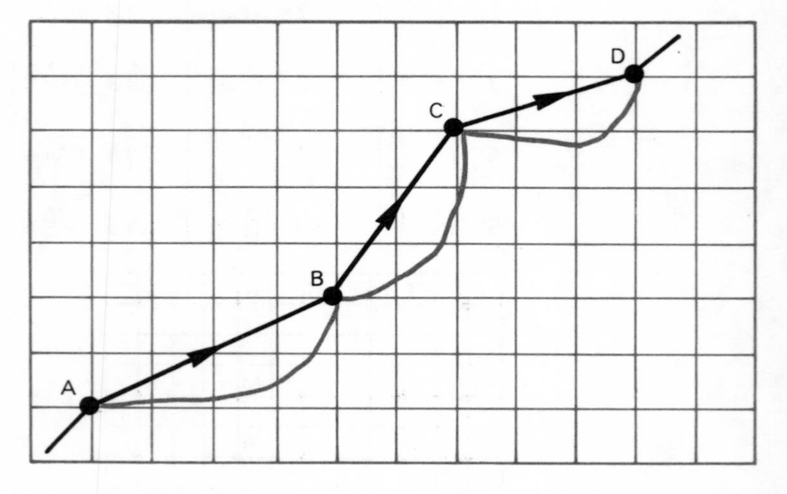

Exercise 2.2

Write down these journeys in the same way.

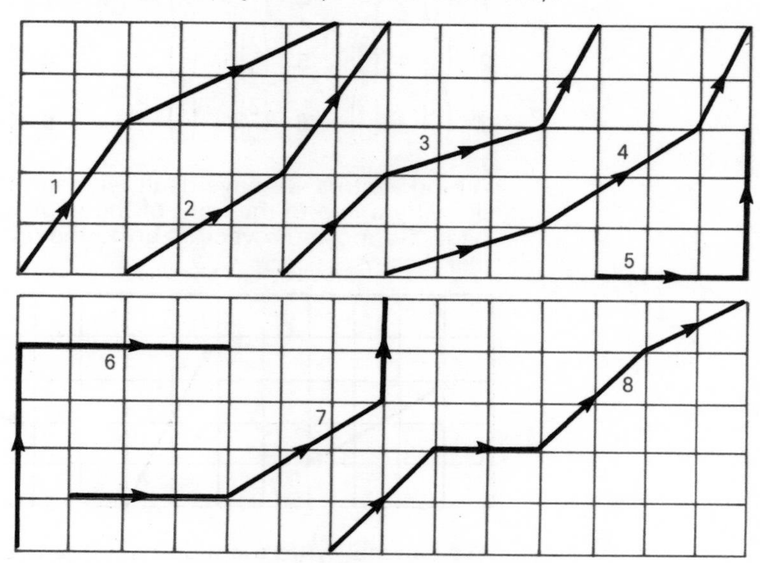

Vectors are a kind of **matrix**. What happens if we add them up?

$$\begin{pmatrix}4\\2\end{pmatrix} + \begin{pmatrix}2\\3\end{pmatrix} + \begin{pmatrix}3\\1\end{pmatrix} = \begin{pmatrix}9\\6\end{pmatrix}$$

What does this new vector stand for?
If we start at A and go 9 along and 6 up we end up at D.

The new vector we get is the vector $\overrightarrow{AD}$.

Exercise 2.3

In each of these examples

a Draw the vectors on squared paper.
b Add the vectors together to get a single vector.
c Draw this single vector starting from the same place as in (a) using a dotted line.
d Make sure that you have ended up at the same place as in (a).

One example has been done for you.

Example

$$\begin{pmatrix}3\\2\end{pmatrix} + \begin{pmatrix}1\\3\end{pmatrix} = \begin{pmatrix}4\\5\end{pmatrix}$$

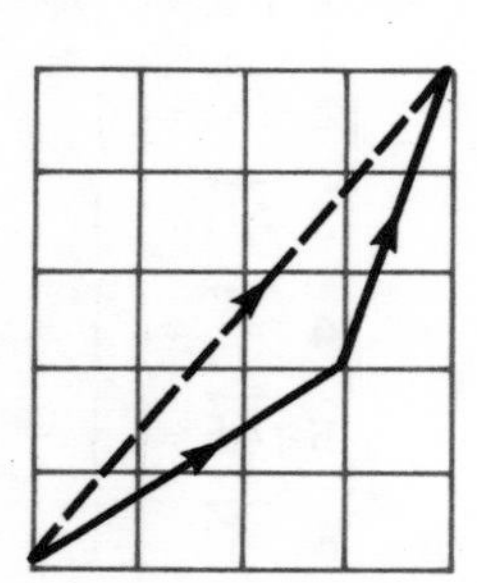

1 $\begin{pmatrix}3\\1\end{pmatrix}+\begin{pmatrix}2\\2\end{pmatrix}$ 4 $\begin{pmatrix}4\\1\end{pmatrix}+\begin{pmatrix}2\\2\end{pmatrix}+\begin{pmatrix}1\\2\end{pmatrix}$ 7 $\begin{pmatrix}0\\2\end{pmatrix}+\begin{pmatrix}3\\2\end{pmatrix}$

2 $\begin{pmatrix}1\\3\end{pmatrix}+\begin{pmatrix}3\\2\end{pmatrix}$ 5 $\begin{pmatrix}2\\4\end{pmatrix}+\begin{pmatrix}4\\1\end{pmatrix}$ 8 $\begin{pmatrix}3\\1\end{pmatrix}+\begin{pmatrix}1\\2\end{pmatrix}+\begin{pmatrix}0\\1\end{pmatrix}$

3 $\begin{pmatrix}4\\2\end{pmatrix}+\begin{pmatrix}2\\3\end{pmatrix}$ 6 $\begin{pmatrix}3\\0\end{pmatrix}+\begin{pmatrix}2\\2\end{pmatrix}$ 9 $\begin{pmatrix}1\\3\end{pmatrix}+\begin{pmatrix}2\\1\end{pmatrix}+\begin{pmatrix}4\\1\end{pmatrix}$

All the vectors we have dealt with so far have meant drawing a line to the right of the page and then up the page. Suppose the vectors go to the left or down the page like these?

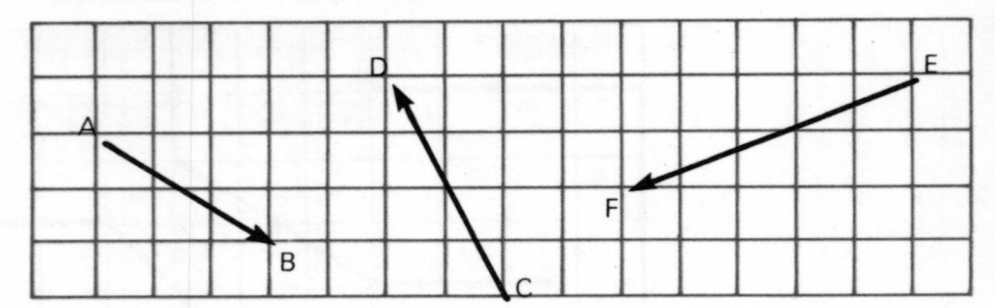

We can still write them down in the usual way if we use minus numbers to show movement to the left or down.

From A to B is 3 *to the right*, but *2 down.*

We can write $\overrightarrow{AB}=\begin{pmatrix}3\\-2\end{pmatrix}$

From C to D is *2 to the left*, and *4 up.*

We can write $\overrightarrow{CD}=\begin{pmatrix}-2\\4\end{pmatrix}$

From E to F is *5 to the left*, and *2 down.*

We can write $\overrightarrow{EF}=\begin{pmatrix}-5\\-2\end{pmatrix}$

Remember that for any movement to the left or down we must use a minus number.

Exercise 2.4

Copy and complete these examples.

1 $\overrightarrow{AB}=\begin{pmatrix}\ \\\ \end{pmatrix}$ 8 $\overrightarrow{AI}=\begin{pmatrix}\ \\\ \end{pmatrix}$ 15 $\overrightarrow{AP}=\begin{pmatrix}\ \\\ \end{pmatrix}$

2 $\overrightarrow{AC}=\begin{pmatrix}\ \\\ \end{pmatrix}$ 9 $\overrightarrow{AJ}=\begin{pmatrix}\ \\\ \end{pmatrix}$ 16 $\overrightarrow{AQ}=\begin{pmatrix}\ \\\ \end{pmatrix}$

3 $\overrightarrow{AD}=\begin{pmatrix}\ \\\ \end{pmatrix}$ 10 $\overrightarrow{AK}=\begin{pmatrix}\ \\\ \end{pmatrix}$ 17 $\overrightarrow{AR}=\begin{pmatrix}\ \\\ \end{pmatrix}$

4 $\overrightarrow{AE}=\begin{pmatrix}\ \\\ \end{pmatrix}$ 11 $\overrightarrow{AL}=\begin{pmatrix}\ \\\ \end{pmatrix}$ 18 $\overrightarrow{AS}=\begin{pmatrix}\ \\\ \end{pmatrix}$

5 $\overrightarrow{AF}=\begin{pmatrix}\ \\\ \end{pmatrix}$ 12 $\overrightarrow{AM}=\begin{pmatrix}\ \\\ \end{pmatrix}$ 19 $\overrightarrow{AT}=\begin{pmatrix}\ \\\ \end{pmatrix}$

6 $\overrightarrow{AG}=\begin{pmatrix}\ \\\ \end{pmatrix}$ 13 $\overrightarrow{AN}=\begin{pmatrix}\ \\\ \end{pmatrix}$ 20 $\overrightarrow{AU}=\begin{pmatrix}\ \\\ \end{pmatrix}$

7 $\overrightarrow{AH}=\begin{pmatrix}\ \\\ \end{pmatrix}$ 14 $\overrightarrow{AO}=\begin{pmatrix}\ \\\ \end{pmatrix}$

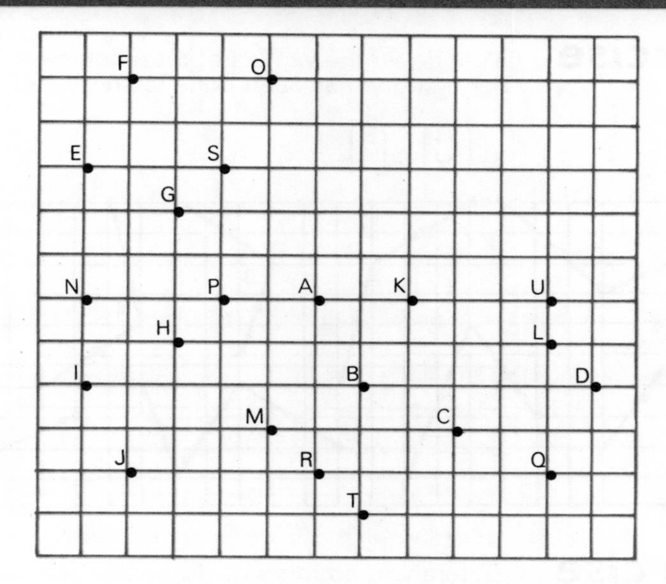

Mark a point O on some squared paper and draw these vectors. The first one has been done to show you what to do.

21 $\overrightarrow{OA} = \begin{pmatrix} 2 \\ -3 \end{pmatrix}$

22 $\overrightarrow{OB} = \begin{pmatrix} 2 \\ -1 \end{pmatrix}$

23 $\overrightarrow{OC} = \begin{pmatrix} 4 \\ -3 \end{pmatrix}$

24 $\overrightarrow{OD} = \begin{pmatrix} 5 \\ -2 \end{pmatrix}$

25 $\overrightarrow{OE} = \begin{pmatrix} -2 \\ 3 \end{pmatrix}$

26 $\overrightarrow{OF} = \begin{pmatrix} -2 \\ 1 \end{pmatrix}$

27 $\overrightarrow{OG} = \begin{pmatrix} -4 \\ 1 \end{pmatrix}$

28 $\overrightarrow{OH} = \begin{pmatrix} -4 \\ 3 \end{pmatrix}$

29 $\overrightarrow{OI} = \begin{pmatrix} -2 \\ -1 \end{pmatrix}$

30 $\overrightarrow{OJ} = \begin{pmatrix} -4 \\ -1 \end{pmatrix}$

31 $\overrightarrow{OK} = \begin{pmatrix} -4 \\ -3 \end{pmatrix}$

32 $\overrightarrow{OL} = \begin{pmatrix} 0 \\ -2 \end{pmatrix}$

33 $\overrightarrow{OM} = \begin{pmatrix} -2 \\ 0 \end{pmatrix}$

34 $\overrightarrow{ON} = \begin{pmatrix} -5 \\ 2 \end{pmatrix}$

35 $\overrightarrow{OP} = \begin{pmatrix} 4 \\ -1 \end{pmatrix}$

36 $\overrightarrow{OQ} = \begin{pmatrix} 0 \\ -3 \end{pmatrix}$

37 $\overrightarrow{OR} = \begin{pmatrix} -2 \\ -3 \end{pmatrix}$

38 $\overrightarrow{OS} = \begin{pmatrix} -5 \\ -2 \end{pmatrix}$

39 $\overrightarrow{OT} = \begin{pmatrix} -4 \\ 0 \end{pmatrix}$

40 $\overrightarrow{OU} = \begin{pmatrix} 0 \\ -1 \end{pmatrix}$

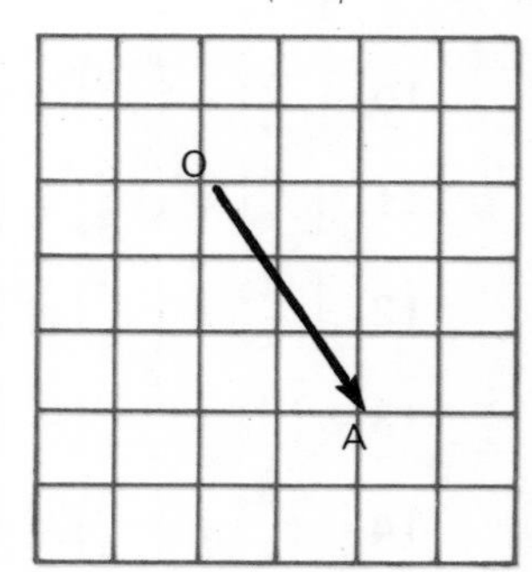

Exercise 2.5

Write down the vectors of these journeys.
The first one has been done to show you what to do.

$\begin{pmatrix} 4 \\ -2 \end{pmatrix} + \begin{pmatrix} 3 \\ 3 \end{pmatrix}$

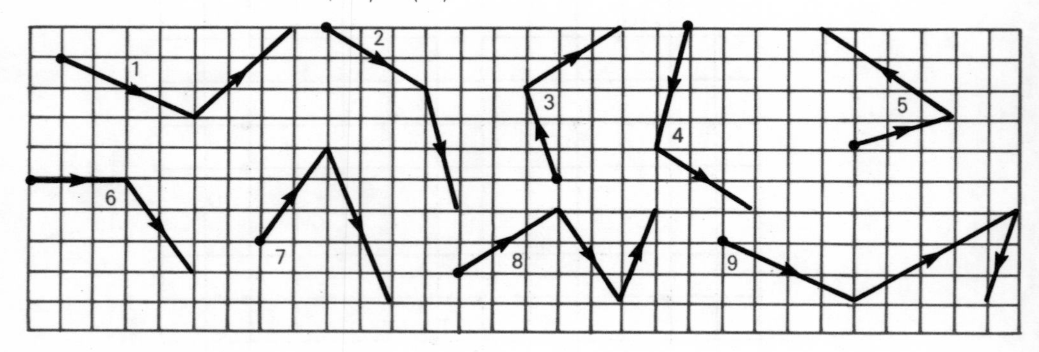

Exercise 2.6

(Before you do this exercise make sure you know how to add minus numbers. This work is done is exercise 6·1.)
In each of these examples:

a Draw the vectors on squared paper.
b Add the vectors together to get a single vector.
c Draw this single vector starting from the same place as in **(a)** using a dotted line.
d Make sure that you end up in the same place as in **(a)**

One example has been done for you.

Example

$\begin{pmatrix} 4 \\ 3 \end{pmatrix} + \begin{pmatrix} 1 \\ -2 \end{pmatrix} = \begin{pmatrix} 5 \\ 1 \end{pmatrix}$

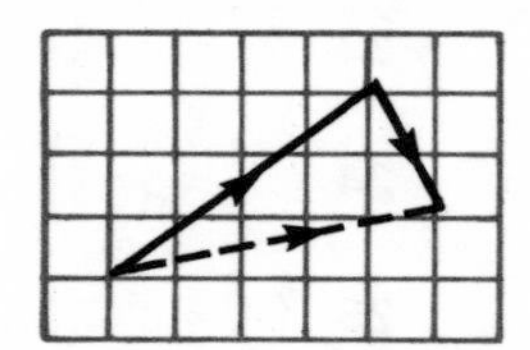

1 $\begin{pmatrix} 2 \\ 4 \end{pmatrix} + \begin{pmatrix} 2 \\ -3 \end{pmatrix}$

2 $\begin{pmatrix} 2 \\ 3 \end{pmatrix} + \begin{pmatrix} 4 \\ -2 \end{pmatrix}$

3 $\begin{pmatrix} 2 \\ -3 \end{pmatrix} + \begin{pmatrix} 3 \\ 4 \end{pmatrix}$

4 $\begin{pmatrix} 3 \\ -2 \end{pmatrix} + \begin{pmatrix} 2 \\ 3 \end{pmatrix}$

5 $\begin{pmatrix} -2 \\ -3 \end{pmatrix} + \begin{pmatrix} 6 \\ 4 \end{pmatrix}$

6 $\begin{pmatrix} -2 \\ -3 \end{pmatrix} + \begin{pmatrix} 4 \\ 1 \end{pmatrix}$

7 $\begin{pmatrix} 3 \\ -2 \end{pmatrix} + \begin{pmatrix} -4 \\ -2 \end{pmatrix}$

8 $\begin{pmatrix} 5 \\ 2 \end{pmatrix} + \begin{pmatrix} -3 \\ -4 \end{pmatrix}$

9 $\begin{pmatrix} -2 \\ -3 \end{pmatrix} + \begin{pmatrix} -3 \\ -1 \end{pmatrix}$

10 $\begin{pmatrix} 0 \\ -4 \end{pmatrix} + \begin{pmatrix} -3 \\ -1 \end{pmatrix}$

11 $\begin{pmatrix} 4 \\ -3 \end{pmatrix} + \begin{pmatrix} 2 \\ 3 \end{pmatrix} + \begin{pmatrix} -3 \\ 0 \end{pmatrix}$

12 $\begin{pmatrix} 4 \\ -2 \end{pmatrix} + \begin{pmatrix} 2 \\ 5 \end{pmatrix} + \begin{pmatrix} -4 \\ -1 \end{pmatrix}$

13 $\begin{pmatrix} 4 \\ 0 \end{pmatrix} + \begin{pmatrix} 0 \\ -4 \end{pmatrix} + \begin{pmatrix} -4 \\ 0 \end{pmatrix}$

14 $\begin{pmatrix} 3 \\ 4 \end{pmatrix} + \begin{pmatrix} 3 \\ -4 \end{pmatrix} + \begin{pmatrix} -3 \\ 0 \end{pmatrix}$

Exercise 2.7

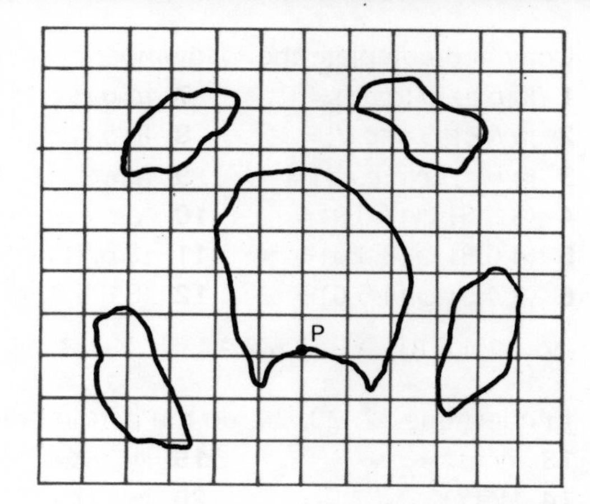

1 Make an accurate copy of this map on squared paper. Starting at P draw the journey around the large island shown by these vectors.

$$\begin{pmatrix} 2 \\ -2 \end{pmatrix} + \begin{pmatrix} 1 \\ 5 \end{pmatrix} + \begin{pmatrix} -3 \\ 3 \end{pmatrix} + \begin{pmatrix} -4 \\ -3 \end{pmatrix} + \begin{pmatrix} 3 \\ -5 \end{pmatrix} + \begin{pmatrix} 1 \\ 2 \end{pmatrix}$$

You can see if you have done it correctly because your journey should end at P, and none of the vectors should cross any of the islands.

2 Draw your own journey around the large island, and write down the vectors of the journey. Give these vectors to a friend to draw and then check to see if your drawing and his are the same.

revision 2 Sets and Venn diagrams

Exercise R 2

Copy and complete the following

1 $\{a,b,c\} \cup \{d,e,f\} =$

2 $\{a,b,c\} \cup \{b,c,d\} =$

3 $\{a,b,c\} \cup \{d,e,f,g\} =$

4 $\{3,5,6\} \cup \{7,8,9\} =$

5 $\{4,6,8\} \cup \{8,4,6\} =$

6 $\{3,4,5\} \cup \{4,5,6\} =$

7 $\{p,q,r\} \cap \{s,t,u,p,r\} =$

8 $\{p,q,r\} \cap \{q,r,s\} =$

9 $\{p,r,t\} \cap \{q,r,s,t\} =$

10 $\{p,q,r\} \cap \{s,t,u\} =$

11 $\{3,5,7\} \cap \{5,7,3\} =$

12 $\{3,8,5,0\} \cap \{2,4,6,8,10\} =$

$W = \{2,4,6,8\}$, $X = \{3,5,7,9,11\}$, $Y = \{3,4,6,8\}$, $Z = \{4,7,10\}$

In questions 22, 23, 24, do the part in brackets first.

13 $W \cup X =$

14 $W \cup Y =$

15 $X \cup Z =$

16 $X \cap Y =$

17 $W \cap Z =$

18 $W \cap Y =$

19 $W \cap X =$

20 $W \cup Y \cup Z =$

21 $X \cup Y \cup Z =$

22 $W \cup (X \cap Y) =$

23 $X \cap (Y \cup Z) =$

24 $(X \cap Y) \cup Z =$

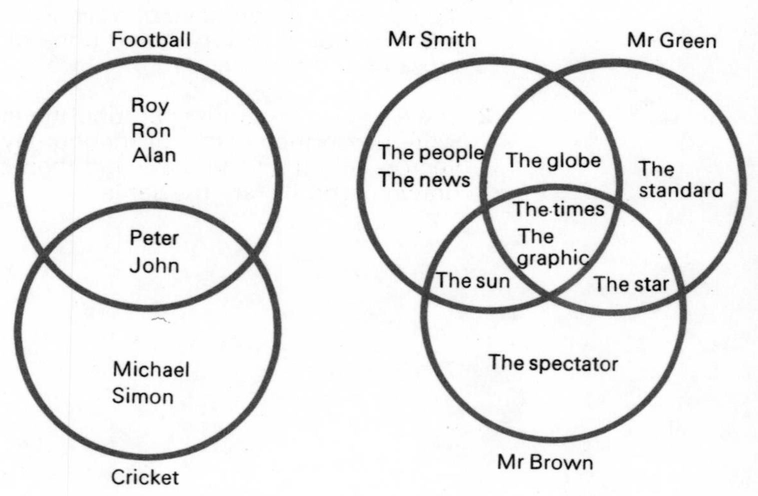

The first Venn diagram shows which boys in a class play football and cricket.

25 Which boys play only football?

26 Which boys play football?

27 Which boys play cricket only?

28 Which boys play cricket?

29 Which boys play both games?

30 Which boys play only one game?

The second Venn diagram shows the papers read by Mr Smith, Mr Green, and Mr Brown.

31 What papers does Mr Smith read?

32 Who reads *The Star*?

33 What papers are read by all three?

34 Who reads the most papers? How many does he read?

35 If all the papers cost 3p each, how much would Mr Brown's papers cost?

36 What is the total amount spent by Mr Green, Mr Brown, and Mr Smith?

Draw Venn diagrams to show the following information.

37 Bill, James, and Alan play only football; John, Simon, and Adrian play only cricket; Peter and Ron play football and cricket.

38 Phillip and Robert only swim; Paul and Andrew only play tennis; Gary, John, and Glenn swim and play tennis.

39 Joseph, Cliff, Keith, and Andrew play football; Savvas, Panos, Leo, Keith, and Andrew play cricket.

40 Brenda, Violet, Jill, and Mary swim. Joy, Jill, and Mary play tennis.

41 Only Mr Smith has *The News*, only Mr Green has *The Standard*, only Mr Brown has *The Globe* and *The Graphic*. Only Mr Smith and Mr Green have *The Sun*, only Mr Smith and Mr Brown have *The Spectator*, only Mr Brown and Mr Green have *The Times*. Mr Smith, Mr Green, and Mr Brown all have *The People* and *The Star*.

You will need to draw a Venn diagram like this for the next question.

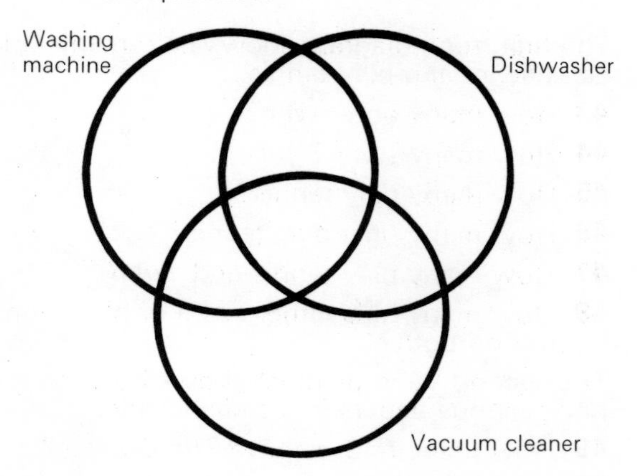

42 Mrs Atkins has only a washing machine, Mrs Smith and Mrs Jones have only a dishwasher, Mrs Hewish has only a vacuum cleaner. Mrs Wright has only a dishwasher and a washing machine, Mrs Brown has only a dishwasher and a vacuum cleaner, Mrs Austin and Mrs Nair have only a washing machine and a vacuum cleaner. Mrs Evans and Mrs Thomas have a dishwasher, a washing machine, and a vacuum cleaner.

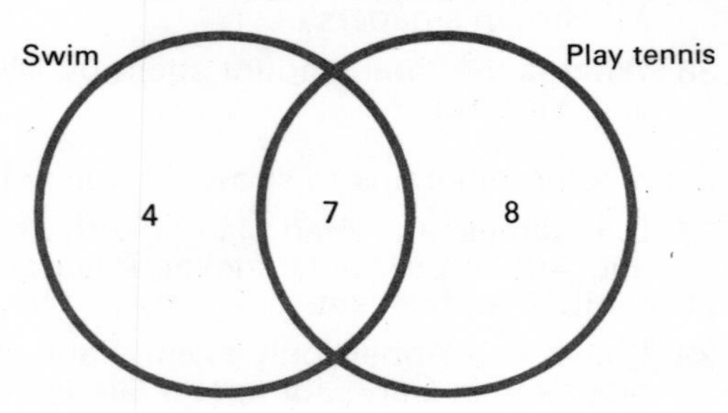

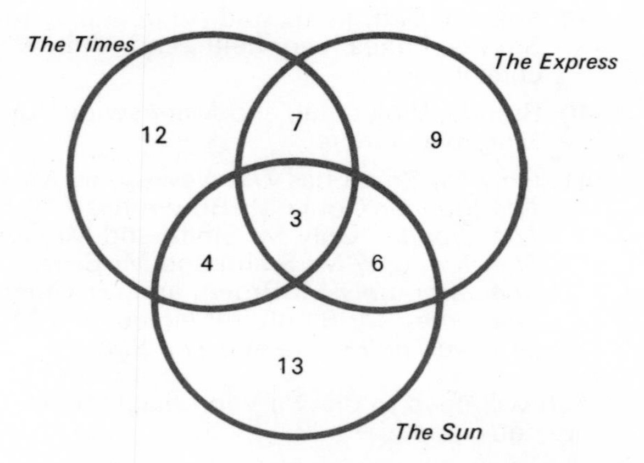

The first Venn diagram shows how many girls in a class swim and play tennis.

43 How many only swim?

44 How many swim?

45 How many play tennis?

46 How many only play tennis?

47 How many play tennis and swim?

48 How many girls either swim or play tennis, but do not do both?

The second Venn diagram shows how many houses have various papers in a small village.

49 How many houses have *The Express*?

50 How many houses have *The Express* and *The Sun*?

51 How many houses have all three papers?

52 How many houses have only one paper?

53 How many houses have two papers?

54 How many houses have three papers?

55 How many papers are sold?

Draw Venn diagrams to show the following information.

56 In a class two girls only swim, six only play tennis, five swim and play tennis.

57 In a class four boys only play football, seven only play cricket, nine play both.

58 In a class seven girls swim, eight play tennis, three swim and play tennis.

59 In a class four boys play football, twelve play cricket, four play both.

60 In a small village seven houses have only *The Times*, eight have only *The Express*, four have only *The Sun*, six have only *The Times* and *The Express*, two have only *The Express* and *The Sun*, none have *The Times* and *The Sun*, and three have all three papers.

unit 3 Sets and Venn diagrams

An advertising firm carries out a survey in a village to see how many people read the two local newspapers, *The Globe* and *The Gazette*. The results of the survey are shown below.

The number who said they read only *The Globe*	10
The number who said they read only *The Gazette*	20
The number who said they read *The Globe* and *The Gazette*	15
The number who said they read *The Globe*	25
The number who said they read *The Gazette*	33

Using the first three lines of the results we can draw a Venn diagram.

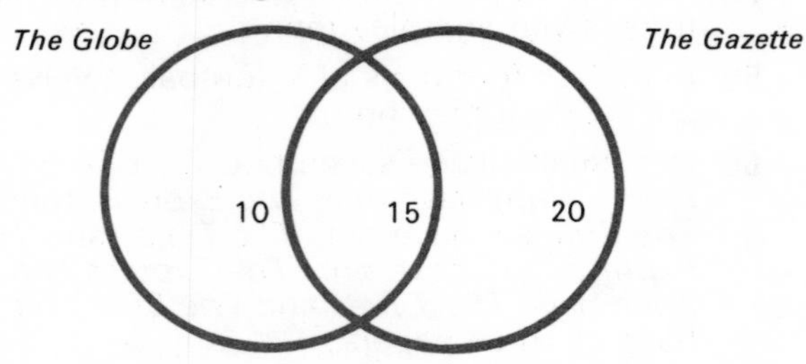

From the Venn diagram we can now see that the number who read *The Globe* is $10+15=25$. This agrees with the fourth line of the survey.

From the Venn diagram we can see that the number who read *The Gazette* is $15+20=35$. This does not agree with the fifth line of the survey.

We can see that some people did not answer all the questions correctly.

It is possible in a survey to see if people are answering the questions correctly, by asking other questions. The answers can then be checked against each other using a Venn diagram.

Exercise 3.1

	A	B	C	D	E	F
The number who said they read only *The Globe*	20	15	8	13	10	5
The number who said they read only *The Gazette*	20	10	5	14	0	11
The number who said they read *The Globe* and *The Gazette*	15	20	7	8	8	10
The number who said they read *The Globe*	35	35	15	19	18	14
The number who said they read *The Gazette*	32	30	12	22	8	20

1 The same survey was carried out in six other villages **A**, **B**, **C**, **D**, **E**, and **F** (see table on previous page). Use Venn diagrams to see in which villages all the people seemed to answer the questions correctly.

2 A garage repairs four makes of car. The types of car that each of their mechanics can repair is shown in the diagram below. This is an example of a Venn diagram which shows the intersection of four sets.

a How many mechanics are there?
b Which mechanics can repair both Ford and Renault cars?
c Which mechanics can repair both Ford, Renault, and Citroen cars?
d Which mechanics can repair all four kinds of car?
e Which mechanics can repair only two makes of car?
f Which mechanics can repair only three makes of car?

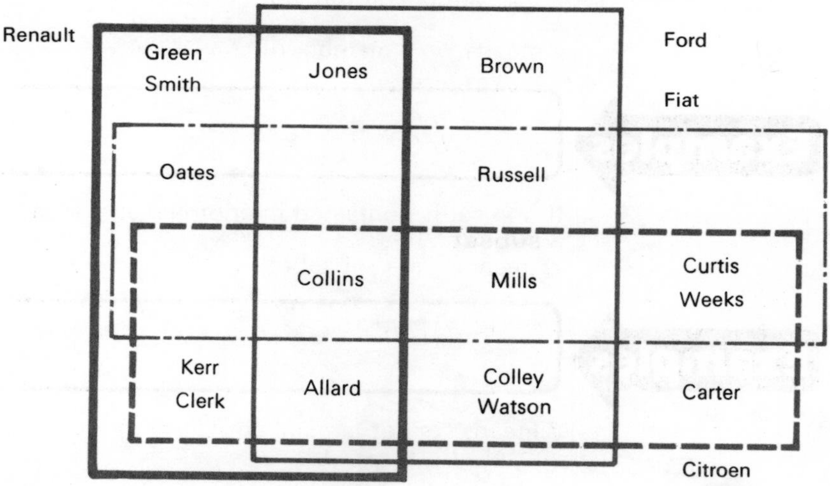

Renault
Green
Smith
Jones
Brown
Ford
Fiat
Oates
Russell
Collins
Mills
Curtis
Weeks
Kerr
Clerk
Allard
Colley
Watson
Carter
Citroen

More about sets

When we talk about sets of things we should state what collection of things the sets are taken from. This is called the **universal set**, and we shall use the letter E to stand for it.
Examples of the universal set are

$E = \{\text{All the girls in a school}\}$
$E = \{\text{The whole numbers from 1 to 10}\}$

There are other signs we sometimes use when talking about sets. To explain them we will use these sets.

$E = \{1,2,3,4,5,6,7,8,9,10\}$
$A = \{1,2,3,4\}$
$B = \{3,4,5,6,7,8\}$
$C = \{7,8\}$

We can show all these sets on a Venn diagram. It is usual to use a **rectangle** for the Universal Set

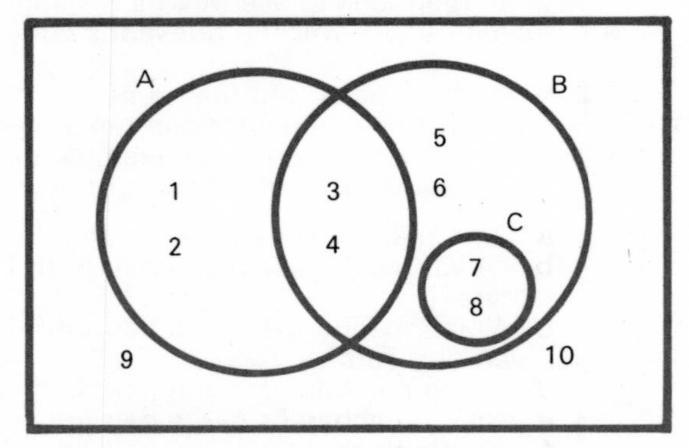

Look carefully at the Venn diagram again.

$\in$ means "is a member of"

Examples

$3 \in E$
$7 \in C$

If one set is contained in another set, we say that it is a **subset**.
$\subset$ means "is a subset of"

Examples

$C \subset B$
$\{a,b,c\} \subset \{a,b,c,d,e\}$

$\notin$ means "is not a member of" $\not\subset$ means "is not a subset of"

Examples

$6 \notin A$
$C \not\subset A$

S′ means "the set containing all the elements not in S". It is called the **complement** of S.

Examples

$A' = \{5,6,7,8,9,10\}$
$B' = \{1,2,9,10\}$

n(S) mean "the number of elements in S"

Examples

$n(A) = 4$
$n\{a,b,c,d,e,f\} = 6$

A set with no elements in is called the **empty set**.
We can write it as { } or ϕ

Examples

$A \cap C = \{\ \}$ or ϕ
$B \cap \{12, 13, 14\} = \{\ \}$ or ϕ

Exercise 3.2

$E=\{p,q,r,s,t,u,v,w,x,y\}$; $A=\{p,q,r,s\}$; $B=\{r,s,t,u,v,w\}$; $C=\{v,w\}$

Use this Venn diagram to help you answer the questions underneath.

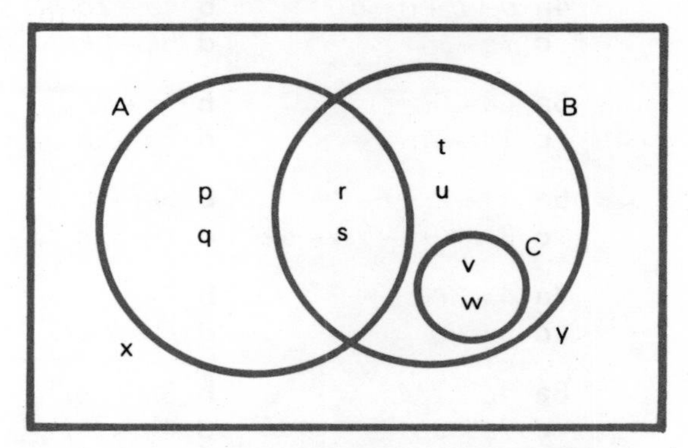

Copy and complete these examples. (Some have more than one answer.)

1 $p \in$ | **9** $p \notin$ | **17** $A' =$
2 $r \in$ | **10** $r \notin$ | **18** $B' =$
3 $t \in$ | **11** $t \notin$ | **19** $C' =$
4 $v \in$ | **12** $v \notin$ | **20** $E' =$
5 $x \in$ | **13** $x \notin$ | **21** $n(A) =$
6 $C \subset$ | **14** $A \not\subset$ | **22** $n(B) =$
7 $B \subset$ | **15** $B \not\subset$ | **23** $n(C) =$
8 $A \subset$ | **16** $C \not\subset$ | **24** $n(E) =$

Exercise R 3

Simplify the following where this is possible. Where they cannot be simplified write the letter N.

1a $a+a+a+a$ **b** $p+q+p+q$
c $l+l+m+m$ **d** $z+z+z+z+z+z$

2a $c+c+d$ **b** $c+d+c+d+c$
c $2a+5b$ **d** $5f+8f+2f$

3a $P \times Q$ **b** $C \times D$
c $6 \times g$ **d** $4 \times A \times B$

4a $d+d+d-d$ **b** $4a-2b$
c $7s-s$ **d** $6k-6k$

5a $6a-6b$ **b** $7z+4z$
c $14t-4t$ **d** $2a+3b+4c$

6a $4s+2t+t+3s$ **b** $3m+4n-n$
c $3w+9x-2w-4x$ **d** $x+3y+7y-x$

7a $4 \times 3b$ **b** $4 \times 5t$
c $9 \times pq$ **d** $P \times R \times Q$

8a $4L \times 7M$ **b** $3h \times 7m$
c $7a \times 5b$ **d** $4m \times 3n$

9a $12w \div 3$ **b** $8a \div 2$
c $10p \div 2$ **d** $15z \div 5$

10a $\frac{12b}{4}$ **b** $\frac{18u}{3}$
c $\frac{16y}{4}$ **d** $\frac{20q}{5}$

11a $\frac{7y}{y}$ **b** $\frac{24z}{z}$
c $\frac{27h}{5}$ **d** $\frac{7d}{d}$

12a $\frac{12bc}{4}$ **b** $\frac{36ab}{c}$
c $\frac{24gh}{8g}$ **d** $\frac{8pq}{2pq}$

Exercise R 4

Write the following without using indices
(Example $n^3 = n \times n \times n$)

	a	b	c	d	e	f
1	b^2	c^3	a^2	d^5	e^3	v^4
2	t^2	w^5	q^1	t^6	z^4	s^1

Simplify the following
(Example $M \times M \times M \times M = M^4$)

3	$a \times a \times a$	$c \times c \times c \times c$	$j \times j \times j$
	$n \times n$	$h \times h \times h \times h \times h$	$y \times y \times y$
4	$b \times b \times b \times b$	$A \times A \times A \times A \times A$	$D \times D$
	$e \times e \times e$	$t \times t$	$P \times P \times P \times P \times P \times P$

Find the value of the following

5	3^2	3^1	3^3
	2^4	2^3	2^2
6	2^1	3^4	5^1
	5^2	2^5	5^3
7	$2^2 + 1^2$	$4^2 + 3^2$	$4^2 + 7^2$
	$6^2 + 5^2$	$5^2 - 4^2$	$3^2 - 2^2$
8	$7^2 - 4^2$	$8^2 - 7^2$	$4^2 + 2^4$
	$5^3 + 2^4$	$1^5 + 5^1$	$2^2 + 3^2 + 4^2$

Find the value of the following if $a = 1$, $b = 2$, $c = 5$, $d = 8$

9	c^2	d^2	b^2
	a^2	b^3	a^3
10	d^3	$a^2 + b^2$	$c^2 + d^2$
	$b^2 + d^2$	$c^2 - b^2$	$c^2 - a^2$
11	$2a^2$	$4d^2$	$2a^2 + 3b^2$
	$4c^2 - 2b^2$	c^b	$2a^2b^2$

12 Write down the area of **a** and **b** using indices.

13 Write down the volume of **c** and **d** using indices.

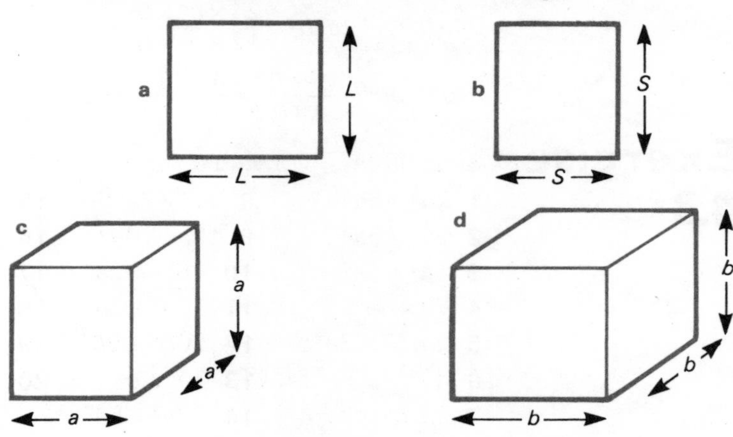

unit 4 More on indices

Is it possible to simplify formulae with indices? Can we simplify something like $a^2 \times a^3$?

If we remember that $a^2 = a \times a$, and $a^3 = a \times a \times a$, we can write

$$a^2 \times a^3 = a \times a \times a \times a \times a = a^5$$

In the same way $b^3 \times b^4 = b \times b \times b \times b \times b \times b \times b = b^7$

If you notice that $2+3=5$ and $3+4=7$ you will see that there is a quicker way of doing these than using the method above.

Exercise 4.1

Simplify the following

1 $a^2 \times a^4$
2 $b^4 \times b^3$
3 $c^5 \times c^4$
4 $d^2 \times d^2$
5 $e^3 \times e^5$
6 $f^4 \times f^4$
7 $g^3 \times g^1$
8 $g^3 \times g$
9 $h^1 \times h^4$
10 $h \times h^4$
11 $j^6 \times j^9$
12 $k^{10} \times k^{11}$
13 $m^4 \times m^5 \times m^3$
14 $n \times n^2 \times n^3$
15 $p^{10} \times p^{11} \times p^{12}$

Look at this example $a^5 \div a^3 = \frac{a^5}{a^3} = \frac{a \times a \times a \times a \times a}{a \times a \times a} = a \times a = a^2$

Here we cancel a three times just as we cancel ordinary numbers.

Notice that $5-3=2$. Can you see a quicker way of doing the example?

Exercise 4.2

Simplify the following

1 $a^5 \div a^2$
2 $b^7 \div b^3$
3 $c^8 \div c^4$
4 $d^9 \div d^5$
5 $e^7 \div e^4$
6 $f^6 \div f^3$
7 $g^5 \div g^2$
8 $h^{10} \div h^5$
9 $k^8 \div k^4$
10 $m^6 \div m^5$
11 $n^7 \div n^6$
12 $p^5 \div p^5$
13 $q^6 \div q^6$
14 $r^1 \div r^1$
15 $r \div r$

Read these examples and then do the next exercise.

$3a^2 \times 4a^3 = 3 \times 4 \times a^2 \times a^3 = 12a^5$

$12a^6 \div 6a^2 = \frac{12 \times a^6}{6 \times a^2} = 2a^4$

$5ab^3 \times 4a^2b^2 = 5 \times 4 \times a \times a^2 \times b^3 \times b^2 = 20a^3b^5$

$24a^4b^5 \div 12a^2b^2 = \frac{24 \times a^4 \times b^5}{12 \times a^2 \times b^2} = 2a^2b^3$

$a^2b^4 \div ab^6 = \frac{a^2 \times b^4}{a \times b^6} = \frac{a}{b^2}$

Exercise 4.3

Simplify the following

1 $2a^3 \times 5a^4$
2 $3a^2 \times 6a^3$
3 $5a^3 \times 4a^5$
4 $7a \times 3a^4$
5 $2a^3b^2 \times 3a^2b^2$
6 $4a^4b^3 \times 5a^3b^4$
7 $2a^3b^5 \times 7a^4b^5$
8 $ab^3 \times 7a^5b$
9 $12a^6 \div 4a^3$
10 $16a^5 \div 2a^3$
11 $18a^7 \div 3a^5$
12 $20b^4 \div 5b^3$
13 $7a^8 \div a^5$
14 $12a^7 \div 5a^2$
15 $6a^4 \div 5a^7$
16 $18a^6b^2 \div 9a^2b$
17 $24a^8b^6 \div 4a^5b^2$
18 $20d^5e^2 \div 5d^2e^7$
19 $36f^4 \div 9f^3g^7$
20 $20 \div 4a^2b^3$

Is it possible to simplify something like a^4+a^3 or d^4-c^3? The rule here is that you can add or subtract them if they have the same letters with the same indices.

Here are some that can be simplified

$s^2+s^2+s^2=3s^2$ $\quad 14d^5-11d^5=3d^5$

$4a^2+5a^2=9a^2$ $\quad 4a^2b^3+3a^2b^3=7a^2b^3$

Here are some that cannot be simplified. Compare them with the previous ones

$s^2+s^3+s^4$ $\quad 14d^5-11d^4$ $\quad a^2+b^2$

$4a^2+5a^3$ $\quad 4a^2b^3+3a^2b^2$

Exercise 4.4

Simplify the following. Where this cannot be done write the letter N.

1. $d^3+d^3+d^3$
2. $f^2+f^2+f^2+f^2$
3. g^2+g^3
4. $d^2+d^2+d^2-d^2$
5. h^5-h^3
6. $5f^3+4f^3$
7. $7h^2+2h^3$
8. $6k^2-4k^3$
9. $12w^4+2w^4$
10. $5g^2+6g^2-3g^2$
11. b^2+c^2
12. b^3-c^2
13. $5ab^2+2ab^2$
14. $3a^2b^2+2a^2b^2+a^2b^2$
15. $6a^2b^3+4a^3b^2$
16. $24ab^6-12ab^6$
17. $5a^1b^2+5ab^2$
18. $a^2+a^3+a^4$
19. $2a^2+4b^2+2a^2$
20. $27a^3b^{14}-27a^3b^{14}$

Example

The diagram shows a floor covered with tiles. Each tile is a square of side S.

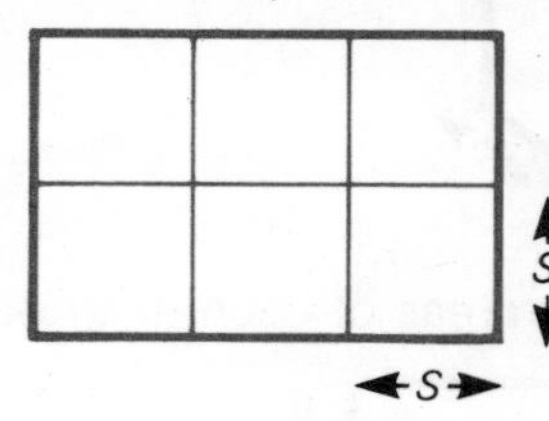

a What is the area of one tile?

b How many tiles are there?

c What is the area of the floor?

a Area of one tile $=S\times S=S^2$

b Number of tiles $=2\times 3=6$

c Area of floor $=6\times S^2=6S^2$

Another way of obtaining the last answer is to say that length $=3S$, width $=2S$
Area $=3S\times 2S=6S^2$

Exercise 4.5

In each of the following examples

a Write down the area of each tile.

b Write down the number of tiles.

c Write down the area of each floor.

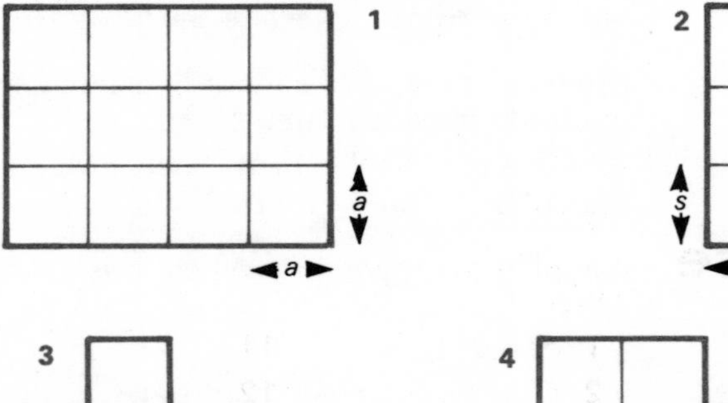

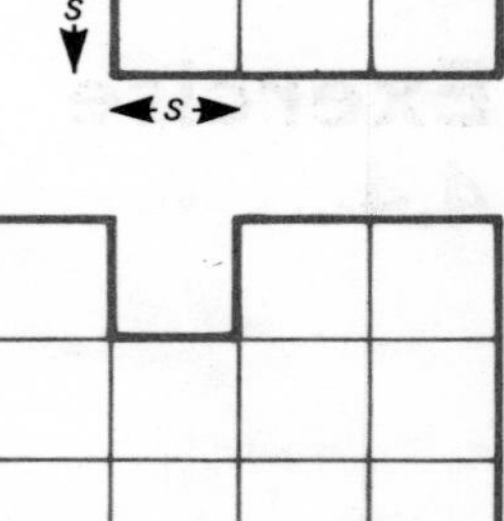

5 **a** What is the area of one of the faces of this cube?

b How many faces has this cube?

c What is the total area of the surface of this cube?

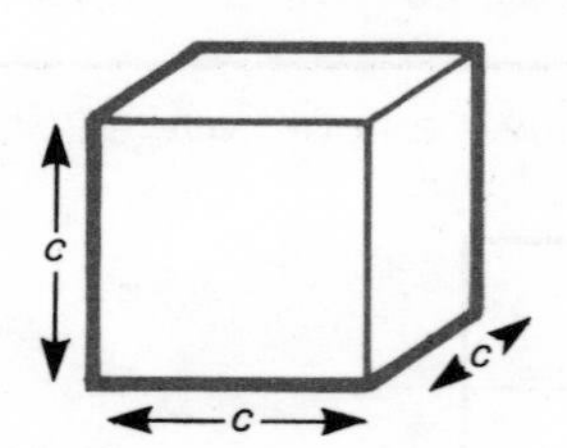

Write down the areas of each of these rectangles.

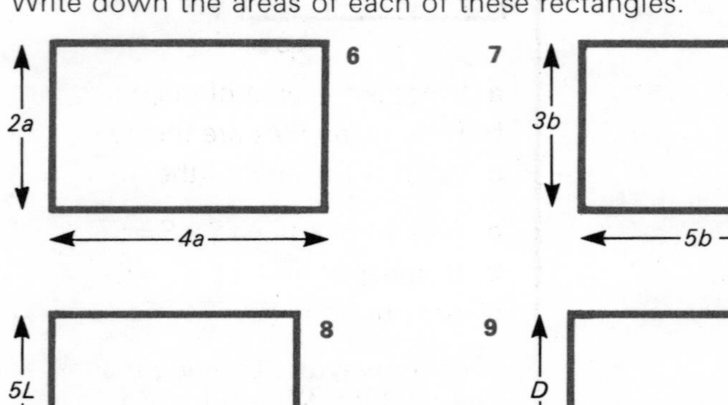

10 A wall is made of bricks shaped like cubes.

a Write down the volume of one cube.

b How many cubes are there?

c What is the total volume of the wall?

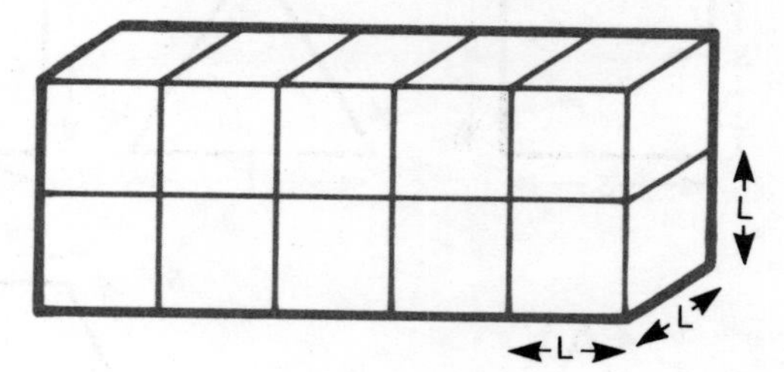

Answer the same questions about these walls.

11

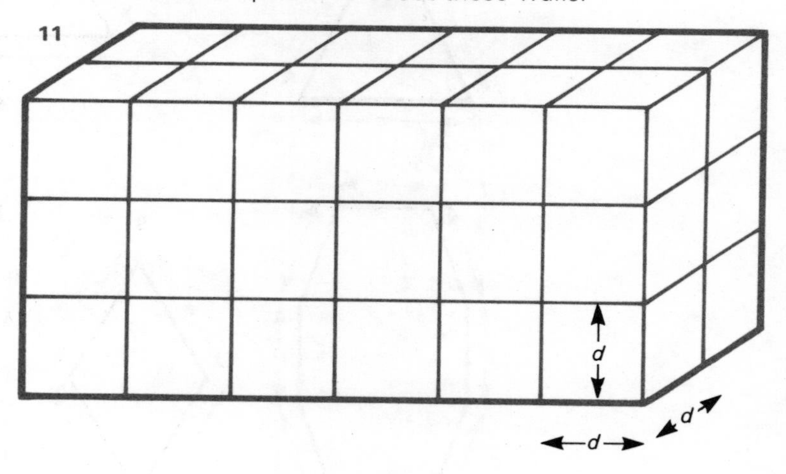

12

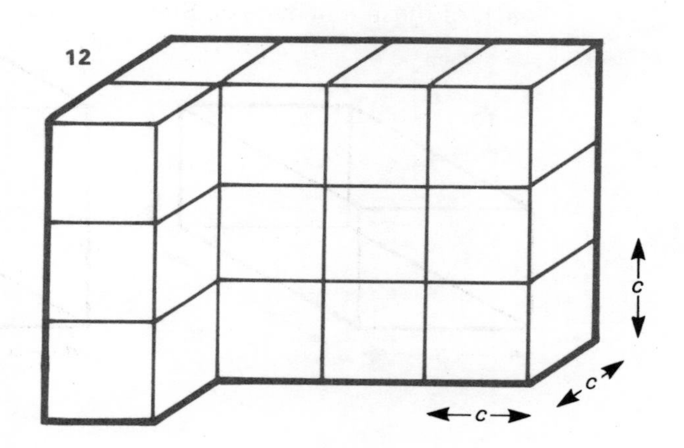

Exercise R 5

Write down formulae for the distance around these shapes. Start each answer with $D=$

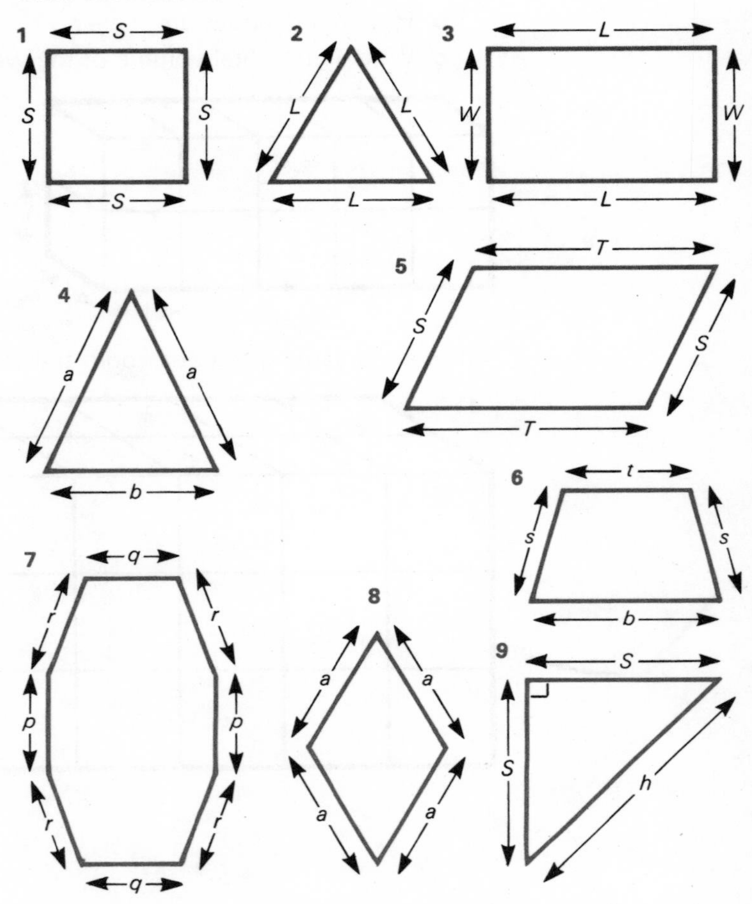

Write down formulae for the lengths of wire needed for each of these frameworks. Start each answer with $W=$

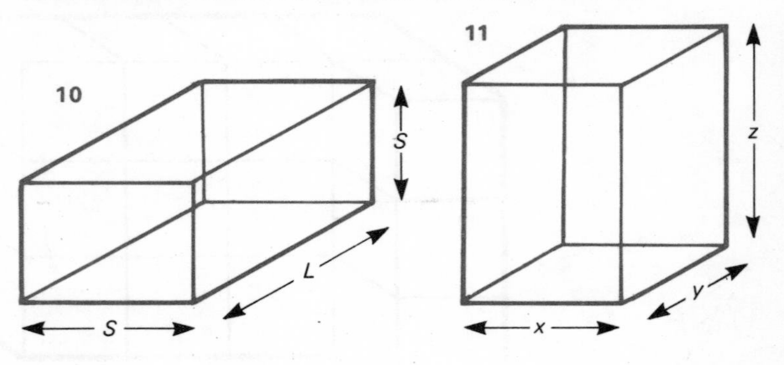

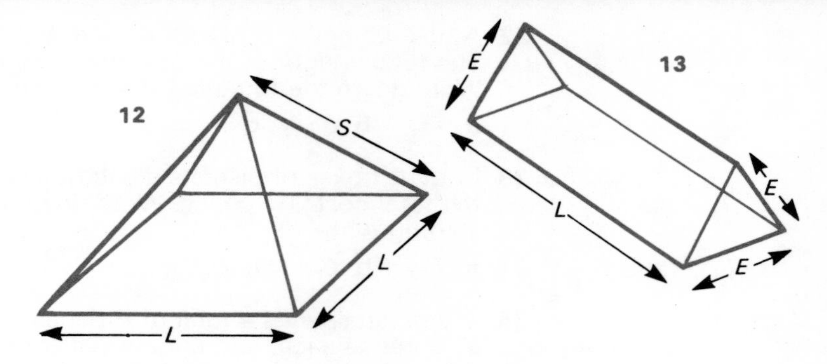

In the next seven questions four answers are given to each question. Two of them are correct and two of them are wrong. Write down the two correct ones.

14 N is the number of pupils in a class. B is the number of boys, and G is the number of girls.

$B=G-N$ $B=N-G$ $N=B+G$ $G=B-N$

15 The number of apple trees in a garden is A. The number of pear trees is P. The total number of trees is T.

$A=P-T$ $T=A+P$ $A=T+P$ $A=T-P$

16 A bottle of jam weighs G grams. The jam weighs J grams, and the empty bottle weighs B grams.

$J=B+G$ $B=J-G$ $G=J+B$ $J=G-B$

17 A man buys a sofa which costs £S and two armchairs which cost £A each. He spends a total of £T.

$T=S+2A$ $T=S+A+2$ $S=T-2A$ $S=2A-T$

18 A man has B boxes of matches. Each box contains M matches. The total number of matches is T.

$M=T-B$ $T=B+M$ $T=BM$ $B=T\div M$

19 There are C coaches on a train. Each coach carries P people. The total number of people is T.

$T=C+P$ $P=C\div T$ $C=T\div P$ $T=CP$

20 A man has C 10p coins. He changes them all into 1p coins and he finds he has Q 1p coins.

$Q=C\div 10$ $C=Q\div 10$ $Q=10C$ $C=10Q$

21 The number of men on a coach is M. The number of women is W. The total number of people on the coach is T.

a Write down the formula starting $T=$

b Write down the formula starting $W=$

c Write down the formula starting $M=$

22 A box weighing B grams is filled with S grams of sugar. The total weight of the box and the sugar is T grams. Write down the formulae starting with:

a $T=$ **b** $B=$ **c** $S=$

23 I buy B boxes of biscuits. Each box costs C pence, and the total cost is T pence. Write down the formulae starting with

a $T=$ **b** $C=$ **c** $B=$

24 A woman spends a total of P pence. She buys C cakes at A pence each, and L loaves at B pence each. Write down the formula starting $P=$

25 Find the total cost of B books at C pence each, and D pencils at E pence each.

26 Find the total cost of B books at C pence each, and B pencils. The pencils cost twice as much each as the books.

If you want to tell someone to add seven and three together you can write down

$$7+3$$

If you want to tell someone to add 7 and 3 together and then multiply the answer by 2 the usual way of writing it is

$$2\times(7+3)$$

The sum inside the brackets is always worked out first, and you would work it out like this

$$2\times(7+3)=2\times10=20$$

Here are some more examples

$3\times(9-7)=3\times2=6$

$4\times(5+2+1)=4\times8=32$

$4(3+2)=4\times5=20$ (Notice that the multiplication sign can be left out)

$\frac{1}{2}6(4+3)=\frac{1}{2}\times6\times7=\frac{1}{2}\times42=21$

$(5-2)(4+3)=3\times7=21$

Exercise 5.1

Find the value of the following.

1a $3\times(7+2)$ **b** $3\times(7-2)$ **c** $5\times(4+2)$
d $5\times(4-2)$ **e** $4\times(3+1)$

2a $4(6+8)$ **b** $3(7-4)$ **c** $4(8+3)$
d $2(9+7)$ **e** $7(6-2)$

3a $5(3+5+8)$ **b** $5(2+3+1)$ **c** $3(5+3+2)$
d $4(6+7-2)$ **e** $4(9-2-3)$

4a $8(3-1-1)$ **b** $6(3+5-2)$ **c** $8(15-5-3)$
d $4(2+5+6-3)$ **e** $3(4-1+6-3)$

5a $(2+3)\times(3+4)$ **b** $(2+1)\times(3+4)$ **c** $(2+3)\times(5-1)$
d $(3+5)\times(6-2)$ **e** $(4-1)\times(3+5)$

6a $(5-2)(7-1)$ **b** $(8-2)(4-2)$ **c** $(6-1)(7-3)$
d $(6+5)(6-1)$ **e** $(12-6)(9-2)$

7a $\frac{1}{2}(6+4)$ **b** $\frac{1}{2}(9-3)$ **c** $\frac{1}{2}(9+7)$
d $\frac{1}{2}(10-4)$ **e** $\frac{1}{2}(7+4)$

8a $\frac{1}{2}(14-9)$ **b** $\frac{1}{2}\times5\times(1+3)$ **c** $\frac{1}{2}\times8\times(2+3)$
d $\frac{1}{2}\times3\times(7+5)$ **e** $\frac{1}{2}\times8\times(9-5)$

9a $\frac{1}{2}\times5\times(2+5)$ **b** $\frac{1}{2}\times7\times(3+2)$ **c** $\frac{1}{2}\times9\times(12-7)$
d $\frac{1}{2}\times3\times(6+1)$ **e** $\frac{1}{2}\times3\times(7-1)$

If $a=7$, $b=3$, $c=4$, $d=6$, $e=1$, $f=0$; find the value of

10a $a(b+c)$ **b** $b(d-c)$ **c** $c(b+a)$
d $c(a-b)$ **e** $a(d+3)$

11a $d(c-1)$ **b** $6(a+b)$ **c** $5(7-b)$
d $d(d+f)$ **e** $f(a+b)$

12a $\frac{1}{2}(a+b)$ **b** $\frac{1}{2}(c+d)$ **c** $\frac{1}{2}(d-2)$
d $\frac{1}{2}(a-e)$ **e** $\frac{1}{2}(a+c)$

13a $\frac{1}{2}c(a+b)$ **b** $\frac{1}{2}c(a-b)$ **c** $\frac{1}{2}d(a+e)$
d $\frac{1}{2}a(b+e)$ **e** $\frac{1}{2}a(b+c)$

14a $\frac{1}{2}b(d+e)$ **b** $\frac{1}{2}a(b+f)$ **c** $(a+b)(c+d)$
d $(a-b)(c+d)$ **e** $(d-e)(a-c)$

15a $3(a+b)(a-b)$ **b** $3(d+c)(d-c)$ **c** $3(a+d)(a-d)$
d $(a+b)(a+b)$ **e** $3{\cdot}1(d+b)(d-b)$

Using formulae with brackets

Many formulae for areas, distances around shapes, etc. use brackets.

Example

The distance around a rectangle is given by the formulae

$$\mathbf{D = 2 \times (L + W)}$$

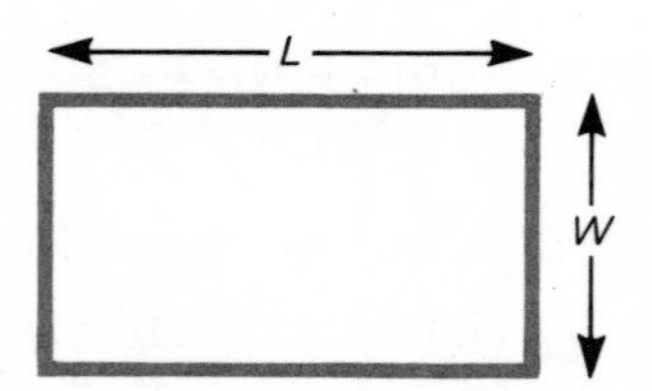

What is the distance around this rectangle?

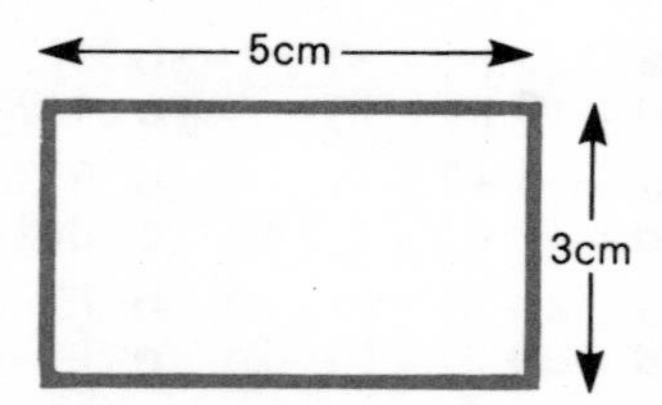

L stands for 5 and W stands for 3, so we can write

$$D = 2 \times (5+3) = 2 \times 8 = 16 \text{ cm}$$

Example

This shape is called an **ellipse**. The distance around it is given approximately by the formula

$$\mathbf{D = 3(a+b)}$$

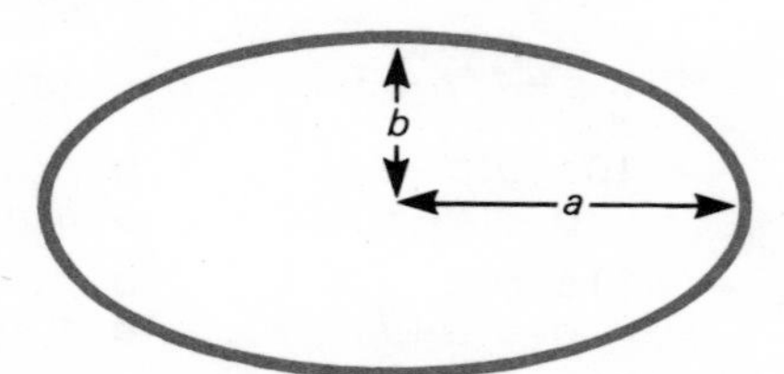

What is the distance around this ellipse?

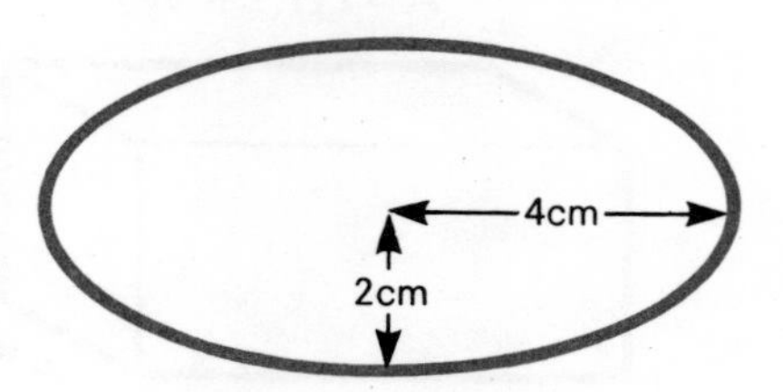

a stands for 4 and b stands for 2 so we can write

$$D = 3(4+2) = 3 \times 6 = 18 \text{ cm}$$

Exercise 5.2

1 Find the distance around these rectangles.

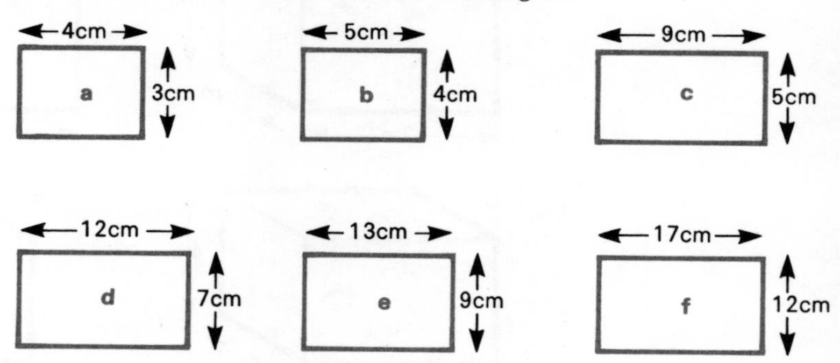

2 Find the distance around these ellipses. The measurements are in cm.

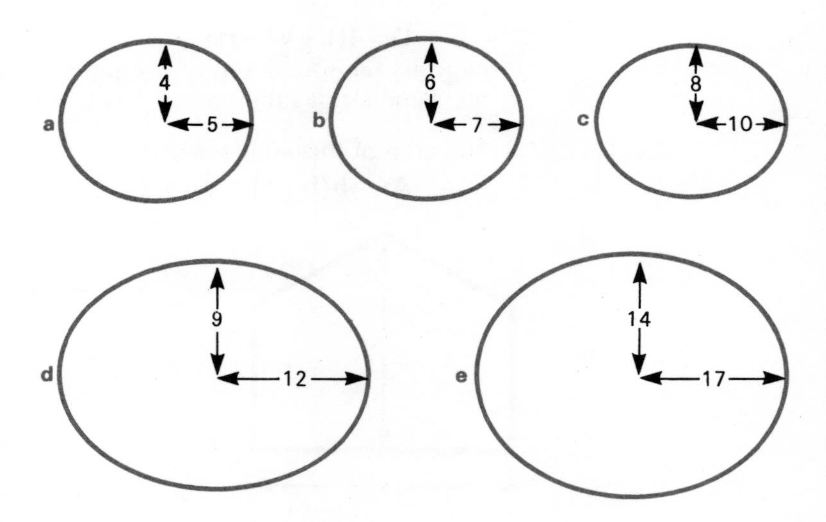

3 The area of the four walls of this room are given by the formula

$$A=2H(L+W)$$

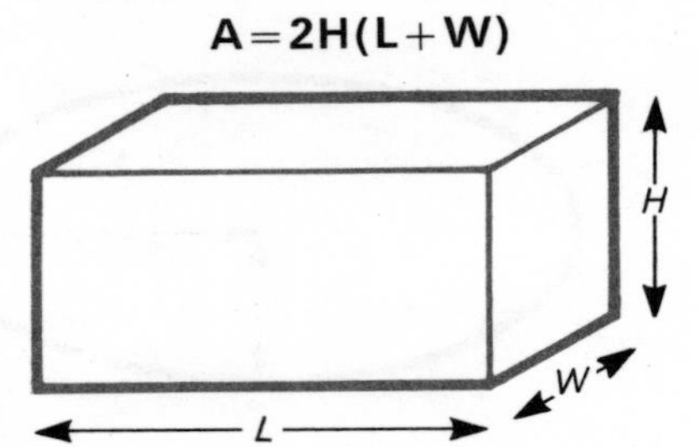

Find the areas of the walls of these rooms.

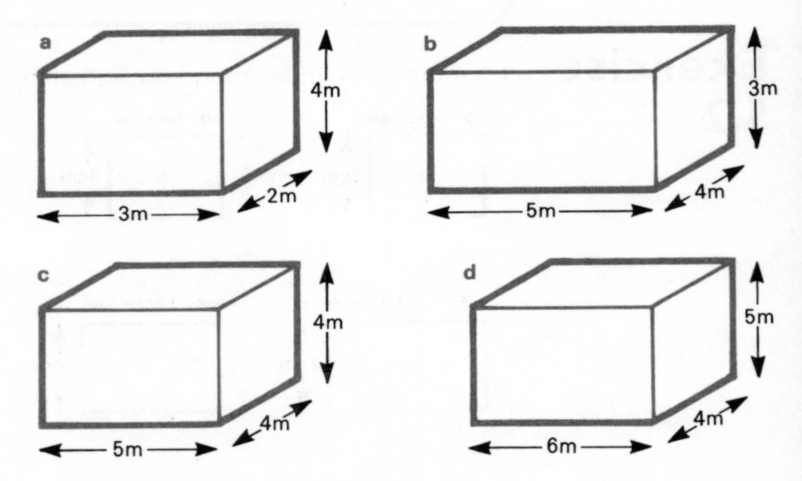

4 The length of wood needed to make a framework the same shape as the room above is given by the formula

$$D=4(L+W+H)$$

Find the lengths of wood needed to make frameworks the same size as the four rooms in question 3.

5 The area of this shape is given by the formula

$$A=\frac{1}{2}b(h+s)$$

h

s

b

Find the areas of these shapes. The measurements are given in metres.

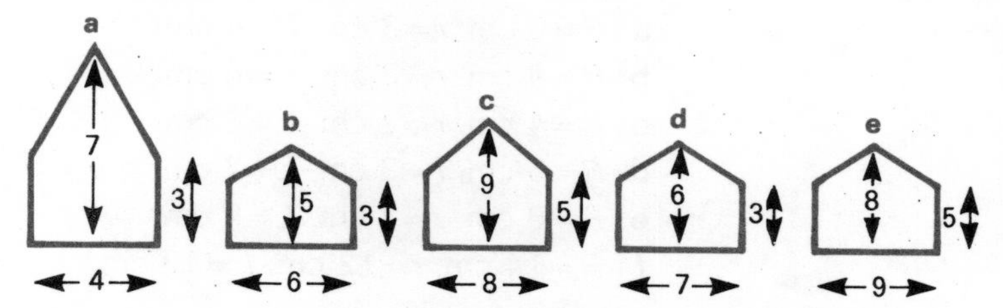

6 The shaded area between these two circles is given approximately by the formula

$$\mathbf{A = 3(c + d)(c - d)}$$

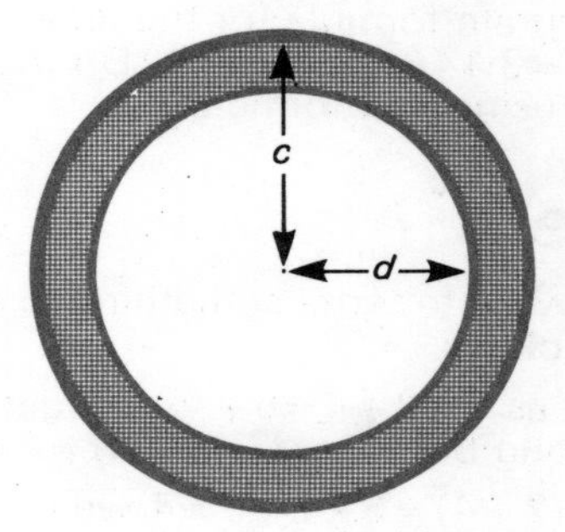

Find the areas between these circles. The measurements are in cm.

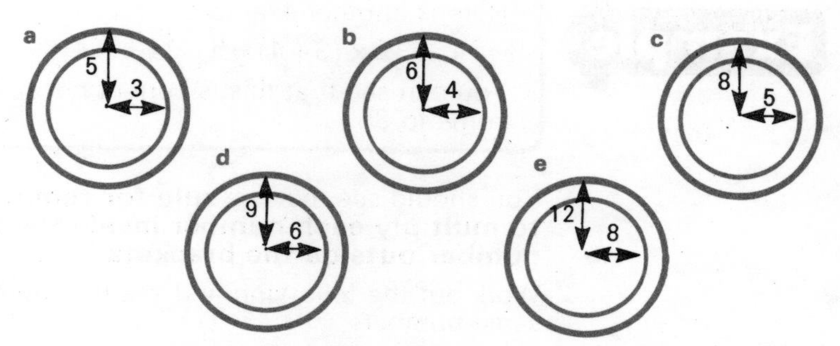

7 The weight of a hollow copper pipe is given by the formula

$$\mathbf{W = 28L(R + r)(R - r)} \text{ grams}$$

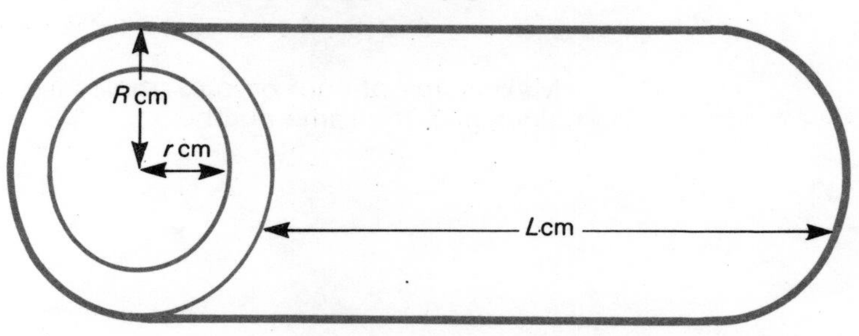

Find the weights of the copper pipes with the following measurements.

a $R=6$ cm, $r=3$ cm, $L=4$ cm.

b $R=6$ cm, $r=3$ cm, $L=5$ cm.

c $R=4$ cm, $r=2$ cm, $L=6$ cm.

d $R=5$ cm, $r=3$ cm, $L=7$ cm.

e $R=9$ cm, $r=4$ cm, $L=9$ cm.

f $R=14$ cm, $r=12$ cm, $L=13$ cm.

8 A more accurate formula for the distance around an ellipse is $D=3{\cdot}14(a+b)$. Do the examples in question **2** using this formula.

9 A more accurate formula for the area between two circles is $A=3{\cdot}14(c+d)(c-d)$. Do the examples in question 6 using this formula.

Removing brackets

Suppose we want to write something like $3\times(2+4)$ **without brackets**.

If you write it as $3\times 2+4=6+4$, you get an answer of 10, but this is wrong because $3\times(2+4)=3\times 6=18$.

If we put $3\times(2+4)=3\times 2+3\times 4$ we can see that this is correct because the right hand side comes to $6+12=18$.

Example

Here is another example

$$5\times(3+4)=5\times 3+5\times 4$$

You can see that this is correct because both sides come to 35.

You should see that the **rule for removing brackets is to multiply each number inside the brackets by the number outside the brackets**.

Work out the following and see that both sides give the same number.

a $3\times(4+2)\ =3\times 4+3\times 2$

b $5\times(7-3)\ =5\times 7-5\times 3$

c $6(2+3)\ =6\times 2+6\times 3$

d $2(3+4+5)=2\times 3+2\times 4+2\times 5$

Make some of your own examples and check that both sides give the same number.

You may also **remove brackets** in **formulae**.

Examples

a $4 \times (r+s) = 4r+4s$
b $3 \times (a-b) = 3a-3b$
c $a(b+c) = ab+ac$
d $2a(3b+d) = 6ab+2ad$
e $c(c+d) = c^2+cd$

Exercise 5.3

Write the following without using brackets.

1a $4(p+q)$ **b** $5(a+b)$
c $2(x+y)$ **d** $7(m+n)$

2a $3(a-b)$ **b** $6(t-s)$
c $4(f-e)$ **d** $5(p-q)$

3a $3(a+b+c)$ **b** $2(r+s+t)$
c $7(x+y-z)$ **d** $5(b-d+c)$

4a $b(b+d)$ **b** $r(s-t)$
c $e(f-e)$ **d** $a(p+q)$

5a $7(x+2)$ **b** $3(d+3)$
c $2(a+3b)$ **d** $4(a-3)$

6a $2(v-2w)$ **b** $2(a+2b-3c)$
c $4(s-t)$ **d** $6(2+f)$

7a $2a(a+8b)$ **b** $6v(a+2b+3c)$
c $ab(d-2e)$ **d** $5z(z+5x)$

8a $3a(2a+b)$ **b** $5t(2c-5n)$
c $ac(3t+7r)$ **d** $4w(1-e-3f)$

Factorization

$3(a+b)$ can be written instead as $3a+3b$

If we start with $3a+3b$ we can put the brackets in again like this

$$3a+3b=3(a+b)$$

Putting brackets in again is called **factorizing**. Here are some more examples.

a $5p+5q = 5(p+q)$
b $2r+2s+2t = 2(r+s+t)$
c $ae+af = a(e+f)$
d $3d+18 = 3(d+6)$
e $x^2+4x = x(x+4)$

You can check that these are correct because if you multiply out the brackets on the right hand side you should get the left hand side.

Here are some examples which cannot be factorized:
$2a+7b$ $ab-cd$ $8+3z$

Exercise 5.4

Factorize the following. Where this is not possible say so.

	a	b	c	d
1	$3a+3b$	$6p+6q$	$4x+4y$	$5m+5n$
2	$7x-7y$	$2s-2t$	$8w-8v$	$3t-3s$
3	$4a+4b+4c$	$9p+9q+9r$	$4d+4e-4f$	$5a+5b+5c$
4	$ab+ad$	$st+sr$	$de+d^2$	a^2-ae
5	$5a+5$	$5a+10$	$5a+10b$	$3x+12$
6	x^2+3x	$2x+4y+8z$	x^2-4x	$7x+2y$
7	$2ab+4a$	$6a-12ab$	$2ab+2ac$	$3n-8mn$
8	$xy+y^2$	$3a+4b$	$4+5x$	$3s-6t+3r$
9	$7v+2z$	$2a+4b+6c$	$ab-4a^2$	$2+3y$
10	$a-ab+ac$	$b-ab+ac$	$3+3y+xy$	$x+y+z$

Making up your own formulae

In **Exercise 2** you did examples using formulae that you were given. In the next exercise you will be making up your own formulae.

Example

Write down the formula for the distance around this shape. Start your answer with $D=$, and factorize your answer.

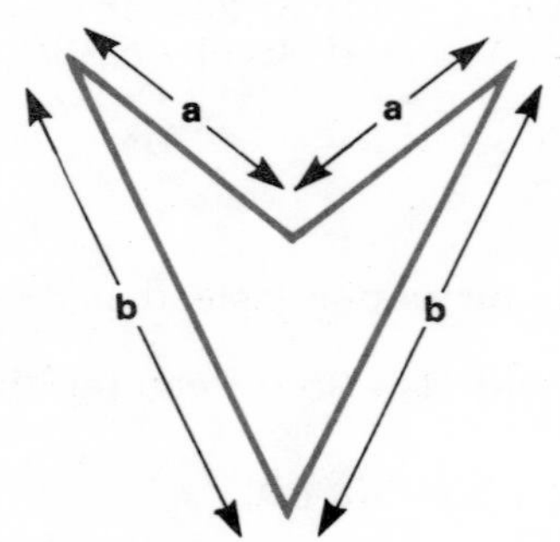

$D=a+a+b+b=2a+2b=2(a+b)$

Example

There are M men and W women on a coach. Their fares are 7p each. T pence is the total cost of all these fares. Which one of these is the correct formula?

$7=T(M+W)$ $\qquad$ $M=T(7+W)$

$T=7(M+W)$ $\qquad$ $T=7(M-W)$

T pence = cost of all fares = cost of men's fares + cost of women's fares $=7\times M+7\times W=7(M+W)$.
You can see that the third formula is the correct one.

Example

I sell A papers on Monday, B papers on Tuesday, C papers on Wednesday. The papers are sold for P pence each. If the total selling price is T write down the formula starting $T=$, and factorize your answer.

Cost of Monday's papers $= A \times P$
Cost of Tuesday's papers $= B \times P$
Cost of Wednesday's papers $= C \times P$
Total cost $= AP + BP + CP = P(A + B + C)$

Answer $T = P(A + B + C)$

Exercise 5.5

Write down formulae for the distances around these shapes. Start each formula with $D=$, and factorize your answers.

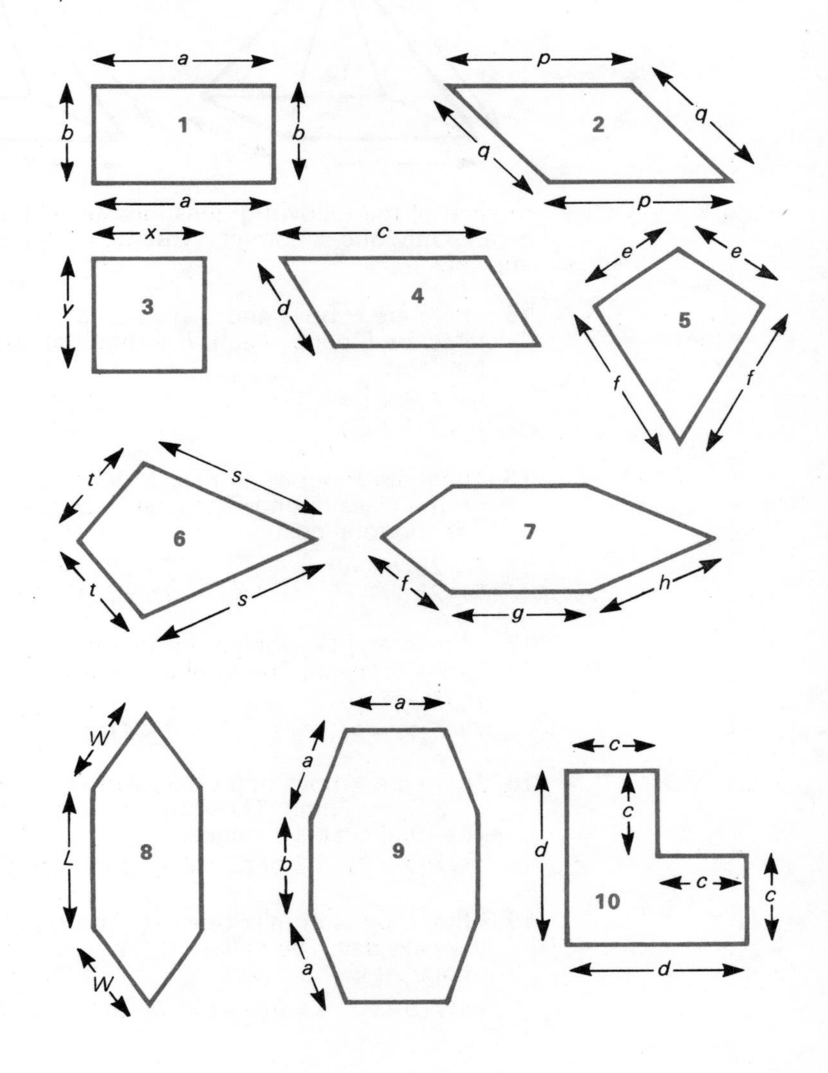

Write down formulae for the length of wire needed to make each of these frameworks. Start each formula with $D=$, and factorize your answers.

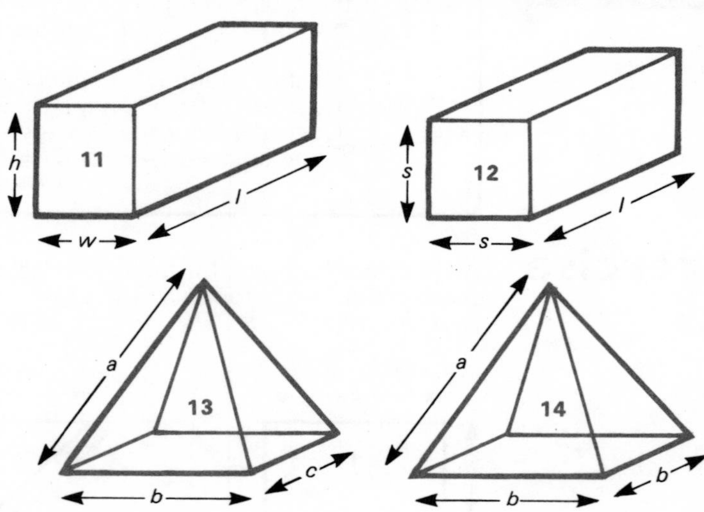

In each of the following questions several answers are given. Only one is correct. Write down the *correct* answer.

15 There are B boys and G girls on a coach. Their fares are 5 pence each. T is the total cost of all the fares.

$5=T(B+G)$ $\quad$ $T=5(B+G)$
$B=T(5+G)$ $\quad$ $G=5(T+B)$

16 There are P pupils in a class. N of them do not go on the class outing. The cost is 25 pence each. T is the total cost.

$T=25(P-N)$ $\quad$ $T=25(P+N)$
$T=25(N-P)$ $\quad$ $25=T(N+P)$

17 M men and W women go on a train journey. The fare is £F each. The total cost is £T.

$T=F(M-W)$ $\quad$ $F=T(M+W)$
$T=F(W-M)$ $\quad$ $T=F(M+W)$

18 There are Y boys in a class, Z of them do not go on a form outing. Those that go pay P pence each. The total cost is T pence.

$T=P(Z-Y)$ $\quad$ $T=P(Z+Y)$ $\quad$ $T=P(Y-Z)$ $\quad$ $P=T(Y-Z)$

19 I buy d pints of milk one day, and b pints of milk the next day. The milk cost 9 pence per pint. The total cost is T pence.

$9=T(d+b)$ $\quad$ $T=9(d-b)$ $\quad$ $T=9(b-d)$ $\quad$ $T=9(d+b)$

20 I buy X pints of milk one day, and Y pints of milk the next day. The milk costs m pence per pint. The total cost is T pence.

$T=m(X-Y)$ $T=m(X+Y)$
$m=X(T+Y)$ $T=m(Y-X)$

21 A rectangular field has a length of L metres, and a width of W metres. The distance around the field is D metres.

$D=2(L+W)$ $D=2(L-W)$
$D=2(W-L)$ $2=D(W+L)$

22 A shopkeeper has A kg of apples. He throws away B kg because they are bad. He sells the rest at P pence per kg. He sells them for a total of T pence.

$T=P(A+B)$ $T=P(A-B)$
$A=T(B-P)$ $B=T(A-P)$

23 There are B boys and G girls on a coach. Their fares are 7 pence each. The total of all the fares is T pence.

a What is the total number of boys and girls on the coach.

b Write down a formula for T. (Factorize your answer.)

24 There are M men and W women on a coach. The fare is 6 pence each. The total cost of all the fares is T.

a What is the cost of all the men's fares?

b What is the cost of all the women's fares?

c Write down a formula for T. (Factorize your answer.)

25 I buy A loaves of bread on one day, and B loaves of bread the following day. They cost C pence each, and the total cost is D pence.

a What is the cost of the loaves I buy on the first day?

b What is the cost of the loaves I buy on the second day?

c Write down a formula for D. (Factorize your answer.)

26 A shopkeeper has K kg of apples. He throws away J kg because they are bad. He sells apples for L pence per kg. The value of the remaining apples he sells is M pence.

a What is the value of the K kg of apples?

b What is the value of the apples that he throws away?

c Write down a formula for M. (Factorize your answer.)

27 I sell P newspapers on Monday, Q on Tuesday, R on Wednesday. I sell them for S pence each, and the total selling price is T pence.

a How many newspapers do I sell altogether?

b Write down a formula for T. (Factorize your answer.)

28 H men and G women go to the theatre. The tickets cost P pence plus T pence tax. The total cost of all the tickets is C pence.

a How many people go to the theatre?

b What is the total cost of each ticket?

c Write down the formula for C.

29 There are P boys in a class. N of them decide not to go on the form outing. The cost for each boy is C pence for the train fare, and M pence for a meal. The total cost of the form outing is T pence.

a How many boys go on the outing?

b What is the cost for each boy?

c Write down the formula for T.

30 A man buys P litres of paint and the shop agrees to take back what he does not use. He returns R litres to the shop. The cost of each litre of paint is C pence plus T pence tax. The total cost of the paint he uses is Y pence.

a How much paint does he use?

b What is the cost of one litre of paint?

c Write down the formula for Y.

31 A plane takes N first-class passengers and N second-class passengers. Each first-class passenger pays A pounds, and each second-class passenger pays B pounds. T is the total cost of all the tickets.

a What is the total cost of all the first-class tickets?

b What is the total cost of all the second-class tickets?

c Write down the formula for T and factorize your answer.

32 A lawn is L metres long and W metres wide. A strip S metres wide is cut from each side of the lawn. The area remaining is A square metres.

a What is the width of the lawn after the strips have been cut off?

b What is the length of the lawn after the strips have been cut off?

c Write down the formula for A.

33 I buy 12 rulers at r pence each, and 6 pencils at p pence each.

a What is the total cost of the rulers?

b What is the total cost of the pencils?

c If T pence is the total cost of the pencils and rulers, write down a formula for T and factorize your answer.

34 buy B books at 6p each, P pens at 12p each, and R rulers at 8p each.

a What is the total cost of the books?

b What is the total cost of the pens?

c What is the total cost of the rulers?

d If T pence is the total cost of the books, pens, and rulers; write down a formula for T and factorize your answer.

35 What length of wire is needed to make the framework for a box L metres, by W metres, by H metres? Factorize your answer.

revision 6 Number patterns

Exercise R 6

Find the next two numbers in each of these sequences.

1a 3, 5, 7, 9 **b** 2, 4, 8, 16
c 15, 20, 25, 30 **d** 1, 3, 9, 27

2a 16, 21, 26, 31 **b** 3, 6, 12, 24
c 0, 7, 14, 21 **d** 1, 10, 100

3a 48, 24, 12 **b** 3, 10, 17, 24
c 20, 17, 14, 11 **d** 32, 16, 8, 4

4a 56, 53, 50, 47 **b** 4, 12, 36, 108
c 9·6, 8·9, 8·2, 7·5 **d** 121·5, 40·5, 13·5, 4·5

5a 2, 3, 5, 8, 12 **b** 1, 6, 12, 19, 27
c 4, 5, 7, 10, 14 **d** 5, 9, 14, 20, 27

6a 0, 2, 6, 12, 20 **b** 28, 27, 25, 22, 18
c 4, 7, 13, 22, 34 **d** 27, 18, 11, 6

7a 2, 7, 17, 32, 52 **b** 53, 50, 45, 38
c $2\frac{1}{2}$, 3, 4, $5\frac{1}{2}$, $7\frac{1}{2}$ **d** 0·1, 10·1, 30·1, 60·1, 100·1

8a 59, 49, 36, 20, 1 **b** 2, 12, 27, 47, 72
c 1, 2, 2, 3, 3, 3 **d** 8, 9, 7, 10, 6

9 Read the lines underneath

$$
\begin{array}{ll}
5 & = 5 = 5 \times 1 \times 1 \\
5+15 & = 20 = 5 \times 2 \times 2 \\
5+15+25 & = 45 = 5 \times 3 \times 3 \\
5+15+25+35 & = 80 = 5 \times 4 \times 4
\end{array}
$$

Now copy and complete these lines

a $5+15+25+35+45=125=$

b $5+15+25+35+45+55=$

c Add up the numbers $5+15+$ . . . . . $+55+65$ using the number pattern above.

d Add up the numbers $5+15+$ . . . . . $+65+75$ using the number pattern above.

e Add up the numbers $5+15+25+$ . . . . . $+95+105$ using the number pattern.

f Add up the numbers $5+15+$ . . . . . $+295+305$ using the number pattern.

revision 7 Negative numbers

Exercise R 7

1a $4-6$	**b** $7-9$	**c** $10-8$	**d** $1-5$
e $3-9$	**f** $5-7$	**g** $8-10$	
2a $9-10$	**b** $5-5$	**c** $6-2$	**d** $2-3$
e $9-13$	**f** $0-8$	**g** $2-20$	
3a $11-4-3$	**b** $3-4-4$	**c** $10-5-5$	**d** $1-6-7$
e $16-7-4$	**f** $1-3-5$	**g** $0-6-4$	
4a $9-14+6$	**b** $9-13-5$	**c** $11-15+3$	**d** $11-15+7$
e $1-6-8$	**f** $4+7-14$	**g** $0-15+24$	
5a $-9+6$	**b** $-9-6$	**d** $-7+12$	**d** $-8+7$
e $-8-9$	**f** $-6-8-9$	**g** $-9+8+2-6$	

A man starts with nothing in the bank. What is his final balance in the following examples?

6 £10 in, £12 in, £14 out, £16 out.

7 £14 out, £12 in, £6 in.

8 £12 out, £12 out, £20 in, £2 in.

9 £8 in, £10 out, £12 in, £8 out.

10 £3 out, £4 out, £8 out.

unit 6 More on negative numbers

A number like $^{-}7$ we call a negative number.
A number like 8 we call a positive number.
Sometime we write a positive number like this: $^{+}8$ instead of 8.

Instead of writing $5-8=^{-}3$ we can write

$$^{+}5-^{+}8=^{-}3$$

Instead of writing $2+4=6$ we can write

$$^{+}2+^{+}4=^{+}6$$

There is an important difference in the way we use the + and − sign.

When it is written above the number like this $^{-}6$, or $^{+}11$ it tells us what **kind** of number we have, whether it is positive or negative.

When it is written between the numbers it tells us what to **do** with the numbers, whether to add them or take them away.

Adding negative numbers

So far you have learnt how to do sums like this

$$^{+}8+^{+}3=^{+}11$$

and $^{+}5-^{+}9 \quad =^{-}4$

Is it possible to do sums like this, and what do they mean?

$$^{+}6+^{-}2$$

If you have £6 in one bank account, and an overdraft of £2 in another, and if you add the two balances together you have a final balance of £4.

We can say $^{+}6+^{-}2=^{+}4$

Adding a negative number is the same as subtracting a positive number

Here are some more examples

$^{-}2+^{+}5=^{+}3$ $\qquad$ $^{-}4+^{+}1=^{-}3$
$^{+}4+^{-}9=^{-}5$ $\qquad$ $^{-}3+^{-}2=^{-}5$

Another way of doing examples like this is to use the number line.

Example

$^{-}8+{}^{+}6+{}^{-}7$

If you start at 0 on the number line, go 8 spaces to the left, 6 to the right, and 7 to the left you will end up at $^{-}9$.

$$^{-}8+{}^{+}6+{}^{-}7={}^{-}9$$

Exercise 6.1

You can do these examples using a number line if you want.

1a $^{+}8+{}^{-}5$ **b** $^{+}9+{}^{-}3$ **c** $^{+}7+{}^{-}4$
d $^{+}9+{}^{-}9$ **e** $^{+}6+{}^{-}7$ **f** $^{+}5+{}^{-}8$

2a $^{+}3+{}^{-}8$ **b** $^{+}1+{}^{-}4$ **c** $^{+}2+{}^{-}9$
d $^{+}7+{}^{-}10$ **e** $^{-}2+{}^{+}7$ **f** $^{-}3+{}^{+}9$

3a $^{-}5+{}^{+}8$ **b** $^{-}9+{}^{+}12$ **c** $^{-}5+{}^{+}4$
d $^{-}8+{}^{+}2$ **e** $^{-}6+{}^{+}3$ **f** $^{-}10+{}^{+}5$

4a $^{-}6+{}^{-}3$ **b** $^{-}6+{}^{-}4$ **c** $^{-}8+{}^{-}3$
d $^{-}7+{}^{-}3$ **e** $^{-}8+{}^{-}2$ **f** $^{-}9+{}^{-}0$

5a $^{+}5+{}^{+}7+{}^{-}9$ **b** $^{+}5+{}^{-}4+{}^{-}8$ **c** $^{-}2+{}^{+}5+{}^{-}12$
d $^{-}6+{}^{-}4+{}^{+}9$ **e** $^{-}13+{}^{-}8+{}^{-}7$ **f** $^{-}23+{}^{+}24+{}^{-}9$

6 A man has £23 and £17 in two bank accounts, and an overdraft of £10 in another account. What is his final balance?

7 A man has overdrafts of £5 and £15 in two bank accounts, but £11 in another account. What is his final balance?

8 A man has overdrafts of £14, £13, and £23 in three bank accounts. What is his final balance?

9 A man has £26 in a bank account, and overdrafts of £14 and £12 in two other accounts. What is his final balance?

10 A man has nothing in one bank account, £25 in a second account, and an overdraft of £25 in a third account. What is his final balance?

Taking away negative numbers

Is it possible to do sums like this? What do they mean?

$$^{+}8-{}^{-}6$$

Suppose you have an overdraft of £6 at your bank and the manager tells you that he is going to take away this overdraft. This means that he is taking away £$^{-}6$. But you are now £6 better off, so what he has really done is to give you £6.

Taking away $^{-}6$ is the same as adding $^{+}6$, so we can say

$$^{+}8-{}^{-}6={}^{+}8+{}^{+}6={}^{+}14$$

To subtract a negative number add the same positive number

Here are some more examples:

Examples

$^{+}9-{}^{-}6={}^{+}9+{}^{+}6={}^{+}15$ $\quad$ $^{+}7-{}^{-}5={}^{+}7+{}^{+}5={}^{+}12$

$^{-}7-{}^{-}9={}^{-}7+{}^{+}9={}^{+}2$ $\quad$ $^{-}10-{}^{-}7={}^{-}10+{}^{+}7={}^{-}3$

Exercise 6.2

1a $^{+}10-{}^{-}7$ **b** $^{+}8-{}^{-}6$ **c** $^{+}7-{}^{-}3$
d $^{+}5-{}^{-}5$ **e** $^{+}8-{}^{-}2$ **f** $^{+}9-{}^{-}3$

2a $^{-}3-{}^{-}8$ **b** $^{-}4-{}^{-}7$ **c** $^{-}5-{}^{-}9$
d $^{-}3-{}^{-}9$ **e** $^{-}1-{}^{-}6$ **f** $^{-}5-{}^{-}5$

3a $^{-}6-{}^{-}3$ **b** $^{-}8-{}^{-}2$ **c** $^{-}9-{}^{-}2$
d $^{-}6-{}^{-}1$ **e** $^{-}4-{}^{-}2$ **f** $^{-}7-{}^{-}1$

Read this example

$$^{+}7-{}^{+}5-{}^{-}9={}^{+}7-{}^{+}5+{}^{+}9={}^{+}11$$

4a $^{+}9-{}^{+}3-{}^{-}10$ **b** $^{+}7-{}^{+}6-{}^{-}9$ **c** $^{+}7-{}^{-}6+{}^{+}8$
d $^{-}8-{}^{-}7+{}^{+}6$ **e** $^{-}8+{}^{-}7-{}^{+}3$ **f** $^{-}9+{}^{-}5+{}^{-}6$

5a $^{-}9+{}^{-}4+{}^{+}9$ **b** $^{-}6+{}^{-}5+{}^{+}9$ **c** $^{-}4-{}^{+}5-{}^{-}5$
d $^{+}9+{}^{+}6-{}^{-}5$ **e** $^{+}8-{}^{+}7+{}^{-}4$ **f** $^{+}8-{}^{+}4-{}^{+}7$

Multiplying and dividing negative numbers

Obviously $^{+}3\times{}^{+}5={}^{+}15$ or 15

But what is the answer to $^{+}3\times{}^{-}5$?

If you have an overdraft of £5, you can say that you have £$^{-}5$ in the bank. The manager might say that he will allow you an overdraft three times that size. He will allow you to have $3\times$£$^{-}5$. But this overdraft three times as big means that you can have £$^{-}15$ in the bank.

We can say

$$^{+}3\times{}^{-}5={}^{-}15$$

In other words

A negative number times a positive number gives a negative number

Here are some more examples

Examples

$^{+}6 \times {}^{-}3 = {}^{-}18$ $\quad$ $^{-}6 \times {}^{+}5 = {}^{-}30$ $\quad$ $^{+}13 \times {}^{-}1 = {}^{-}13$

Can we now give a meaning to something like $^{-}5 \times {}^{-}4$?

What are the next 3 numbers in this set? $^{-}3 \quad {}^{-}2 \quad {}^{-}1 \quad 0$
They are obviously $^{+}1 \quad {}^{+}2 \quad {}^{+}3$

Now look at this multiplication table:

$^{-}1 \times {}^{+}3 = {}^{-}3$
$^{-}1 \times {}^{+}2 = {}^{-}2$
$^{-}1 \times {}^{+}1 = {}^{-}1$
$^{-}1 \times 0 = 0$
$^{-}1 \times {}^{-}1 =$
$^{-}1 \times {}^{-}2 =$
$^{-}1 \times {}^{-}3 =$

What are the three missing numbers? This is rather like the question that you have answered above, and we can complete the table like this.

$^{-}1 \times {}^{-}1 = {}^{+}1$
$^{-}1 \times {}^{-}2 = {}^{+}2$
$^{-}1 \times {}^{-}3 = {}^{+}3$

The rule that we have just discovered is

A negative number times a negative number gives a positive number

There are other ways of showing that this is true.

Here are some more examples:

Examples

$^{-}3 \times {}^{-}4 = {}^{+}12$ $\quad$ $^{-}5 \times {}^{-}4 = {}^{+}20$ $\quad$ $^{-}1 \times {}^{-}13 = {}^{+}13$

Exercise 6.3

1a $^{-}4 \times {}^{+}3$ **b** $^{-}2 \times {}^{-}3$ **c** $^{+}4 \times {}^{-}6$
d $^{-}3 \times {}^{+}5$ **c** $^{+}7 \times {}^{-}3$ **f** $^{-}7 \times {}^{-}2$

2a $^{+}2 \times {}^{+}7$ **b** $^{-}6 \times {}^{+}3$ **c** $^{-}8 \times {}^{-}3$
d $^{+}2 \times {}^{-}1$ **e** $^{-}4 \times {}^{-}6$ **f** $^{+}5 \times {}^{+}6$

3a $^{+}5 \times {}^{-}6$ **b** $^{-}1 \times {}^{-}8$ **c** $^{-}1 \times {}^{+}8$
d $^{+}1 \times {}^{+}8$ **e** $^{-}9 \times {}^{-}2$ **f** $^{+}5 \times {}^{-}0$

4a $^{-}2 \times {}^{-}3 \times {}^{-}4$ **b** $^{+}2 \times {}^{-}3 \times {}^{-}4$ **c** $^{+}2 \times {}^{-}3 \times {}^{+}4$
d $^{-}4 \times {}^{-}4$ **e** $({}^{-}4)^2$ **f** $({}^{-}6)^2$

5a $^-5\times{}^-6\times{}^-3$ **b** $^+4\times{}^+4\times{}^-2$ **c** $^-5\times{}^-6\times{}^+2$
d $(^-1)^2$ **e** $(^-3)^3$ **f** $(^-2)^3$

It can be shown that similar rules apply to dividing

Dividing a negative number and a positive number gives a negative number

Examples

$^+12\div{}^-4={}^-3$ $^-24\div{}^+3={}^-8$

Dividing a negative number by a negative number gives a positive number

Examples

$^-12\div{}^-3={}^+4$ $^-24\div{}^-3={}^+8$

Exercise 6.4

1a $^-16\div{}^+2$ **b** $^-15\div{}^+5$ **c** $^-18\div{}^-3$
d $^-21\div{}^-3$ **e** $^+6\div{}^-1$ **f** $^-18\div{}^+3$

2a $^+18\div{}^-3$ **b** $^+21\div{}^-3$ **c** $^+16\div{}^-2$
d $^-21\div{}^+3$ **e** $^+15\div{}^-5$ **f** $^+21\div{}^+3$

3a $^+16\div{}^+2$ **b** $^+15\div{}^+5$ **c** $^+18\div{}^+3$
d $^-16\div{}^-2$ **e** $^-8\div{}^-1$ **f** $^-15\div{}^-5$

In the next examples do the part in the brackets first like this example:

Example

$(^+3\times{}^-4)\div{}^-6={}^-12\div{}^-6={}^+2$

4a $(^+4\times{}^+3)\div{}^-2$ **b** $(^+4\times{}^-3)\div{}^+6$ **c** $(^-4\times{}^+3)\div{}^+2$
d $(^-4\times{}^-5)\div{}^-2$ **e** $(^-4\times{}^+5)\div{}^-2$ **f** $(^-4\times{}^-3)\div{}^-2$

5a $(^+8\div{}^+2)\div{}^-2$ **b** $(^+6\div{}^-2)\times{}^+5$ **c** $(^-9\div{}^+3)\times{}^+4$
d $^-2\times(^-4\times{}^-3)$ **e** $^-8\div(^+6\div{}^-2)$ **f** $(^-5\div{}^-1)\div{}^-5$

Read these examples and then do the next exercise.

Find the value of the following if $a=4$, $b={}^-2$, $c=3$, $d={}^-5$, $e={}^-7$.

$3b=3\times{}^-2={}^-6$

$ab = {}^{+}4 \times {}^{-}2 = {}^{-}8$
$a+b = {}^{+}4 + {}^{-}2 = {}^{+}2$
$d+e = {}^{-}5 + {}^{-}7 = {}^{-}12$
$3a+5b = {}^{+}12 + {}^{-}10 = {}^{+}2$
$3a-5b = {}^{+}12 - {}^{-}10 = {}^{+}12 + {}^{+}10 = {}^{+}22$
$(d+e) \div b = {}^{-}12 \div {}^{-}2 = {}^{+}6$
$b^2 - d^2 = ({}^{-}2 \times {}^{-}2) - ({}^{-}5 \times {}^{-}5) = {}^{+}4 - {}^{+}25 = {}^{-}21$

Exercise 6.5

Find the value of the following

1a $4a$ **b** $2b$ **c** $4d$
d $3e$ **e** bc **f** cd

2a de **b** $c+b$ **c** $a+d$
d $e+a$ **e** $a-b$ **f** $c-d$

3a $b-d$ **b** $d-b$ **c** $5a+3b$
d $4a+3e$ **e** $2c+b$ **f** $5c+6b$

4a $3b-2d$ **b** $2e-3b$ **c** $2b-3e$
d $(d+e) \div a$ **e** $(d-e) \times a$ **f** abc

5a cde **b** e^2+d^2 **c** d^2-e^2
d b^3 **e** $ab+cd$ **f** $ad-bc$

Some practical uses of negative numbers

Exercise 6.6

1 If a flare is shot upwards with a starting speed of 50 m per second, its speed after t seconds is given by $S = 50 - 10t$.

For example, its speed after 2 seconds is $S = 50 - (10 \times 2) = 50 - 20 = 30$ m per sec.

a What is its speed after 3 seconds?

b What is its speed after 4 seconds?

c What is its speed after 5 seconds?

d Explain the answer to **c**.

e What is its speed after 7 seconds?

f Explain the answer to **e**.

2 The distance of the same flare above the ground after t seconds is given by $D = 50t - 5t^2$.

For example, its distance above the ground after 2 seconds is $D = (50 \times 2) - (5 \times 2 \times 2) = 100 - 20 = 80$ m.

a What is its distance above the ground after 3 secs?

b What is its distance above the ground after 4 secs?

c What is its distance above the ground after 5 secs?

d What is its distance above the ground after 6 secs?

e What is its distance above the ground after 7 secs?

f Look at the answers to **a** to **e** and explain what is happening to the flare.

g What is its distance above the ground after 10 secs?

h Explain the answer to **g**.

i What is its distance above the ground after 11 secs?

j What does the answer to **i** mean? Assume that the flare has been shot up at the edge of a cliff.

3

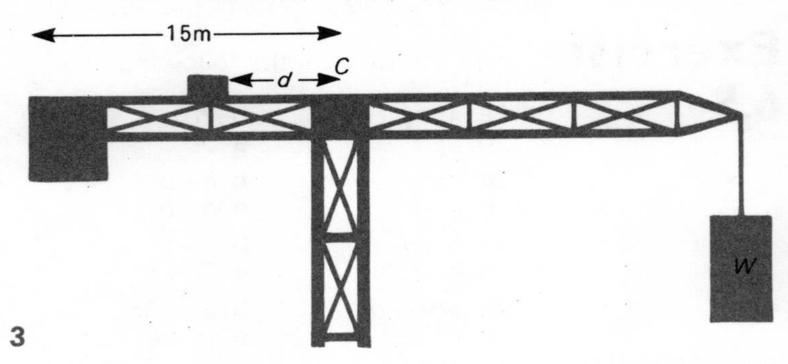

Before the crane shown in the diagram lifts a weight of W tonnes, a counterweight moves a distance d metres from the point C. The distance d is given by the formula

$$d=\tfrac{1}{2}W-5 \text{ metres.}$$

a What is the distance d when the crane has to lift 20 tonnes?

b What is the distance d when the crane has to lift 30 tonnes?

c What is the distance d when the crane has to lift 40 tonnes?

d What is the largest load that the crane can lift?

e What is the distance d when the crane has to lift 10 tonnes?

f What is the distance d when the crane has to lift 4 tonnes?

g Explain the answer you get in **f**.

h Where will the counterweight be when there is no load on the crane? (Put $W=0$ in the formula.)

Exercise R 8

If $a=9$, $b=4$, $c=2$, $d=1$, $e=0$. Find the value of

	a	b	c	d	e
1	$4a$	$5b$	$7d$	$6e$	$9a$
2	$a+b$	$c+d$	$d+e$	$a+e$	$a+d$
3	$b+5$	$7+a$	$6+e$	$d+11$	$b+9$
4	$a-b$	$c-d$	$b-a$	$c-e$	$a-c$
5	$2a+5b$	$4c+3d$	$e-17$	$2e+4d$	$3a+b$
6	$2c-b$	$4c-3d$	$2b+3c$	$2e-4d$	$3b-a$
7	ab	bc	$2b-3c$	bd	be
8	$3ab$	$4bc$	ac	$7cd$	$5ac$
9	$\frac{b+c}{2}$	$\frac{a+8}{b-c}$	$\frac{b-d}{b-c}$	$\frac{2a+4}{b}$	$\frac{3a+d}{c+2d}$
10	$\frac{a+c}{b}$	$\frac{a+b}{c}$	$\frac{2b+c}{a}$	$\frac{a+b+c}{c}$	$\frac{2a+b-c}{c+4}$

11 The cost of carpet for a room with a length of L metres, and a width of W metres is given by the formula cost $=5LW$ pounds. Find the cost of buying carpet for rooms of the following size.

a 2 m by 3 m, **b** 3 m by 4 m, **c** 2 m by 4 m, **d** 2 m by 5 m, **e** 6 m by 7 m.

12 The cost of delivering T tonnes of coal is given by the formula $C=2+20T$. Find the cost of delivering the following weights of coal.

a 1 tonne, **b** 2 tonnes, **c** 3 tonnes, **d** 5 tonnes, **e** 9 tonnes.

13 The area of this shape is given approximately by the formula $A=3ab$.

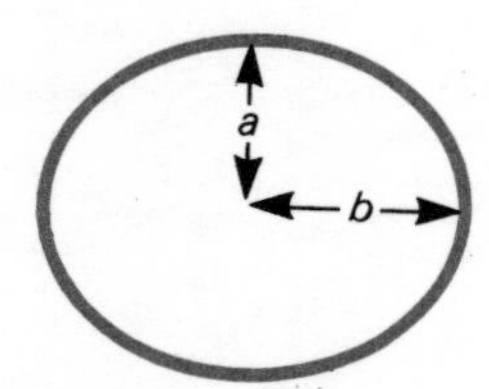

Find the areas of these shapes.

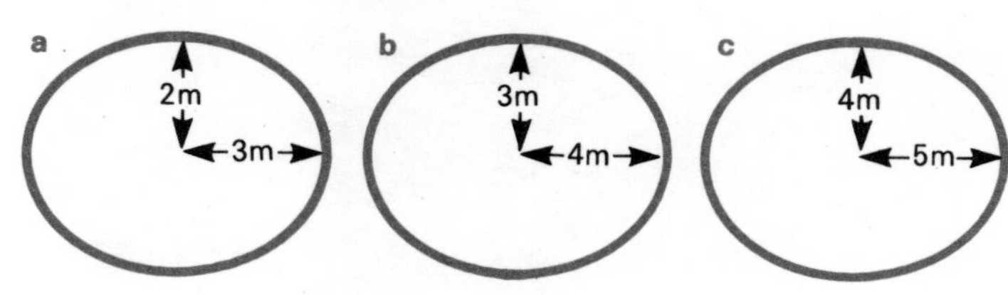

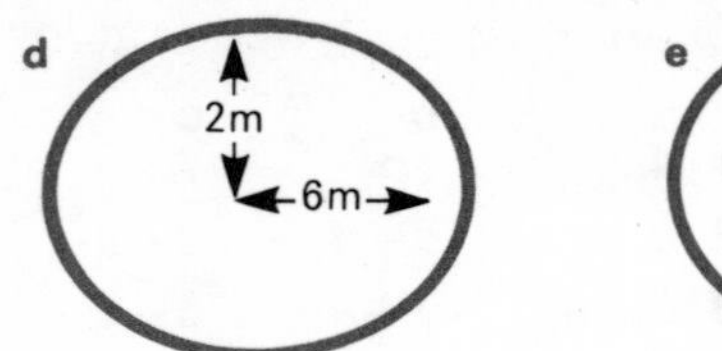

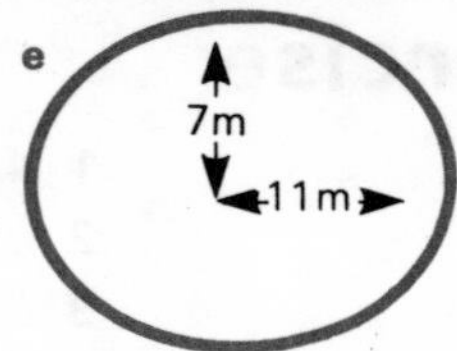

14 The area of this shape is given by the formula area $= 2pq$.

Find the areas of these shapes.

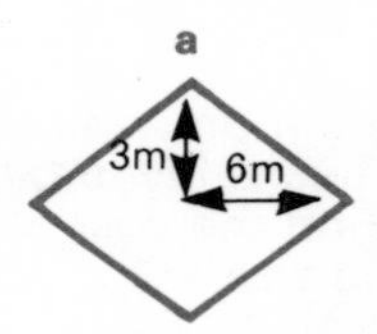

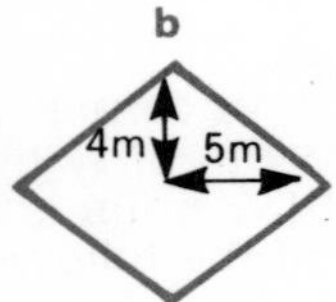

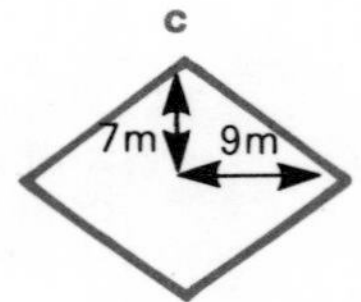

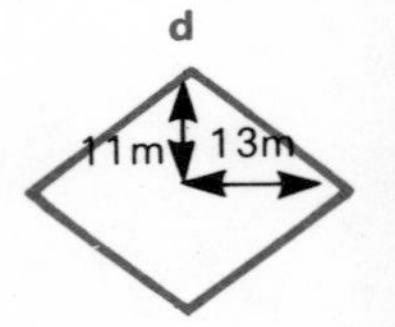

15 The weight of a bar of copper of length L cm, width W cm, and height H cm is given by the formula Weight $= 9LHW$ grams.

Find the weights of the copper bars with the following dimensions.

a 3 cm by 2 cm by 4 cm, **b** 5 cm by 2 cm by 6 cm,
c 5 cm by 10 cm by 12 cm,
d 2 cm by 2·5 cm by 7 cm,
e 3·5 cm by 4·5 cm by 20 cm.

Exercise R 9

1 The graph shows the temperature from 8 a.m. to 8 p.m. during a day in August. Use the graph to answer the following questions.

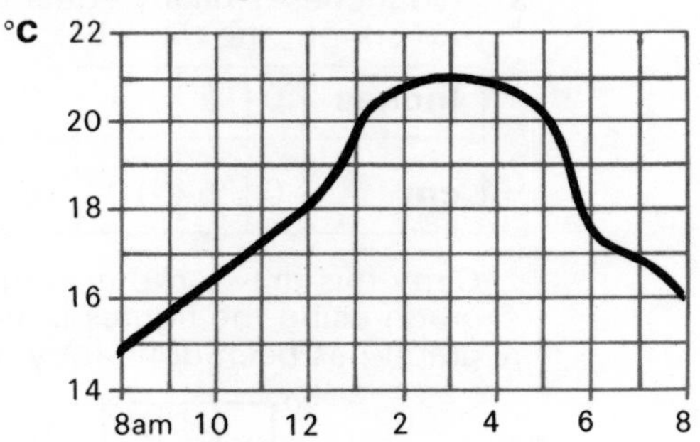

a What was the temperature at 1 p.m.?

b What was the temperature at 7 p.m.?

c At what time was the temperature 15°C?

d At what time was the temperature 18°C?

e What was the maximum (greatest) temperature?

f What was the minimum (lowest) temperature?

2 The graph shows the distance of a train from a station between midday and midnight. Use the graph to answer the following questions.

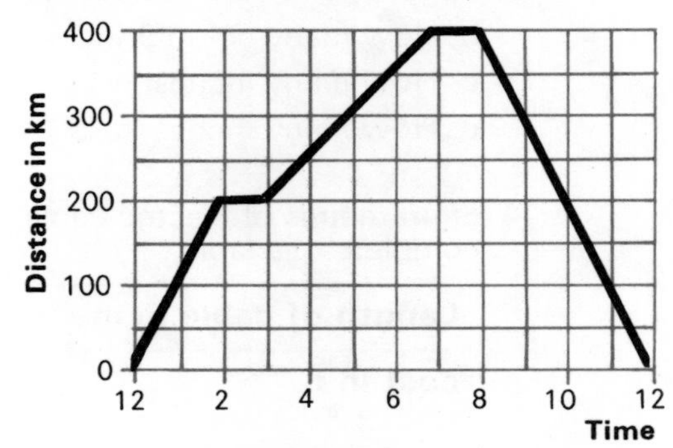

a How far from the station is the train at 2 p.m.?

b How far from the station is the train at 11 p.m.?

c How far is the train from the station at midnight?

d At what time is the train 100 km from the station?

e How far is the train from the station at 2 p.m. and 3 p.m.?

f What is the train doing between 2 p.m. and 3 p.m.?

g What is the train doing between 7 p.m. and 8 p.m.?

h What is the speed of the train between 3 p.m. and 7 p.m.?

i Between what times is the train travelling fastest, and what is its speed?

3 Two inches is nearly equal to 5 cm. This was used to work out the table below.

Inches	0	2	4	6	8	10
cm	0	5	10	15	20	25

Copy this drawing onto squared paper and draw a graph using the figures in the table. Part of the graph has been done for you.

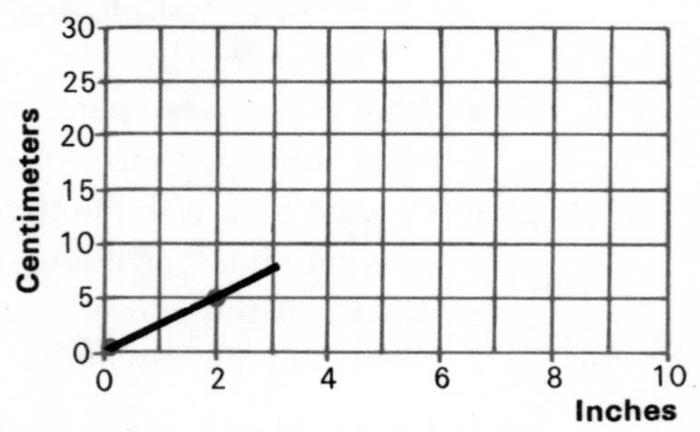

Use your graph to answer the following questions.

a How many cm in 3 inches?

b How many cm in 9 inches?

c How many inches in $12\frac{1}{2}$ cm?

d How many inches in $17\frac{1}{2}$ cm?

4 Eight metres of electric cable costs £3. Copy and complete this table.

Length of cable in m	0	8	16	24	32
cost in £		3	6		

Use the table to draw a graph, and then use your graph to answer the following questions.

a Find the cost of 4 m of cable.

b Find the cost of 20 m of cable.

c How much cable can be bought for £4·50?

d How much cable can be bought for £10·50?

e Find the cost of 11 m of cable.

f Find the cost of 17 m of cable.

g How much cable can be bought for £10?

unit 7 Graphs of formulae

You may be given a formula and told to draw its graph.

For example the distance around a circle is given approximately by the formula:

$$C = 3D$$

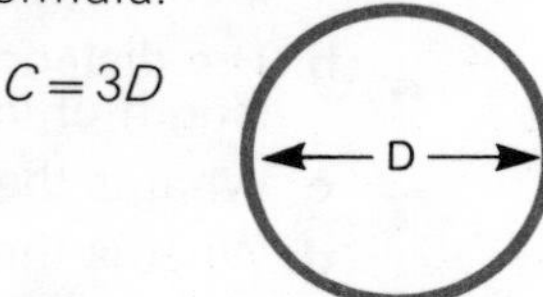

We will draw the graph for values of D from 0 cm to 8 cm.

The first thing we have to do is to make a table of values like this.

D cm	0	2	4	6	8
C cm	0	6	12	18	24

We then mark the values of D up to 8 cm along the horizontal axis, and the values of C up to 24 cm along the vertical axis. (To save space we have used one square for 2 cm on the vertical axis.) We then plot the points, and join them with a straight line.

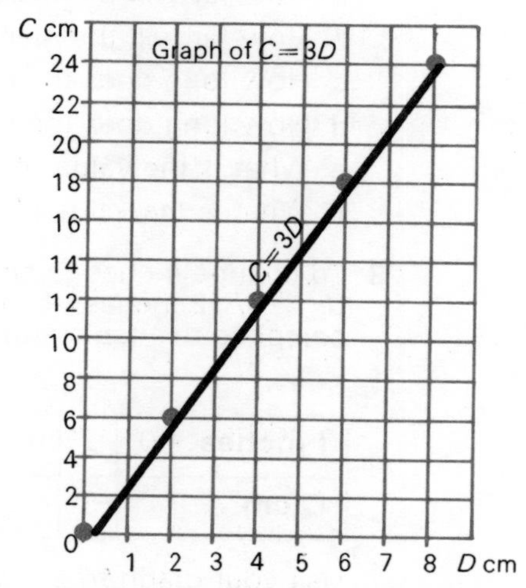

We have now drawn the graph of $C = 3D$. The formula of the graph is usually written on the graph near the line.

Exercise 7.1

1 The distance around a square is given by the formula $D = 4S$. Copy and complete this table and use it to draw the graph of $D = 4S$.

S cm	0	2	4	6	8
D cm	0	8			

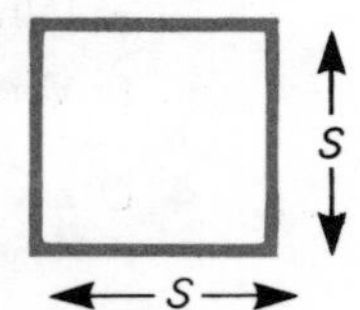

Use your graph to answer the following questions.

a What is the distance around a square with a side of 3 cm?

b The distance around a square is 20 cm. What is the length of its side?

c What is the value of D when $S=7$?

d What is the value of S when $D=14$?

e What is the value of D when $S=3$?

f What is the value of S when $D=18$?

2 The distance a man walks in km if he walks for h hours is given by the formula $D=5h$.
Copy and complete this table and use it to draw the graph of $D=5h$.

h hours	0	1	2	3	4	5	6	7
D km	0		10					

Use your graph to answer the following questions.

a How far will the man walk in $2\frac{1}{2}$ hours?

b How far will the man walk in $3\frac{1}{2}$ hours?

c How long does the man take to walk $7\frac{1}{2}$ km?

d How long does the man take to walk $22\frac{1}{2}$ km?

e What is the value of D when $h=5\frac{1}{2}$?

f What is the value of h when $D=32\frac{1}{2}$?

3 To change inches to cm we can use the formula $C=2{\cdot}5l$ (2 inches = 5 cm approximately). Copy and complete this table and use it to draw the graph of $C=2{\cdot}5l$.

l inches	0	2	4	6	8	10
C cm		5				

Use your graph to answer the following questions.

a How many cm in 5 inches?

b Change 7 inches to cm.

c Change 8 cm to inches.

d How many inches in 12 cm?

e What is the value of C when $l=3$?

f What is the value of l when $C=18$?

4 Draw the graph giving the distance around a circle, but instead of using $C=3D$, use either $C=3{\cdot}1D$ or $C=3{\cdot}14D$.

Use your graph to answer the following questions.

a What is the distance around a circle with a diameter of 5 cm?

b What is the distance around a circle with a diameter of 3 cm?

c The distance around a circle is 15 cm. What is its diameter?

d The distance around a circle is 13 cm. What is its diameter?

e What is the value of C when $D=7$?

f What is the value of D when $C=16$?

5 Copy and complete this table and use it to draw the graph of $y=5x$.

x	0	2	4	6	8	10
y						

Use your graph to answer the following questions.

a Find the value of x when $y=22$.

b Find the value of x when $y=18$.

c Find the value of x when $y=43$.

So far the formulae you have dealt with have all had straight line graphs. We will now deal with formulae with curved graphs. When you have marked the points on your graph you must not join them with straight lines. You must carefully draw a curve through them.

Exercise 7.2

1 The area of a square is given by the formula $A=s^2$ where s is the length of the side.

Copy and complete this table, and use it to draw the graph of $A=s^2$.

s cm	0	1	2	3	4	5	6	7	8	9	10
A sq cm				9		25	36				

Use your graph to answer the following questions.

a What is the area of a square with a side of 4·5 cm?

b What is the area of a square with a side of 7·5 cm?

c Find the area of a square with a side of 6·5 cm.

d The area of a square is 45 sq cm. Find the length of its side.

e The area of a square is 70 sq cm. Find the length of its side.

f The area of a square is 30 sq cm. What is the length of its side?

g What is the value of A when $s=8{\cdot}5$ cm?

h What is the value of s when $A=40$?

2 The area of a circle is given approximately by the formula $A=3R^2$.

Copy and complete this table, and use it to draw the graph of $A=3R^2$.

R cm	0	1	2	3	4	5
A sq cm	0					75

Use your graph to answer the following questions.

a What is the area of a circle with a radius of 1·5 cm?

b What is the area of a circle with a radius of 2·5 cm?

c What is the area of a circle with a radius of 4·5 cm?

d What is the radius of a circle with an area of 50 sq cm?

e What is the radius of a circle with an area of 60 sq cm?

f What is the radius of a circle with an area of 85 sq cm?

3 Repeat question **2** using either the formula $A=3{\cdot}1R^2$, or the formula $A=3{\cdot}14R^2$.

4 If a stone is dropped from the top of a cliff, the distance it has fallen after t seconds is given approximately by the formula $D=5t^2$.

Copy and complete this table and use it to draw the graph of $D=5t^2$.

t seconds	0	1	2	3	4	5	6
D metres				45			

Use your graph to answer the following questions.

a How far will the stone fall in 2·5 secs?

b How far will the stone fall in 4·5 secs?

c How long will the stone take to fall 150 m?

d How long will the stone take to fall 100 m?

e A stone is dropped from the top of a tower. It is seen to hit the ground 3·5 secs later. What is the height of the tower?

5 Copy and complete this table and use it to draw the graph of $y=\sqrt{x}$. (See page 92 for square roots.)

x	0	1	4	9	16	25	36	49	64	81	100
y	0	1		3							

Use your graph to find

a $\sqrt{20}$, **b** $\sqrt{30}$, **c** $\sqrt{70}$, **d** $\sqrt{35}$,
e $\sqrt{55}$, **f** $\sqrt{75}$, **g** $\sqrt{39}$, **h** $\sqrt{53}$

6 Copy and complete this table and use it to draw the graph of $y=\frac{36}{x}$

x	2	3	4	6	9	12
y		12				

Copy and complete this next table and use your graph to fill in the missing values for **a**, **b**, **c**, **d**, **e**, and **f**.

x	a	b	c	5	8	10
y	15	11	7	d	e	f

Example

A sand dealer charges £5 per tonne plus a delivery charge of £5 whatever the weight. We can write this in the formula $C=5+5t$ where t is the weight in tonnes, and C is the cost in £.

The formula gives the following table

t tonnes	0	2	4	6	8	10
C £	5	15	25	35	45	55

The graph can now be drawn for $C=5+5t$.

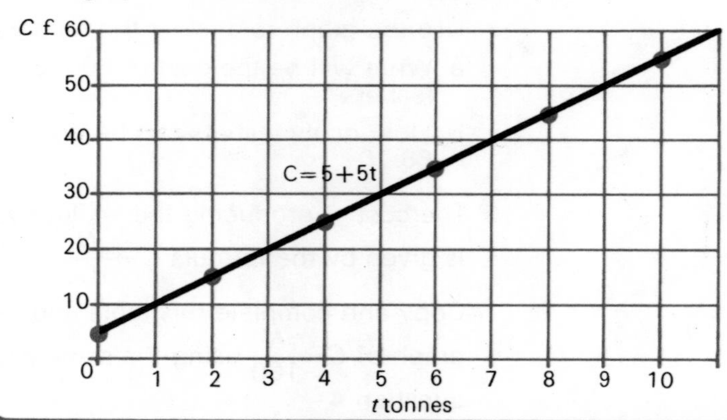

Make sure that you know how to use the graph to get the following answers.

a 7 tonnes of sand cost £40.

b For £50 the dealer will deliver 9 tonnes of sand.

Exercise 7.3

1 If the sand dealer charges £10 for delivery and £5 per tonne the formula is $C=10+5t$.

Work out a table of values and then draw the graph of $C=10+5t$.

Use your graph to answer the following questions.

a How much will 3 tonnes of sand cost?

b How much will 9 tonnes of sand cost?

c How much will $7\frac{1}{2}$ tonnes of sand cost?

d How much sand will be delivered for £15?

e How much sand will be delivered for £25?

f How much sand will be delivered for £33?

2 Repeat question **1** using the formula $C=4+3t$.

3 Repeat question **1** using the formula $C=7+4t$.

4 The cost of producing N copies of a leaflet by offset lithography is given by the formula

$C=4+\frac{N}{400}$.

Copy and complete this table and use it to draw the graph of $C=4+\frac{N}{400}$.

Mark up to £12 on the vertical axis as you will need this for the next question.

N	0	400	800	1200
C £		5		

Use the graph to answer the following questions.

a What will be the cost of 600 leaflets, and 700 leaflets?

b How many leaflets can be bought for £4·50 and £6·70?

5 The cost of producing the leaflet by photocopying is given by the formula $C=\frac{N}{100}$.

Copy and complete this table and use it to draw the graph of $C=\frac{N}{100}$ using the same paper as for question **4**.

N	0	400	800	1200
C £		4		

Use the graph to answer the following questions.

a What will be the cost of 200 and 700 photo-copied leaflets?

b How many photocopied leaflets can be got for £3, and for £8.50?

c Over how many copies will it be cheaper to use offset lithography rather than photocopying?

d How much does each photocopied leaflet cost?

e Explain why we cannot ask the same question as in **d** about leaflets produced by offset lithography.

f What is the cheapest method of producing 1000 leaflets, and how much will each leaflet cost?

revision 10 Areas of rectangles and triangles

Exercise R 10

Find the area of, and the distance around, each of these shapes. With the more difficult examples you might find that it will help to draw them on squared paper and count the number of squares to find the area.

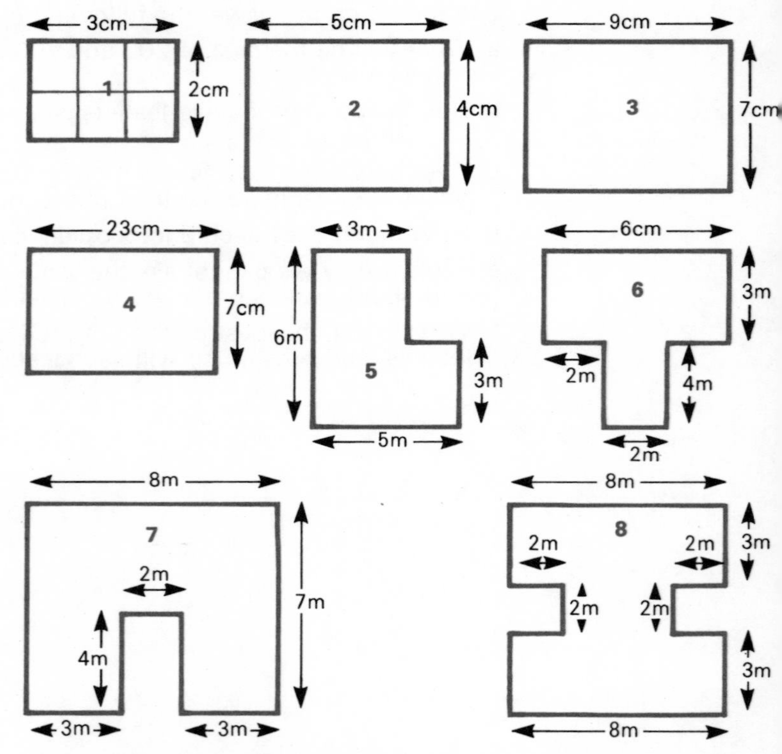

The diagrams in questions 5 to 8 represent plans of rooms. Find out how much it would cost to carpet each room at £3 per square metre, and at £4·37 per square metre.

For the next examples you will need to remember that

Area of triangle = $\frac{1}{2}$ × base × height

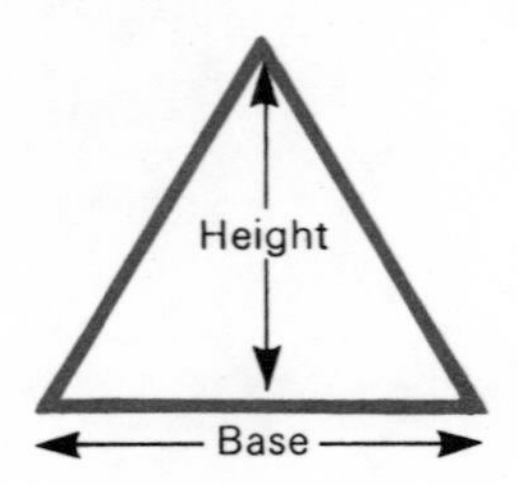

Find the areas of these triangles.

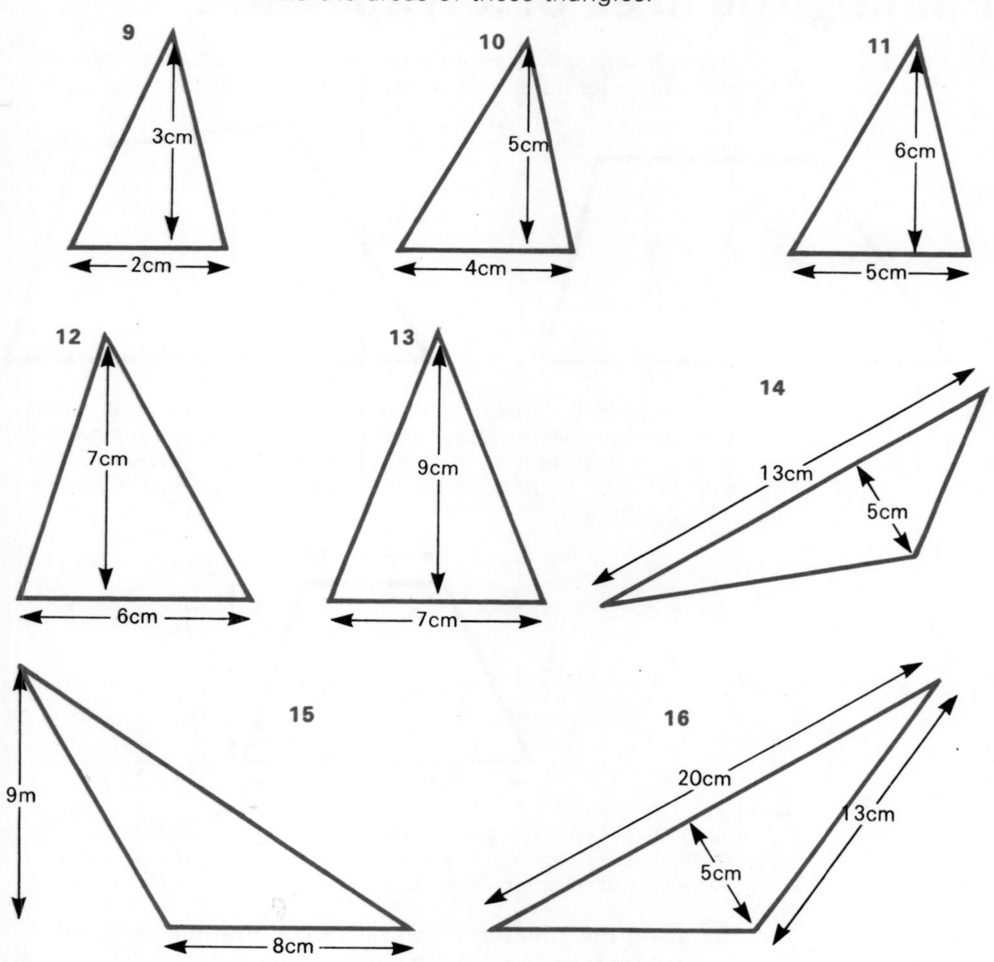

Finding the area of a trapezium

A trapezium is a triangle with the top cut off by a line parallel to its base. A is a trapezium. B is not a trapezium.

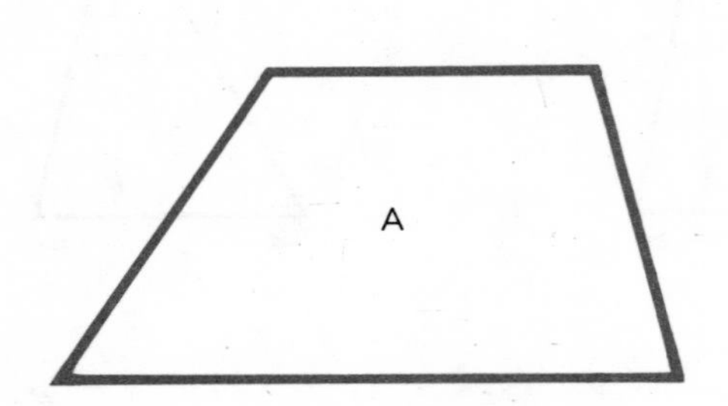

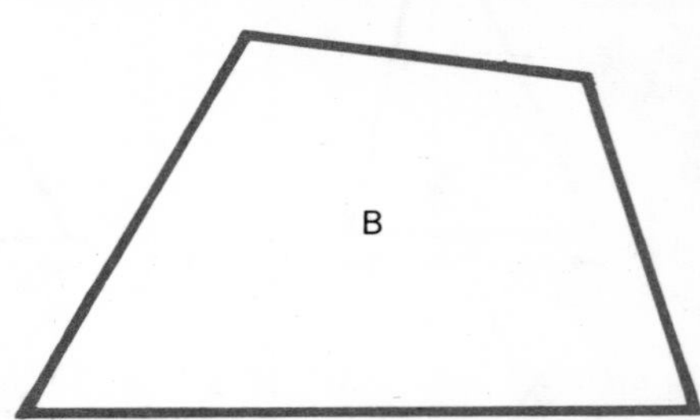

One way of finding the area of a trapezium is to cut it into two triangles which have been marked T_1 and T_2, find the area of each triangle, and add the two areas together. If you turn the page upside down you will see that the base of T_2 is 4cm.

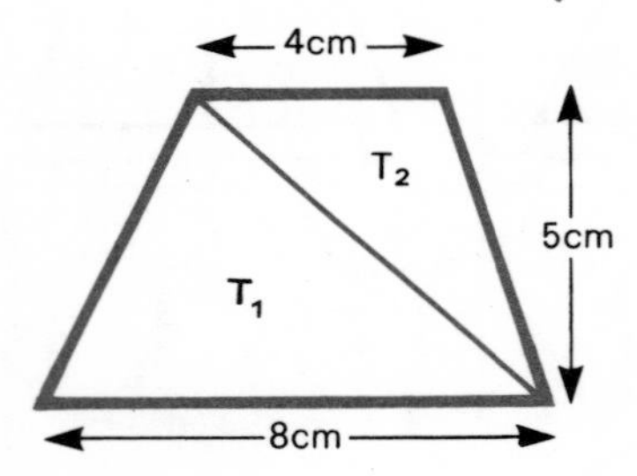

Area of $T_1 = \frac{1}{2} \times 8 \times 5 = \frac{1}{2} \times 40 = 20$ sq cm
Area of $T_2 = \frac{1}{2} \times 4 \times 5 = \frac{1}{2} \times 20 = 10$ sq cm
Area of trapezium $= 20 + 10 = \underline{30 \text{ sq cm}}$

By using this method it can be shown that the area will always be given by the following formula.

Area of trapezium = $\frac{1}{2}$ × (base + top) × height

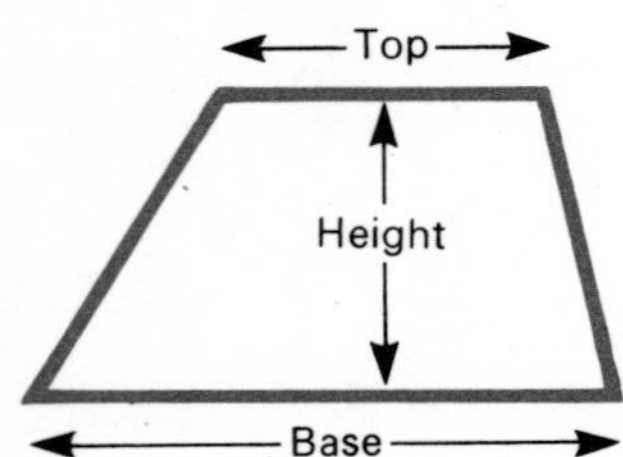

If we use this formula in the example we have just done we get –
Area $= \frac{1}{2} \times (8+4) \times 5 = \frac{1}{2} \times 12 \times 5 = \frac{1}{2} \times 60 = \underline{30 \text{ sq cm}}$

Example

Find the area of this trapezium.

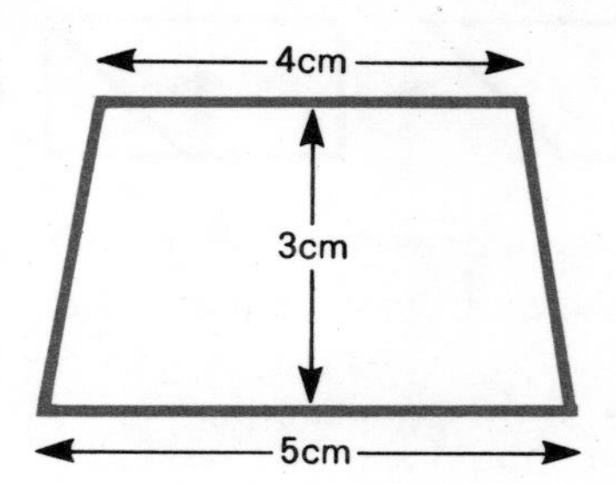

Area $= \frac{1}{2} \times (5+4) \times 3 = \frac{1}{2} \times 9 \times 3 = \frac{1}{2} \times 27 = \underline{13\frac{1}{2} \text{ sq cm}}$
You should note that the two numbers inside the brackets are added together first.

Exercise 8.1

Find the areas of these trapeziums.

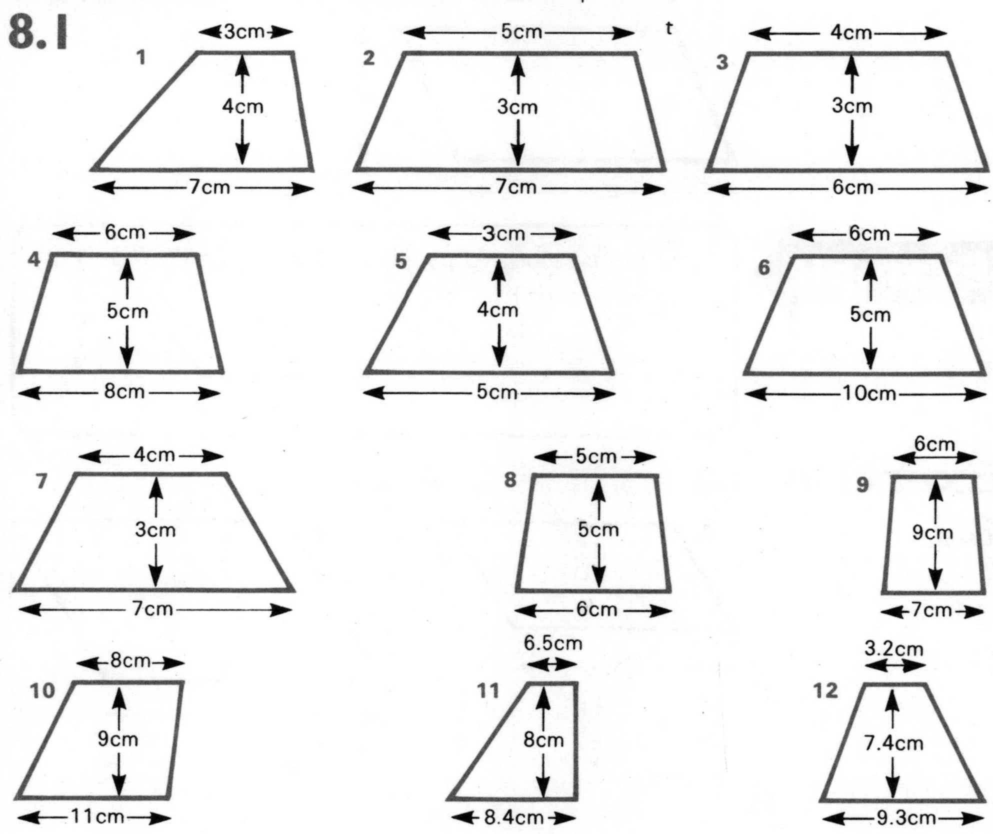

To find the area of a parallelogram

Look at these pictures which show how you can cut up a parallelogram and turn it into a rectangle.

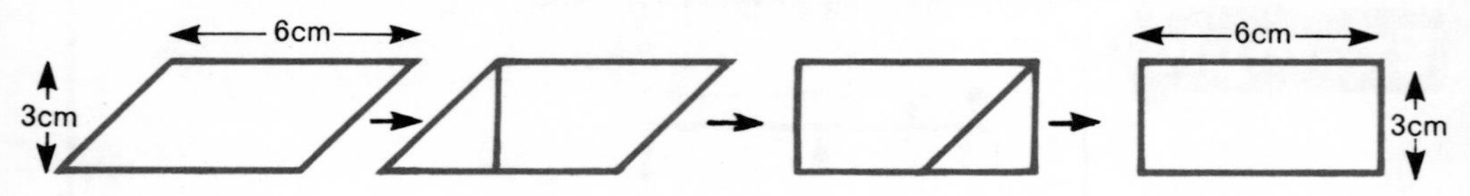

Since the area does not change, we can say:

Area of Parallelogram = area of rectangle
= 6 × 3 = 18 sq cm

The **base** of the parallelogram is 4 cm, and the **height** is 3 cm. You should see that we found the area by multiplying the **base** by the **height**. We can obviously find the area of any parallelogram the same way and so we can write:

Area of parallelogram = base × height

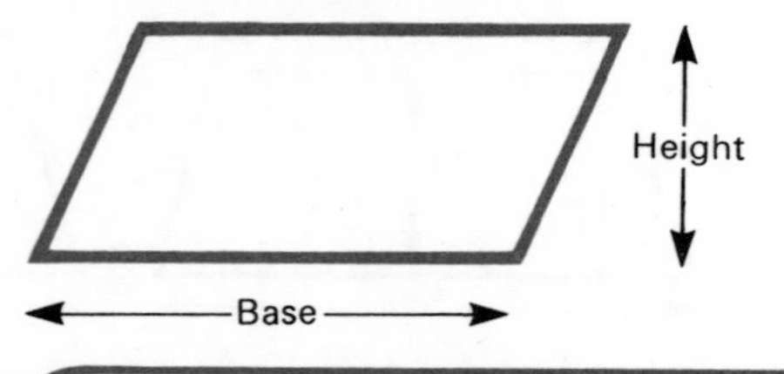

Example

Find the area of this parallelogram.

Area = 7 × 9 = 63 sq cm

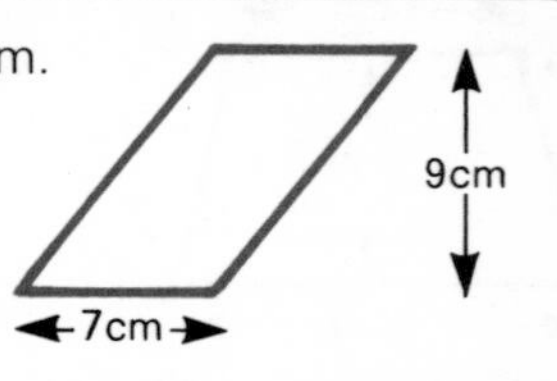

Exercise 8.2

Find the areas of these parallelograms.

1

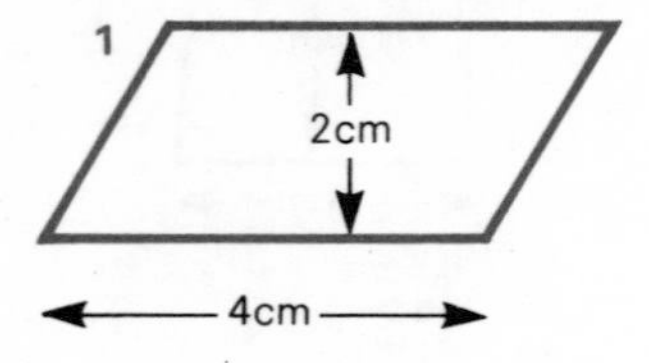

2

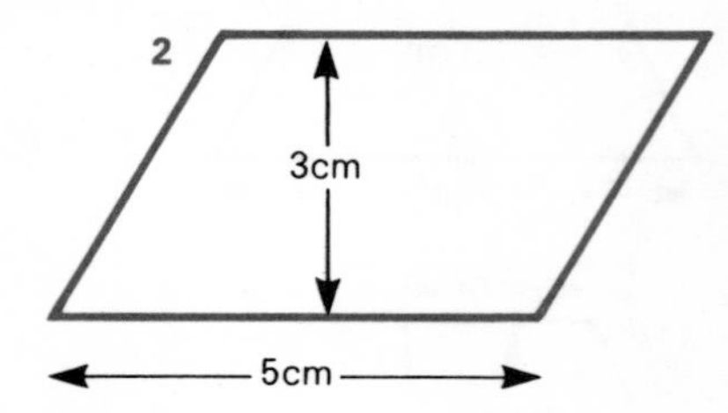

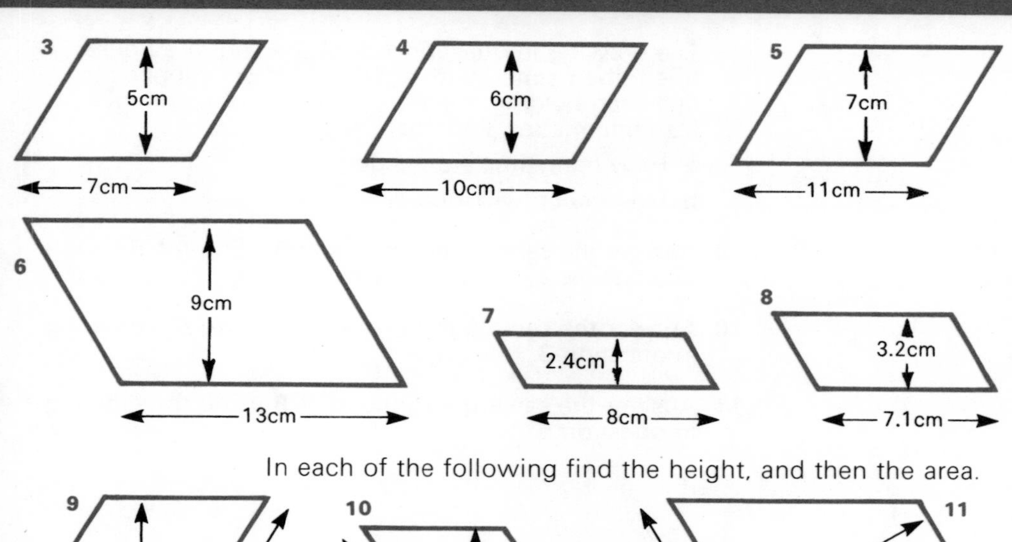

In each of the following find the height, and then the area.

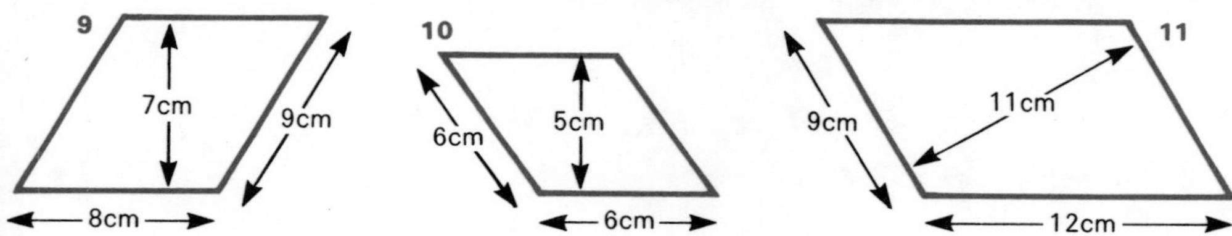

Exercise 8.3

For each of these shapes find the area of part A, the area of part B, and then the total area.

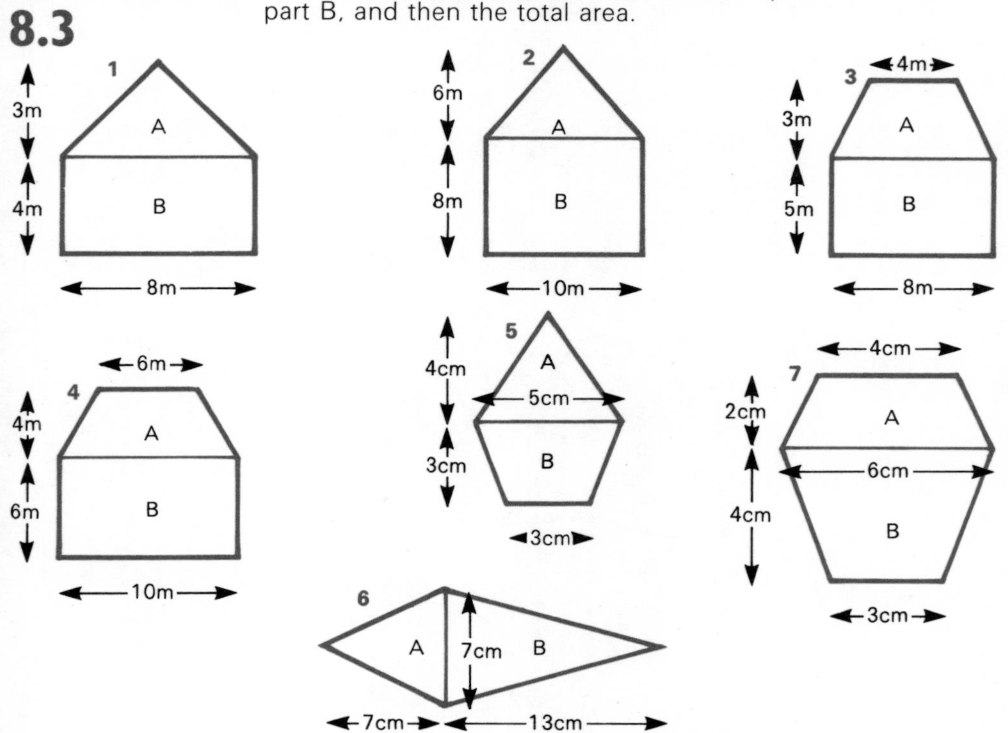

8 The drawing in question **1** is of the end of a house. It is to be painted with a paint which is supplied in tins each holding one litre. Each litre of paint covers 9 square metres, and costs 75p.

a How many tins are needed?

b How much will it cost?

9 Answer the same questions as in **8** using the drawing in question **2**.

10 Answer the same questions as in **8** using the drawing in question **3**.

11 Answer the same questions as in **8** using the drawing in question **4**.

Exercise R 11

In this exercise you can use $\pi = 3$, or 3·1, or 3·14.

Distance around circle = $\pi \times$ diameter

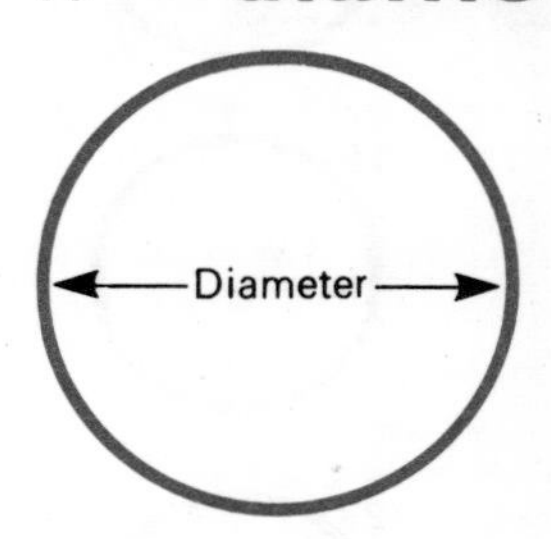

Find the distance around these circles.

1
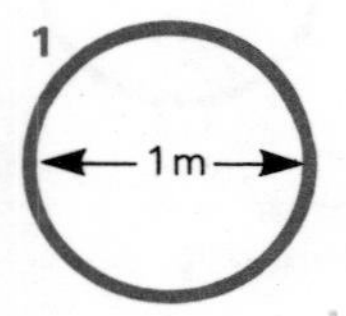

2
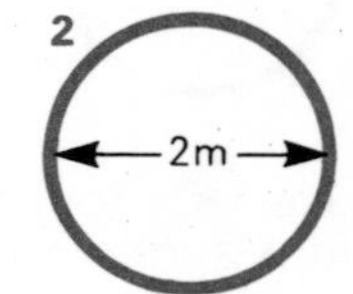

3
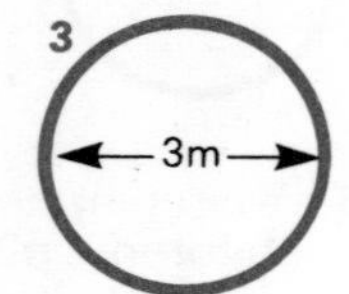

4
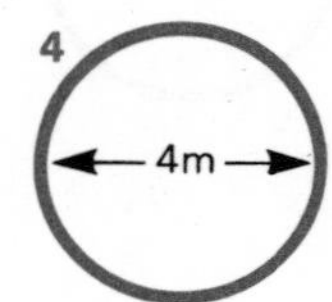

5
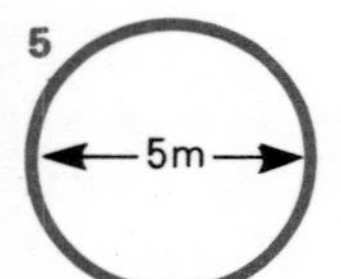

6
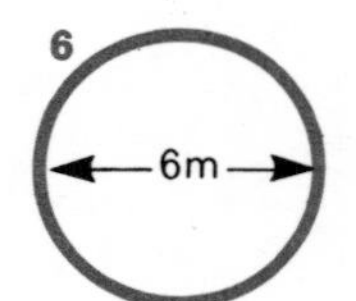

7
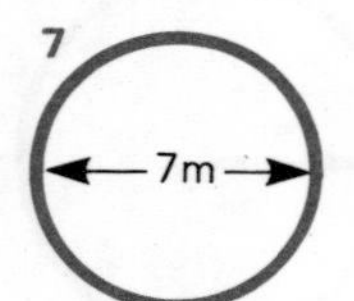

8
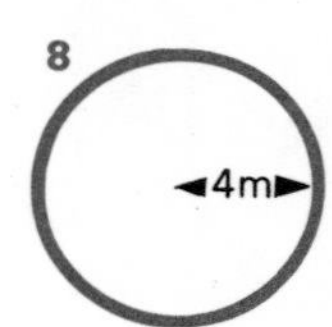

9
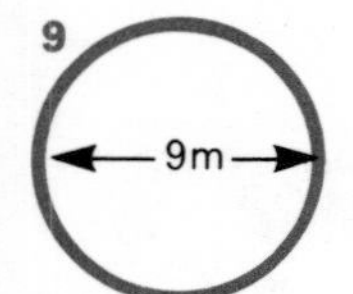

10
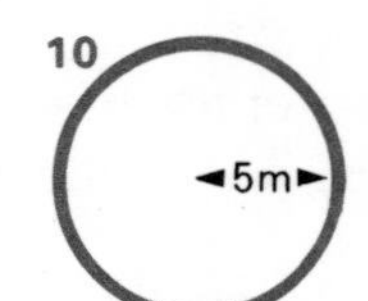

11
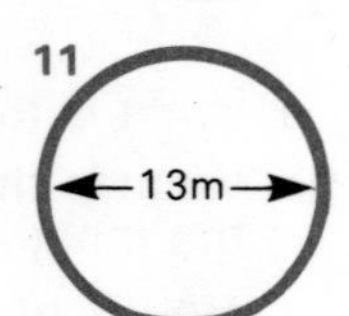

12
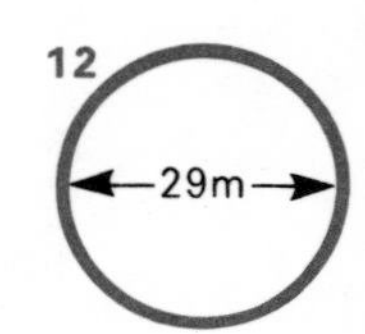

Area of circle = $\pi \times$ radius $\times$ radius

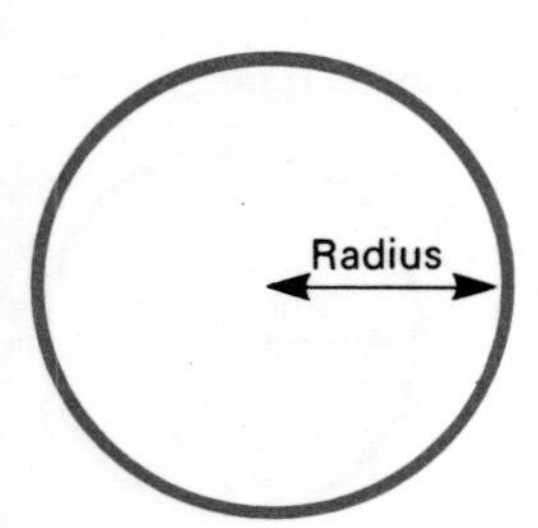

Find the areas of these circles.

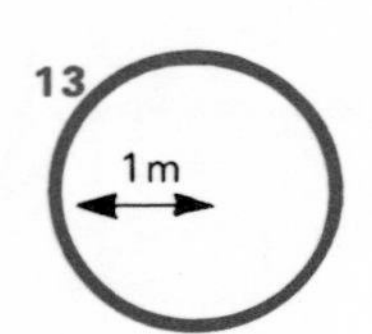

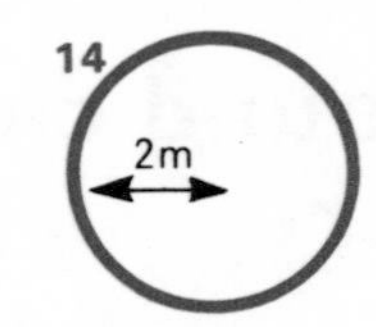

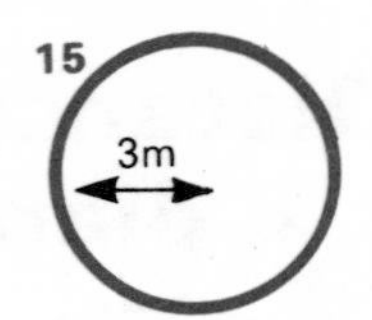

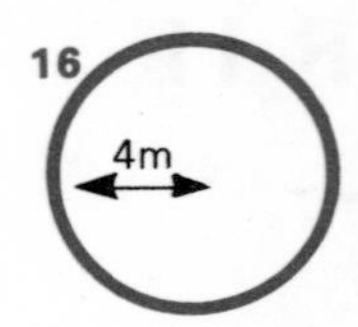

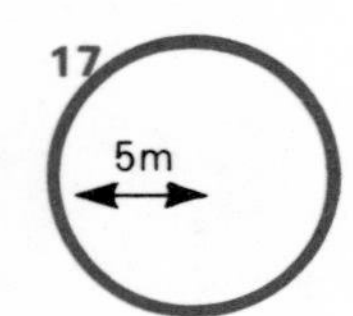

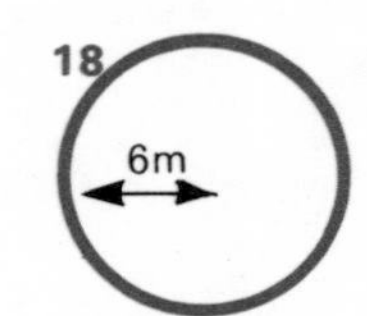

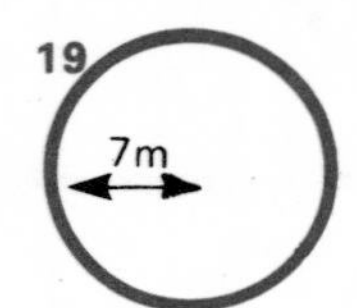

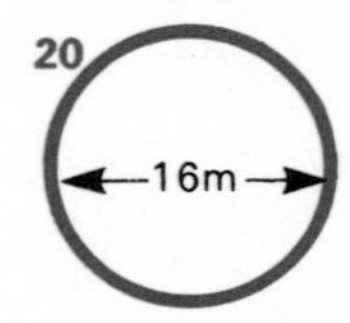

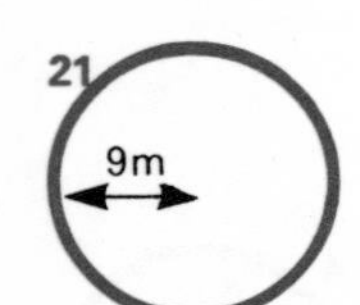

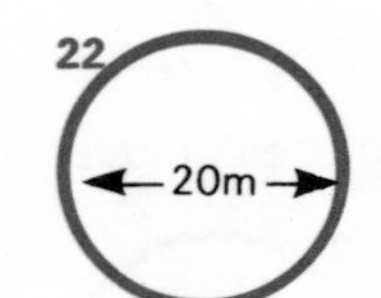

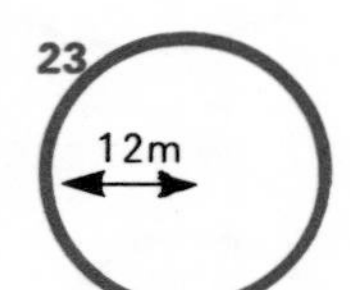

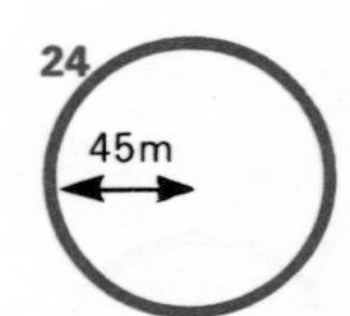

25 A circular concrete base for a roundabout on a children's playing field is surrounded by a fence.

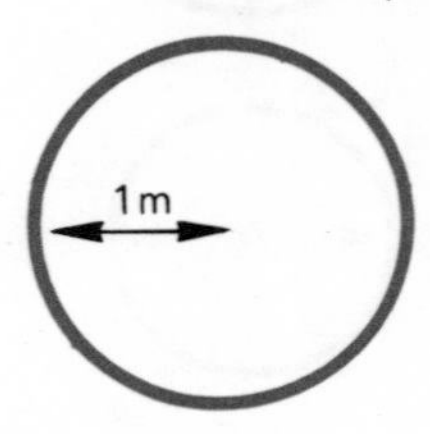

a What is the diameter of the base?

b What is the distance around the circle (the length of the fence)?

c How much will the fence cost at 75p per metre?

d What is the radius of the base?

e What is the area of the base?

f How much will the concrete cost at £4 per square metre?

Repeat question 25 with the following concrete bases.

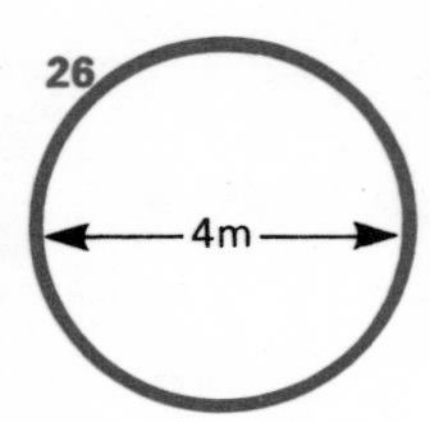

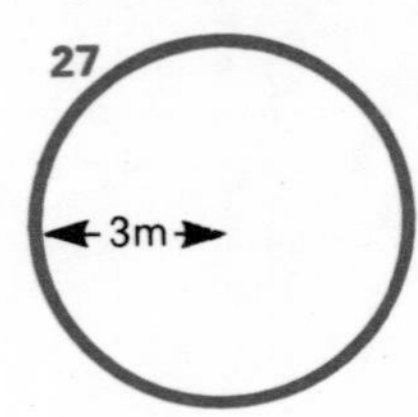

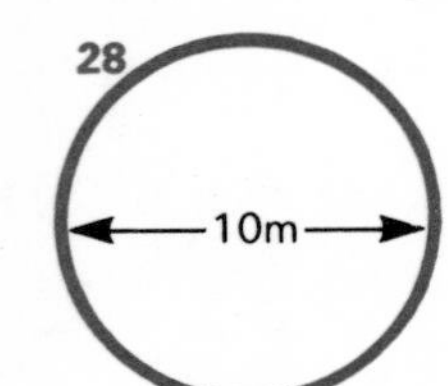

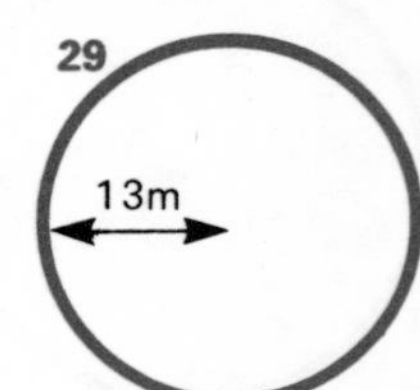

revision 12 Volumes of rectangular blocks

Exercise R 12

Find the volumes and surface areas of these rectangular blocks.

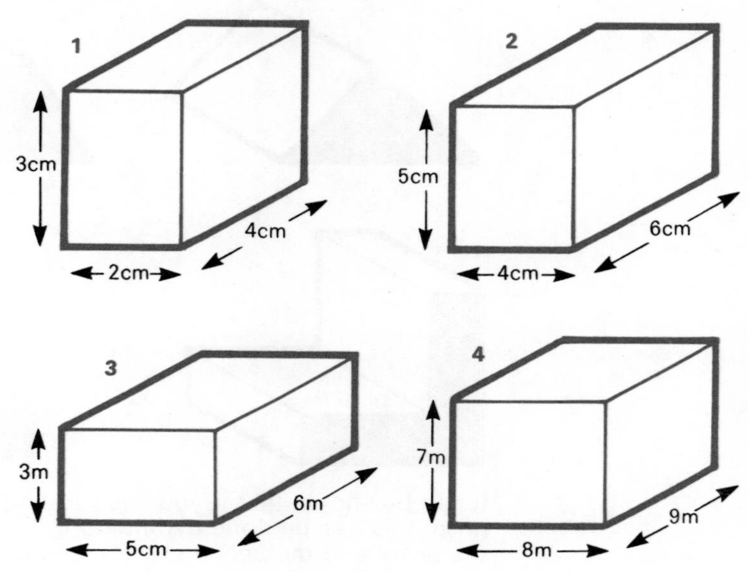

Find the volume and surface area of the rectangular blocks with the following measurements.

5 5 cm by 6 cm by 7 cm.

6 6 m by 7 m by 8 m

7 5 cm by 7 cm by 10 cm

8 3·5 m by 7 m by 8 m

unit 9 More volumes

Look at these shapes. They are called prisms.

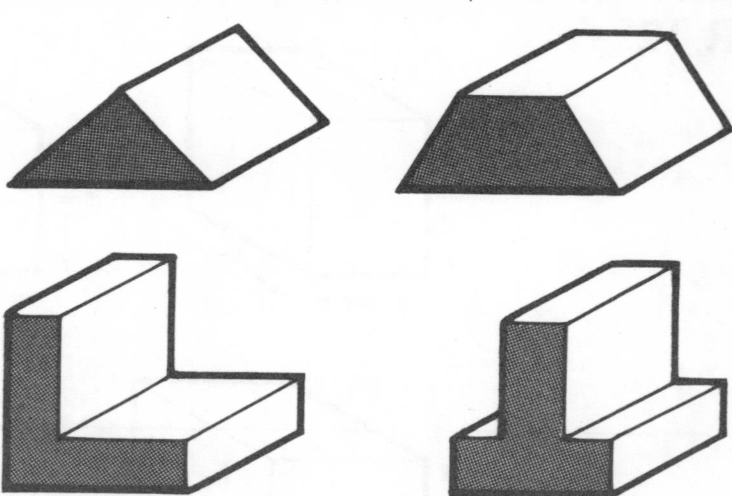

It can be shown that if you need to find the area of a prism this can be done by finding the area of the end of the prism and multiplying by the length.

Volume of prism = area of end × length

Exercise 9.1

Read each of these examples, and then do the examples underneath it in the same way.

Example

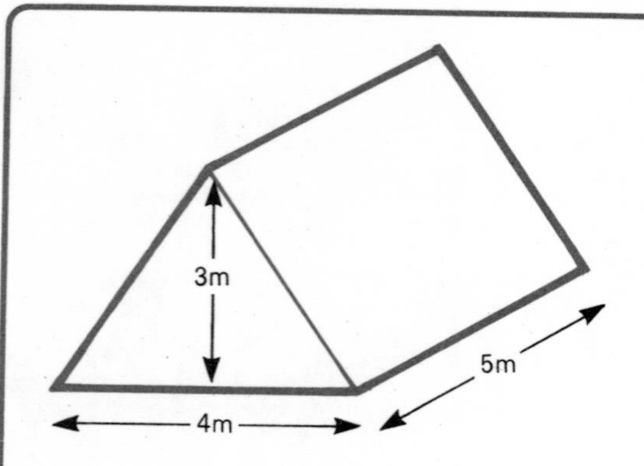

The end is shaped like a triangle so we can say:

Area of end $= \frac{1}{2} \times$ Base $\times$ Height $= \frac{1}{2} \times 4 \times 3 = \frac{1}{2} \times 12$
$= \underline{6 \text{ sq metres}}$

Volume $= 6 \times 5 = \underline{30 \text{ cubic metres}}$

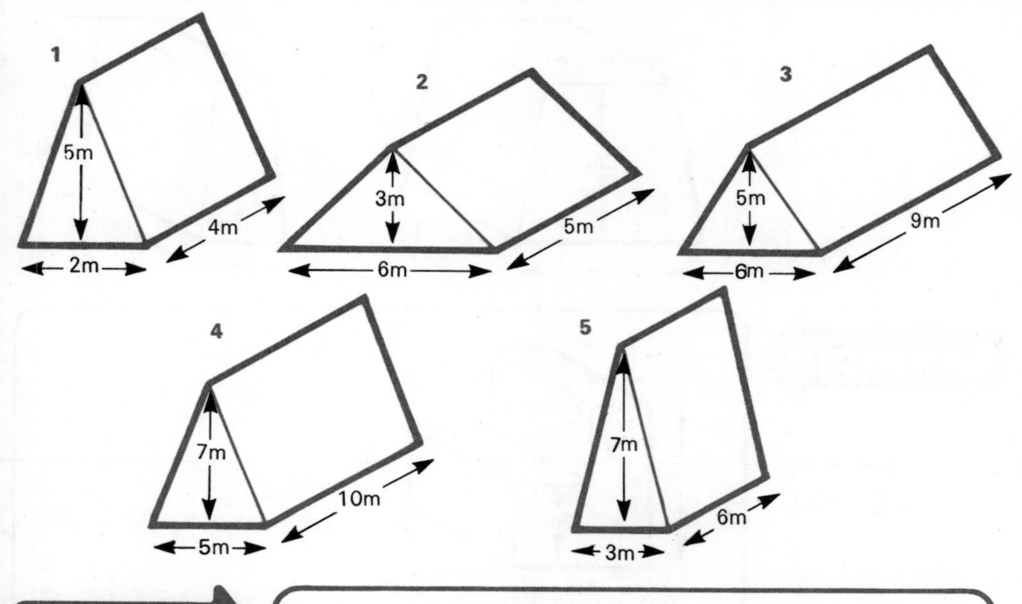

Example

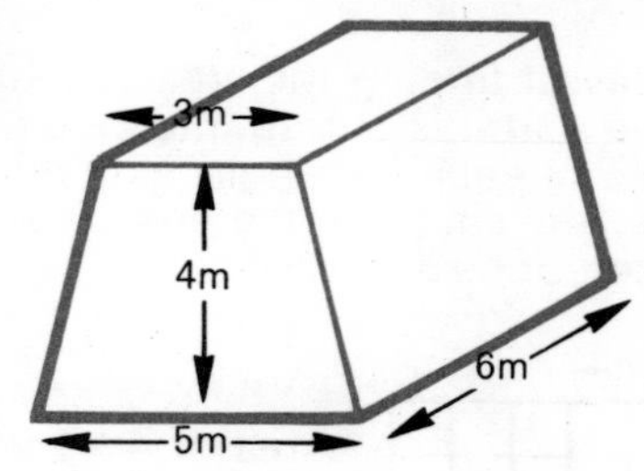

The shape of the end is a trapezium so we can say:

Area of end $= \frac{1}{2} \times (\text{Top} + \text{Bottom}) \times \text{Height}$

$= \frac{1}{2} \times (3+5) \times 4 = \frac{1}{2} \times 8 \times 4 = \underline{16 \text{ sq metres}}$

Volume $= 16 \times 6 = \underline{96 \text{ cubic metres}}$

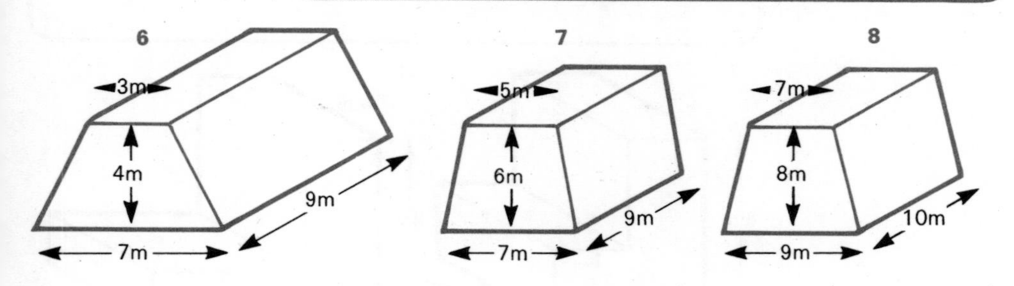

9

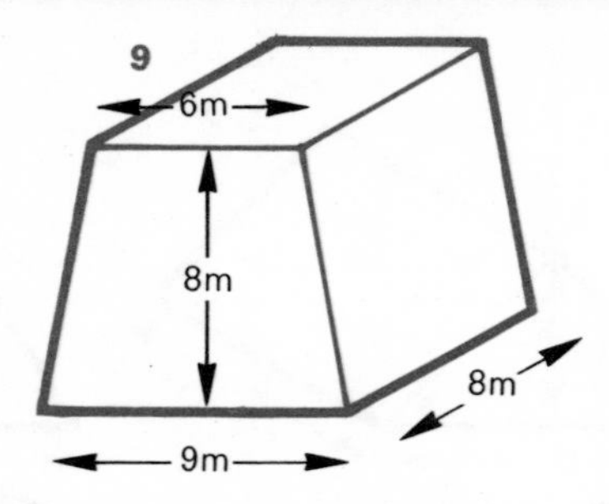

10

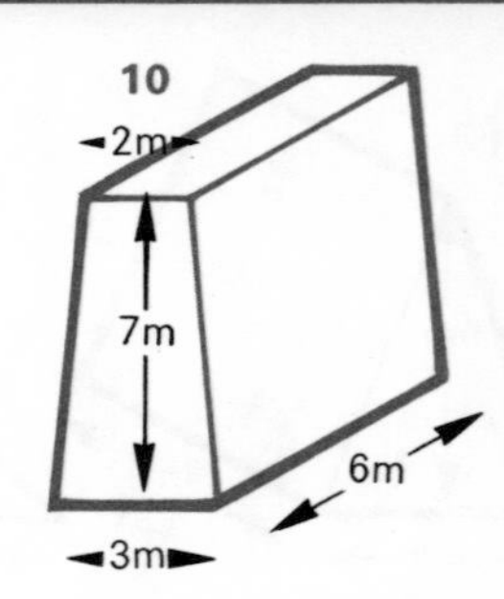

Example

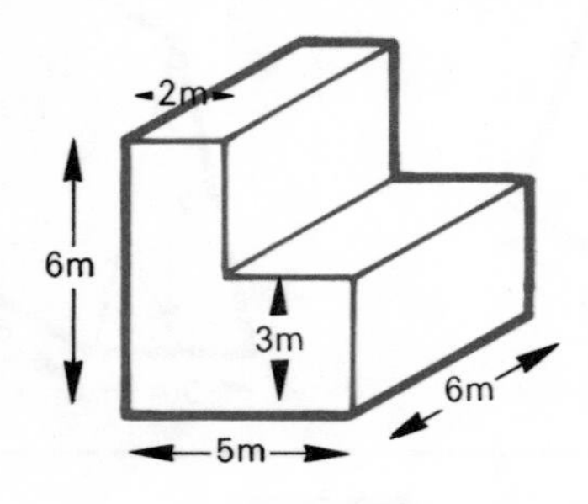

One way of finding the area of the end of this is to imagine that it is cut up into two rectangles, and find the area of each rectangle. You might find that it helps to draw the shape of the end on squared paper. This has been done below.

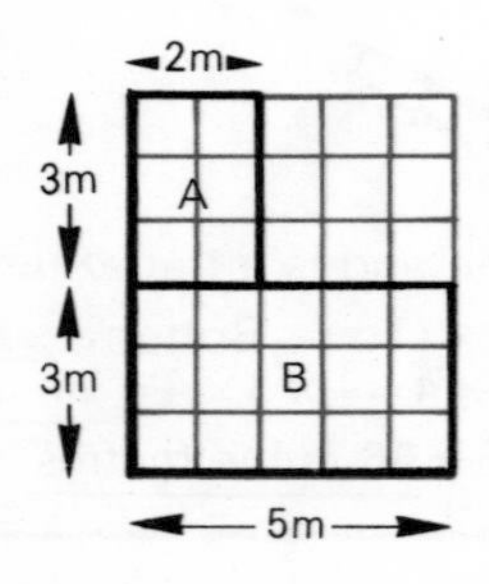

Area of A $= 2 \times 3 =$ 6 sq metres
Area of B $= 3 \times 5 =$ 15 sq metres
Area of end $= 6 + 15 =$ 21 sq metres
Volume $= 21 \times 6 =$ 126 cubic metres

11 **12**

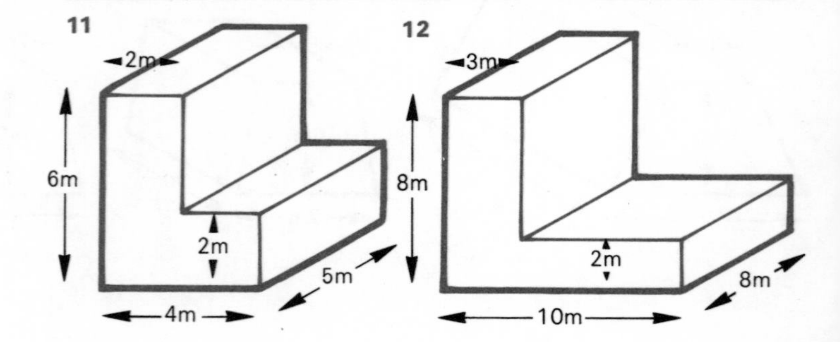

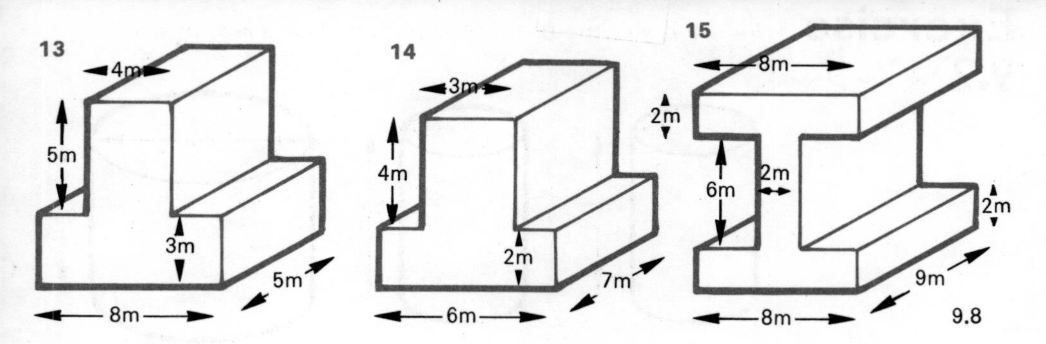

9.8

Volume of a cylinder

A cylinder can be regarded as a prism.
The area of its end is πR^2.
We usually write H for its length, so we can say

Volume of a cylinder = $\pi R^2 H$

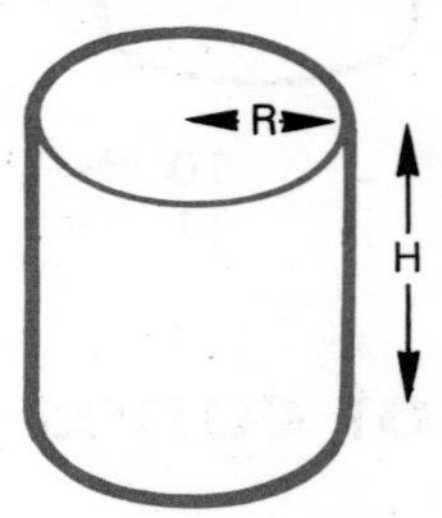

Remember that this is just a short way of writing $\pi \times R \times R \times H$.

Example

Find the volume of this cylinder.

4m
5m

$V = 3{\cdot}1 \times 4 \times 4 \times 5 =$ 248 cubic metres.

Exercise 9.2

Find the volume of these cylinders. Use $\pi = 3$, or 3·1, or 3·14.

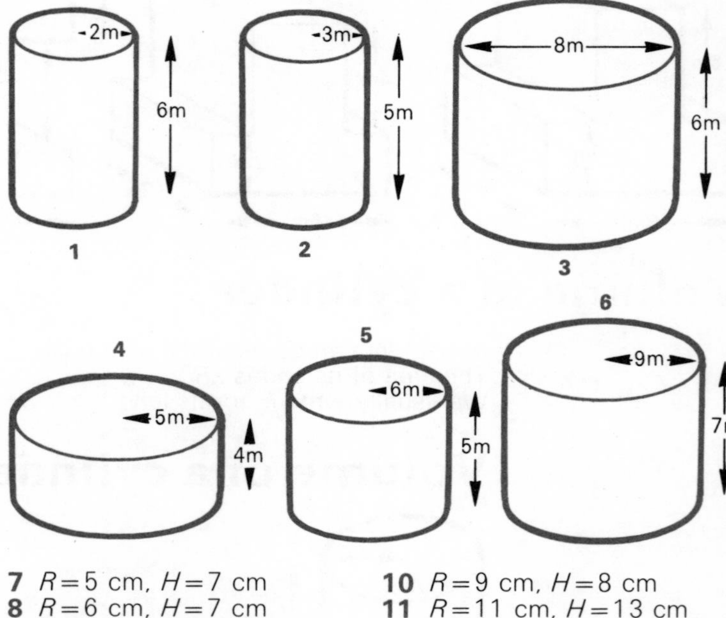

7 $R = 5$ cm, $H = 7$ cm
8 $R = 6$ cm, $H = 7$ cm
9 $R = 7$ cm, $H = 5$ cm
10 $R = 9$ cm, $H = 8$ cm
11 $R = 11$ cm, $H = 13$ cm

Finding the volumes of cones and pyramids

This shape is called a **cone.**

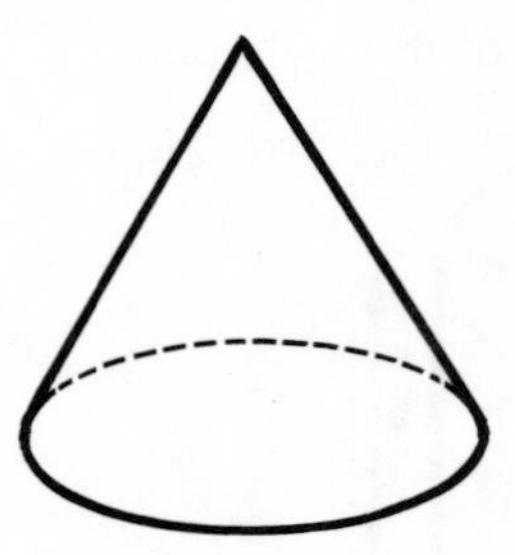

This shape is called a **pyramid.**

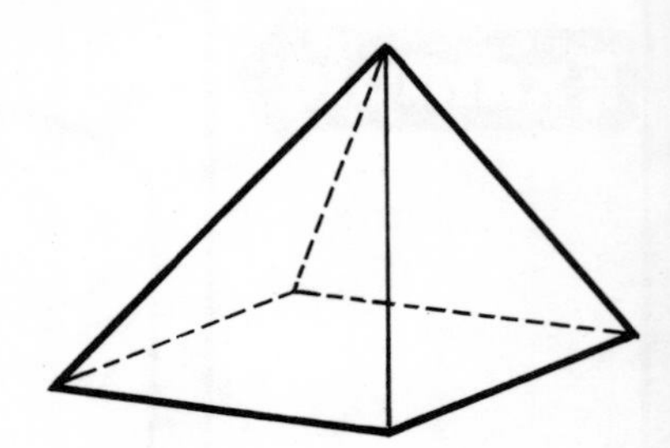

It can be shown that

Volume of cone or pyramid
$= \frac{1}{3} \times$ **area of base** $\times$ **height**

Exercise 9.3

Read each of these examples and then do the examples underneath it in the same way.

Example

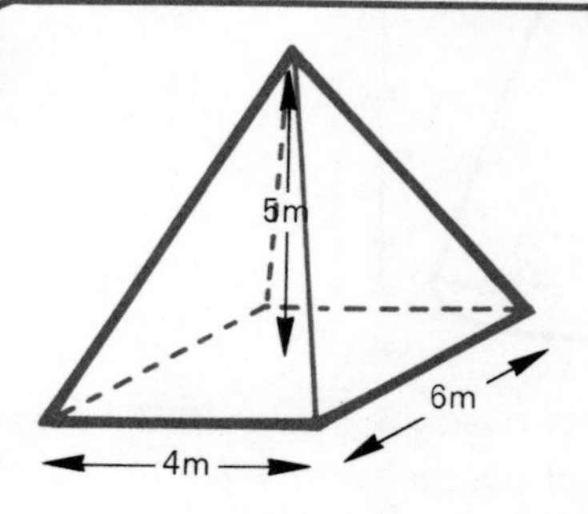

Area of Base $= 4 \times 6 = 24$ sq m

Area of Base $\times$ Height $= 24 \times 5 = 120$ cubic m

Volume $= \frac{1}{3} \times 120 =$ 40 cubic m

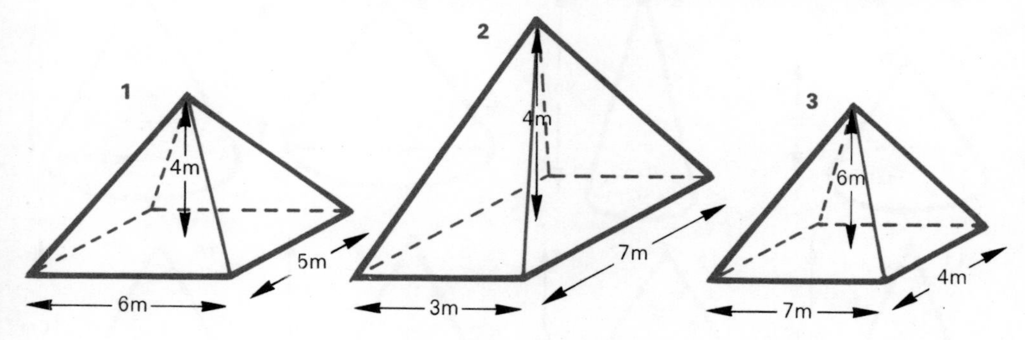

The next three examples do not come out exactly and you will need to do this sort of calculation:

$$\frac{1}{3} \times 20 = 20 \div 3 = 6{\cdot}666$$

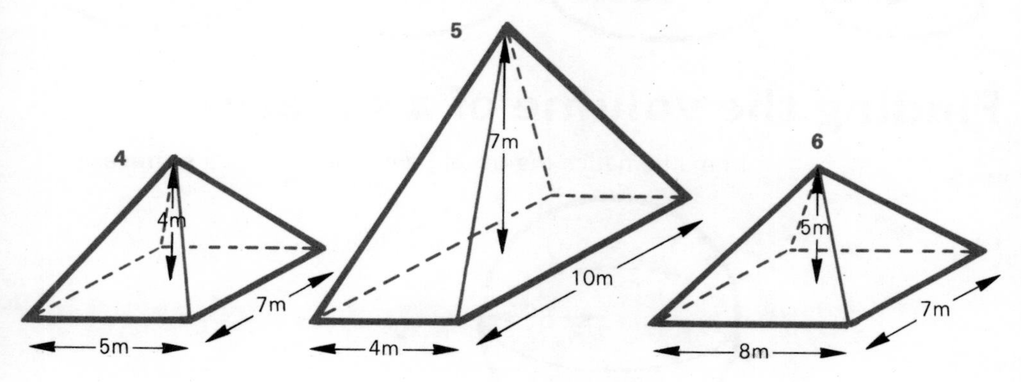

Example

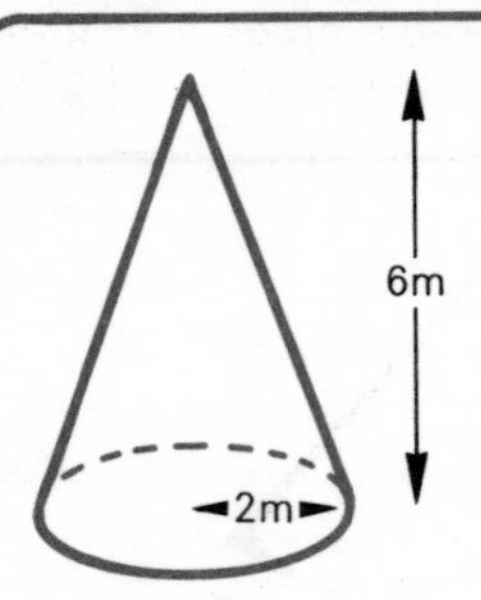

Area of base = 3·1 × 2 × 2 = 3·1 × 4 = 12·4 sq m
Area of base × height = 6 × 12·4 = 74·4 cubic m
Volume = $\frac{1}{3}$ × 74·4 = <u>24·8 cubic m</u>

With the next examples use $\pi = 3$, or 3·1, or 3·14.

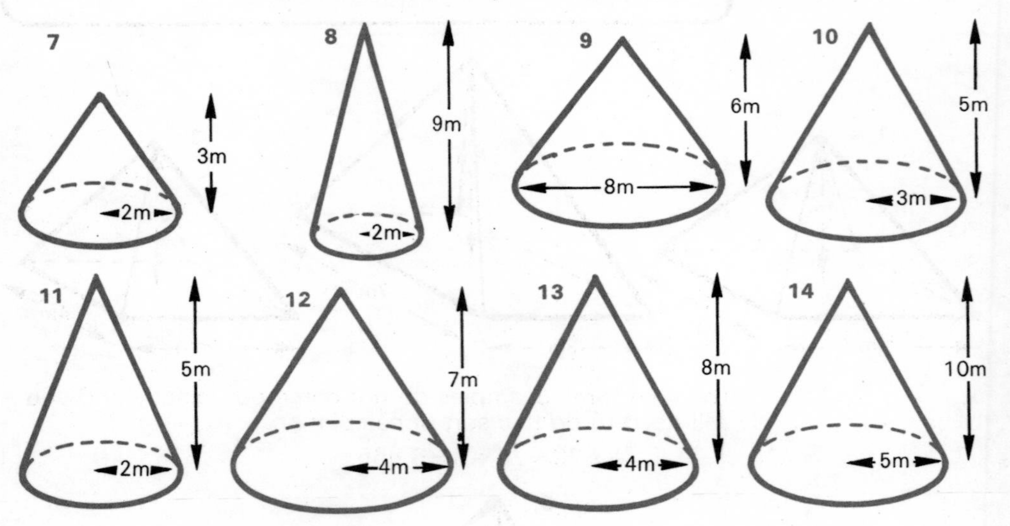

Finding the volume of a sphere

In mathematics the usual name for a ball is a **sphere**.

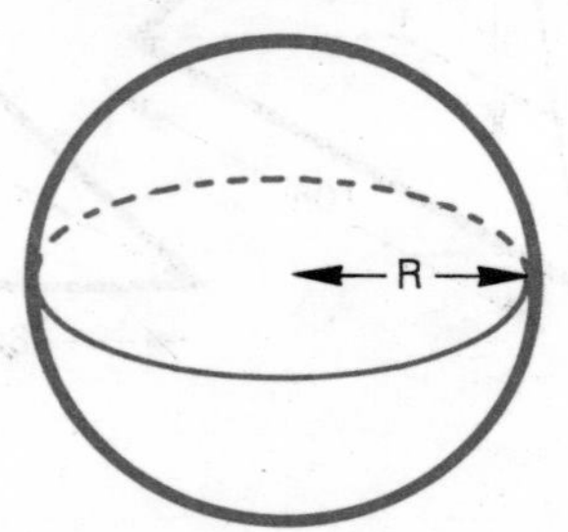

It can be shown that

Volume of sphere = $\frac{4}{3}\pi r^3$

Remember that this is just a short way of writing $4 \times \pi \times R \times R \times R \div 3$.

Exercise 9.4

Read this example, and then do the examples underneath. Use $\pi = 3$, or 3·1, or 3·14.

Example

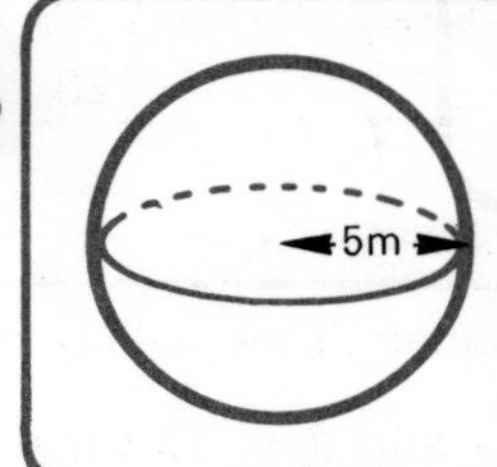

$V = 4 \times 3{\cdot}1 \times 5 \times 5 \times 5 \div 3$
$= 3{\cdot}1 \times 500 \div 3$
$= 1550 \div 3$
$= \underline{516{\cdot}6 \text{ cubic m}}$

1
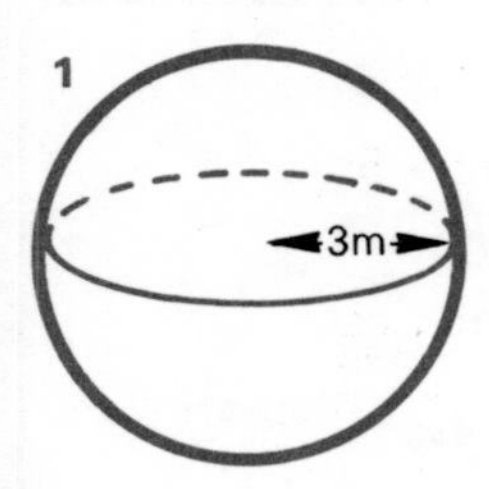

2
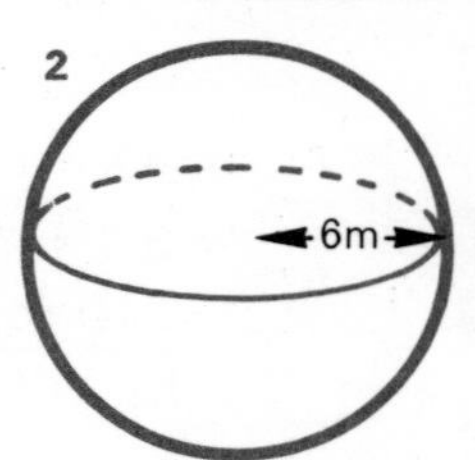

3
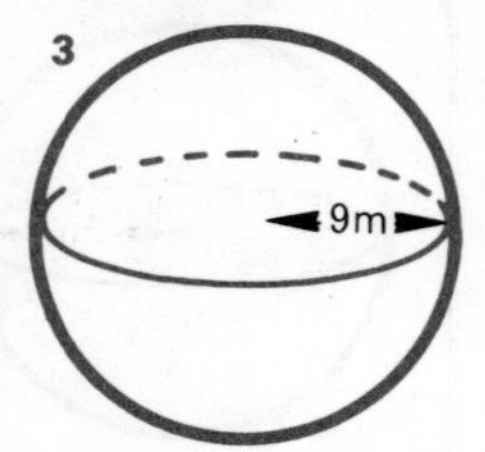

4
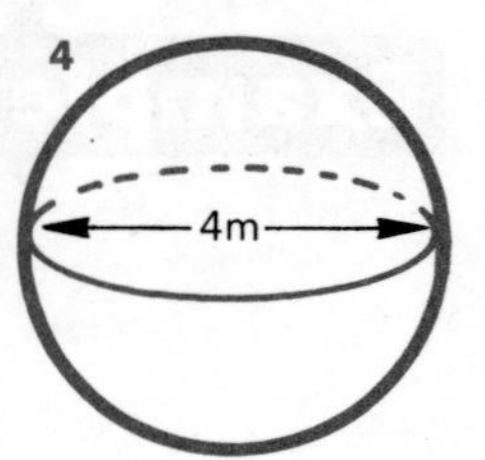

Find the volumes of spheres with the following radii.

5 $R = 4$ m **6** $R = 10$ m **7** $R = 7$ m **8** $R = 8$ m

Finding the surface areas of solids

It is sometimes necessary to find the surface areas of solids – for example, in deciding how much paint is needed to paint a large tank shaped like a cylinder.

Here are the formulae for the surface areas of some solids.

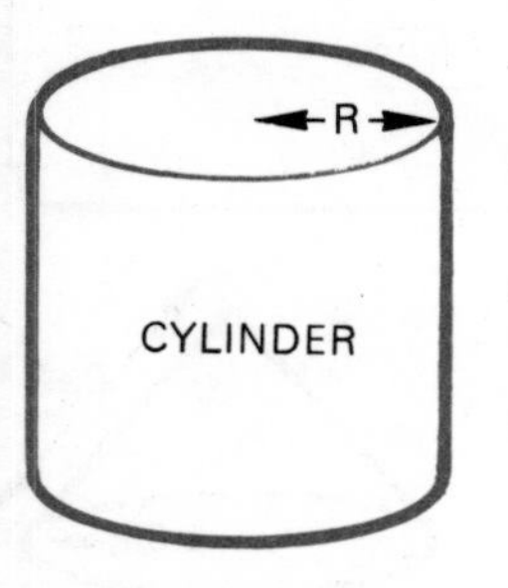

$A = 2\pi R(R + H)$

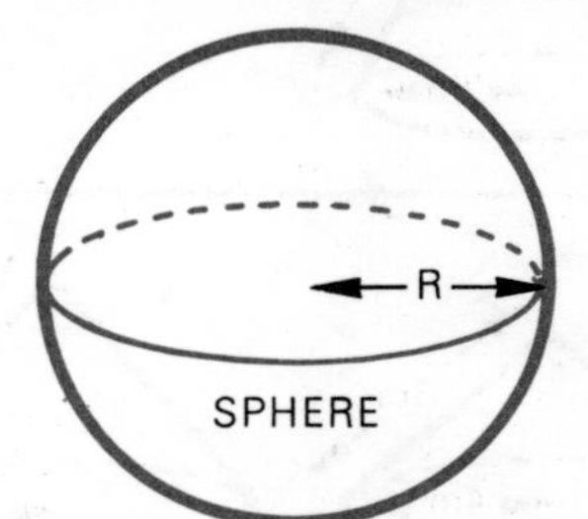

$A = 4\pi R^2$

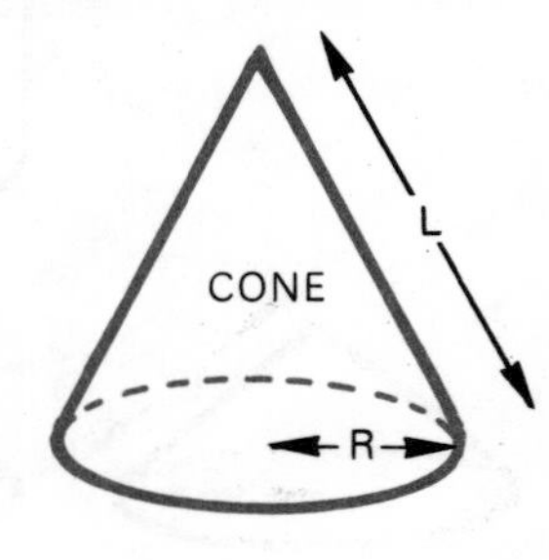

$A = \pi RL$ (Area of curved part)

You should notice that the formula gives the area of the curved surface of the cone only. The total surface area is given by $A = \pi RL + \pi R^2$.

Exercise 9.5

Read this example and then do the examples underneath. Use $\pi = 3$, or 3·1, or 3·14.

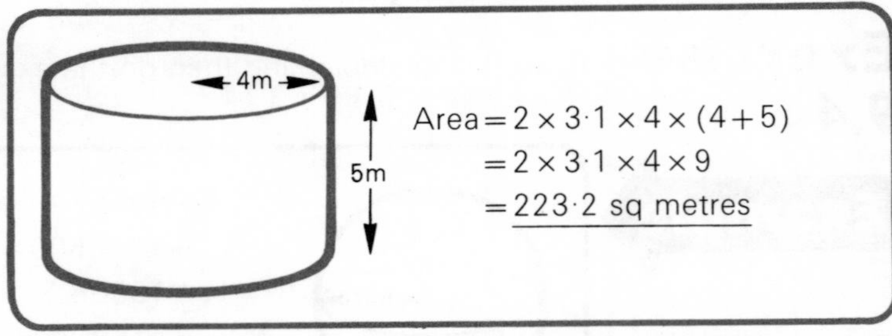

$$\text{Area} = 2 \times 3{\cdot}1 \times 4 \times (4+5)$$
$$= 2 \times 3{\cdot}1 \times 4 \times 9$$
$$= \underline{223{\cdot}2 \text{ sq metres}}$$

A Find the surface areas of the cylinders in Exercise **9.2**.

Read this example and then do the examples underneath.

Example

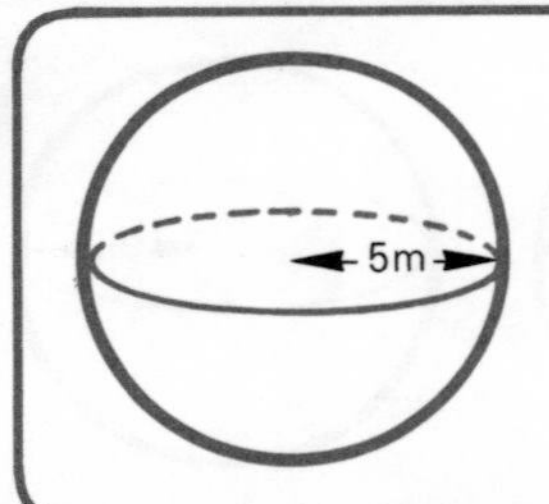

$$\text{Area} = 4 \times 3{\cdot}1 \times 5 \times 5$$
$$= 3{\cdot}1 \times 100$$
$$= \underline{310 \text{ sq metres}}$$

B Find the surface areas of the spheres in Exercise **9.4**.

Exercise 9.6

Read this example and then do the examples underneath. Use $\pi = 3$, or 3·1, or 3·14.

Example

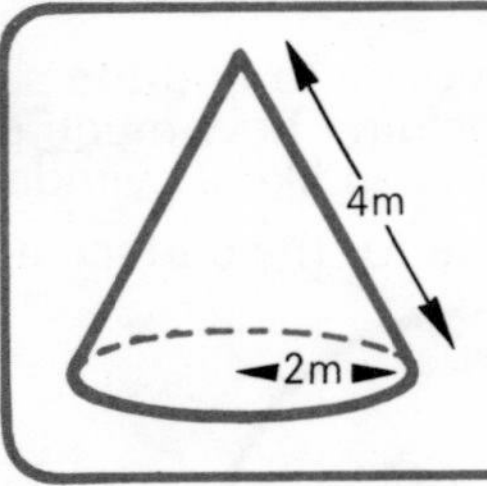

$$\text{Area} = 3{\cdot}1 \times 2 \times 4$$
$$= 3{\cdot}1 \times 8$$
$$= \underline{24{\cdot}8 \text{ sq metres}}$$

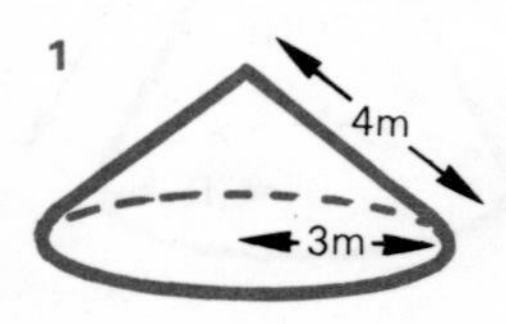

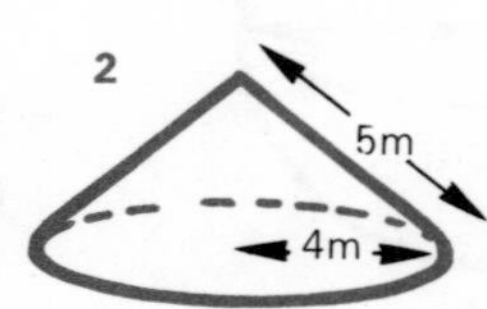

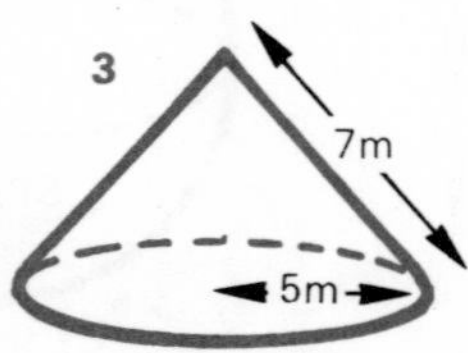

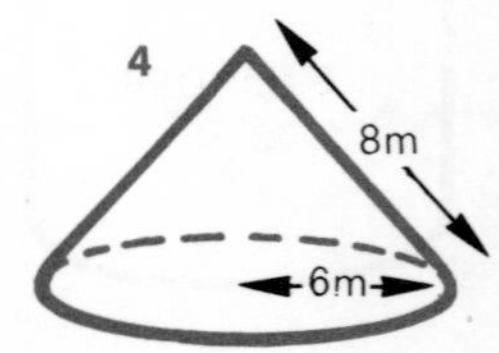

Read this example and then do the examples underneath.

Example

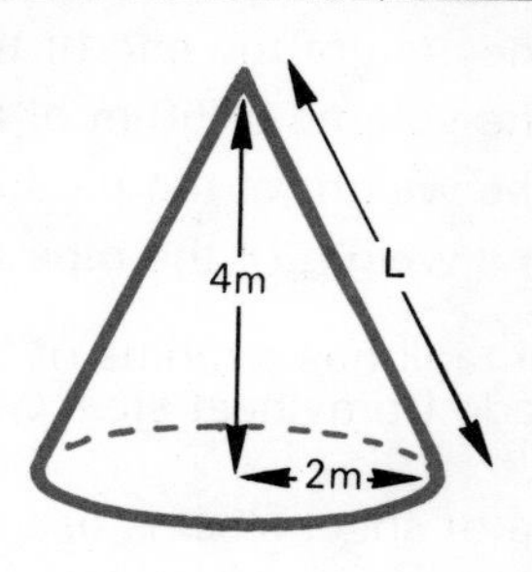

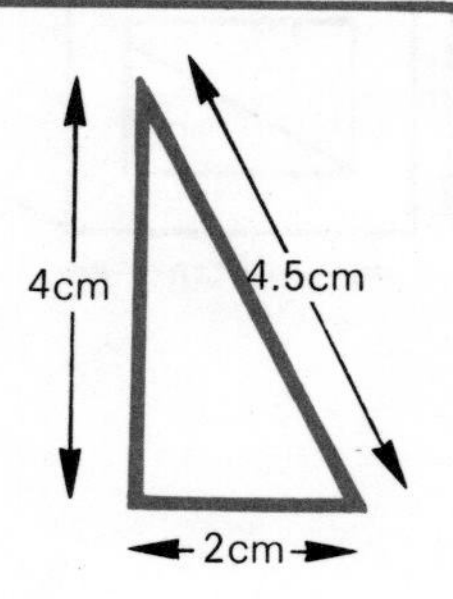

You will see that in this case the height of the cone is given, and we need to find the length L. This can be done by means of a scale drawing (shown on the right). Or by using the Rule of Pythagoras (see Unit 13). This is done underneath the example. An accurate scale drawing will give $L = 4{\cdot}5$ m.

Area $= 3{\cdot}1 \times 2 \times 4{\cdot}5$
$= \underline{27{\cdot}9 \text{ sq metres}}$

$L^2 = 4^2 + 2^2$
$\Rightarrow L^2 = 20$
$\Rightarrow L = \sqrt{20} = \underline{4{\cdot}47 \text{ m}}$

Turn to Exercise **9.3** (Numbers **7** to **14**) and find

a The length L for each cone.

b The area of the curved surface of each cone.

Exercise 9.7

Use $\pi = 3$, or $3{\cdot}1$, or $3{\cdot}14$.

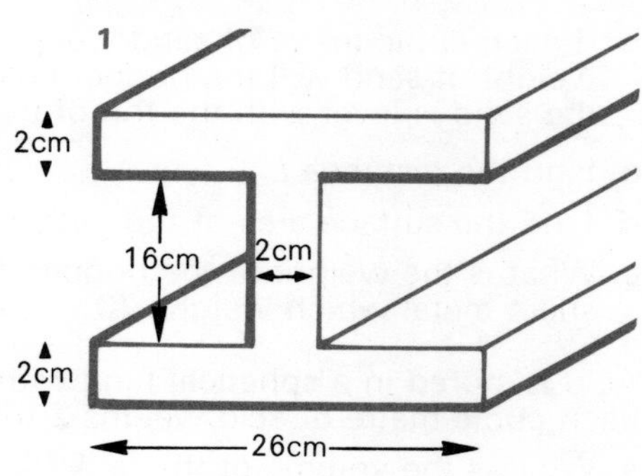

1 The drawing shows a steel girder 4 m long. The density of the steel is 8 grams per cubic cm.

a Find the area of the end of the girder.

b Change 4 m to cm.

c Find the volume of the girder.

d Find the weight of the girder in grams.

e Find the weight of the girder in kg.

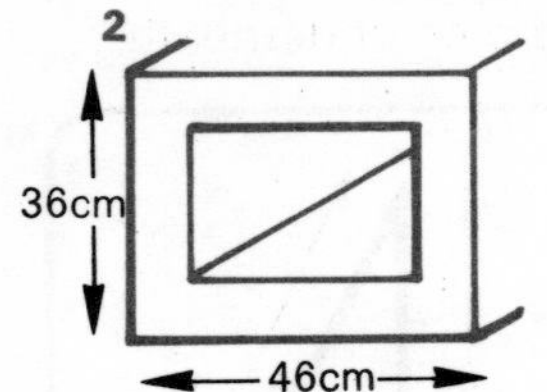

2 The drawing shows a hollow concrete pipe with sides 6 cm thick, and has a density of 6 grams per cubic cm.

a What is the area of the end of the pipe?

b What is the volume of 50 m of this pipe?

c What is the weight of the pipe in grams?

d What is the weight of the pipe in kg?

3 A cylindrical tank has a radius of 2 m and a height of 4 m. It is made from sheet steel which weighs 80 kg per square metre.

a What area of sheet steel is used to make the tank?

b What will be the weight of the steel?

c What is the volume of the cylinder?

The tank contains a liquid which has a density of 2 tonnes per cubic m.

d Find the weight of the liquid.

4 A cylindrical tank has a radius of 3 m and a height of 7 m.

a What is the surface area of the tank?

The tank has to be painted using tins of paint, each one of which will cover 9 square metres.

b How many tins will need to be bought?

c If each tin costs 73p, how much will the paint cost?

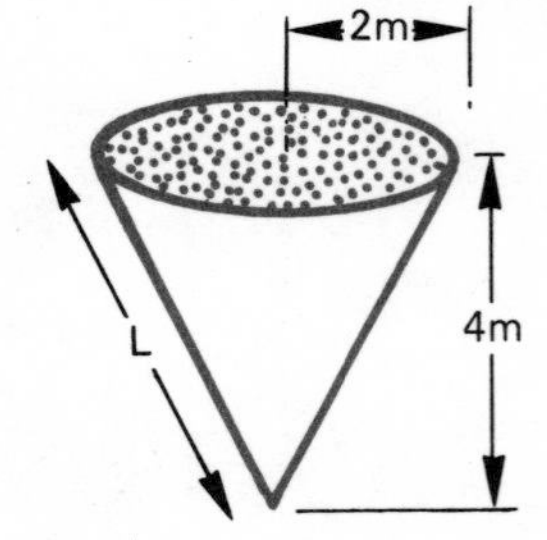

5 Sand is stored in the conical hopper shown with an open t

a What is the volume of the hopper?

b If each cubic metre of sand weighs 6 tonnes, what weight of sand will the hopper hold? (Assume that the sand is level with the top of the hopper.)

c Find the distance L.

d Find the surface area of the outside of the hopper.

e What is the weight of the hopper if it is made from sheet metal which weighs 70 kg per square metre?

6 Acid is stored in a spherical tank with a radius of 11 m. Each cubic metre of acid weighs 2 tonnes.

a What is the volume of the tank?

b What is the weight of acid in the tank?

c What is the area of the surface of the tank?

d How many litre tins of paint will be needed if each litre covers 8 square metres?

e How much will the paint cost if each tin costs £1·23?

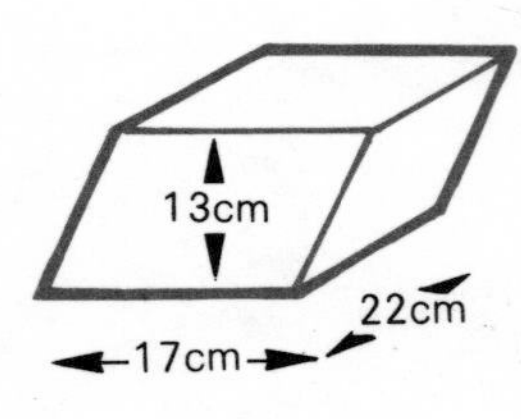

7 a What is the area of the end of this shape?

b What is the volume of this shape?

Exercise R 13

		a	b	c	d	e	f
1	Write 12 in base	4	5	6	7	8	9
2	Write 10 in base	3	4	5	6	7	8
3	Write 16 in base	4	5	6	7	8	9
4	Write 76 in base	3	4	5	6	7	8
5	Write 246 in base	4	5	6	7	8	9

Change these numbers back into base 10.

	a	b	c	d	e	f
6	13_5	18_9	14_6	15_7	11_8	11_4
7	13_6	12_5	14_7	16_9	16_7	12_8
8	24_5	28_9	33_8	30_5	41_6	20_7
9	78_9	36_8	64_7	44_5	37_8	76_9
10	345_6	457_8	343_5	245_7	2233_9	4573_8

The following numbers are in base 2. Change them into base 10.

	a	b	c	d	e	f
11	111	1000	101	110	1001	1011
12	1010	1101	1100	1110	1111	10000
13	11011	10111	10101	100001	1010111	1110111

Change the following numbers into binary (base 2).

	a	b	c	d	e	f
14	14	13	17	23	27	31
15	33	43	48	53	62	64
16	65	68	74	89	108	245

revision 14 Equations

Exercise R 14

1a $a+1=10$ **b** $b+7=12$ **c** $c+10=11$
d $d+2=13$ **e** $e+3=14$

2a $q+6=26$ **b** $r+8=27$ **c** $s+7=25$
d $t+9=30$ **e** $u+10=42$

3a $a-1=5$ **b** $b-3=6$ **c** $c-2=5$
d $d-4=8$ **e** $e-5=7$

4a $q-9=18$ **b** $r-11=20$ **c** $s-10=27$
d $t-12=35$ **e** $u-13=42$

5a $3a=12$ **b** $2b=18$ **c** $6c=18$
d $4d=36$ **e** $5e=35$

6a $8q=56$ **b** $9r=81$ **c** $7s=70$
d $10t=80$ **e** $9u=54$

7a $\frac{a}{2}=3$ **b** $\frac{b}{3}=2$ **c** $\frac{c}{4}=6$
d $\frac{d}{4}=5$ **e** $\frac{e}{2}=5$

8a $\frac{q}{10}=10$ **b** $\frac{r}{9}=11$ **c** $\frac{s}{10}=12$
d $\frac{t}{9}=10$ **e** $\frac{u}{8}=13$

9a $3a+4=25$ **b** $7u+2=30$ **c** $3n+8=29$
d $2r+6=12$ **e** $9t+5=23$

10a $7b-1=20$ **b** $8m-3=37$ **c** $9p-7=20$
d $6r-6=12$ **e** $6h-3=63$

11a $2a=3$ **b** $5b=27$ **c** $6c=25$
d $7d=20$ **e** $8e=31$

12a $2p+6=11$ **b** $3q+8=22$ **c** $5r+7=33$
d $8s+7=26$ **e** $9t+10=53$

13a $5v-3=24$ **b** $7w-9=30$ **c** $4x-10=37$
d $8y-6=55$ **e** $13z-7=1$

14a $5-a=8$ **b** $b+7=2$ **c** $5+2c=42$
d $23-6d=14$ **e** $3e+9=6$

The area of a rectangular room is given by the formula $A=L\times W$. In each of the following examples write down an equation and then solve it.

15 Find the width of a room with an area of 24 sq m and a length of 8 m.

16 Find the width of a room with an area of 56 sq m and a length of 8 m.

17 Find the length of a room with an area of 36 sq m and a width of 4 m.

18 Find the length of a room with an area of 70 sq m and a width of 8 m.

Find the weights of sugar, butter, and flour needed to make the following weights of pastry. (Write down an equation and then solve it.)

19 48 kg pastry: 1 part sugar, 2 parts butter, 5 parts flour.

20 28 kg pastry: 1 part sugar, 2 parts butter, 4 parts flour.

21 36 kg pastry: 1 part sugar, 3 parts butter, 5 parts flour.

22 200 g pastry: 2 parts sugar, 3 parts butter, 5 parts four.

23 900 g pastry: 2 parts sugar, 3 parts butter, 4 parts flour.

If we know the speed of a car and the time it takes to do a journey, the formula for the distance is given by
distance = speed × time.
Find the speeds in the following examples by first writing down an equation and then solving it.

24 Time = 2 h, Distance = 40 km.

25 Time = 4 h, Distance = 200 km.

26 Time = 7 h, Distance = 350 km.

27 Time = 8 h, Distance = 640 km.

28 Time = 5 h, Distance = 347 km.

29 Time = 13 h, Distance = 579 km.

Squares of numbers

You should remember from Book 1 that 5^2 (read as "5 squared") $= 5 \times 5 = 25$. What is the value of 56^2?

Most people cannot do this in their heads, and because finding the squares of numbers is needed often in Mathematics, tables of squares have been prepared. Look at the table of squares on page 222. You will see that next to 56 there is the number 3136. This means that $56^2 = 3136$.

Make sure that you see how the following are done using the tables.

$35^2 = 1225$

$48^2 = 2304$

$79^2 = 6241$

Exercise 10.1

Find the value of the following using the table of squares.

1a 17^2 **b** 23^2 **c** 65^2
d 59^2 **e** 28^2

2a $13^2 + 12^2$ **b** $45^2 + 23^2$ **c** $34^2 - 17^2$
d $56^2 - 45^2$ **e** $45^2 - 44^2$

3a $23^2 + 24^2 + 25^2$ **b** 11^2 **c** 11^3
d 13^2 **e** 13^4

Square roots

7 is called the **square root** of 49 because $7 \times 7 = 49$.

The sign for square root is $\sqrt{}$ and we can write $\sqrt{49} = 7$.

What is the square root of 9?

We can write this as $\sqrt{9} = 3$.

What is the square root of 36?

We can write this as $\sqrt{36} = 6$.

Example

Work out $\sqrt{169}$.

We know that $10 \times 10 = 100$, so the answer must be larger than 10. $20 \times 20 = 400$, so the answer must be less than 20.

Try 15. $15 \times 15 = 225$, so the answer must be less than 15.

Try 13. $13 \times 13 = 169$. The answer is 13. We can write $\sqrt{169} = \underline{13}$.

Exercise 10.2

Find the value of the following.

	a	b	c	d
1	$\sqrt{16}$	$\sqrt{9}$	$\sqrt{1}$	$\sqrt{25}$
2	$\sqrt{81}$	$\sqrt{4}$	$\sqrt{49}$	$\sqrt{64}$
3	$\sqrt{121}$	$\sqrt{100}$	$\sqrt{36}$	$\sqrt{0}$
4	$\sqrt{289}$	$\sqrt{169}$	$\sqrt{144}$	$\sqrt{256}$
5	$\sqrt{529}$	$\sqrt{400}$	$\sqrt{225}$	$\sqrt{196}$
6	$\sqrt{900}$	$\sqrt{441}$	$\sqrt{324}$	$\sqrt{361}$
7	$\sqrt{2809}$	$\sqrt{3364}$	$\sqrt{5776}$	$\sqrt{7396}$

Square roots that do not work out exactly

We know that $\sqrt{16}=4$ and that $\sqrt{25}=5$.
What is $\sqrt{20}$?

It must lie between 4 and 5.
Let us try to guess it.
$4{\cdot}3\times4{\cdot}3=18{\cdot}49$
$4{\cdot}4\times4{\cdot}4=19{\cdot}36$
$4{\cdot}5\times4{\cdot}5=20{\cdot}25$

You should see that $\sqrt{20}$ lies between 4·4 and 4·5, so we can say

$$\sqrt{20}=4{\cdot}4 \quad \text{or} \quad \sqrt{20}=4{\cdot}5$$

Since 20·25 is nearer to 20 than 19·36, it is more accurate to say $\sqrt{20}=4{\cdot}5$.

If we kept on trying to find $\sqrt{20}$ exactly we should find that this cannot be done; very few numbers have square roots that come out exactly, although we can get as near as we want.

For example, if we work out $4{\cdot}472\times4{\cdot}472$ it comes to 19·998 784 which is very close to 20. So we can say that for most purposes

$$\sqrt{20}=4{\cdot}472.$$

Example

Find $\sqrt{46}$ to one decimal place.

Try $6\times6=36$	Too small
Try $7\times7=49$	Too large
Try $6{\cdot}8\times6{\cdot}8=46{\cdot}24$	Too large
Try $6{\cdot}7\times6{\cdot}7=44{\cdot}89$	Too small

We can say $\sqrt{46}=6{\cdot}8$.

Exercise 10.3

Find the value of the following to one decimal place.

	a	b	c	d	e	f
1	$\sqrt{5}$	$\sqrt{7}$	$\sqrt{8}$	$\sqrt{11}$	$\sqrt{17}$	$\sqrt{18}$
2	$\sqrt{23}$	$\sqrt{27}$	$\sqrt{30}$	$\sqrt{34}$	$\sqrt{37}$	$\sqrt{42}$
3	$\sqrt{45}$	$\sqrt{62}$	$\sqrt{73}$	$\sqrt{89}$	$\sqrt{92}$	$\sqrt{98}$

Using tables of square roots

You will see from the last exercise that finding square roots is rather difficult. Because of this tables of square roots have been prepared. There is a table of square roots on page 222.

You will see from the table that $\sqrt{26}=5{\cdot}10$ and $\sqrt{69}=8{\cdot}31$.

Exercise 10.4

Use the table of square roots to do Exercise **10.3**.

Using square roots to solve equations

Later on you will need to be able to solve more difficult equations like the ones below.

Exercise 10.5

Read the next two examples, and then do the exercise below.

$$x^2=36 \quad\Rightarrow x=\sqrt{36} \quad\Rightarrow \underline{x=6}$$

$$a^2=47 \quad\Rightarrow a=\sqrt{47} \quad\Rightarrow \underline{a=6{\cdot}86}$$

	a	b	c	d	e
1	$z^2=49$	$s^2=64$	$r^2=81$	$d^2=25$	$f^2=16$
2	$y^2=13$	$b^2=29$	$x^2=56$	$m^2=67$	$n^2=93$

Read the next three examples, and then do the exercise below.

$$3b^2=75 \quad\Rightarrow b^2=75\div 3 \quad\Rightarrow b^2=25 \quad\Rightarrow b=\sqrt{25} \quad\Rightarrow \underline{b=5}$$

$$5y^2=35 \quad\Rightarrow y^2=35\div 5 \quad\Rightarrow y^2=7 \quad\Rightarrow y=\sqrt{7} \quad\Rightarrow \underline{y=2{\cdot}65}$$

$$4c^2=57 \quad\Rightarrow c^2=57\div 4 \quad\Rightarrow c^2=14{\cdot}2 \quad\Rightarrow c=\sqrt{14{\cdot}2} \quad\Rightarrow \underline{c=3{\cdot}74}$$

In the third example we have had to look up $\sqrt{14}$ instead of $\sqrt{14{\cdot}2}$ because the tables do not give $\sqrt{14{\cdot}2}$. If you used more accurate square root tables you would find that a more accurate answer was 3·77.

	a	b	c	d	e
3	$5a^2=45$	$3b^2=48$	$5c^2=20$	$4d^2=100$	$2c^2=98$
4	$2p^2=20$	$3q^2=18$	$4r^2=32$	$5s^2=100$	$7t^2=77$
5	$5x^2=26$	$4y^2=34$	$3z^2=25$	$8w^2=95$	$9v^2=100$

Read the next two examples and then do the exercise below.

$$\begin{aligned} & z^2+11=36 \\ \Rightarrow\ & z^2=36-11 \\ \Rightarrow\ & z^2=25 \\ \Rightarrow\ & z=\sqrt{25} \\ \Rightarrow\ & \underline{z=5} \end{aligned} \qquad \begin{aligned} & x^2-23=11 \\ \Rightarrow\ & x^2=11+23 \\ \Rightarrow\ & x^2=34 \\ \Rightarrow\ & x=\sqrt{34} \\ \Rightarrow\ & \underline{x=5{\cdot}83} \end{aligned}$$

6a $a^2+9=25$ **b** $b^2+14=50$ **c** $c^2+15=79$
d $d^2+21=70$ **e** $e^2+1=10$

7a $p^2-4=12$ **b** $q^2-5=20$ **c** $r^2-13=23$
d $s^2-45=4$ **e** $t^2-14=86$

8a $v^2+23=47$ **b** $w^2-9=10$ **c** $x^2+17=68$
d $y^2-23=42$ **e** $z^2-47=0$

unit 11 Using equations to solve area and volume problems

Exercise 11.1

Read this and then do the examples underneath.

Example

A builder wishes to put a rectangular water tank on a base 4 m long and 3 m wide. The tank has to hold 24 cubic m. What must be the height of the tank?

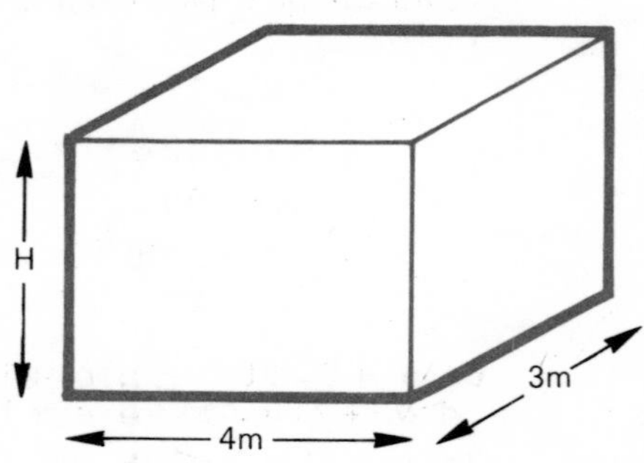

We know that Volume = Length × Width × Height

Put in the numbers we know $24 = 4 \times 3 \times H$

We can write this equation as $12H = 24$

$\Rightarrow H = 24 \div 12$

$\Rightarrow \underline{H = 2\text{ m}}$

1 In each of the following examples first write an equation down and then solve it to find the height of the tank.

	a	b	c	d	e	f	g	h	i	j	k	l	m	n	o	p	q
Volume in cubic m	18	16	45	48	54	32	90	126	99	96	55	42	20	39	55	42	106
Length in m	3	2	3	4	3	2	5	3	3	4	5	3	4	3	5	7	9
Width in m	2	2	3	2	2	2	2	2	3	2	1	3	2	2	2	3	5

2 A petrol tank has to be built in a space 2 m wide and 5 m high. What must the length of the tank be if it is to hold 55 cubic m?

3 A petrol tank has to be built in a space 4 m wide and 7 m high. What must the length of the tank be if it is to hold 120 cubic m?

4 A store room has to be constructed with a volume of 250 cubic m. The height between the floor and the ceiling is 4 m, and the room has to be built in a space 12 m wide. What will be the length of the room?

Exercise 11.2

Read this example and then do the exercise underneath.

Example

The distance around a wheel is 12 m. What is its diameter?

We know that Distance around = $\pi \times$ Diameter

Example

Put in the numbers we know $12 = 3 \times D$
We can write this equation as $3D = 12$

$$\Rightarrow D = 12 \div 3$$
$$\Rightarrow \underline{D = 4 \text{ m}}$$

To get a more accurate answer we can use $\pi = 3{\cdot}1$ and we would get:

$$D = 12 \div 3{\cdot}1 = 120 \div 31 = \underline{3{\cdot}87 \text{ m}}$$

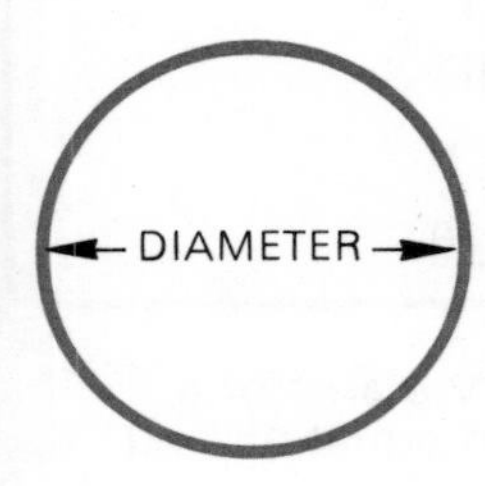

In each of the following examples you can use $\pi = 3$, or $3{\cdot}1$, or $3{\cdot}14$.

1a The distance around a circle is 6 m. Find its diameter.
b The distance around a circle is 15 m. Find its diameter.
c The distance around a circle is 18 m. Find its diameter.
d The distance around a circle is 21 m. Find its diameter.

Find the diameters of a circle if the distance around the circle is **e** 24 m, **f** 27 m, **g** 7 cm, **h** 11 cm, **i** 10 cm.

2 Some printing machines work by rolling a cylinder containing the type over the sheet of paper.

a What must the diameter of the cylinder be if the length of the paper is 33 cm?

b What must the diameter of the cylinder be if the length of the paper is 42 cm?

What must be the diameter of the cylinder if the length of the paper is:

c 72 cm, **d** 80 cm, **e** 250 cm?

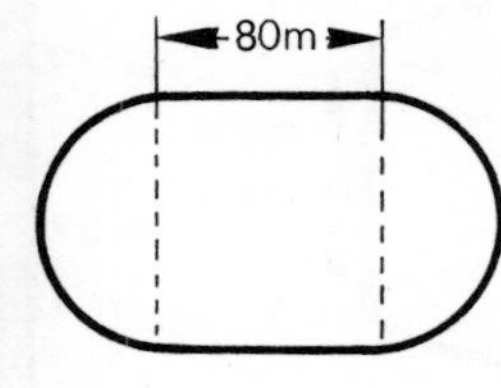

3 The diagram shows a running track made up of two 80 m straights, and two half circles. The total length of the running track is 400 m.

a What is the distance around the circular part of the track?

b What is the diameter of the circular ends?

4 Repeat question **3** for a running track with straights of 65 m.

5 Repeat question **3** for a running track with straights of 70 m.

6 Repeat question **3** for a running track with straights of 100 m.

Exercise 11.3

Read this example and then do the exercise underneath.

Example

A gardener wishes to plant 48 square metres of grass to make a circular lawn. What will be the radius of the lawn?

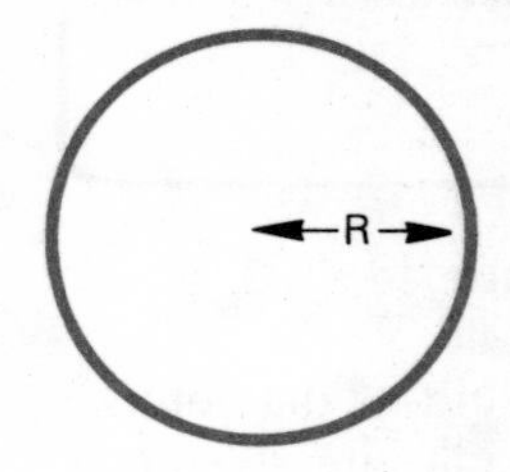

We know that Area $=\pi R^2$

Put in the numbers we know $48=3R^2$

We can write this equation as $3R^2=48$

$\Rightarrow R^2=48\div 3$

$\Rightarrow R^2=16$

$\Rightarrow R=\sqrt{16}$

$\Rightarrow \underline{R=4\text{ m}}$

In each of these examples you are given the area of a circle. Find the radius by writing down an equation and then solving it. $\pi=3$, or 3·1, or 3·14.

1 12 sq m
2 27 sq m
3 75 sq m
4 108 sq m
5 192 sq m
6 147 sq m
7 243 sq m
8 300 sq m
9 363 sq m
10 432 sq m
11 30 sq cm
12 42 sq cm
13 60 sq km
14 45 sq cm
15 71 sq cm
16 12·3 sq km

Exercise 11.4

Read this example and then do the exercise underneath.

Example

A cylindrical storage tank is to be built to hold 300 cubic m. What will the height of the tank be if the radius of the base is 5 m?

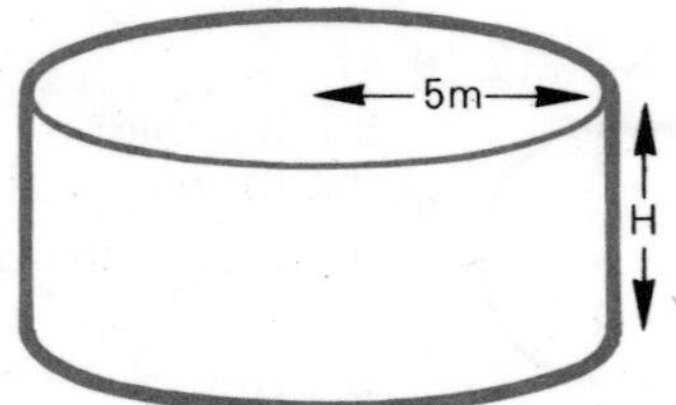

We know that Volume $=\pi R^2 H$

Put in the numbers we know $300=3\times 5\times 5\times H$

We can write this equation as $75H=300$

$\Rightarrow H=300\div 75$

$\Rightarrow \underline{H=4\text{ m}}$

Here is another way of solving this equation.

$$3\times 5\times 5\times H=300$$

Divide each side by 3 ⇒ $5 \times 5 \times H = 100$
Divide each side by 5 ⇒ $5 \times H = 20$
Divide each side by 5 ⇒ $\underline{H = 4 \text{ m}}$

Find the height of these cylindrical storage tanks. Use $\pi = 3$, or 3·1, or 3·14.

	a	b	c	d	e	f	g	h	i	j	k	l	m
Volume in cubic m	36	54	48	144	150	60	72	216	324	162	243	268	965
Radius in m	2	3	2	4	5	2	2	6	6	3	5	6	7

Exercise 11.5

Read this example and then do the exercise underneath.

Example

A cylindrical storage tank is to be built to hold 135 cubic m. What will the radius be if the height is 5 m?

We know that Volume $= \pi R^2 H$

Put in the numbers we know $135 = 3 \times R^2 \times 5$

We can write this equation as $15R^2 = 135$

⇒ $R^2 = 135 \div 15$

⇒ $R^2 = 9$

⇒ $R = \sqrt{9}$

⇒ $\underline{R = 3 \text{ m}}$

Find the radii of these cylindrical storage tanks. Use $\pi = 3$, or 3·1, or 3·14.

	a	b	c	d	e	f	g	h	i	j	k	l
Volume in cubic m	36	54	60	96	144	300	450	48	90	90	198	570
Height	3	2	5	2	3	4	6	2	3	2	3	4

unit 12 Clock arithmetic

Look at this sum $10+5=3$. It is obviously wrong. Sometimes, however, a sum that looks like this can be correct.

Suppose you leave home at 10 a.m. on a journey that will take you one hour. What time will you arrive?

Since $10+1=11$, you will arrive at 11 a.m.

Suppose the journey takes 5 hours. When will you arrive? $10+5=15$, but since 15 is not marked on most clocks you will have to say that you arrive at 3 p.m.

This is an example where it makes sense to say $10+5=3$.

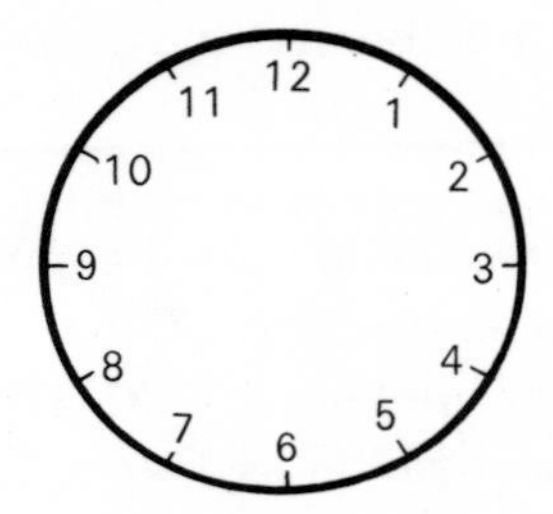

Exercise 12.1

Examples like this can be done easily by using a drawing of a clock face. To work out $10+5$ start at 10 and then count around 5 hours. You will see that you end up at 3.

1. I start a 7 hour journey at 10 a.m. What time will I arrive?
2. I start a 5 hour journey at 11 a.m. What time will I arrive?
3. I start an 8 hour journey at 7 a.m. What time will I arrive?
4. I start a 6 hour journey at 9 a.m. What time will I arrive?
5. I start a 9 hour journey at 5 p.m. What time will I arrive?
6. I start a 4 hour journey at 10 p.m. What time will I arrive?
7. I start a 10 hour journey at 5 p.m. What time will I arrive?
8. I start a 14 hour journey at 7 p.m. What time will I arrive?

Example

I want to get to Cardiff by 4 p.m. The journey takes 7 hours. What time must I start?

If you start at 4 and go 7 hours back you will end up at 9 a.m.

In clock arithmetic we can say that $4-7=9$.

Exercise 12.2

1. I want to end a 5 hour journey at 2 p.m. What time must I start?
2. I want to end an 8 hour journey at 5 p.m. What time must I start?
3. I want to end a 6 hour journey at 4 p.m. What time must I start?
4. I want to end a 7 hour journey at 1 p.m. What time must I start?
5. I want to end a 4 hour journey at 3 a.m. What time must I start?
6. I want to end a 9 hour journey at 4 a.m. What time must I start?
7. I want to end a 10 hour journey at 2 a.m. What time must I start?
8. I want to end a 14 hour journey at 8 a.m. What time must I start?

Another name for Clock Arithmetic is **Finite Arithmetic**. The clock arithmetic we have been doing is **modulo 12**. If there were only 10 hours in the day we would have to use a finite arithmetic **modulo 10**.
In finite arithmetics it is usual to start the numbers with 0 like this:

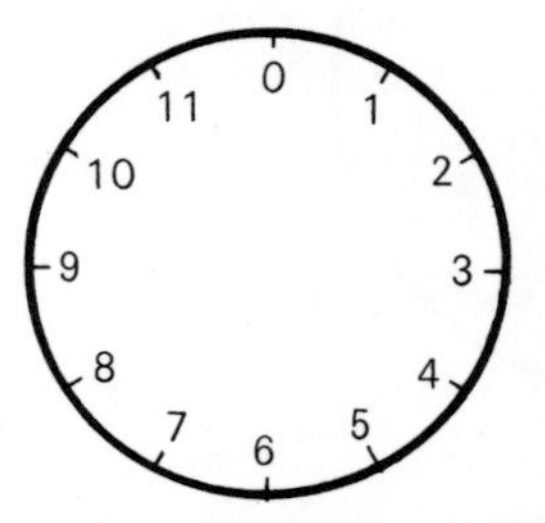

Example

I start a journey at 10 p.m. I travel 3 hours by train, and then 2 hours by boat, and then another 5 hours by train. What time will I arrive?

This is the same as working out $10+3+2+5$ and if you use the clock drawing you will see that $10+3+2+5=8$. Answer: 8 a.m.

Example

Work out $5+3-9$ (modulo 12).

If you use the clock face and start at 5, go 3 forwards, and then 9 back, you will end up at 11. We can say $5+3-9=11$ (modulo 12).

Exercise 12.3

1 I start a journey at 11 p.m. I travel 5 hours by train, 3 hours by boat, and then another 2 hours by train. What time will I arrive?

2 I start a journey at 8 a.m. I travel 2 hours by car, 7 hours by plane, and then 3 hours by car. What time will I arrive?

Now try these sums (modulo 12).

3 $4+6+8$
4 $6+9+6$
5 $2-9$
6 $4-8$
7 $1-9$
8 $1-8$
9 $3+7+9+7$
10 $5+8+9+4$
11 $4+5-9$
12 $5+8-9-7$
13 $6-8+9-3$
14 $4-6-9-7$
15 $3-7-9$
16 $9+8-7-7$
17 $6-9-8$
18 $6+7+8+9$

Modulo 24

Many travel organisations use a 24-hour clock. Use the drawing to do examples 3 to 18 in modulo 24 (that is, using the 24-hour clock).

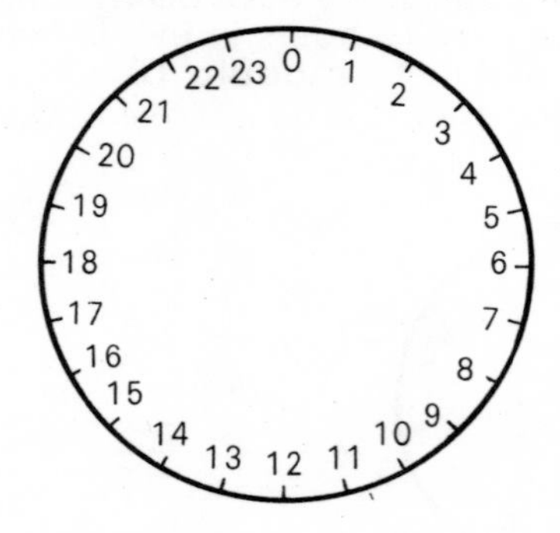

Modulo 10

Imagine that there are only 10 hours in a day. Draw a clock using the numbers 0 to 9 and then do examples 3 to 18 in modulo 10.

Multiplication in clock arithmetic

Can we multiply in clock arithmetic? If you remember that multiplying is just repeated addition you should be able to multiply.

3×7 (Modulo 12)

$3 \times 7 = 7+7+7$ and if we start at 7 and then move around 7 spaces, and then another 7 spaces we should end up at 9. We can say $3 \times 7 = 9$ (modulo 12).

Check that you can do these before you start on the next exercise.

$4 \times 5 = 8$ (modulo 12). $4 \times 6 = 0$ (modulo 12). $5 \times 5 = 1$ (modulo 24).

Exercise 12.4

1 3×7 (modulo 12)
2 5×3 (modulo 12)
3 2×8 (modulo 12)
4 5×6 (modulo 12)
5 2×6 (modulo 12)
6 4×7 (modulo 12)
7 3×6 (modulo 12)
8 3×4 (modulo 12)
9 4×4 (modulo 12)
10 2×7 (modulo 12)
11 5×5 (modulo 24)
12 5×6 (modulo 24)
13 4×7 (modulo 24)
14 4×6 (modulo 24)
15 3×5 (modulo 8)
16 2×7 (modulo 8)

Proper finite arithmetics used by mathematicians do not contain decimals or fractions, only whole numbers, so that sums like the following are not possible.

$2\frac{1}{2} + 3\frac{3}{4} = 6\frac{1}{4}$ $\quad$ $3{\cdot}7 + 4{\cdot}8 = 8{\cdot}5$ $\quad$ $5 \div 2 = 2{\cdot}5$

Exercise 12.5

(In this exercise only whole number answers are allowed.)

1 Copy and complete this multiplication table modulo 6. (A clock with the numbers 0 to 5 will help you.)

	0	1	2	3	4	5
0					0	0
1					4	5
2					2	4
3						
4						
5						

Use the table to answer the following questions.

2 What number $\times 5 = 4$?
3 Solve the equation $5a = 3$.
4 Solve the equation $5b = 2$.
5 Solve the equation $2c = 4$.
6 Solve the equation $2d = 2$.
7 Solve the equation $4e = 4$
8 Solve the equation $4f = 1$.
9 Solve the equation $4g = 3$.

revision 15 Congruence, similarity and enlargements

Exercise R 15

Which of the following pairs are congruent?

1

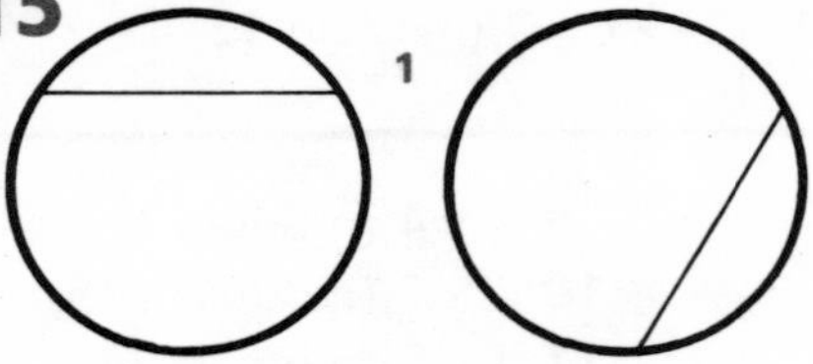

2

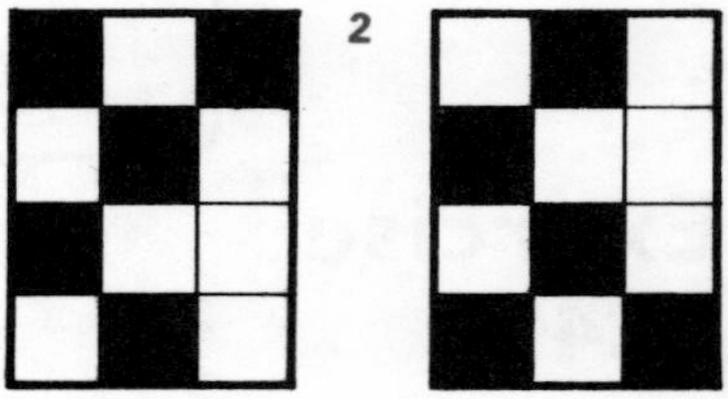

3

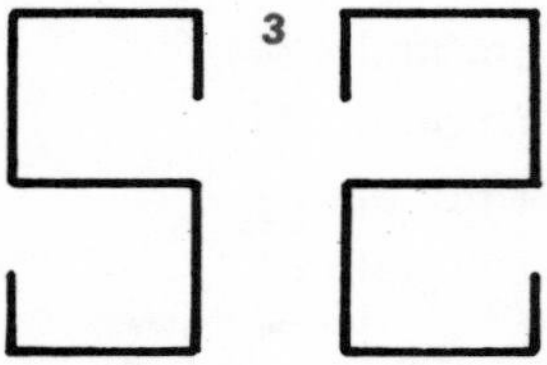

4

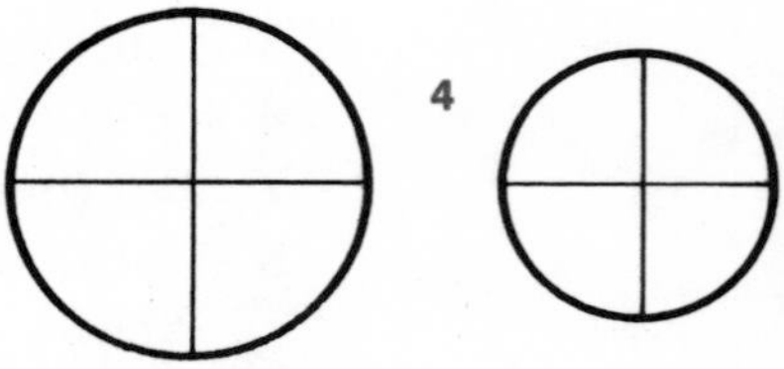

5

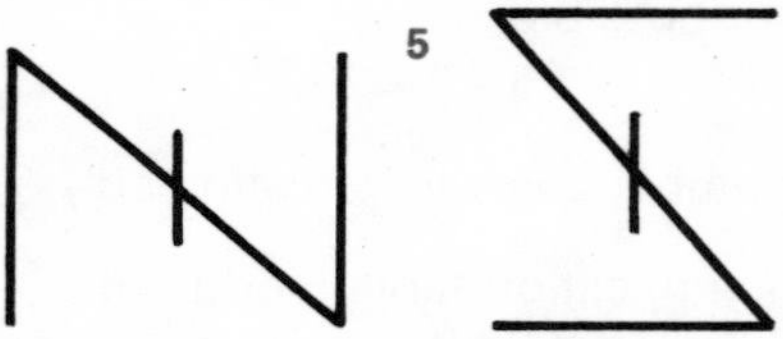

6

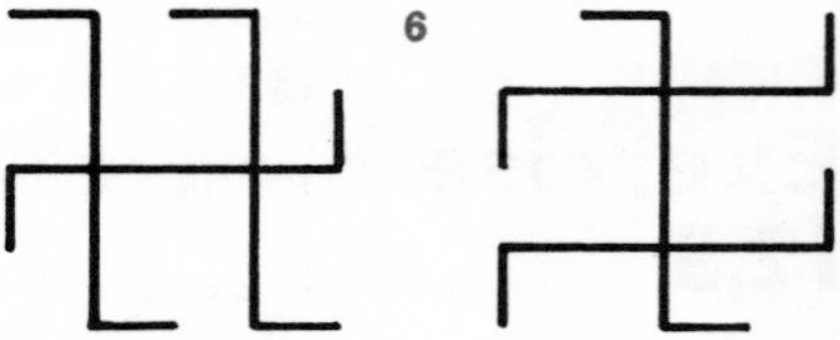

7 **a** On squared paper mark the horizontal axis from 0 to 14 and the vertical axis from 0 to 17.

b Mark in the points $A(3, 4)$, $B(1, 4)$, $C(1, 8)$ and join them to make the letter L.

c With centre of enlargement (0, 0) enlarge the letter L with scale factor 2. Mark the enlargement with the letters A', B', C'.

d Write down the co-ordinates of A', B', C'.

e Copy and complete this matrix multiplication.

$$2 \times \begin{bmatrix} 3 & 4 \\ 1 & 4 \\ 1 & 8 \end{bmatrix} = \begin{bmatrix} \quad & \quad \\ & \\ & \end{bmatrix}$$

f Check that the answers to **d** are the same as the numbers in the matrix answer to **e**.

8 **a** On squared paper mark the horizontal axis from 0 to 14 and the vertical axis from 0 to 17. (Or use the axis marked for question 7.)

b Mark in the points $P(2, 1)$, $Q(4, 1)$, $R(2, 3)$, $S(4, 3)$ and join them to make the shape Σ.

c With centre of enlargement (0, 0) enlarge the shape with scale factor 3. Mark the enlargement with the letters P', Q', R', S'.

d Write down the co-ordinates of P', Q', R', S'.

e Copy and complete this matrix multiplication.

$$3 \times \begin{bmatrix} 2 & 1 \\ 4 & 1 \\ 2 & 3 \\ 4 & 3 \end{bmatrix} = \begin{bmatrix} & \\ & \\ & \\ & \end{bmatrix}$$

f Check that the answers to **d** are the same as the numbers in the matrix answer to **e**

In each of the following examples find the scale factor, and then find the lengths marked by the letters.

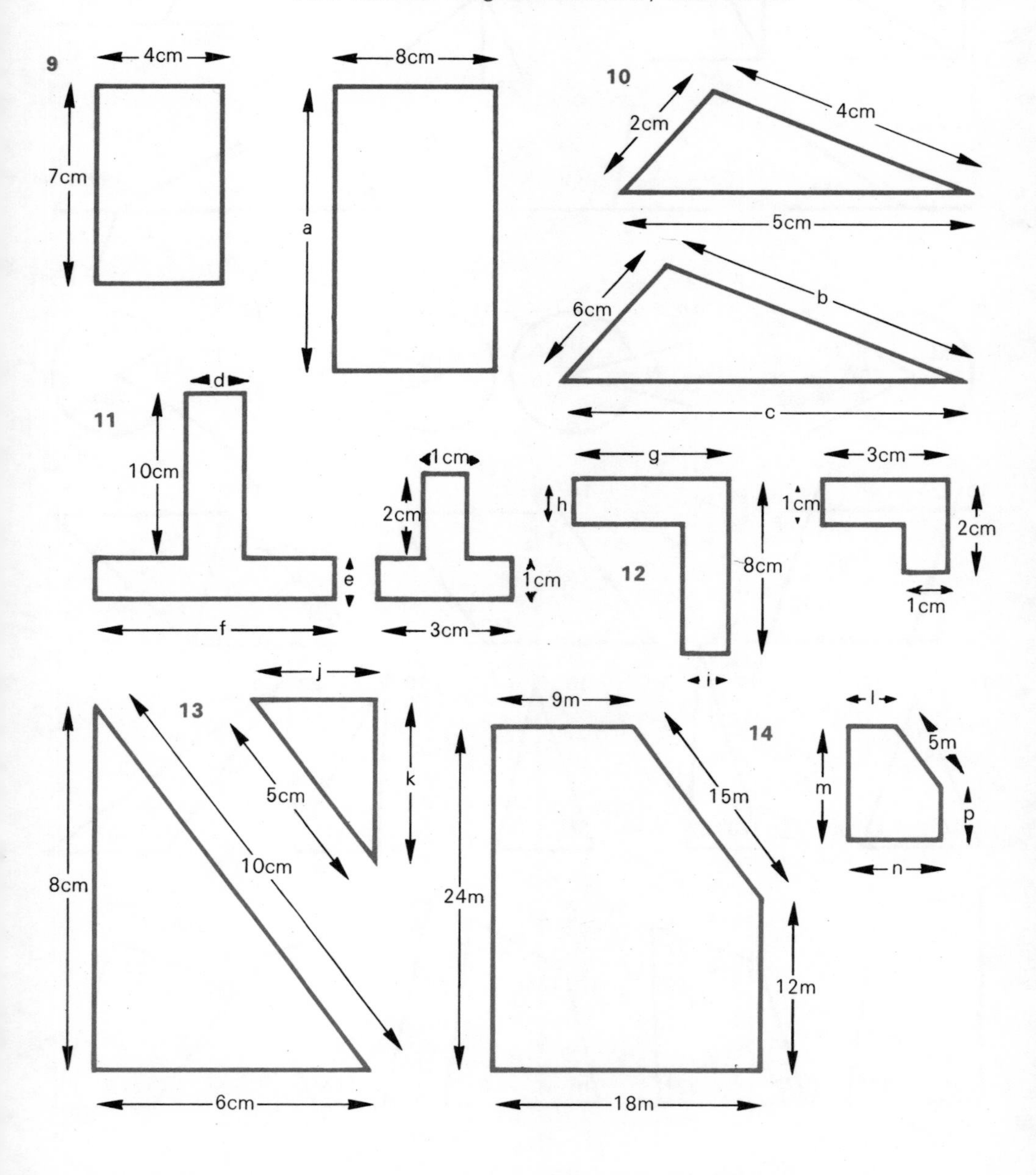

revision 16 Angles

Exercise R 16

To do the examples in this exercise you will need to remember that the angles of a triangle add up to 180°, and the angles of a four-sided shape add up to 360°.

Find the size of the angles marked by the letters.

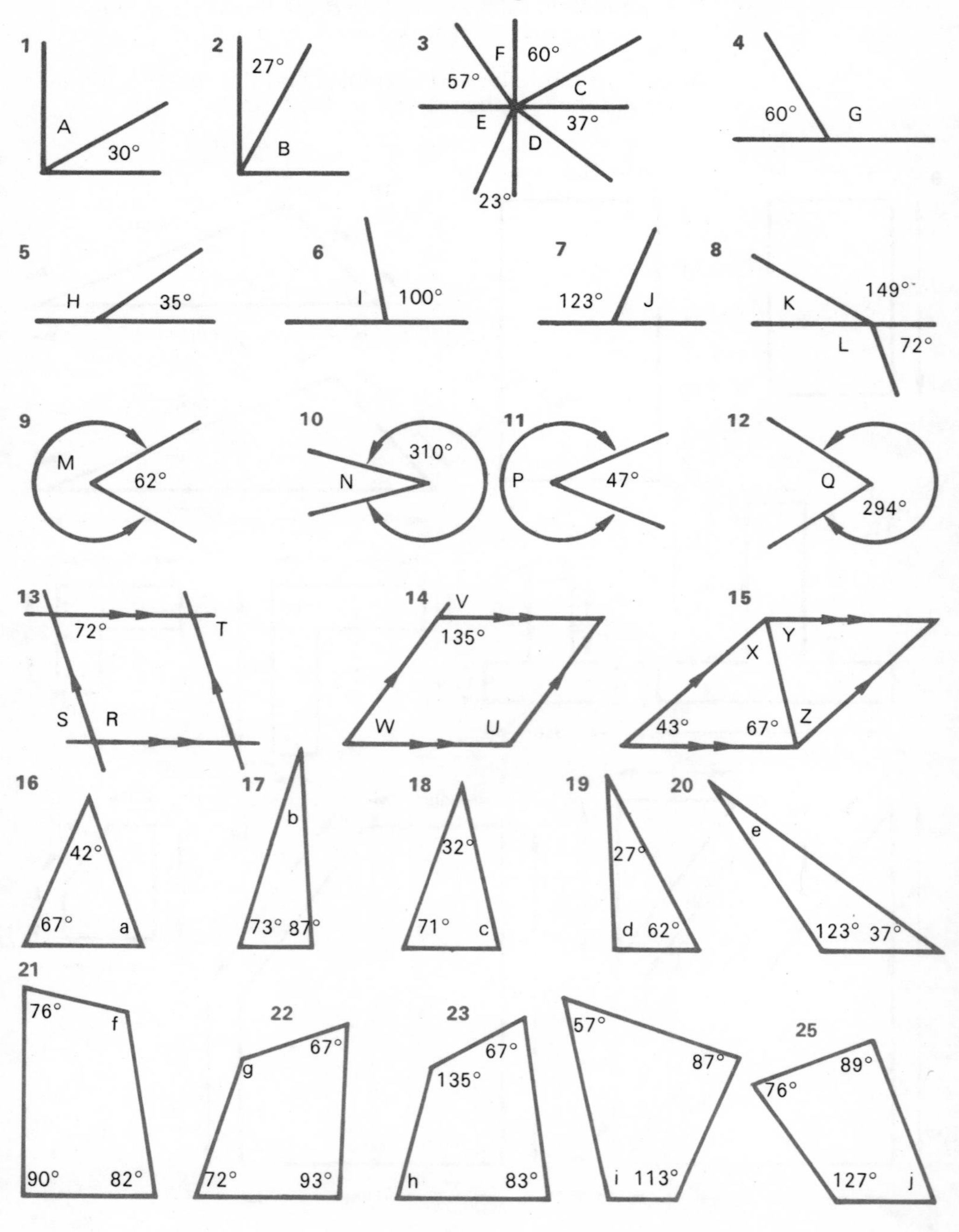

unit 13 Finding the sides of right-angled triangles

There are many problems involving right angled triangles that can be solved by means of scale drawings.

Example

What is the length of the longest side of this triangle?

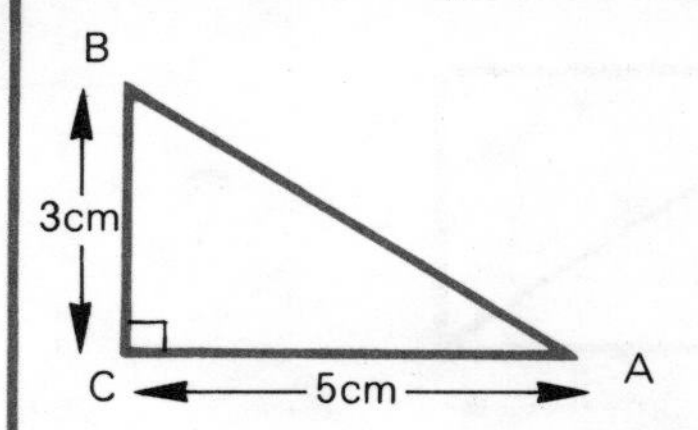

If the drawing is done on squared paper there is no need to worry about how to draw an accurate right angle.
The first thing to do is mark the point *C*, and then from *C* draw two lines one 5 cm long, and the other 3 cm long. Mark the ends of the line *A* and *B*. Now carefully draw a line from *A* to *B* and measure it.
The drawing below has been done accurately. If you measure it you will find that $AB = 5{\cdot}8$ cm.

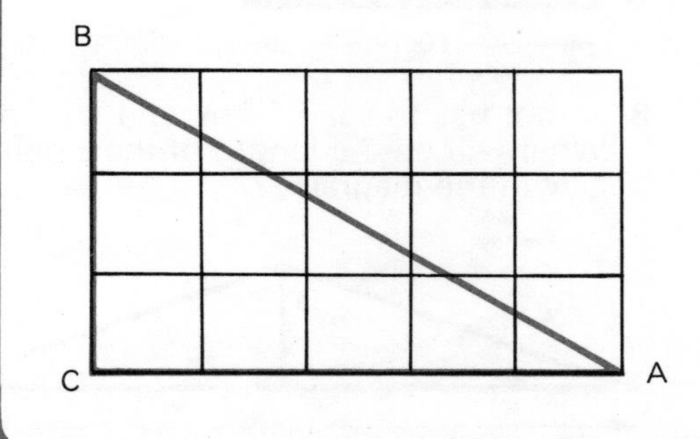

Exercise 13.1

Make scale drawings of these triangles and measure the lengths of the longest sides.

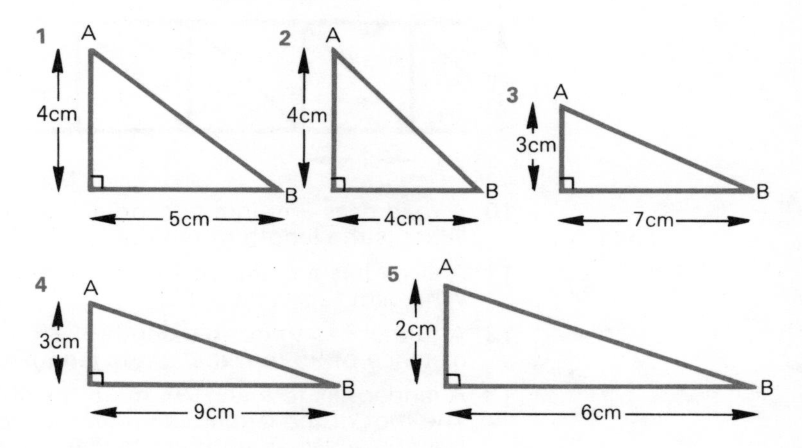

6 A man wishes to strengthen a gate which is 2 m high and 3 m wide, by placing a metal bar across from *A* to *B*. Make a drawing of the triangle *ABC* and measure the length of the bar. (Let 1 cm stand for one metre.)

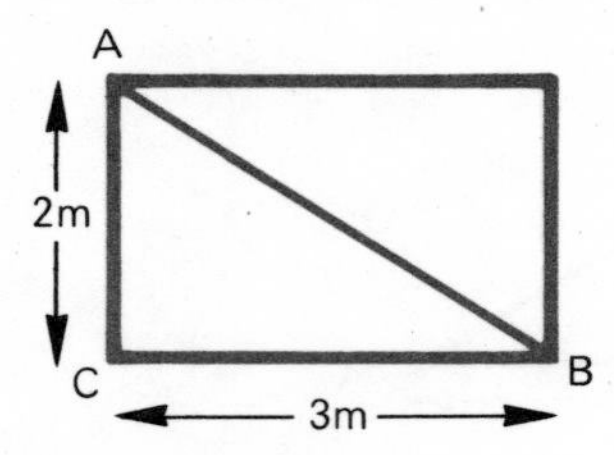

7 A clothes line is to be placed from one corner of a garden to the other. The garden is 6 m by 10 m. What length of line will be needed?

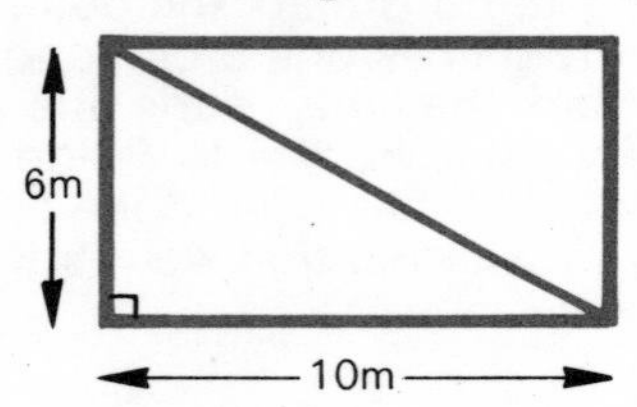

8 A roof has to span 14 m, and it is to be 2 m high. What will be the length of the sloping beams (marked *MN* in the diagram)?

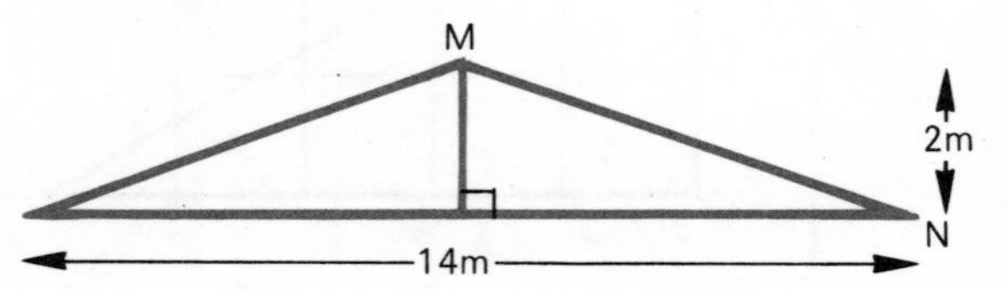

9 What is the length of the sloping girders of the bridge shown in the diagram (marked *PQ*)?

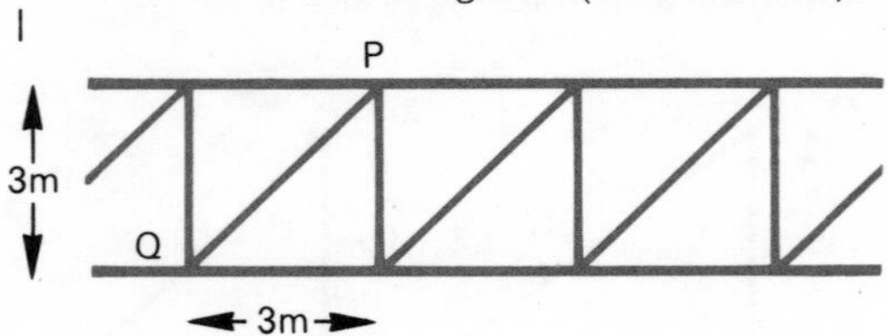

10 A road rises 3 m in a horizontal distance of 10 m. What is the length of the road?

11 A cone has a radius of 2·5 cm, and a height of 3·7 cm. What is its slant height?

12 A tunnel has to go to a depth of 35 m in a horizontal distance of 95 m. What is the length of the tunnel?

13 A ladder has to reach 7·5 m up the side of a house. The foot of the ladder can be no nearer than 1·5 m to the house. What length of ladder will be needed?

Finding the length of one of the short sides

Example

What is the length of the side *BC* of this triangle?

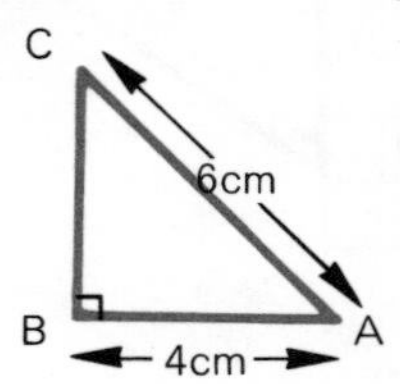

This type of problem is best done on squared paper to get an accurate right angle.
The first thing to do is to draw a line *BA* 4 cm long. Then open a compass to a radius of 6 cm, put the point on *A*, and draw part of the circle as shown below. Mark the point *C* where the circle cuts the vertical line above *B*. Draw the lines *BC* and *AC*. Measure the line *BC*. You will find that *BC* = 4·5 cm.

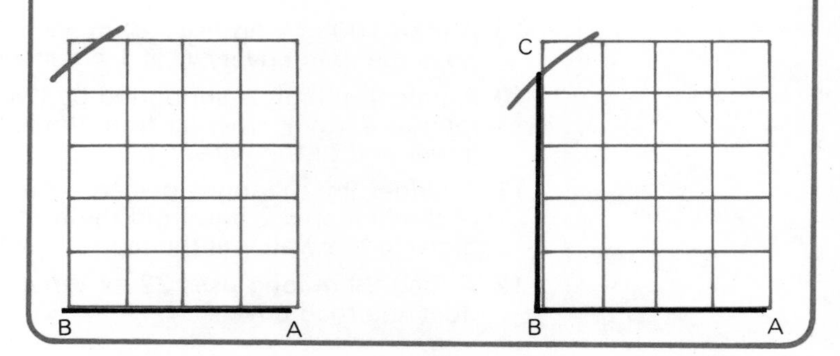

Exercise 13.2

Make scale drawings of these triangles and measure the lengths of the sides marked by the letters.

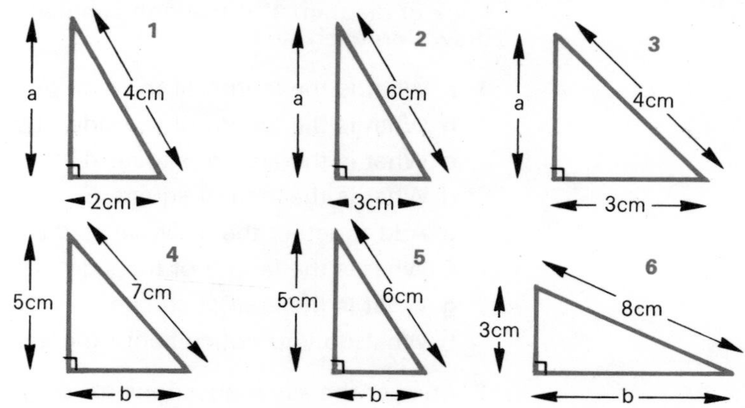

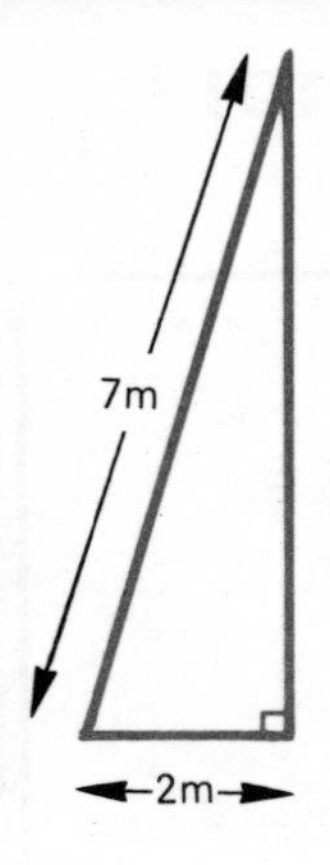

7 A ladder 7 m long is placed against a wall. The foot of the ladder is 2 m from the foot of the wall. How high up the wall will the ladder reach? (Let 1 cm stand for 1 metre.)

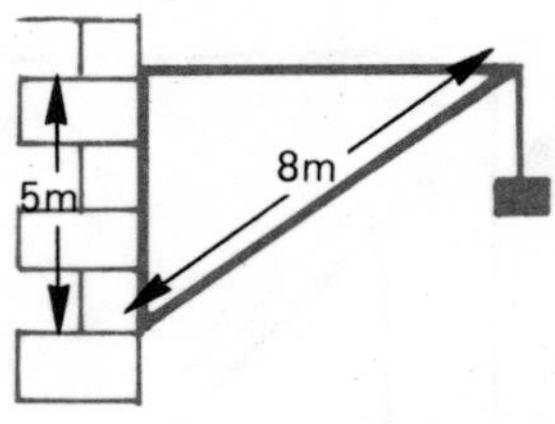

8 A crane is fixed to a wall so that the top of the jib is 5 m above the ground. The jib is 8 m long. How far out from the foot of the wall does the jib reach?

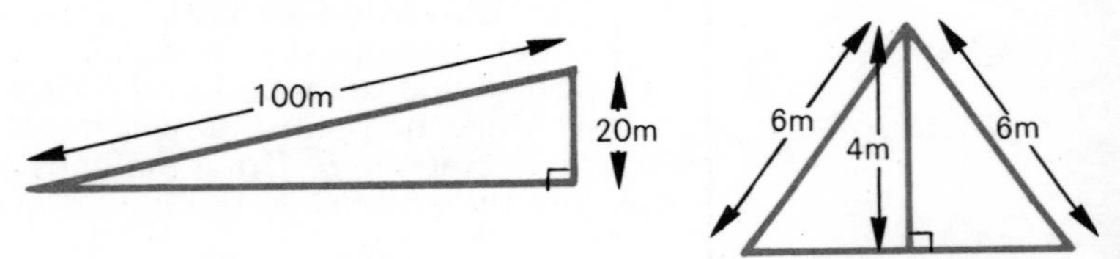

9 A road 100 m long rises 20 m. What horizontal distance does the road cover? (Let 1 cm stand for 10 m.)

10 A pole 4 m high is supported by two wires each 6 m long as shown. How far from the foot of the pole is the lower end of the wire?

11 A ladder 9·5 m long is placed against a wall. The foot of the ladder is 2·5 m from the foot of the wall. How high up the wall will the ladder reach?

12 A road 89 m long rises 22 m. What horizontal distance does the road cover?

Exercise 13.3

Use the table of squares on page 222 to help you answer these questions.

Look at diagram **1**. It is a right-angled triangle with squares drawn on each side.

1 **a** What is the length of the side of square *A*?
b What is the length of the side of square *B*?
c What is the area of square *A*?
d What is the area of square *B*?
e Add together the answers to **c** and **d**.
f What is the length of the side of square *C*?
g What is the area of square *C*?
h What do you notice about the answers to **e** and **g**?

2 Answer the same questions for diagram **2**.

3 Answer the same questions for diagram **3**.

4 Answer the same questions for diagram **4**.

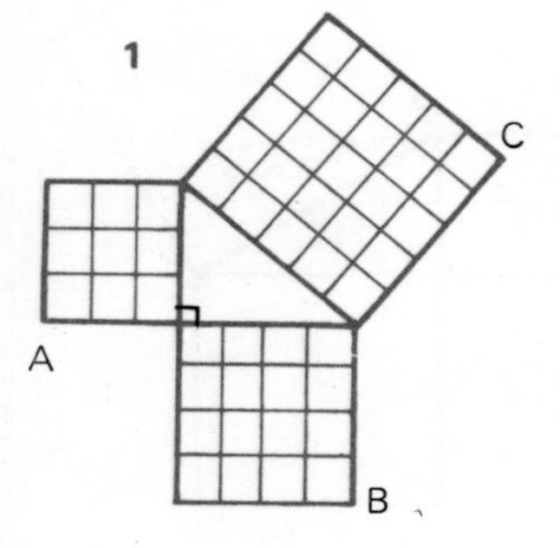

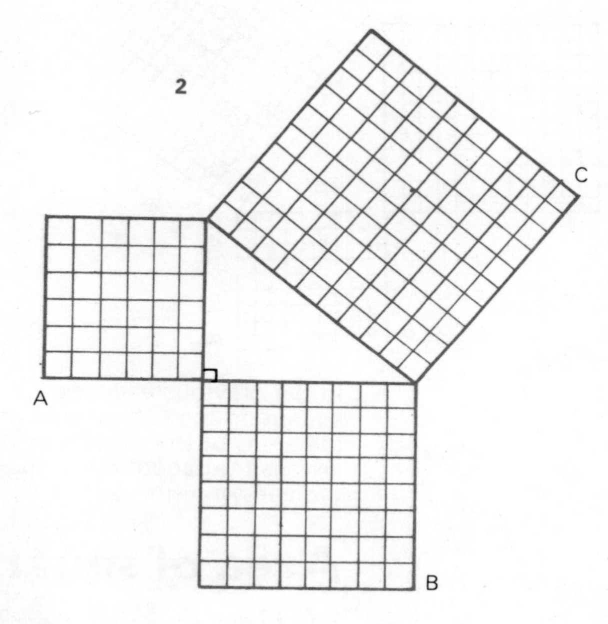

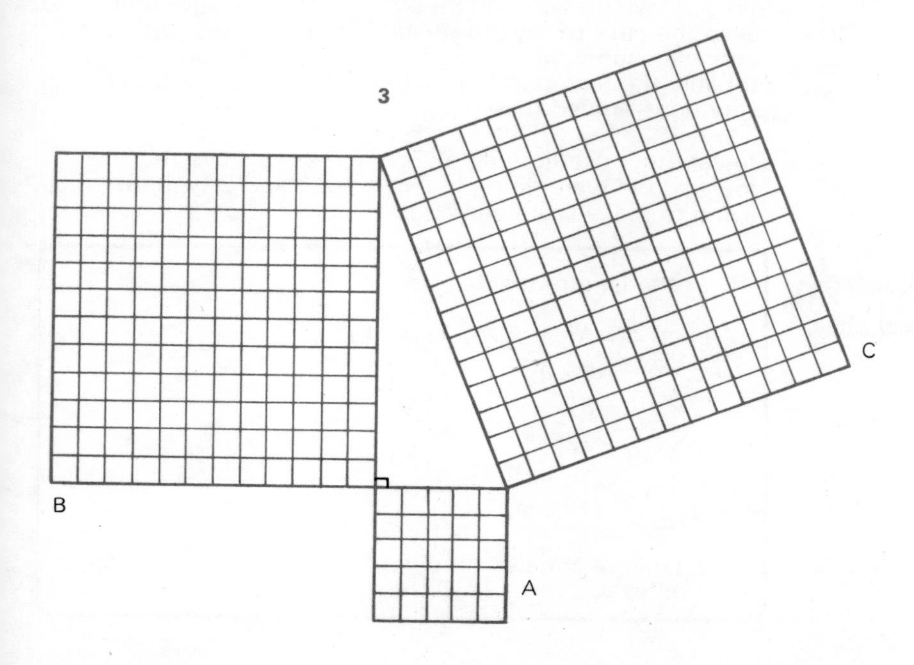

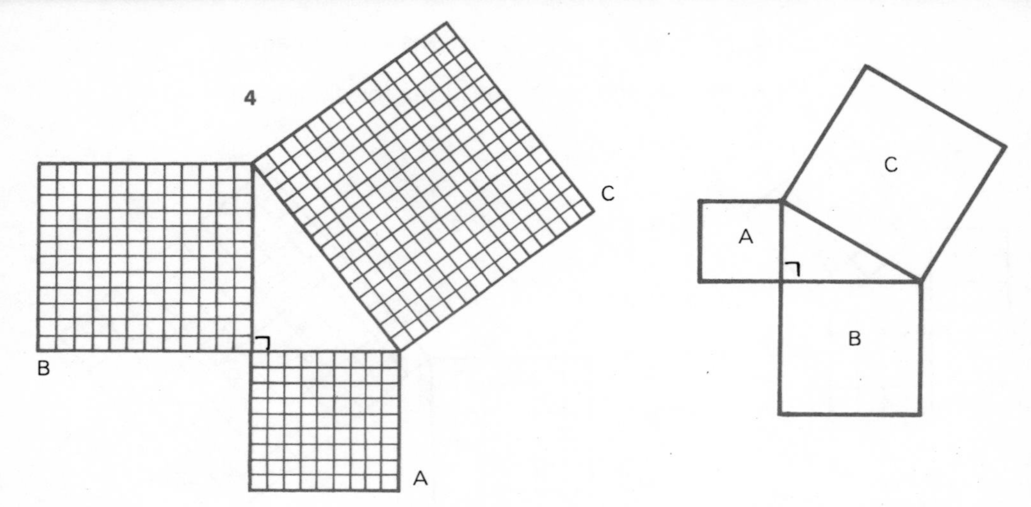

In the previous exercise you have found that the area of the square on the largest side of a right-angled triangle equals the areas of the other two squares added together. This is always true about right-angled triangles.
We can write this as:

Area of square A + Area of square B = Area of square C

It is not known who discovered this fact about right-angled triangles, which was known over 2500 years ago. It is called the **rule of Pythagoras**. Pythagoras was a Greek mathematician who was born about 569 B.C., and he was probably the first man to prove that this rule was true for every right-angled triangle.

We can now use the rule of Pythagoras to solve the problems we were doing in Exercises 1 and 2 without having to make scale drawings.

Example

Find the length of side c.

Area of C = Area of A + Area of B

$\Rightarrow c \times c = 9 \times 9 + 12 \times 12$

$\Rightarrow c \times c = 81 + 144$

$\Rightarrow c \times c = 225$

$\Rightarrow c \times c = 15 \times 15$

$\Rightarrow \underline{c = 15 \text{ cm}}$

The table of squares on page 222 will help you to do examples like this.

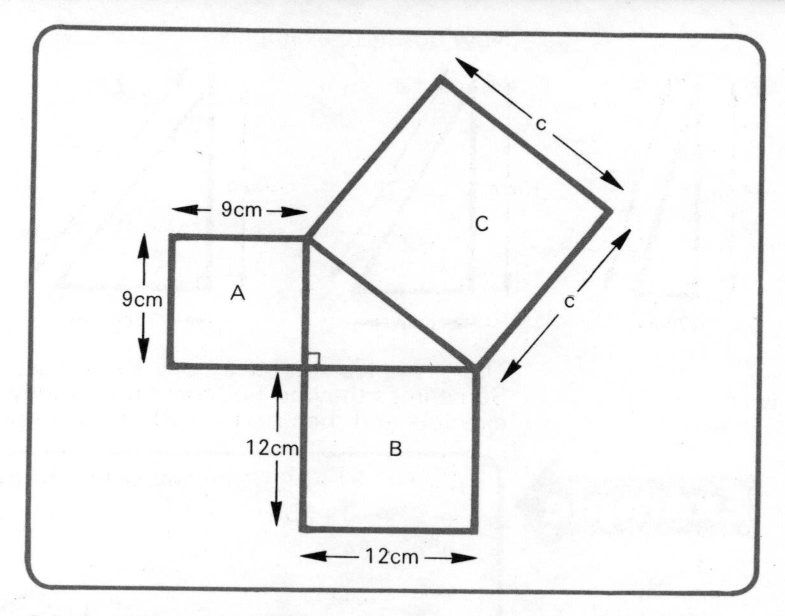

Exercise 13.4

Copy and complete these examples. Use the tables of squares on page 222 to help you.

1

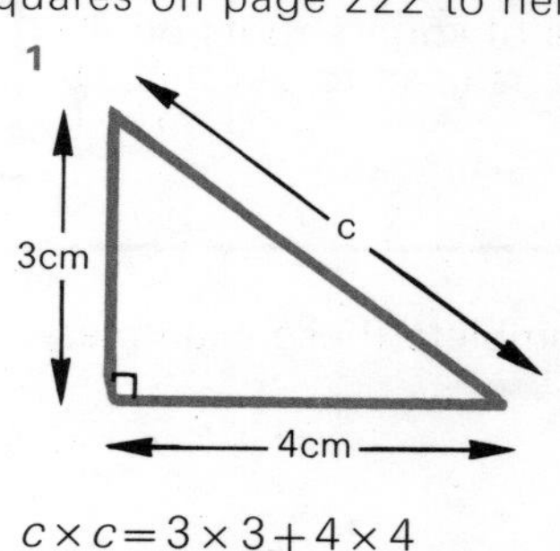

$c \times c = 3 \times 3 + 4 \times 4$

$\Rightarrow c \times c = \ 9 + 16$

$\Rightarrow c \times c = 25$

$\Rightarrow c \times c = \ 5 \times$

$\Rightarrow \underline{c \quad = \quad \text{cm}}$

2

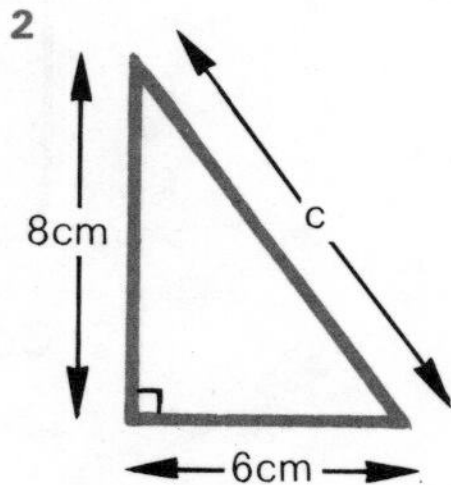

$c \times c = 8 \times 8 + 6 \times 6$

$\Rightarrow c \times c = 64 + 36$

$\Rightarrow c \times c =$

$\Rightarrow c \times c =$

$\Rightarrow \underline{c \quad = \quad \text{cm}}$

3

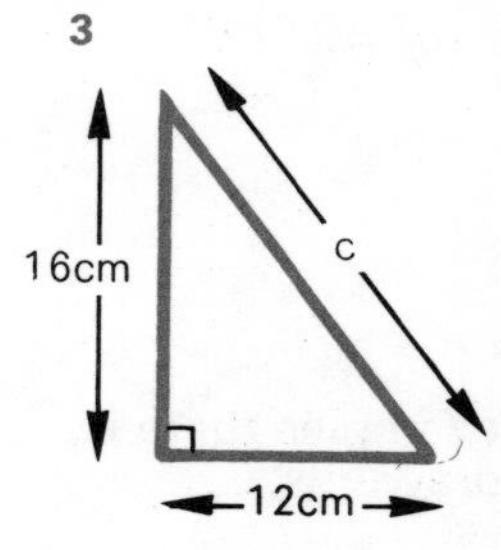

$c \times c = 12 \times 12 + 16 \times 16$

4

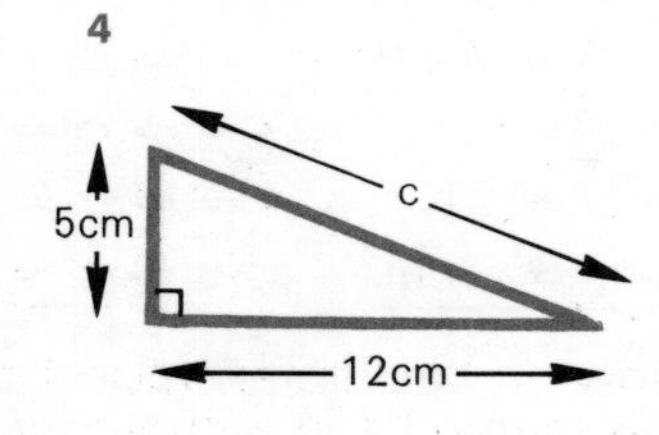

$c \times c =$

Now do these examples.

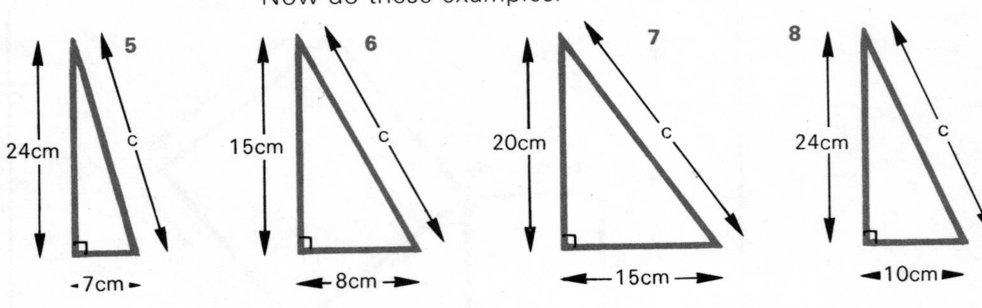

The examples you have done so far have come out exactly. Sometimes they do not come out exactly. Read the next example and then do the rest of this exercise.

Example

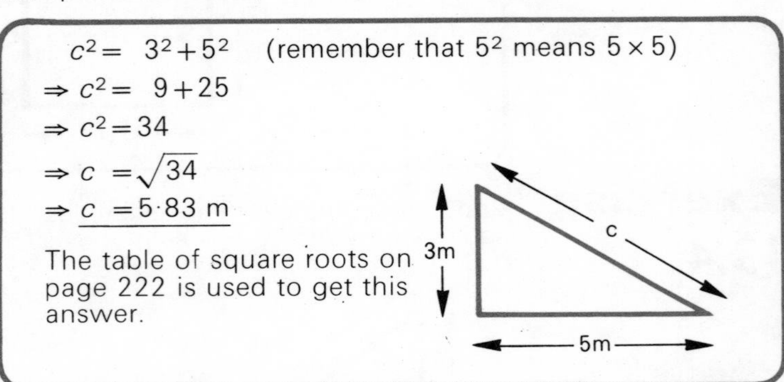

$c^2 = 3^2 + 5^2$ (remember that 5^2 means 5×5)

$\Rightarrow c^2 = 9 + 25$

$\Rightarrow c^2 = 34$

$\Rightarrow c = \sqrt{34}$

$\Rightarrow \underline{c = 5{\cdot}83 \text{ m}}$

The table of square roots on page 222 is used to get this answer.

Copy and complete these examples.

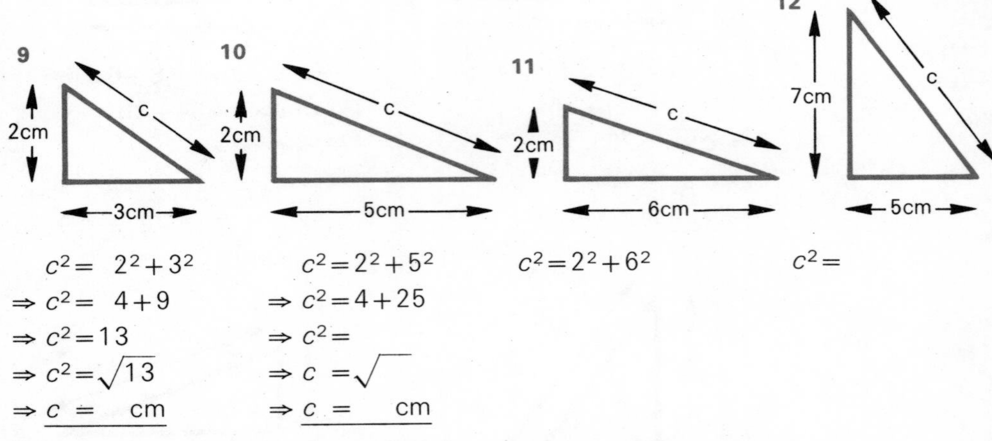

9

$c^2 = 2^2 + 3^2$

$\Rightarrow c^2 = 4 + 9$

$\Rightarrow c^2 = 13$

$\Rightarrow c^2 = \sqrt{13}$

$\Rightarrow \underline{c = \quad \text{cm}}$

10

$c^2 = 2^2 + 5^2$

$\Rightarrow c^2 = 4 + 25$

$\Rightarrow c^2 =$

$\Rightarrow c = \sqrt{\quad}$

$\Rightarrow \underline{c = \quad \text{cm}}$

11

$c^2 = 2^2 + 6^2$

12

$c^2 =$

Now go back to exercise 1 and do the examples there by this method and without making scale drawings.

In Exercise **4** we found the length of the longest side when the lengths of the two shortest sides were known. It is also possible to find the length of one of the short sides if the lengths of the other two sides are known.

Example

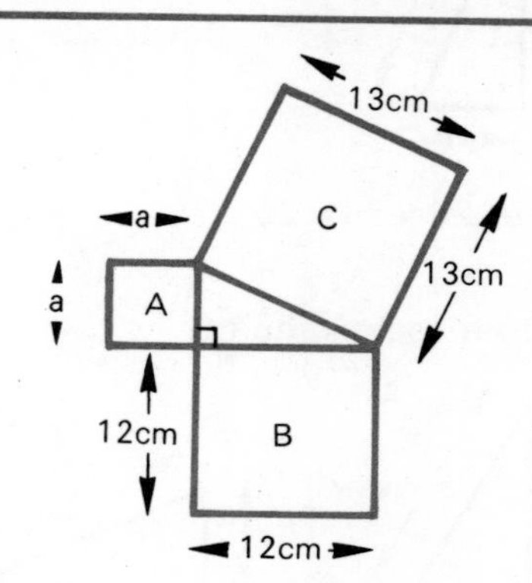

Find the length of side a.

Area of A + Area of B = Area of C

$\Rightarrow a\times a + 12\times 12 = 13\times 13$

$\Rightarrow a\times a + 144 = 169$

$\Rightarrow a\times a = 169-144$

$\Rightarrow a\times a = 25$

$\Rightarrow a\times a = 5\times 5$

$\Rightarrow a = 5$ cm

Notice that on the first line of the equation we always put the longest side of the triangle by itself to start with, whether we know it or not.

Exercise 13.5

Copy and complete these examples.

1

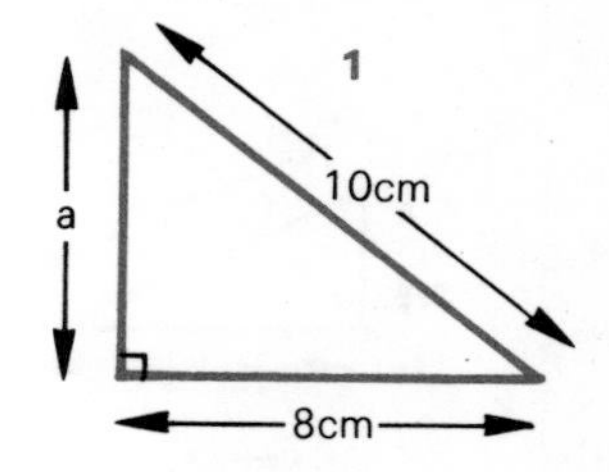

$a\times a + 8\times 8 = 10\times 10$

$\Rightarrow a\times a + 64 = 100$

$\Rightarrow a\times a = 100-64$

$\Rightarrow a\times a = 36$

$\Rightarrow a\times a = 6\times$

$\Rightarrow a =$ cm

2

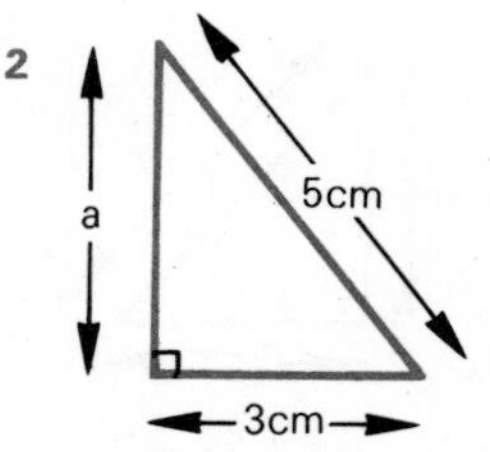

$a\times a + 3\times 3 = 5\times 5$

$\Rightarrow a\times a + 9 = 25$

$\Rightarrow a\times a = 25-9$

$\Rightarrow a\times a =$

$\Rightarrow a\times a =$

$\Rightarrow a =$ cm

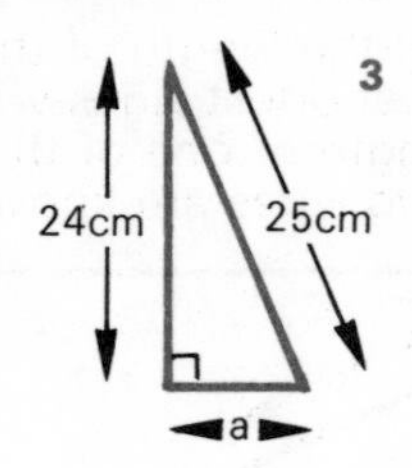

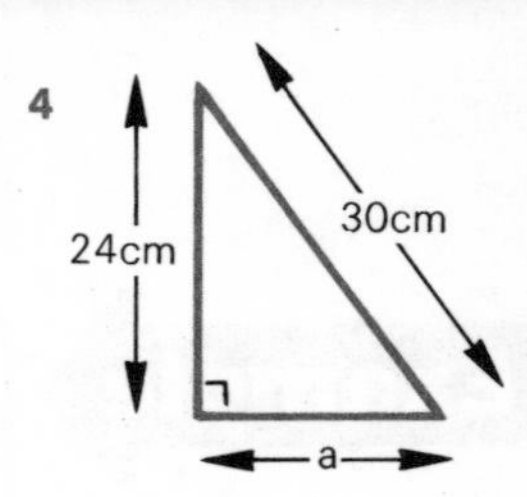

$a \times a + 24 \times 24 = 25 \times 25$ $\quad$ $a \times a + 24 \times 24 =$

$\Rightarrow$ $\quad$ $\Rightarrow$

Now do these examples.

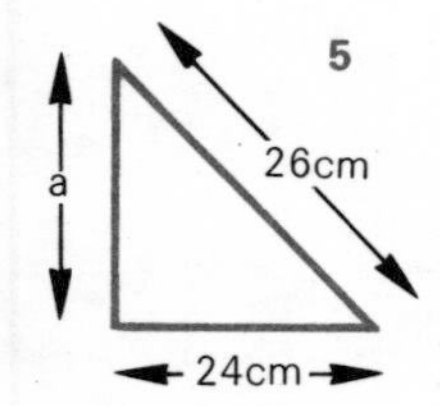

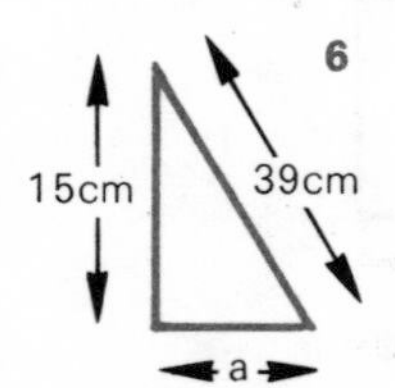

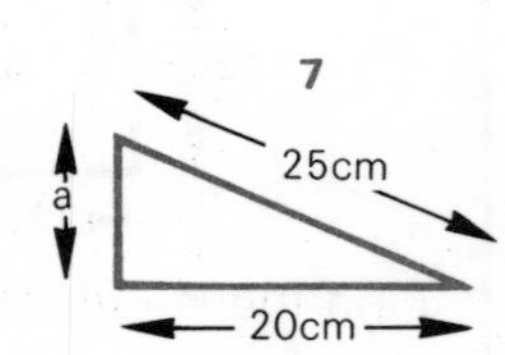

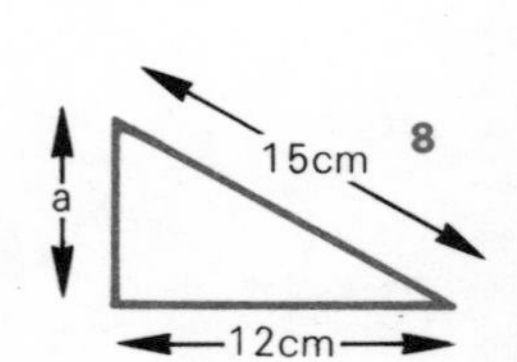

Read the next example and then do the rest of this exercise.

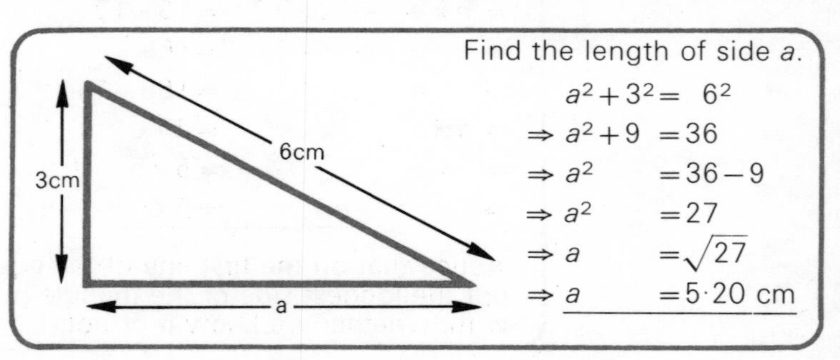

Find the length of side a.

$$
\begin{aligned}
& a^2 + 3^2 = 6^2 \\
\Rightarrow\ & a^2 + 9 = 36 \\
\Rightarrow\ & a^2 = 36 - 9 \\
\Rightarrow\ & a^2 = 27 \\
\Rightarrow\ & a = \sqrt{27} \\
\Rightarrow\ & \underline{a = 5{\cdot}20 \text{ cm}}
\end{aligned}
$$

Copy and complete these examples.

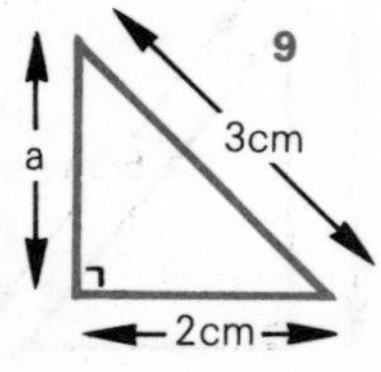

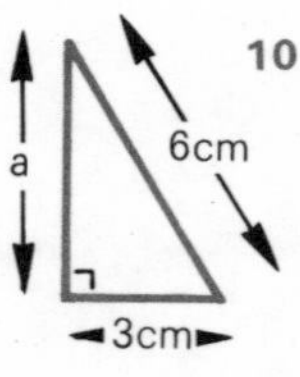

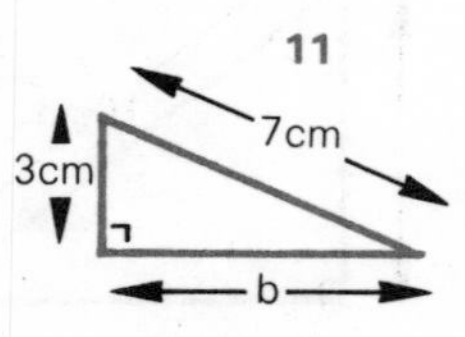

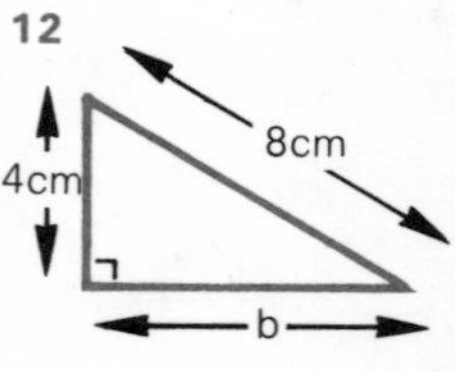

$$
\begin{aligned}
& a^2 + 2^2 = 3^2 \\
\Rightarrow\ & a^2 + 4 = 9 \\
\Rightarrow\ & a^2 = 9 - 4 \\
\Rightarrow\ & a^2 = 5 \\
\Rightarrow\ & a = \sqrt{5} \\
\Rightarrow\ & \underline{a = \quad \text{cm}}
\end{aligned}
$$

$$
\begin{aligned}
& a^2 + 3^2 = 6^2 \\
\Rightarrow\ & a^2 + 9 = 36 \\
\Rightarrow\ & a^2 = 36 - 9 \\
\Rightarrow\ & a^2 = \\
\Rightarrow\ & a = \sqrt{\quad} \\
\Rightarrow\ & \underline{a = \quad \text{cm}}
\end{aligned}
$$

$b^2 + 3^2 = 7^2$

$b^2 + 4^2 =$

Now go back to **13.2** (p. 109) and do the examples there by this method and without making scale drawings.

Exercise 13.6

Find the lengths of the lines marked by the letters.

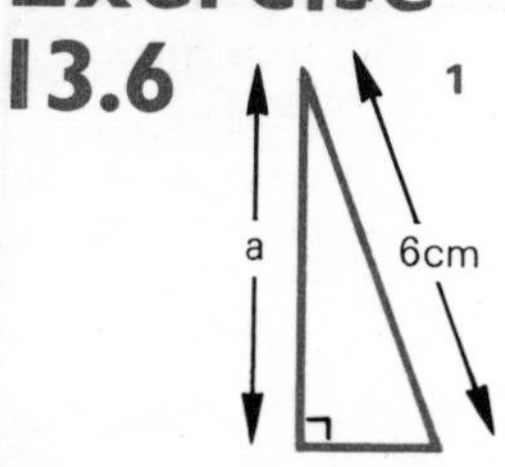

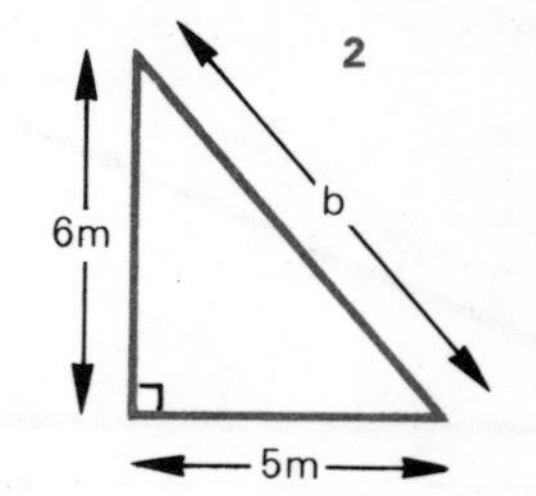

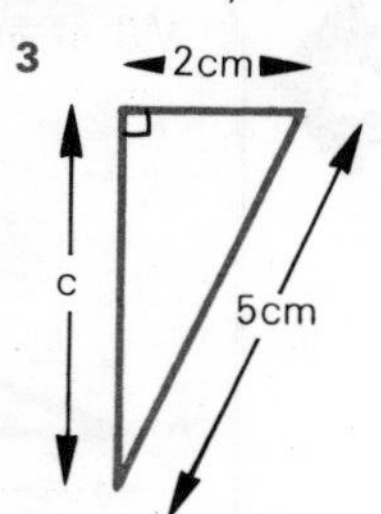

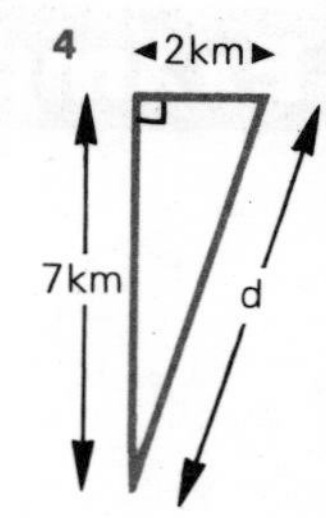

5 The framework of a large gate is made up of two vertical bars 2 m long, three horizontal bars 2 m long, and a diagonal bar. Find:

a The length of the diagonal bar.

b The total length of the other five pieces.

c The total length of all six pieces.

d The total cost of the bars at 32p per metre.

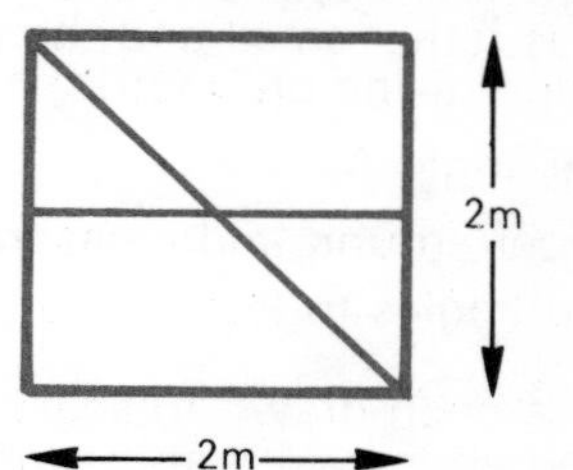

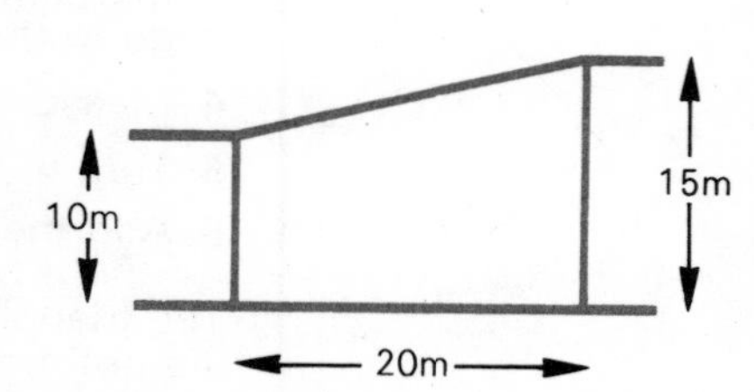

6 Two buildings are 20 m apart. One of the buildings is 10 m high, and the other is 15 m high. What length of wire is needed to stretch from the top of one building to the top of the other? What is the cost of this wire at 11p per metre?

7 A television mast 10 m high is supported by four wires. The foot of each wire is 5 m from the foot of the mast.

a Find the length of one of the wires.

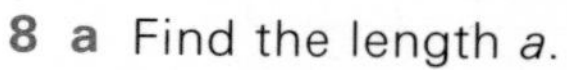

b Find the length of all four wires.

c Find the total cost of all the wire at 9p per metre.

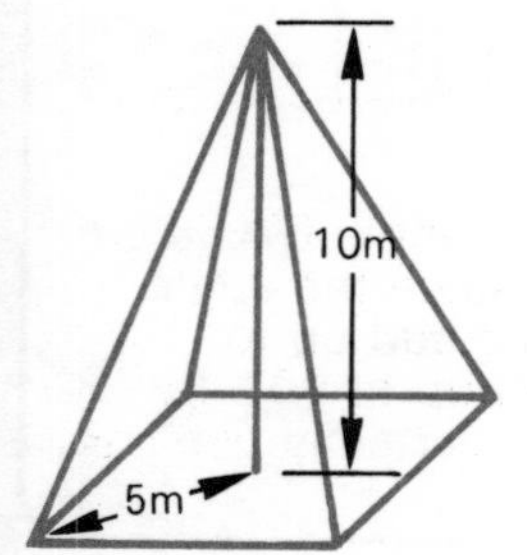

8 **a** Find the length *a*.

b Find the length *b*.

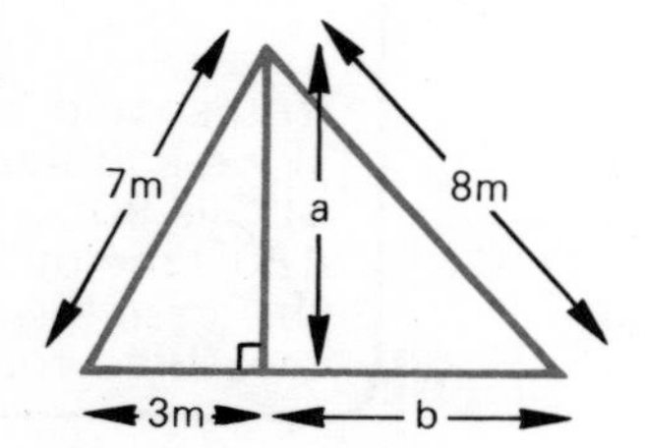

unit 14 The angles of right-angled triangles

There are many practical situations where you may need to find the angles of a right-angled triangle. This can be done easily by using scale drawings.

Example

What are the sizes of angles A and B in this triangle?

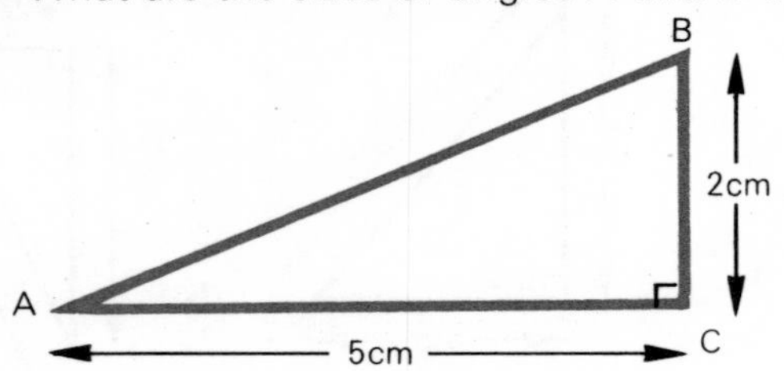

This problem is best done on squared paper to get an accurate right angle at C.

1 Mark the point C, and then draw the line CA 5 cm long.
2 Draw the line CB 2 cm long.
3 Draw the line AB. (You will see that we have continued the line beyond A and B. This is because the triangle is rather small and the line AB may not go to the edge of the protractor.)
4 Measure the angle A.
5 Turn the paper around and measure the angle B.
6 Add the two angles together as a check. Why?

The triangle has been drawn to scale below. If you are not drawing your own triangle measure the angles of this one and see what you get.

You should find that angle A=22° and angle B=68°. If we add them together we get 90°. Since the angle at C is 90°, and the angles of a triangle add up to 180°, the other two angles should add up to 90°. In this case they do, so they have almost certainly been measured correctly.

Exercise 14.1

Make scale drawings of these triangles, and measure the sizes of angles A and B. Add them together to make sure that they come to 90°. (Since it is difficult to measure angles exactly, you can count your answers correct if when you add them together you get between 88° and 92°.)

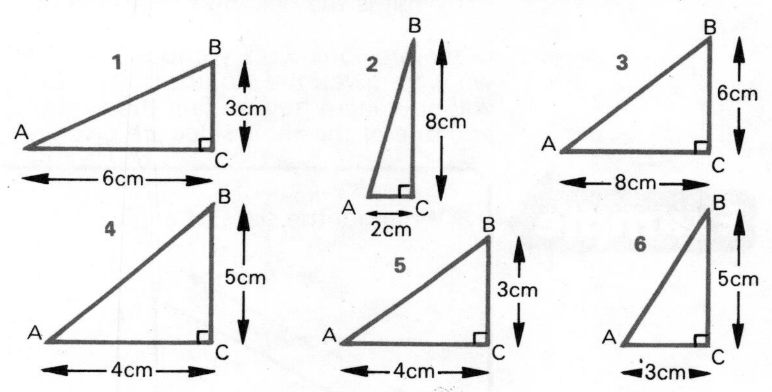

7 A road rises 3 m over a horizontal distance of 10 m. What angle does the road make with the horizontal? Let 1 cm stand for 1 m.

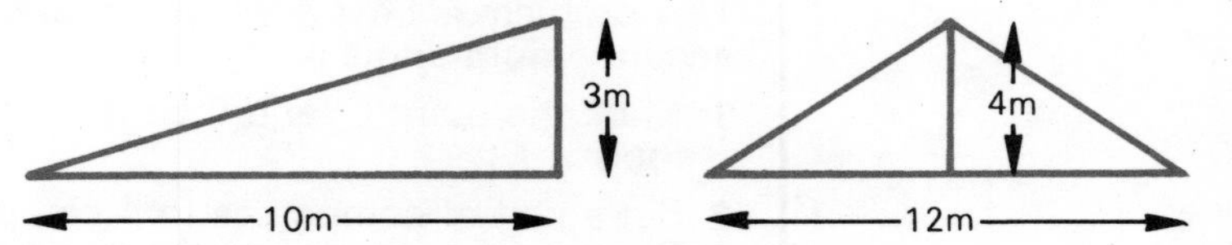

8 A roof is 4 m high, and has to span 12 m. What angle does the slope of the roof make with the horizontal? Let 1 cm stand for 1 m.

9 From the point where a plane leaves the ground the top of a hill 100 m high is 500 m away (measured horizontally). What is the least angle of climb of the plane? Let 1 cm stand for 100 m.

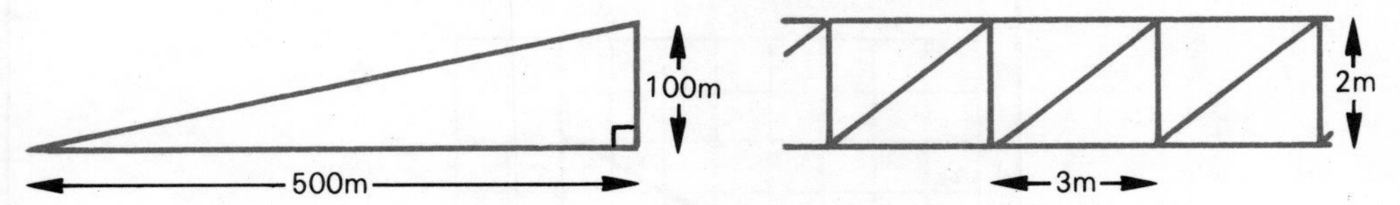

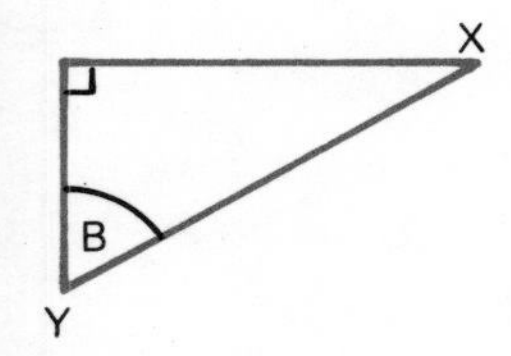

10 What angle do the sloping girders make with the horizontals in the bridge shown in the diagram?

11 A town X is 30 km North and 70 km East of town Y. What is the bearing of X from Y? Let 1 cm stand for 10 km. (You have to measure the angle B.)

12 A road rises 4 m over a horizontal distance of 11 m. What angle does the road make with the horizontal?

13 From the point where a plane leaves the ground the top of a hill 150 m high is 600 m away (measured horizontally). What is the least angle of climb of the plane?

14 A town P is 35 km North and 80 km East of town Q. What is the bearing of P from Q?

In the previous work we found the angles of the triangle if we were given the lengths of its two shortest sides. You will now learn how to find the angles if the longest side and one of the short sides are given.

Example

What are the sizes of angles A and B in this triangle?

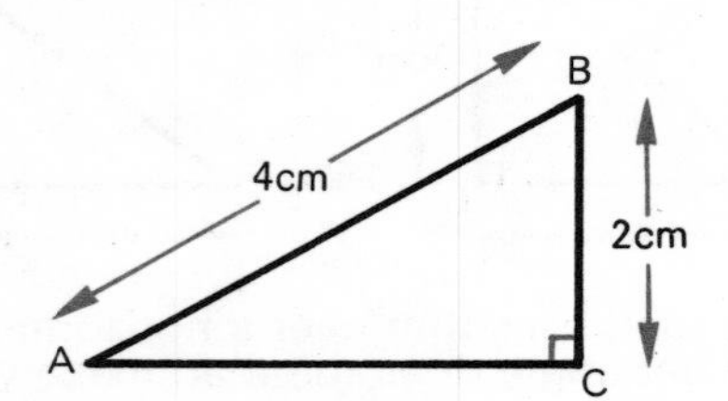

This problem is best done on squared paper to get an accurate right angle at C.

1 Mark the point C, and then draw the line CB 2 cm long.

2 Set a pair of compasses to 4 cm, put the point on B, and draw the part of the circle that cuts the horizontal line from C.

3 Mark the point A, and draw the line AC.

4 Measure the angles A and B.

5 Add them together to check that they come to 90°.

Your drawing should look like this.

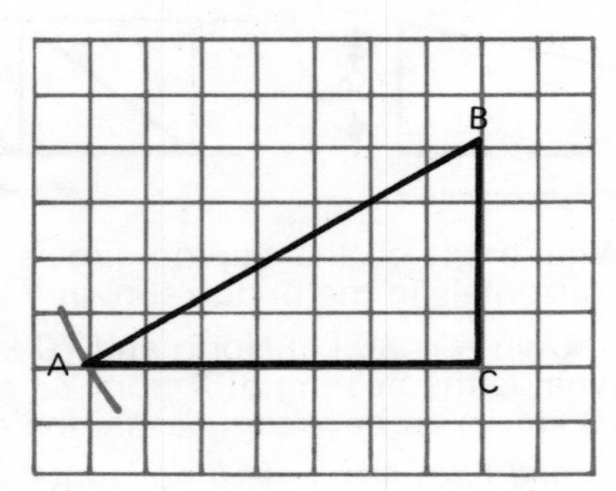

If you measure the angles you will find that A = 30°, and B = 60°.

Exercise 14.2

Make scale drawings of these triangles, and measure the size of angles A and B. Add them together to make sure that they come to 90°. (Since it is difficult to measure angles exactly, you can regard your answers as correct if when you add them together you get between 88° and 92°.)

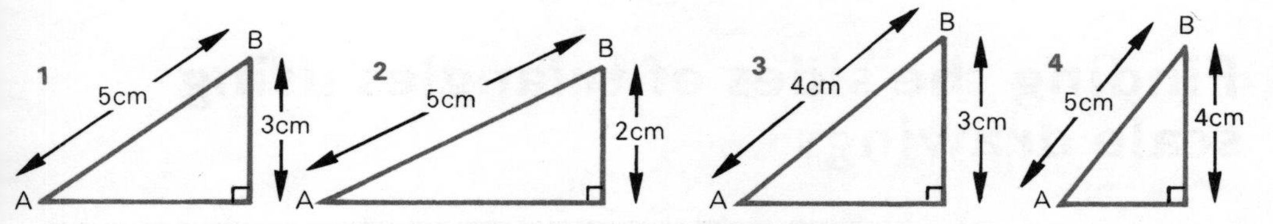

If you turn this page sideways you will see that the next four examples are like the four that you have just done.

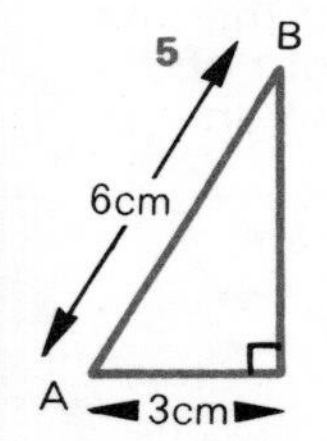

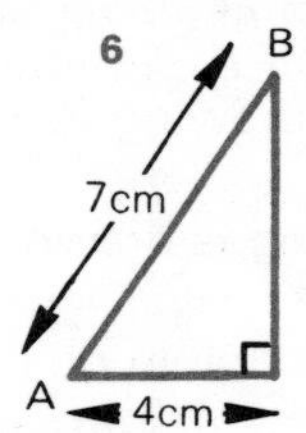

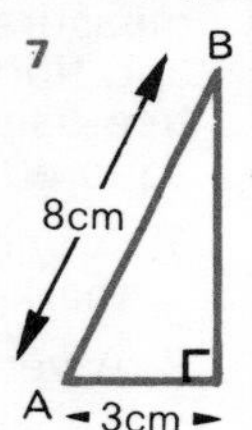

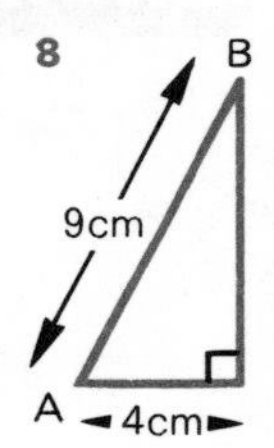

9 A road 8 m long rises 2 m. What angle does the road make with the horizontal? (Let 1 cm stand for 1 m.)

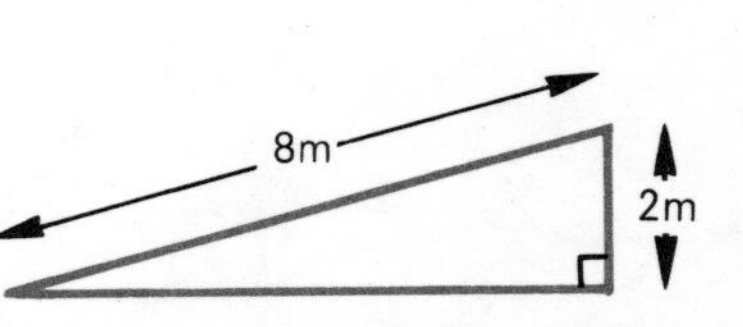

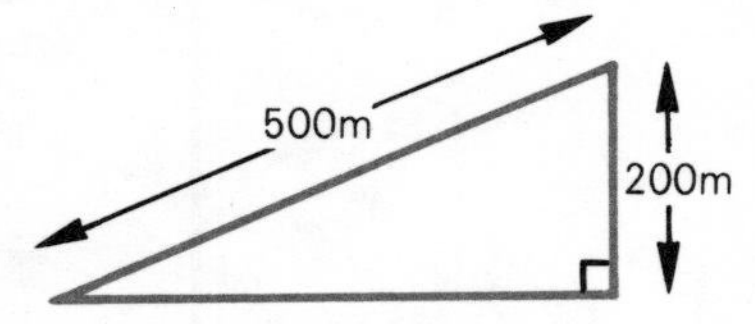

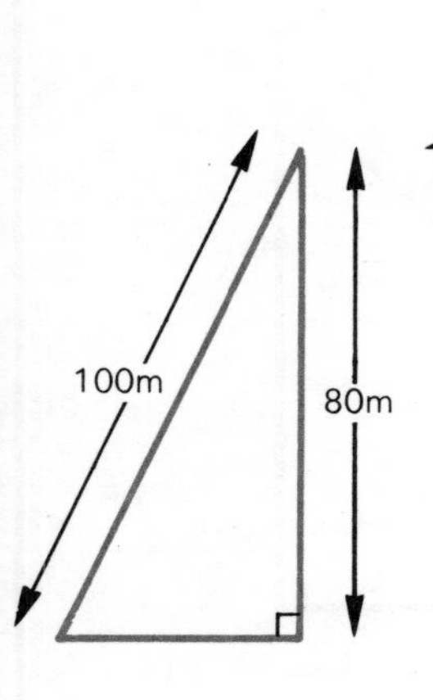

10 A plane after flying a distance of 500 m is 200 m above the ground. What is its angle of climb? (Let 1 cm stand for 100 m.)

11 A television tower 80 m high is supported by wire stays 100 m long. What angle do these stays make with the horizontal?

12 A tunnel 150 m long is driven from the surface to a point 60 m below the surface. What angle does the tunnel make with the horizontal?

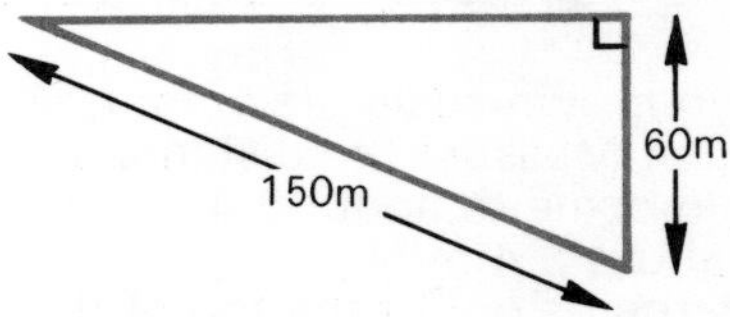

13 A road 12 m long falls 3 m. What angle does the road make with the horizontal?

14 A plane after flying a distance of 600 m is 200 m above the ground. What is its angle of climb?

15 A television tower 70 m high is supported by wire stays 90 m long. What angle do these stays make with the horizontal?

16 A tunnel 70 m long is driven from a point 30 m under the surface of the earth to a point 50 m under the surface of the earth. What angle does the tunnel make with the horizontal?

Finding the sides of triangles using scale drawings

Example

A man who wants to find the height of a tree goes to a point 10 m from the foot of the tree and then measures the angle of elevation of the top of the tree and finds it is 40°. He can now find the height of the tree using a scale drawing. (This is best done on squared paper.)

1 A line 10 cm long is drawn and the ends are marked S and B.

2 A vertical line is drawn going from B.

3 An angle of 40° is drawn at S and where the line cuts the vertical line the point T is marked. This represents the top of the tree on the scale drawing.

4 The line BT is measured to get the height of the tree.

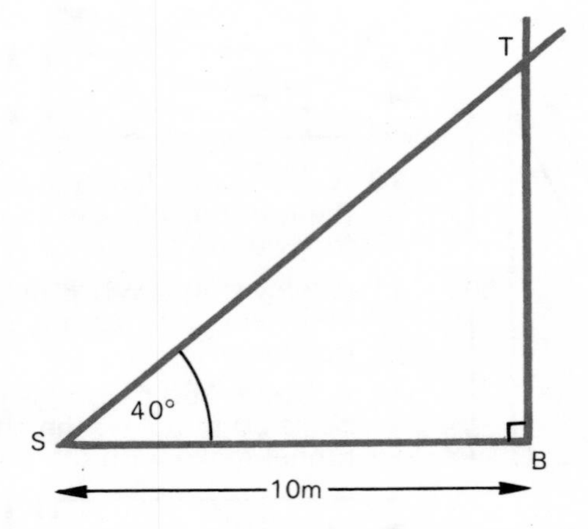

Make your own scale drawing. It should look like this, but larger. Measure the distance BT.
It should come to about 8·4 cm. This means that the tree is about 8·4 m high.
(This is the height of the top of the tree above the level of his eye. To get the true height of the tree we should add on the height of the man's eye above the ground. (About 1·5 m.).)

Exercise 14.3

Make scale drawings of these triangles, and measure the distances marked by the letter *d*.

1

60° d 3cm

2

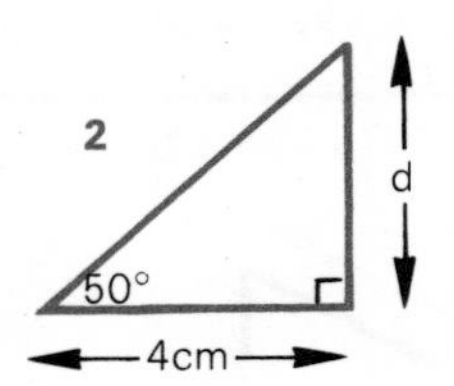

3

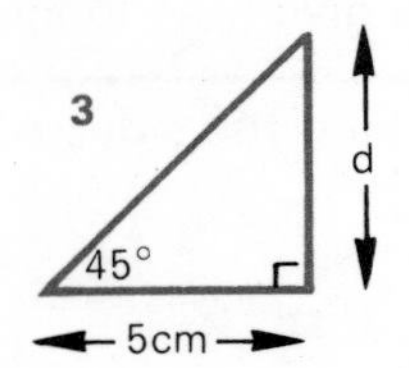

4

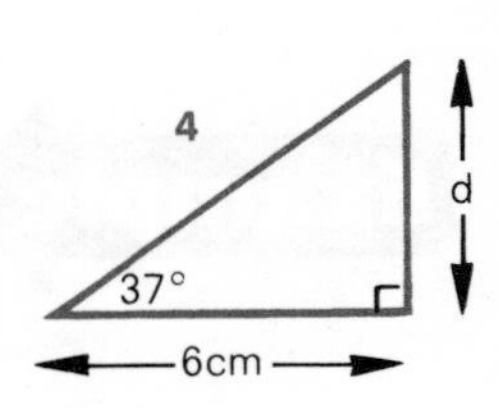

If you turn this page sideways you will see that the next four examples are like the four that you have just done.

5

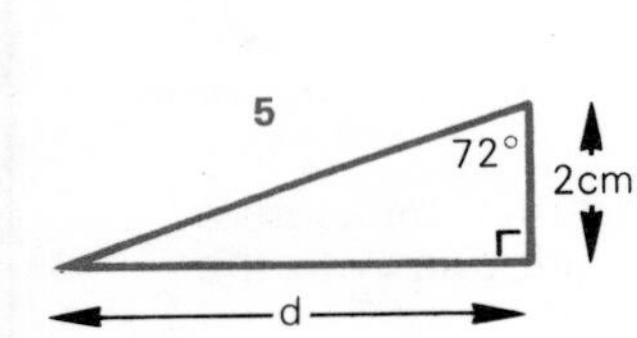

6

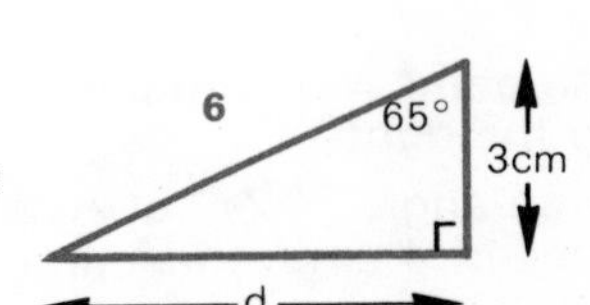

7

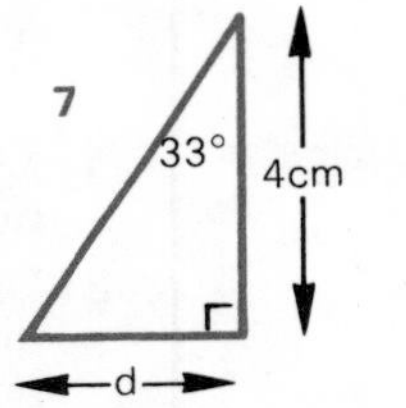

8

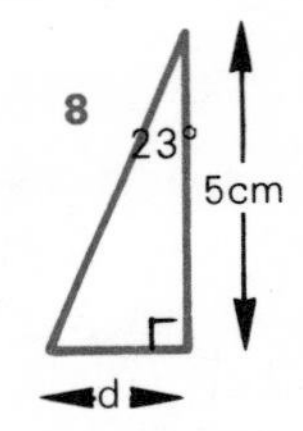

9 The angle of elevation of the top of a tree from a point 8 m from the foot of the tree is 48°. What is the height of the tree? (Ignore the height of the man taking the measurements.)

10 The angle of elevation of the top of a cliff from a point 120 m from its foot is 35°. What is the height of the cliff? (Let 1 cm stand for 10 m.)

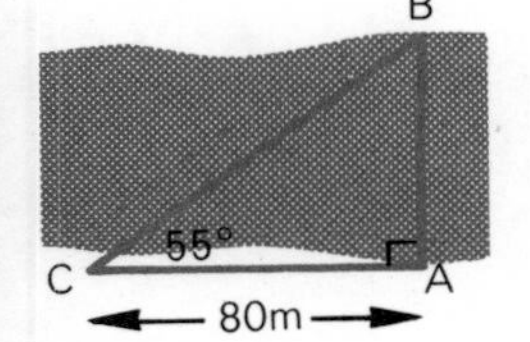

11 A man stands directly opposite a tree B on the other bank of a river. He walks along the bank from A to C. The distance AC = 80 m, the angle ACB = 55°, and the angle CAB is a right angle. Find AB, the distance across the river. (Let 1 cm stand for 10 m.)

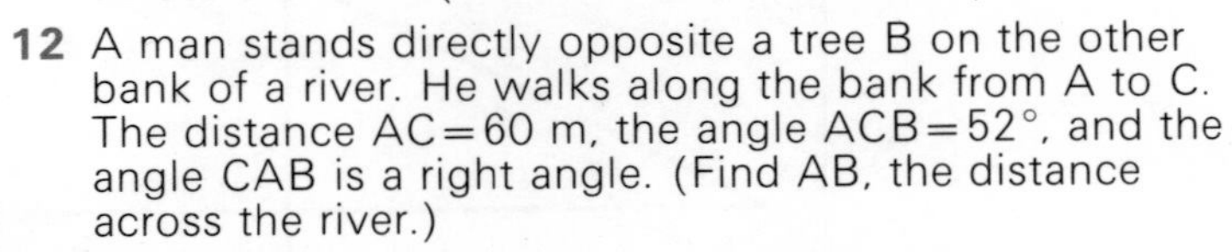

12 A man stands directly opposite a tree B on the other bank of a river. He walks along the bank from A to C. The distance AC = 60 m, the angle ACB = 52°, and the angle CAB is a right angle. (Find AB, the distance across the river.)

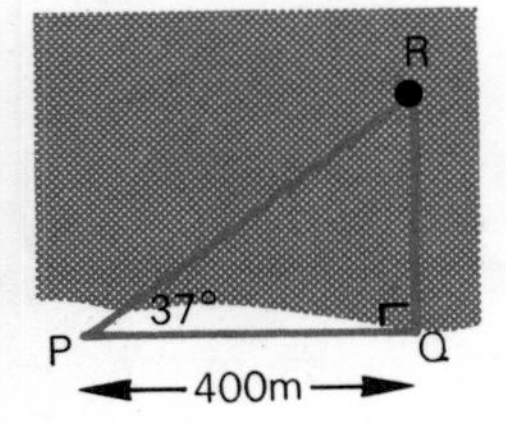

13 To find the distance of an island R from the shore, a man starts from a point Q opposite R and walks to P, 400 m along the shore, so that the angle PQR is a right angle. He finds that the angle QPR = 37°. Find RQ, the distance of the island from the shore. (Let 1 cm stand for 100 m.)

14 Do question **13** again with the angle equal to 26°, and the distance equal to 500 m.

15 The angle of elevation of a satellite is 49°. At the same time the satellite is directly over a point 400 km away. What is the height of the satellite? (Let 1 cm stand for 100 km.)

16 Repeat question **15** with the distance equal to 350 km, and the angle equal to 38°.

In the previous examples you were given one of the short sides of the triangle, and you had to find the other short side. You are now going to find the two short sides if you are given the longest side and one of the angles.

Example

Find the sides AB and BC in this triangle.

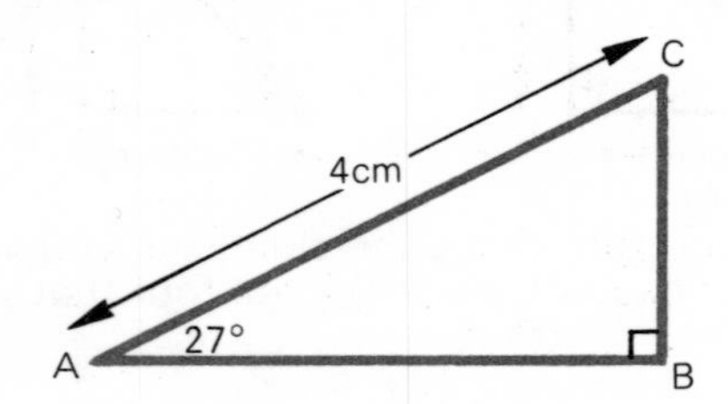

1 Mark a point A on the paper and draw a line going out to the right.
2 Draw an angle of 27° at A, and then measure a distance of 4 cm from A and mark this with the point C.
3 Draw a vertical line down from C and where it crosses the line from A mark in the letter B.
4 Measure AB and BC.

Your drawing should look like this. If you do your own scale drawing you will find that AB = 3·6 cm, and BC = 1·8 cm.

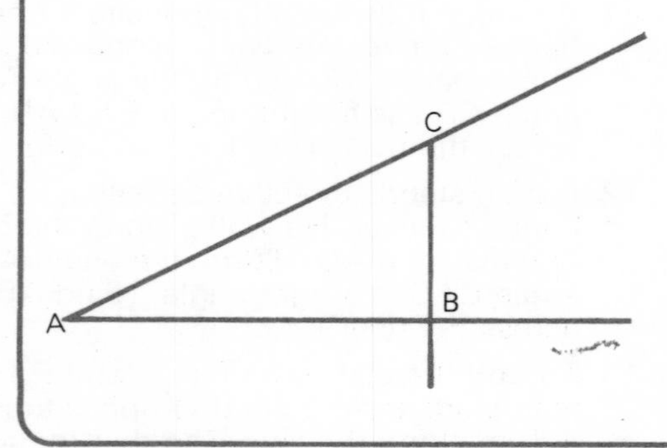

Exercise 14.4

Make scale drawings of these triangles, and measure the distances marked by the letters *a* and *b*.

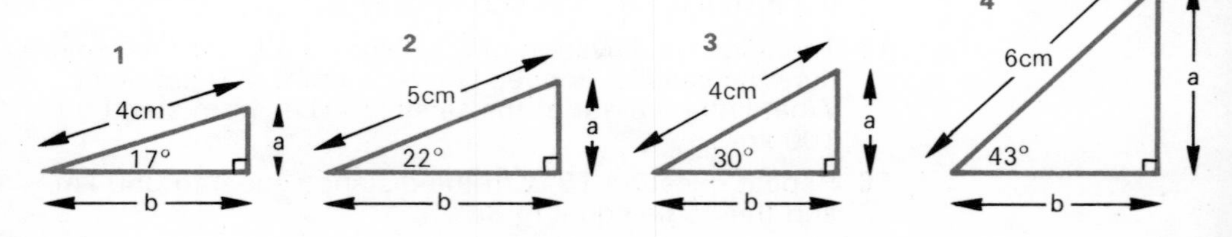

If you turn this page sideways you will see that the next four examples are like the four that you have just done.

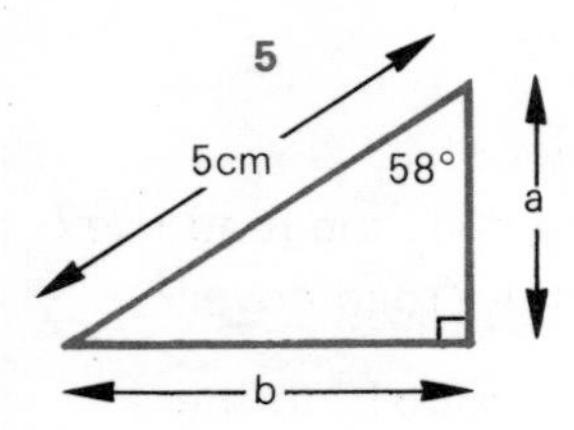

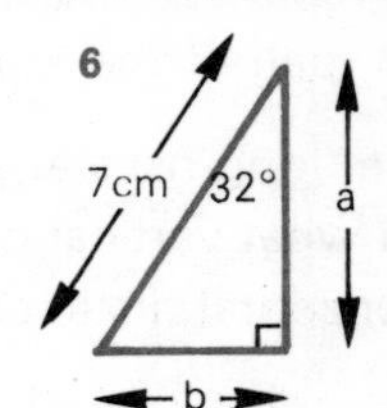

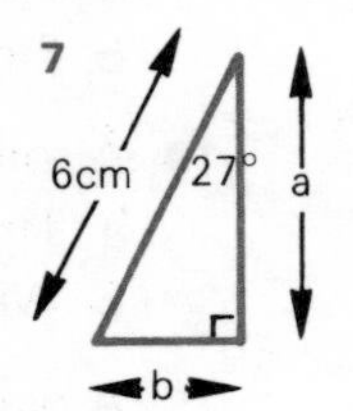

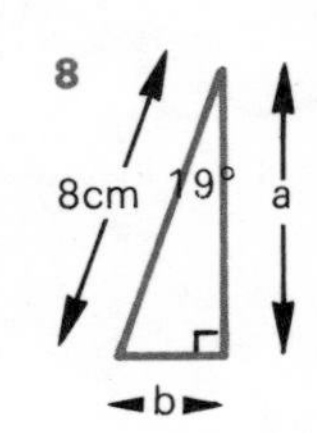

9 A roof 6 m long rises at an angle of 32°.

a Find the height of the roof (the distance *a*).

b Find the span of the roof (the distance *b*).

Let 1 cm stand for 1 m.

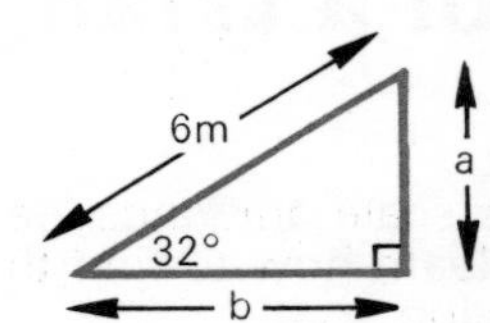

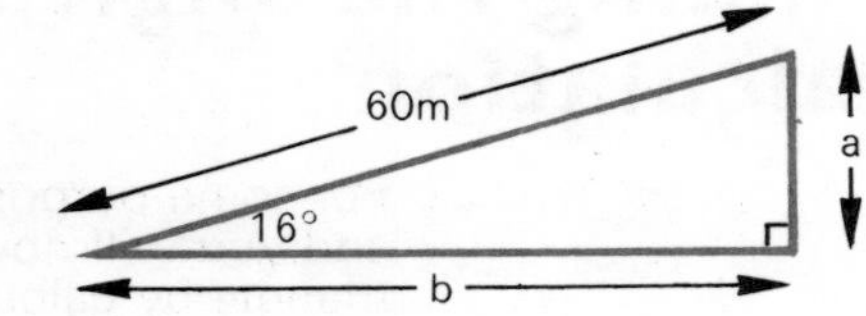

10 A road 60 m long rises at an angle of 16°.

a Through what vertical distance does the road rise (the distance *a*)?

b What horizontal distance does the road cover (the distance *b*)?

Let 1 cm stand for 10 m.

11 After taking off a plane flies for 500 m at an angle of 23° to the horizontal.

a Through what vertical distance has it risen (the distance *a*)?

b Through what horizontal distance has it flown (the distance *b*)?

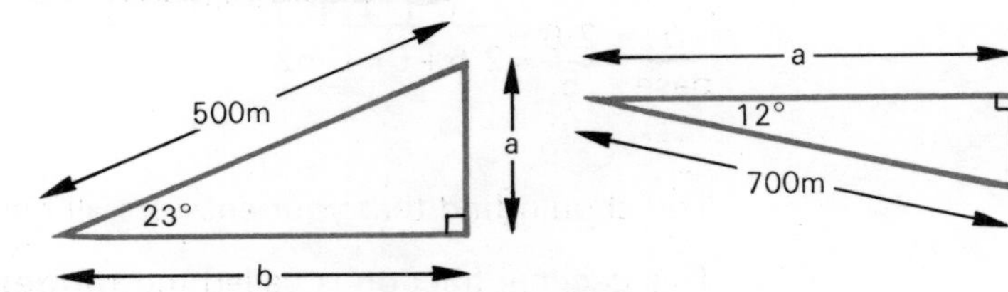

12 A tunnel 700 m long is driven underground at an angle of 12°.

a What horizontal distance has it covered (the distance *a*)?

b What is the depth of the end of the tunnel (the distance *b*)?

13 A roof 5 m long rises at an angle of 32°.

a Find the height of the roof.

b Find the span of the roof.

14 A road 50 m long rises at an angle of 13°.

a Through what vertical distance does the road rise?

b What horizontal distance does the road cover?

15 A plane flies for 700 m at an angle of 31° to the horizontal after taking off.

a Through what vertical distance has it risen?

b Through what horizontal distance has it flown?

Finding the angles of a triangle by calculation

For some purposes scale drawings are not accurate enough, and you will now learn how to find the angles of a triangle by calculation.

Draw some right-angled triangles with the base *a* whole number of cm long, and an angle of 27°. Measure the distance *h*, and work out as a decimal the fraction $\frac{h}{\text{base}}$. One example is done below.

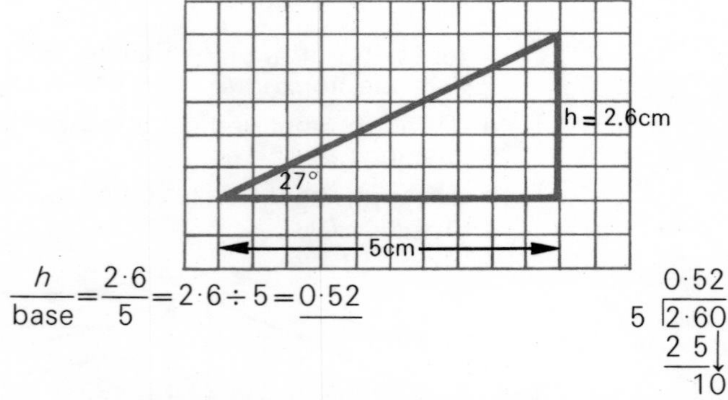

$$\frac{h}{\text{base}} = \frac{2{\cdot}6}{5} = 2{\cdot}6 \div 5 = \underline{0{\cdot}52}$$

$$\begin{array}{r} 0{\cdot}52 \\ 5\,\overline{)\,2{\cdot}60} \\ 2\;5\downarrow \\ \hline 10 \end{array}$$

You should find that your answers all come to about 0·52.

This decimal fraction is called the **tangent** of the angle, and it depends only on the angle, not on the lengths of the sides.

If you did the same thing with angles of 28°, 29°, 30° and so on and made a list of your results they would look like this.

ANGLE	TANGENT
27°	0·51
28°	0·53
29°	0·55
30°	0·58

A complete table of tangents is on page 223.

To find out the angle of a right-angled triangle all you need to do is to work out the fraction $\frac{\text{height}}{\text{base}}$ and then see what angle gives this fraction using the table of tangents on page 223.

Example

Find the angles of this triangle.

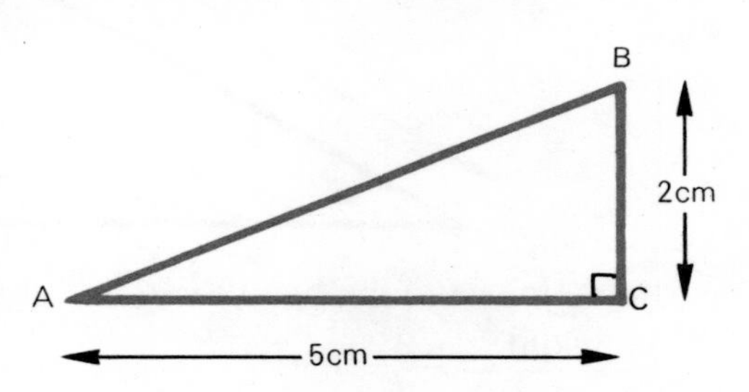

a Tangent of $A=\frac{2}{5}=0{\cdot}40$

$\underline{A=22°}$

b To find angle B we need to turn the triangle on its side so that 2 is now the base and 5 the height.

Tangent of $B=\frac{5}{2}=2{\cdot}50$

$\underline{B=68°}$ Check: $68°+22°=90°$

To do these examples you need to remember that:

Tangent of Angle = Height/Base

Sometimes you will find that the tangent you get cannot be found exactly in the tables. In this case you get as near as you can.

For example if you get tangent of A = 0·71, you will find that the nearest you can get is 0·70. You will then get A = 35°.

Exercise 14.5

Go back and do Exercise 14.1 using the method you have just been shown.

In some problems you are given the longest side of a right-angled triangle and one of the other sides, and you have to calculate the angles.

If you draw a right-angled triangle with an angle of 30° and work out the fraction $\frac{\text{height}}{\text{longest side}}$ you will always find that it comes to about 0·50. An example has been done below.

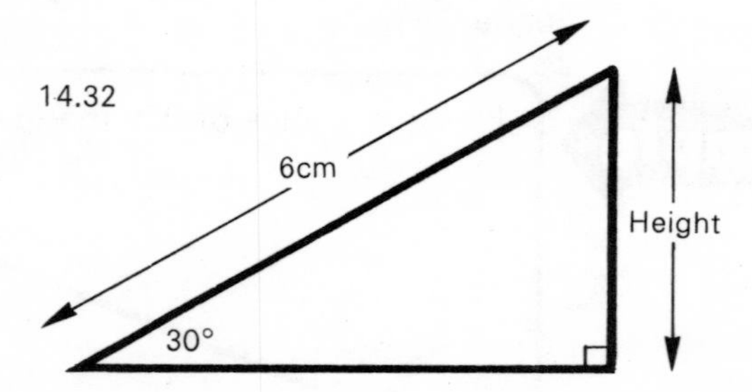

If you measure the height you find that it comes to 3 cm.

$$\frac{\text{height}}{\text{longest side}} = \frac{3}{6} = 0{\cdot}50$$

If you do this for all right-angled triangles with angles from 1° to 89° you get what is called a table of **sines**. Part of a table of sines is shown below.

ANGLE	SINE
35°	0·57
36°	0·59
37°	0·60

To find out the angle of a right-angled triangle all you need to do is to work out the fraction $\frac{\text{height}}{\text{longest side}}$, and then see what angle gives this fraction using the table of sines on page 223.

Example

Find the angles of this triangle.

B
7cm
2cm
A
C

a Sine of $A = \frac{2}{7} = 0{\cdot}28$

$\Rightarrow \underline{A = 16°}$

b We now use the fact that $A + B = 90°$
So we get $\underline{B = 74°}$

To do examples like this you need to remember that

$$\textbf{Sine of angle} = \frac{\textbf{height}}{\textbf{longest side}}$$

Exercise 14.6

Go back and do Exercise **14.2** using the method you have just been shown.

Finding the sides of a triangle by calculation

If you are given one of the angles of a right-angled triangle and one of the short sides it is possible to calculate the other short side as follows.

Example

Find the height of this triangle.

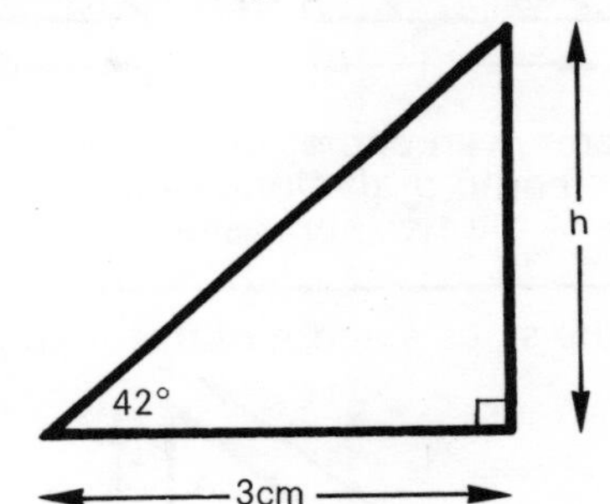

We know that $\frac{\text{height}}{\text{base}}$ = tangent of angle.

We also know that the angle = 42° and the base = 3 cm, so we can write

$$\frac{h}{3} = \text{tangent of } 42°$$

If we look in the tables we see that the tangent of 42° = 0·90

So we can write $\frac{h}{3} = 0{\cdot}90$.

This is a simple equation to solve and we can now write

$$h = 3 \times 0{\cdot}90$$

$$\Rightarrow \underline{h = 2{\cdot}70 \text{ cm}}$$

Exercise 14.7

Go back and do Exercise **14.3** using the method you have just been shown. Then try the examples underneath.

In each of these examples you have to find the angle A, and then use it to find the height or the side marked by the letter *h*.

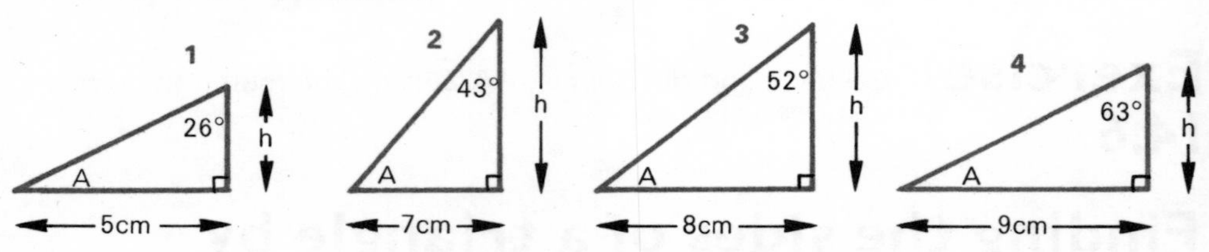

If you turn the page around you will see that the next four examples are done in the same way as the first four.

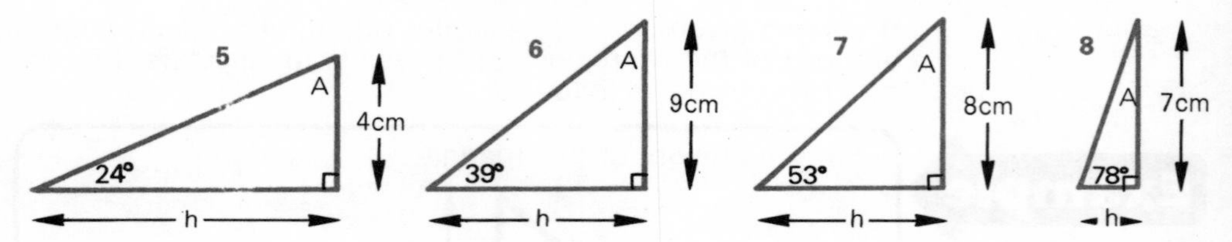

If you are given one of the angles of a right-angled triangle and the length of the longest side it is possible to find the lengths of the two short sides as follows.

Example

Find the sides *a* and *b* of this triangle.

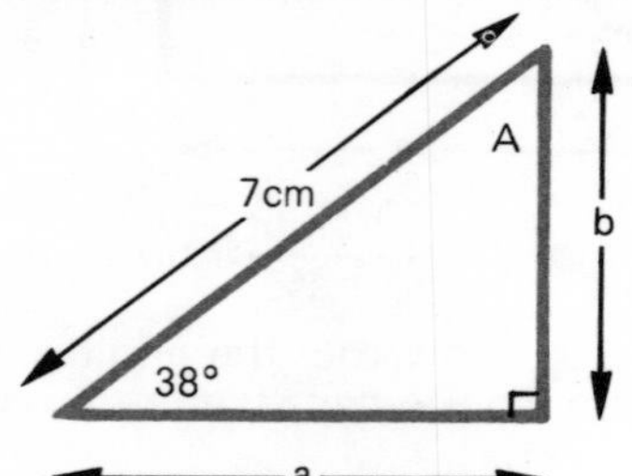

We know that $\frac{\text{height}}{\text{longest side}} = \text{sine of angle}$

So we can write $\frac{b}{7} = \text{sine of } 38°$

$$\Rightarrow \frac{b}{7} = 0{\cdot}62$$

$$\Rightarrow b = 7 \times 0{\cdot}62$$

$$\Rightarrow \underline{b = 4{\cdot}34 \text{ cm}}$$

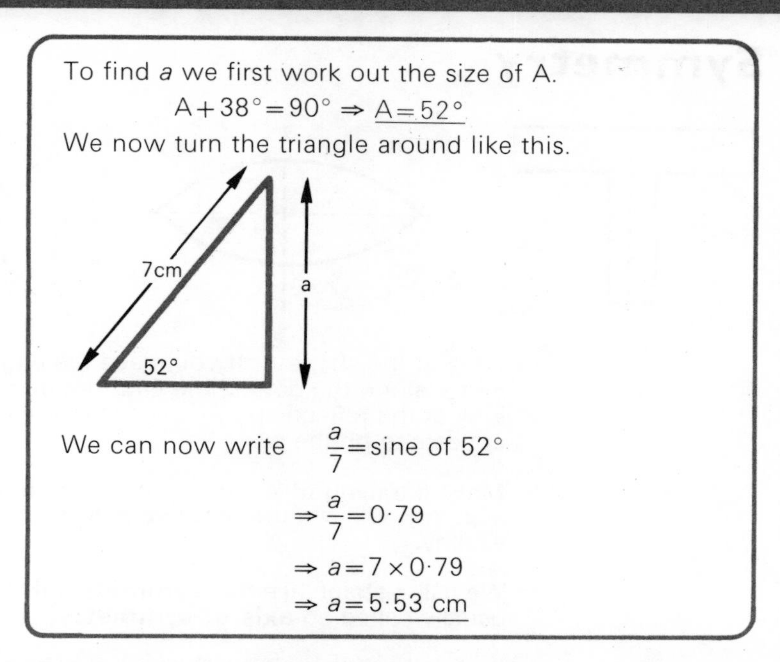

To find a we first work out the size of A.

$$A + 38° = 90° \Rightarrow \underline{A = 52°}$$

We now turn the triangle around like this.

We can now write $\frac{a}{7} = \text{sine of } 52°$

$$\Rightarrow \frac{a}{7} = 0{\cdot}79$$

$$\Rightarrow a = 7 \times 0{\cdot}79$$

$$\Rightarrow \underline{a = 5{\cdot}53 \text{ cm}}$$

Exercise 14.8

Go back and do Exercise **14.4** using the method you have just been shown.

Exercise 14.9

Find the sides and angles marked by the letters. You can find the answers either by calculation or by scale drawing.

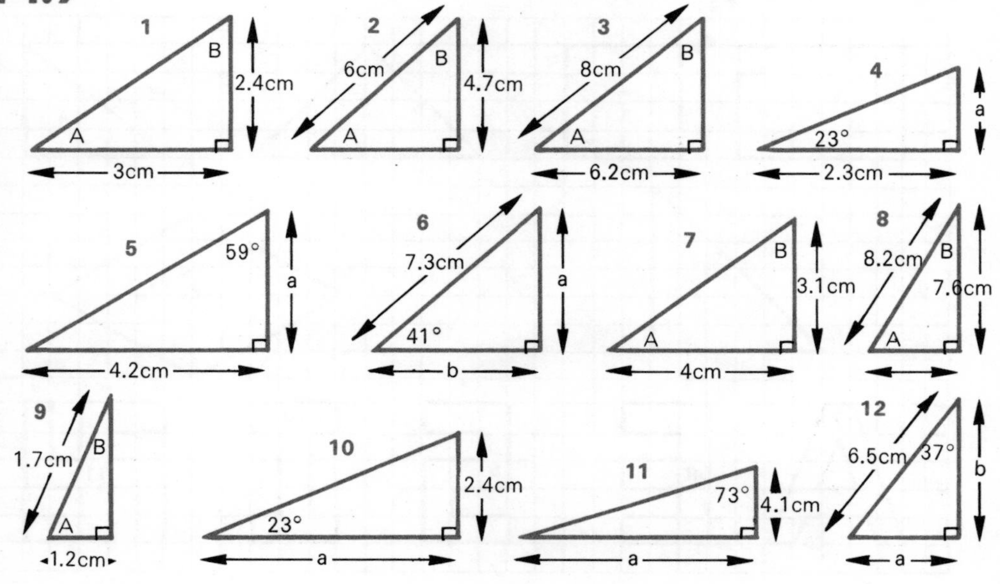

Symmetry

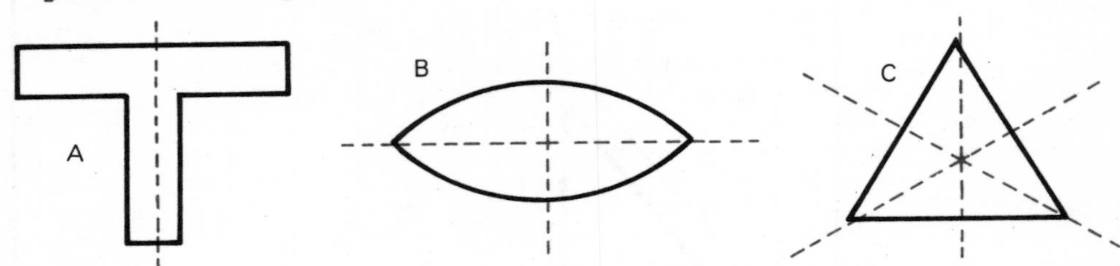

Look at the shape A. If you place the edge of a small mirror along the dotted line with the mirror vertical and look at the reflection you will see that the shape still appears to be the same.

Make a tracing of A, cut it out, and fold it along the dotted line. You will see that the two halves fit over each other exactly.

We call a shape like this **symmetrical**. The dotted line is usually called an **axis of symmetry**.

You will see that shape B has 2 axes of symmetry, and that shape C has 3 axes of symmetry.

Exercise 15.1

Make a copy of each of these shapes on squared paper and mark in all the axes of symmetry with dotted lines. Some of the shapes have no axis of symmetry. Do not copy them, but just write down the number of the shape and "No axis of symmetry".

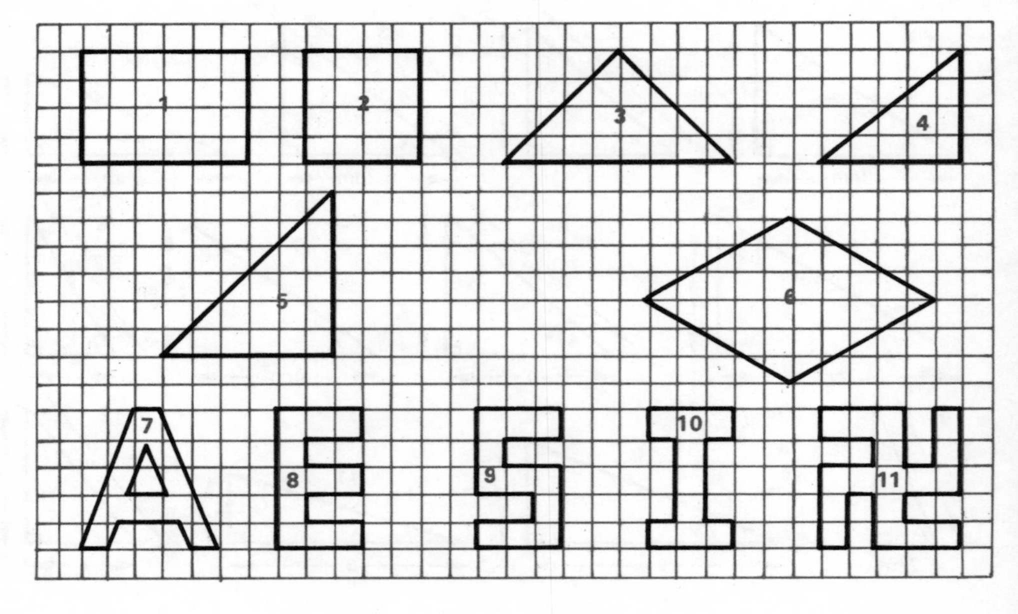

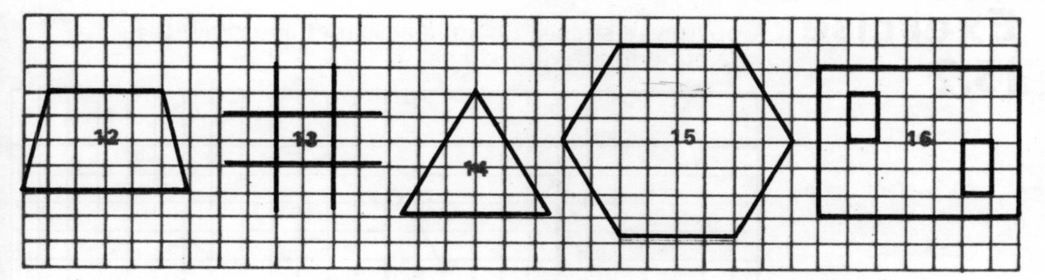

Copy these half-shapes and their axes of symmetry on squared paper and draw the other half of the shape.

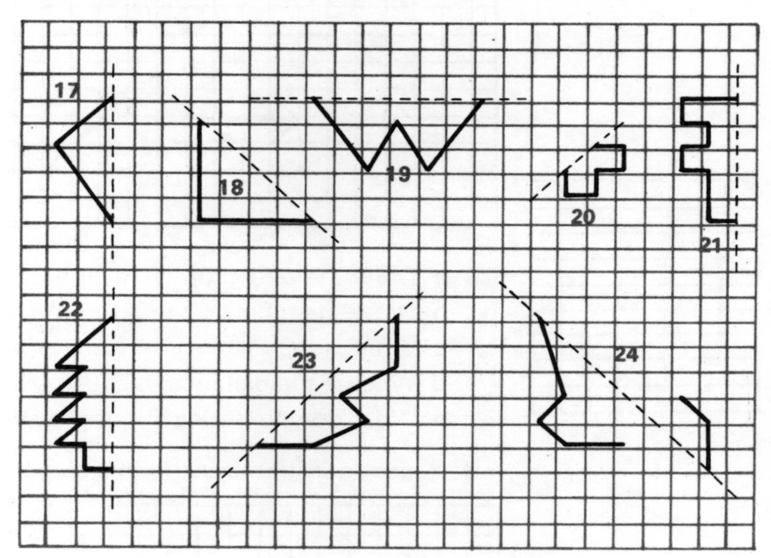

Reflections

Look at the two arrows A_1 and A_2. If you place the edge of a small mirror along the dotted line you will see that the reflection of one of the arrows appears just where the other arrow would be.

We say that A_2 is the **reflection** of A_1 in the line. You will also see that the dotted line is now a line of symmetry.

Exercise 15.2

Copy these shapes on squared paper and draw their reflection in the dotted line.

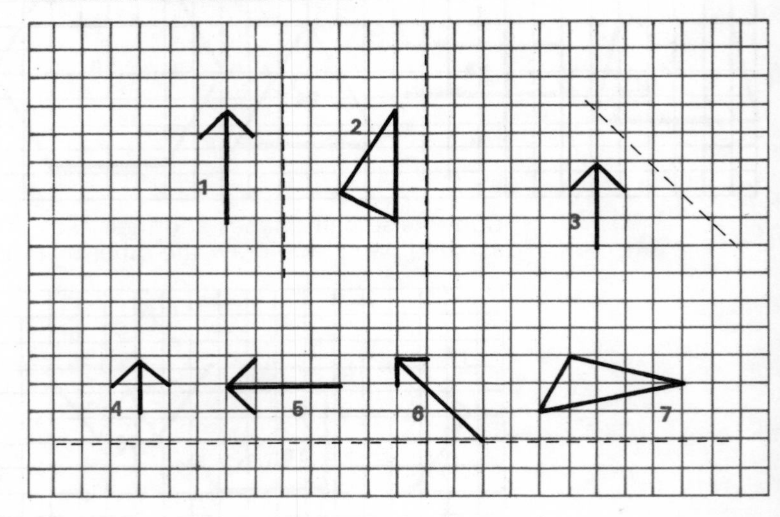

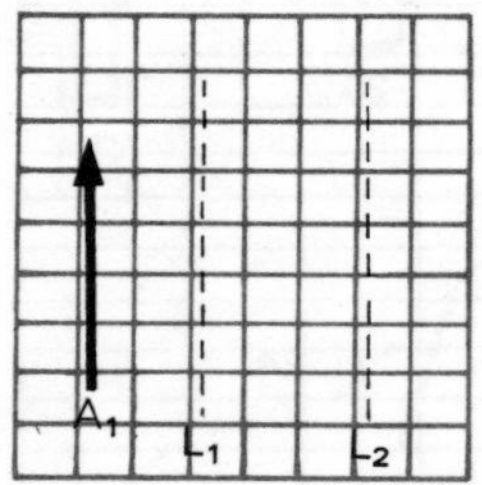

8 **a** Draw the reflection of A_1 in the line L_1. Label it A_2.
b Draw the reflection of A_2 in the line L_2. Label it A_3.
c What is the distance in squares between A_1 and A_3?
d What is the distance in squares between L_1 and L_2?
e What do you notice about the answers to **c** and **d**?

9 Repeat question **8** using this diagram.

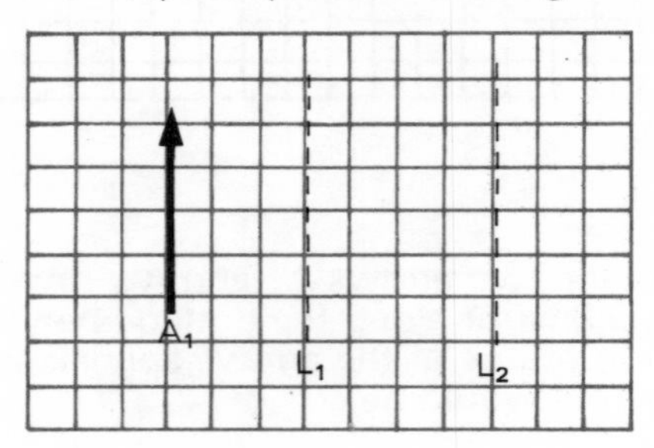

10 Repeat question **8** using this diagram.

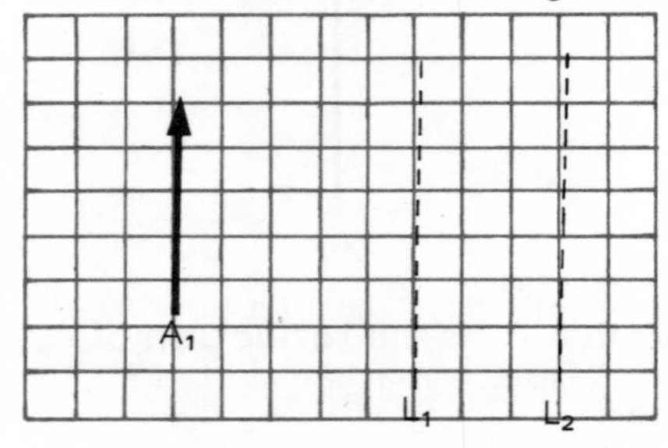

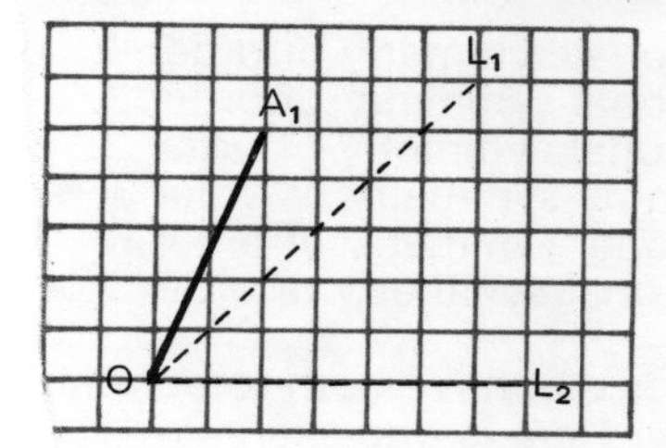

11 **a** Draw the reflection of the point A_1 in the line L_1 and label it A_2.

b Draw the reflection of the point A_2 in the line L_2 and label it A_3. Draw a line from A_3 to 0.

c Measure the angle between the lines L_1 and L_2.

d Measure the angle between lines from 0 to A_1 and from 0 to A_2.

e What do you notice about the answers to **c** and **d**?

12 Repeat question **11** with the point A_1 in a different place.

Rotations

Look at these shapes and see whether they have any axes of symmetry.

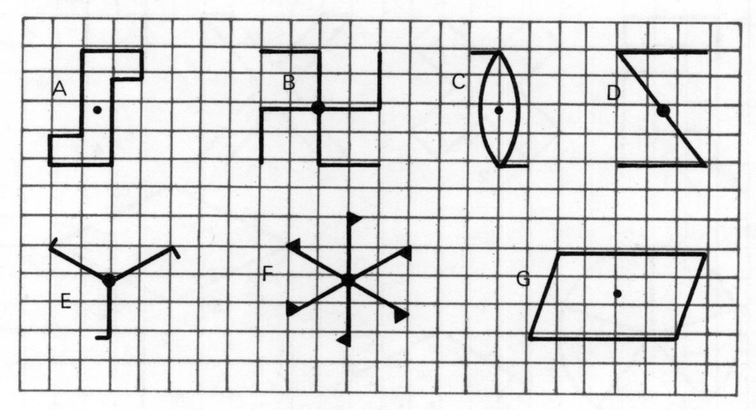

You should have seen that none of them had any axes of symmetry. However, they still seem to be symmetrical in some way. Why is this?

Copy A on squared paper and cut it out. Put a pin through the dot and then turn it through half a turn. The shape will now look the same. Another half turn will again make it look the same and it will be back where it started. We say that A has **rotational symmetry**, and because there are 2 positions where it looks the same we say that the rotational symmetry has **order 2**.

B has rotational symmetry of order 4.
C has rotational symmetry of order 2.
D has rotational symmetry of order 2.
E has rotational symmetry of order 3.
F has rotational symmetry of order 6.
G has rotational symmetry of order 2.

Some shapes can have both rotational symmetry and axes of symmetry. Here is one example.

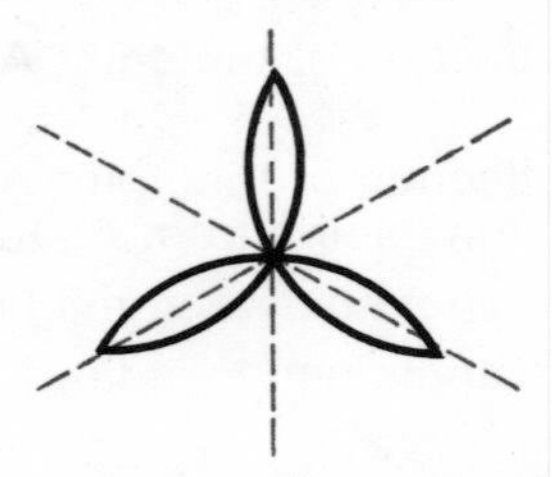

Exercise 15.3

Answer these questions about each of these shapes.

a Has it got rotational symmetry?

b If the answer to **a** is yes, what is the order?

c Has it got any axes of symmetry?

d If the answer to **c** is yes, how many axes of symmetry are there?

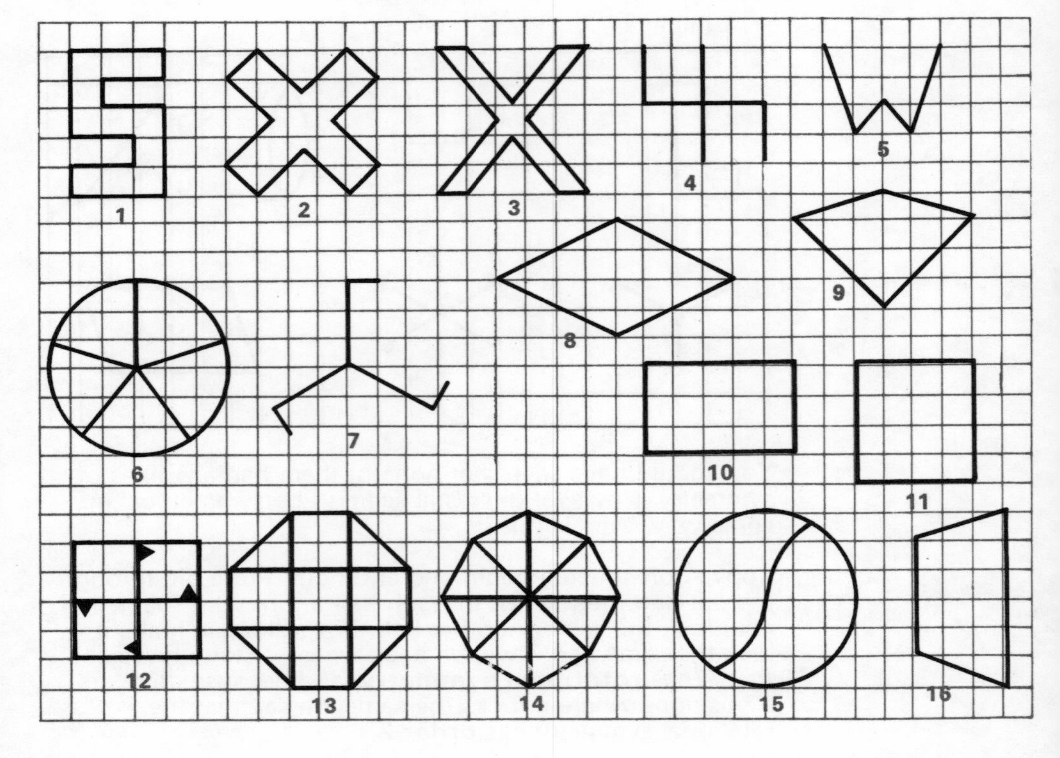

Symmetry of solid shapes

Look at this shape (it is called a triangular prism). Is it possible to cut it into two equal halves? One way of doing this is shown.

The three sides of the triangle at the end are all the same length.

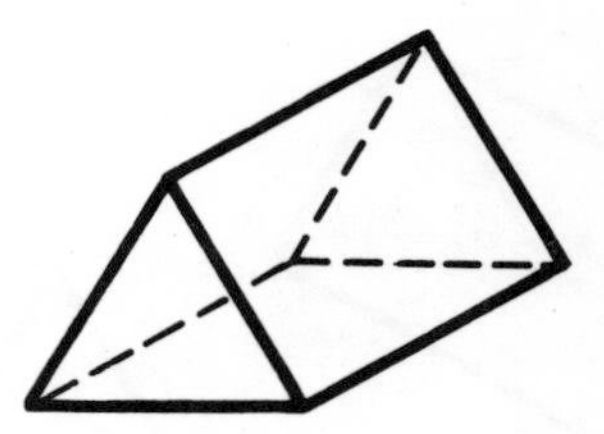

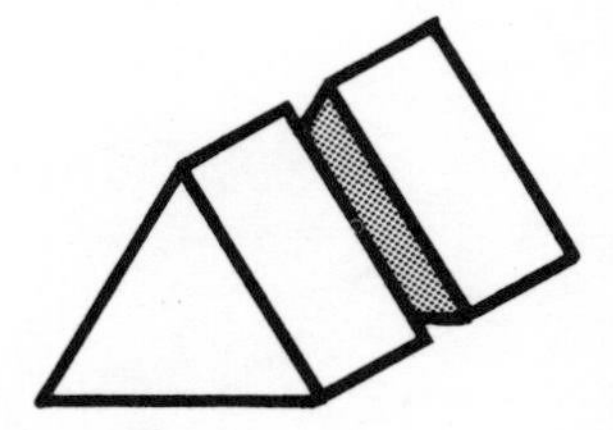

If you put one of the halves against a mirror where you have made the cut, the shape will appear to be whole again like this.

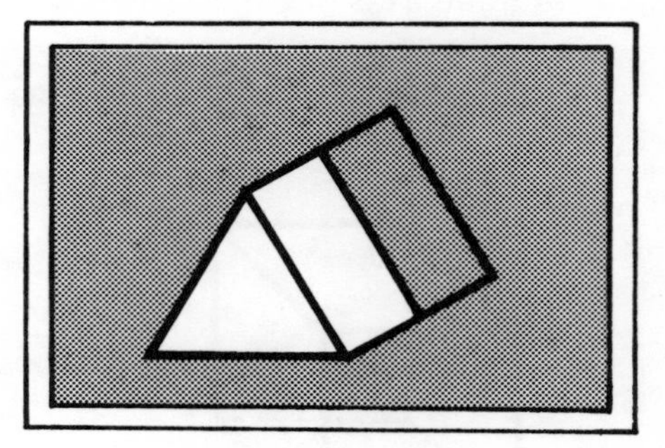

There is another way of cutting it into two equal halves. This is shown below.

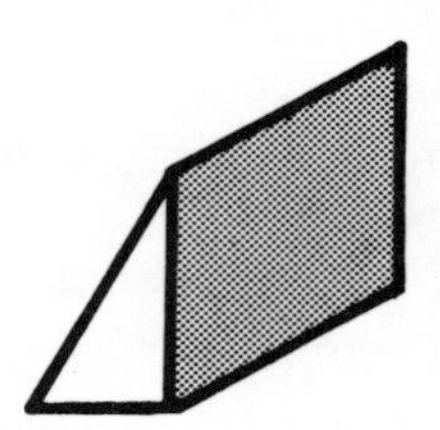

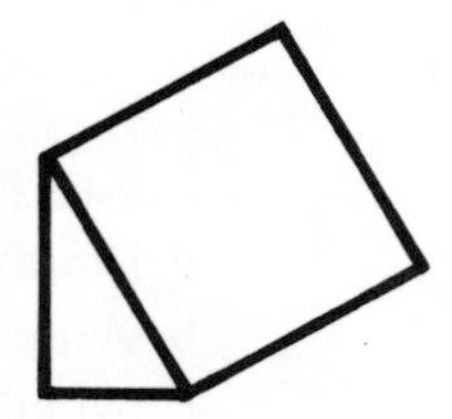

If you put one of the halves against a mirror where you have made the cut, the shape will appear to be whole again. If you cut it this way there are three ways that the cut can be made. These are shown below as dotted lines.

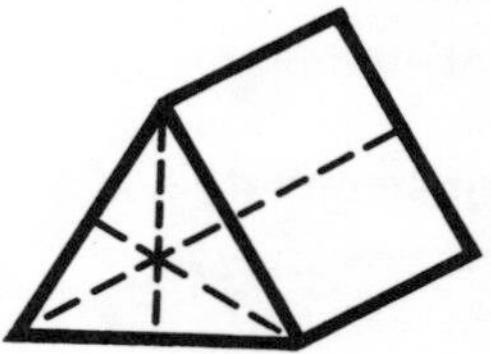

Cuts like this are called **planes of symmetry**. This shape has four planes of symmetry.

It also has four axes of symmetry. Some of them are shown below. If we make half a turn about A the shape will look the same. The order of rotation about A is 2. If we make one third of a turn about B the shape will look the same. The order of rotation about B is three.

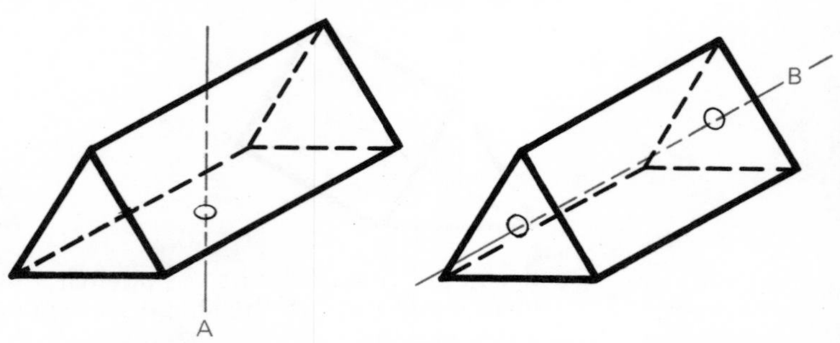

There are three axes like this. There is only one axis like this.

Exercise 15.4

1 The drawings show some of the axes of symmetry of a cube.

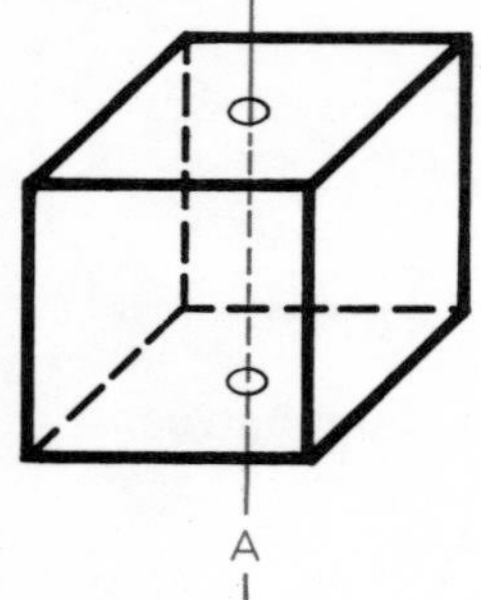

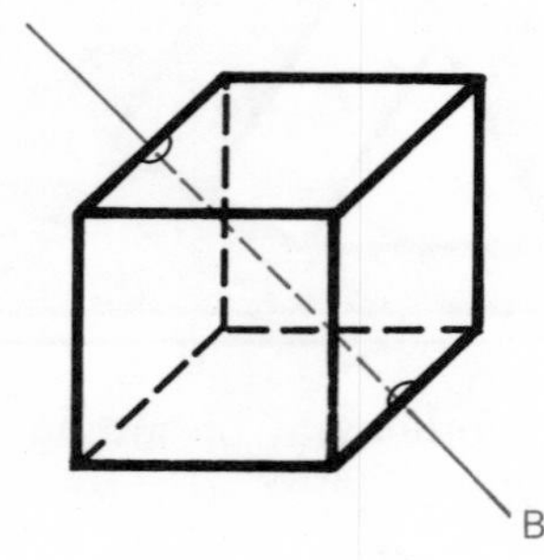

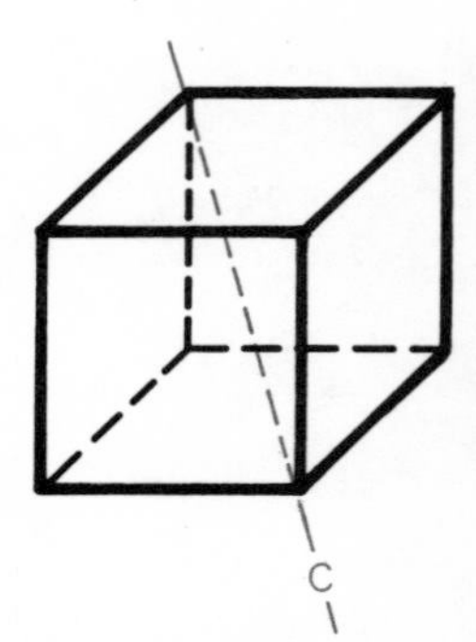

Cardboard models with sticks through them might help you to answer these questions.

a What is the order of symmetry about A?

b What is the order of symmetry about B?

c What is the order of symmetry about C?

d How many axes of symmetry like A are there in a cube?

e How many axes of symmetry like B are there in a cube?

f How many axes of symmetry like C are there in a cube?

g How many axes of symmetry has a cube?

2 The drawings show some of the planes of symmetry of a cube.

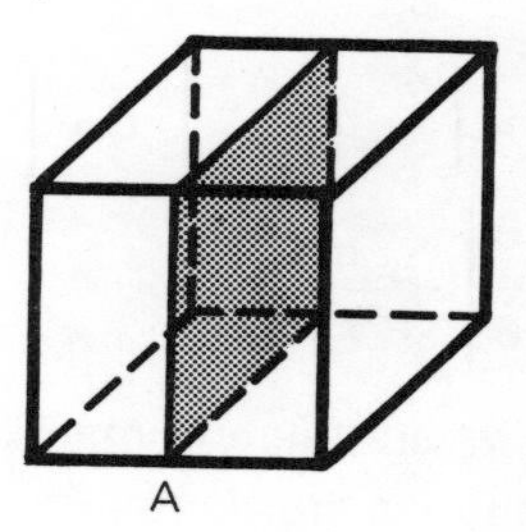

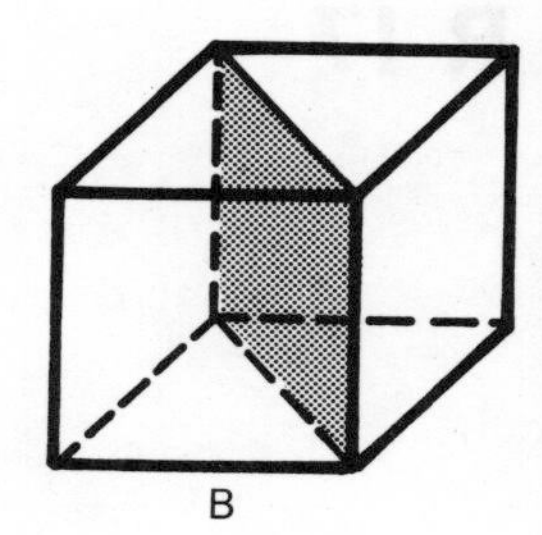

a How many planes of symmetry like A are there in a cube?

b How many planes of symmetry like B are there in a cube?

c How many planes of symmetry has a cube?

3 The drawings show some of the axes of symmetry of a hexagonal prism.

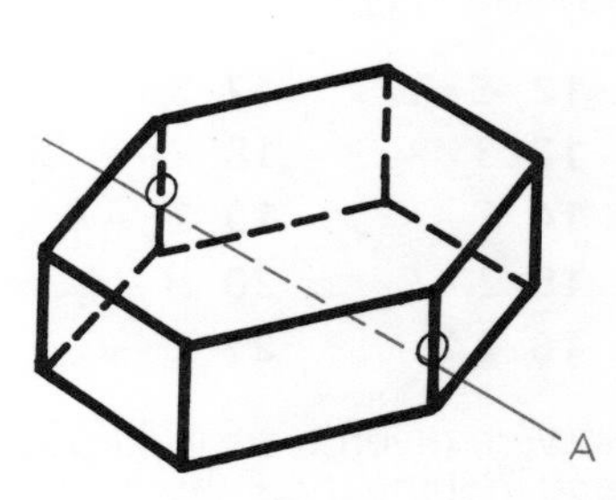

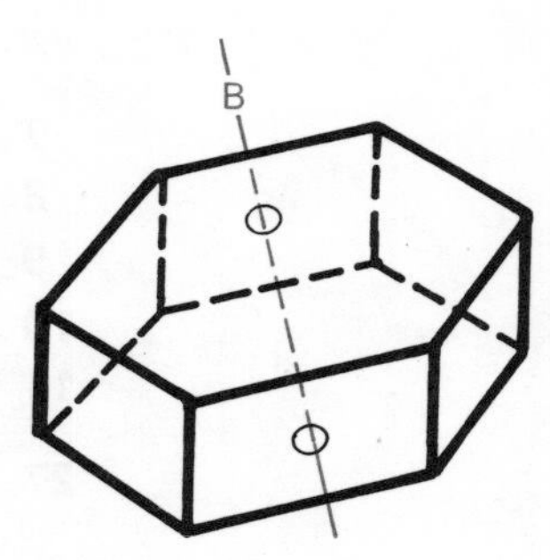

a What is the order of symmetry about A?

b What is the order of symmetry about B?

c How many axes like A are there?

d How many axes like B are there?

e There is another kind of axis of symmetry. Make a drawing showing it.

f What is the order of the axis of symmetry in **e**?

g How many axes of symmetry has this shape?

4 The drawings show some of the planes of symmetry of a hexagonal prism.

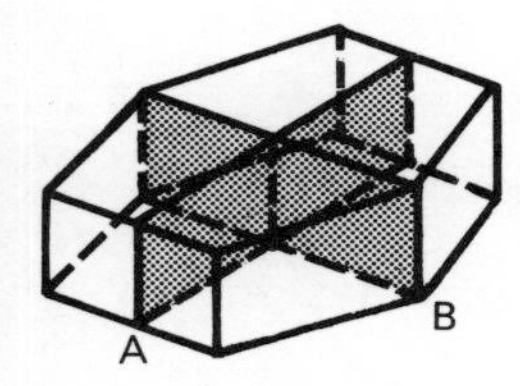

a How many planes of symmetry like A are there?

b How many planes of symmetry like B are there?

c There is another kind of plane of symmetry. Make a drawing about it.

d How many planes of symmetry has this shape?

Exercise R 17

Some of these questions have no answer. In these cases write "No answer".

$A=\begin{bmatrix}4 & 5 & 6\\ 3 & 7 & 8\end{bmatrix}$, $B=\begin{bmatrix}2 & 4 & 2\\ 3 & 6 & 4\end{bmatrix}$, $C=\begin{bmatrix}11 & 15 & 13\\ 14 & 16 & 24\end{bmatrix}$,

$D=\begin{bmatrix}4 & 8\\ 3 & 2\end{bmatrix}$, $E=\begin{bmatrix}6\\ 4\\ 6\end{bmatrix}$, $F=\begin{bmatrix}4\\ 0\\ 5\end{bmatrix}$

1 **a** How many rows are there in matrix A?

b How many columns are there in matrix A?

c What is the size or order of matrix A?

d How many elements are there in matrix A?

e Add together the elements of the second column A.

f What number is in the second row and third column of A?

Questions **2** to **6**. Answer the same questions about the matrices B, C, D, E, F.

Work out the following.

7 $A+B$	**12** $C-D$	**17** $3B$	**22** $4C$
8 $C+D$	**13** $3\times A$	**18** $2A+3B$	**23** $2D$
9 $B+D$	**14** $E-F$	**19** $2E-F$	**24** $4C-2D$
10 $A-B$	**15** $2\times F$	**20** $B-A$	**25** $\frac{1}{2}D$
11 $D+E$	**16** $2A$	**21** $D-C$	**26** $F-E$

27 Four boys were given tests in English, Mathematics and Science in January, February and March. Their marks are given in the matrices below. The marks are out of ten. Some of the marks for March were accidentally torn from the book.

	Peter	John	Brian	Victor
$J=$ English	7	8	3	4
Maths	5	9	4	5
Science	6	7	5	5

$F=\begin{bmatrix}8 & 9 & 4 & 5\\ 4 & 10 & 5 & 6\\ 7 & 6 & 6 & 4\end{bmatrix}$ $\quad M=\begin{bmatrix}8 & 6 & & \\ 4 & & & \\ & & & \end{bmatrix}$

The marks for the whole term (January, February and March) are given in the matrix below

$$T=\begin{bmatrix}23 & 23 & 11 & 14\\ 13 & 28 & 15 & 16\\ 18 & 19 & 15 & 15\end{bmatrix}$$

a What mark did Brian get on his Science test in February?

b What mark did Peter get on his English test in March?

c Who did best in the January exams?

d Who did worst in the February exams?

e If $A=J+F$, write down the matrix A. What does the matrix A tell us?

f $M=T-A$. Write down the matrix M. What does the matrix M tell us?

g If $B=10J$, write down the matrix B.

h Copy and complete this sentence. "Matrix B gives the marks out of . . ."

i What percentage did Victor get in his Science mark in January?

j Who did worst in the exams for the whole term?

k Copy and complete this sentence. "Matrix T gives the marks out of . . ."

l What number have we to divide T by to get the marks out of 10?

unit 16 Multiplying matrices

You know how to add and subtract matrices, and to multiply a matrix by a number. Is it possible to multiply two matrices together? This is possible if the two matrices are the right size.

Suppose a baker delivers three sorts of bread to three cafés. He could write out each day's delivery in the form of a matrix like this.

	Brown Loaves	White Loaves	Rolls
Railway Café	7	6	20
Fred's	5	10	30
Corner Café	1	8	10

He could also write out the cost of the loaves and the rolls in a matrix like this.

Brown Loaves	8p
White Loaves	9p
Roll	2p

How much must he charge each café?

The Railway Café will owe

$$7 \times 8p + 6 \times 9p + 20 \times 2p = 56 + 54 + 40 = 150p$$

In the same way you will find that
Fred's café owes 190p
and that the Corner Café owes 100p

One way of writing this out is shown below

$$\begin{bmatrix} 7 & 6 & 20 \\ 5 & 10 & 30 \\ 1 & 8 & 10 \end{bmatrix} \times \begin{bmatrix} 8 \\ 9 \\ 2 \end{bmatrix} = \begin{bmatrix} 150 \\ 190 \\ 100 \end{bmatrix}$$

You will see that the **first** number in the answer is got by multiplying the **first** row in the first matrix by the prices, the **second** number is got by multiplying the **second** row by the prices, the **third** number is got by multiplying the **third** row by the prices.

This is an example of how matrices can be multiplied together. You should notice that there are **three** columns in the first matrix, and **three** rows in the second matrix.

Example

$$\begin{bmatrix} 3 & 5 & 4 \\ 4 & 6 & 7 \\ 3 & 2 & 1 \end{bmatrix} \times \begin{bmatrix} 4 \\ 2 \\ 3 \end{bmatrix} = \begin{bmatrix} 34 \\ 49 \\ 19 \end{bmatrix} \leftarrow \begin{aligned} &3 \times 4 + 5 \times 2 + 4 \times 3 \\ &= 12 + 10 + 12 = 34 \end{aligned}$$

You are shown how the first number in the answer is worked out. See whether you can get the other two numbers in the same way.

Exercise 16.1

1 A baker delivers bread to three houses in Green Road. His deliveries for a complete week are shown below. The prices are those given on page 142

	Brown Loaves	White Loaves	Rolls
No. 4	3	7	9
No. 7	1	4	4
No. 9	8	2	6

Copy and complete this matrix multiplication, which will tell you how much each house will have to pay. The first one has been done for you.

$$\begin{bmatrix} 3 & 7 & 9 \\ 1 & 4 & 4 \\ 8 & 2 & 6 \end{bmatrix} \times \begin{bmatrix} 8 \\ 9 \\ 2 \end{bmatrix} = \begin{bmatrix} 105 \\ \\ \end{bmatrix} \leftarrow \begin{array}{l} 3\times8+7\times9+9\times2 \\ =24+63+18=105 \end{array}$$

2 Do question **1** again using these matrices.

a $\begin{bmatrix} 3 & 4 & 7 \\ 2 & 5 & 4 \\ 5 & 2 & 3 \end{bmatrix} \times \begin{bmatrix} 7 \\ 4 \\ 3 \end{bmatrix}$

b $\begin{bmatrix} 3 & 4 & 5 \\ 1 & 5 & 3 \\ 5 & 7 & 6 \end{bmatrix} \times \begin{bmatrix} 4 \\ 2 \\ 3 \end{bmatrix}$

c $\begin{bmatrix} 2 & 1 & 6 \\ 3 & 0 & 2 \\ 1 & 1 & 3 \end{bmatrix} \times \begin{bmatrix} 5 \\ 4 \\ 8 \end{bmatrix}$

d $\begin{bmatrix} 6 & 9 & 1 \\ 2 & 8 & 6 \\ 0 & 0 & 5 \end{bmatrix} \times \begin{bmatrix} 4 \\ 3 \\ 6 \end{bmatrix}$

3 Copy and complete these matrix multiplications. The stars tell you how many numbers in each answer. If any of them cannot be done write N.

a $\begin{bmatrix} 4 & 5 & 3 \\ 2 & 6 & 7 \end{bmatrix} \times \begin{bmatrix} 2 \\ 3 \\ 1 \end{bmatrix} = \begin{bmatrix} * \\ * \end{bmatrix}$

b $\begin{bmatrix} 2 & 3 & 6 & 2 \\ 1 & 2 & 3 & 4 \end{bmatrix} \times \begin{bmatrix} 3 \\ 5 \\ 2 \\ 6 \end{bmatrix} = \begin{bmatrix} * \\ * \end{bmatrix}$

c $\begin{bmatrix} 4 & 7 \\ 3 & 8 \\ 2 & 3 \end{bmatrix} \times \begin{bmatrix} 5 \\ 3 \end{bmatrix} = \begin{bmatrix} * \\ * \\ * \end{bmatrix}$

d $\begin{bmatrix} 3 & 5 & 7 \\ 2 & 4 & 6 \\ 3 & 1 & 7 \end{bmatrix} \times \begin{bmatrix} 5 \\ 1 \end{bmatrix} = \begin{bmatrix} \\ \end{bmatrix}$

4 A farmer delivers eggs, potatoes and apples. His deliveries for one week are shown in the matrix below. The numbers in the potato and apple columns stand for kilograms.

	Potatoes	Apples	Eggs
Mrs Brown	5	2	10
Mrs Green	6	0	6
Mrs Smith	3	1	0
Mrs Jones	4	5	8

	Prices
Potatoes	6p
Apples	8p
Eggs	3p

Multiply the two matrices together to get a matrix which will show how much each the women will have to pay.

5 Do question **4** again using these matrices.

a $\begin{bmatrix} 1 & 5 & 8 \\ 2 & 2 & 9 \\ 3 & 6 & 0 \\ 4 & 7 & 1 \end{bmatrix} \times \begin{bmatrix} 7 \\ 5 \\ 2 \end{bmatrix}$

b $\begin{bmatrix} 3 & 9 & 7 \\ 0 & 4 & 8 \\ 6 & 5 & 1 \\ 8 & 6 & 7 \end{bmatrix} \times \begin{bmatrix} 8 \\ 4 \\ 4 \end{bmatrix}$

c $\begin{bmatrix} 2 & 8 & 7 \\ 9 & 3 & 6 \\ 0 & 4 & 1 \\ 5 & 3 & 2 \end{bmatrix} \times \begin{bmatrix} 7 \\ 5 \\ 4 \end{bmatrix}$

d $\begin{bmatrix} 4 & 1 & 9 \\ 3 & 7 & 0 \\ 5 & 8 & 2 \\ 4 & 6 & 5 \end{bmatrix} \times \begin{bmatrix} 3\frac{1}{2} \\ 8\frac{1}{2} \\ 2\frac{1}{2} \end{bmatrix}$

6 Do these matrix multiplications. In each example explain what multiplying by the first matrix does to the second matrix.

a $\begin{bmatrix} 1 & 0 & 0 \\ 0 & 1 & 0 \\ 0 & 0 & 1 \end{bmatrix} \times \begin{bmatrix} 1 \\ 2 \\ 3 \end{bmatrix}$

b $\begin{bmatrix} 0 & 0 & 1 \\ 0 & 1 & 0 \\ 1 & 0 & 0 \end{bmatrix} \times \begin{bmatrix} 1 \\ 2 \\ 3 \end{bmatrix}$

c $\begin{bmatrix} 2 & 0 & 0 \\ 0 & 2 & 0 \\ 0 & 0 & 2 \end{bmatrix} \times \begin{bmatrix} 1 \\ 2 \\ 3 \end{bmatrix}$

d $\begin{bmatrix} 1 & 1 & 1 \\ 0 & 0 & 0 \\ 0 & 0 & 0 \end{bmatrix} \times \begin{bmatrix} 1 \\ 2 \\ 3 \end{bmatrix}$

e $\begin{bmatrix} 0 & 0 & 5 \\ 0 & 5 & 0 \\ 5 & 0 & 0 \end{bmatrix} \times \begin{bmatrix} 1 \\ 2 \\ 3 \end{bmatrix}$

7 a A factory makes two sizes of boxes of mixed sweets. The large box contains 10 orange, 6 lemon and 6 lime. The small box contains 5 orange, 3 lemon and 3 lime. Write this in the form of a matrix with three rows and two columns.

b On one day the factory has to produce 200 large boxes and 100 small boxes. Write this as a matrix with two rows and one column.

c Multiply the two matrices together to find the number of orange sweets they will need, the number of lemon sweets and the number of lime sweets.

Changing shapes with matrices

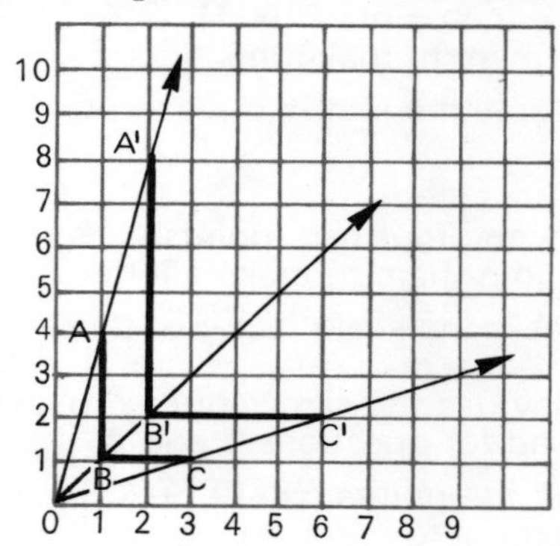

Look at the small letter L on the squared paper.

Its corners are at $A(1, 4)$, $B(1, 1)$, $C(3, 1)$.

We may write them like this $\begin{bmatrix} 1 \\ 4 \end{bmatrix}$ $\begin{bmatrix} 1 \\ 1 \end{bmatrix}$ $\begin{bmatrix} 3 \\ 1 \end{bmatrix}$

Let us multiply each one of these by the matrix $\begin{bmatrix} 2 & 0 \\ 0 & 2 \end{bmatrix}$

$$\begin{bmatrix} 2 & 0 \\ 0 & 2 \end{bmatrix} \times \begin{bmatrix} 1 \\ 4 \end{bmatrix} = \begin{bmatrix} 2 \\ 8 \end{bmatrix} \quad \begin{bmatrix} 2 & 0 \\ 0 & 2 \end{bmatrix} \times \begin{bmatrix} 1 \\ 1 \end{bmatrix} = \begin{bmatrix} 2 \\ 2 \end{bmatrix}$$

$$\begin{bmatrix} 2 & 0 \\ 0 & 2 \end{bmatrix} \times \begin{bmatrix} 3 \\ 1 \end{bmatrix} = \begin{bmatrix} 6 \\ 2 \end{bmatrix}$$

This gives three new points $A'(2, 8)$, $B'(2, 2)$, $C'(6, 2)$.

If we mark these three points on the same sheet of squared paper you see that we still get the letter L, but twice the size. It has been enlarged with centre of enlargement 0, and scale factor 2.

Exercise 16.2

1 **a** Write the numbers from 0 to 20 along the bottom and side of some squared paper.

b Mark the points $A(3, 3)$, $B(1, 3)$, $C(1, 6)$ and join them to make the letter L.

c Multiply the matrix of each point by $\begin{bmatrix} 3 & 0 \\ 0 & 3 \end{bmatrix}$ as in the example above.

d Mark in the three new points with the letters A', B', C' and join them to make a letter L.

e What is the scale factor of this enlargement?

2 You may use the same squared paper for this question as you used for question **1**.

a Mark the points $P(6, 1)$, $Q(4, 1)$, $R(6, 4)$, $S(4, 4)$ and join them to make the letter Z.

b Multiply the matrix of each point by $\begin{bmatrix} 2 & 0 \\ 0 & 2 \end{bmatrix}$

c Mark the four new points with the letters P', Q', R', S' and join them to make a letter Z.

d What is the scale factor of this enlargement?

3 You may use the same squared paper for this question as you used for questions **1** and **2**.

a Mark the points $I(8, 6)$, $J(4, 5)$, $K(3, 6)$, $L(4, 7)$ and join them to make an arrow.

b Multiply the matrix of each point by $\begin{bmatrix} 2 & 0 \\ 0 & 2 \end{bmatrix}$

c Mark the four new points with the letters I', J', K', L' and join them to make an arrow.

d What is the scale factor of this enlargement?

Matrices that reflect

The matrices that you have just dealt with are called enlarging matrices. Are there matrices that do other things to shapes? You will see that there are in the next exercise.

Exercise 16.3

1 **a** Make a copy of the following diagram on squared paper.

b Multiply the matrices of the points A, B, C by the matrix $\begin{bmatrix} 0 & 1 \\ 1 & 0 \end{bmatrix}$

c Mark in the three new points with the letters A', B', C'

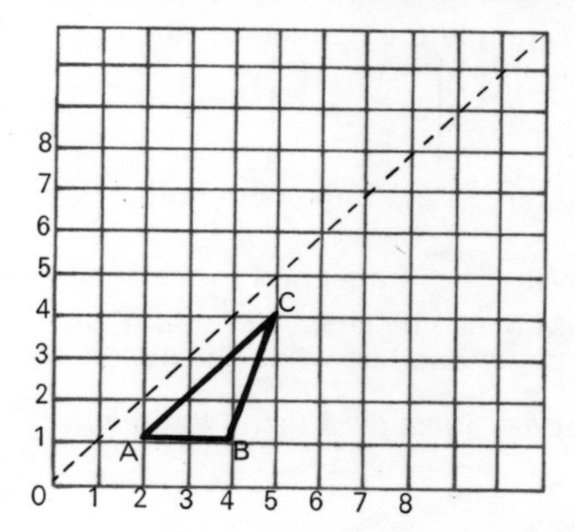

d You should find that the new shape is exactly the same as the old one, except that it is on the other side of the dotted line, and has been turned over. We say that it has been **reflected** in the dotted line.

2 Make some shapes of your own and multiply the matrices of their points by the matrix $\begin{bmatrix} 0 & 1 \\ 1 & 0 \end{bmatrix}$

You should find that they are all reflected in the dotted line.

If you are going to try the rest of this exercise you should revise the work on co-ordinates in exercise **R1**, and the work on negative numbers in Unit **6**. You will need squared paper with the numbers on the horizontal axis marked from -6 to $+6$, and the numbers on the vertical axis marked from -6 to $+6$ as shown below.

3 a Copy the triangle *ABC* and the dotted line on your squared paper.

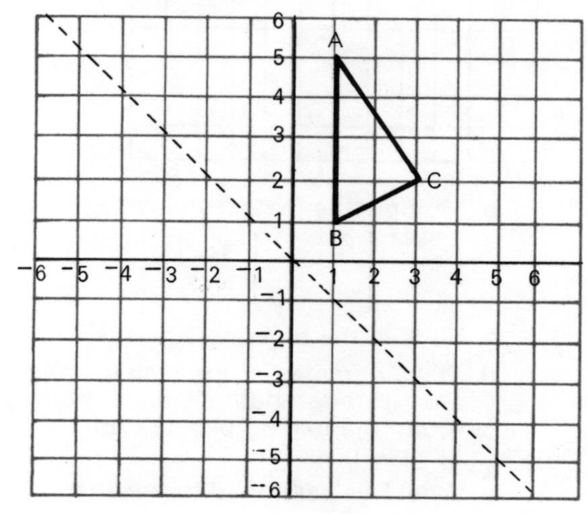

b Multiply the matrices of the points A, B, C by the matrix $\begin{bmatrix} 0 & ^{-}1 \\ ^{-}1 & 0 \end{bmatrix}$

c Mark in the three new points with the letters A', B', C'.

d What has happened to the shape ABC?

e Make some shapes of your own on your squared paper and see what this matrix does to them.

4 Repeat question **3** using the matrix $\begin{bmatrix} 1 & 0 \\ 0 & ^{-}1 \end{bmatrix}$

5 Repeat question **4** using the matrix $\begin{bmatrix} ^{-}1 & 0 \\ 0 & 1 \end{bmatrix}$

Matrices that rotate

Before you do the next exercise make sure that you know what a rotation is.

Exercise 16.4

1 a Make a copy of this drawing on squared paper and mark the numbers on the axis as shown.

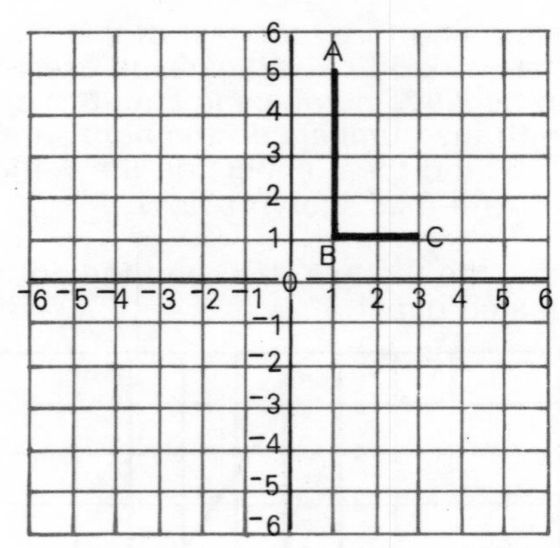

b Multiply the matrices of the points A, B. C, by the matrix $\begin{bmatrix} 0 & ^{-}1 \\ 1 & 0 \end{bmatrix}$

c Mark in the three new points with the letters A', B', C'.

d Make a tracing of the L shape and the point 0. Stick a pin through both sheets of paper at the point 0 and rotate the tracing paper so that shape L moves to its new position.

e What has happened to the shape L?

Repeat question **1** with the following matrices. Questions **2** and **3** can be done on the same sheet of paper as question **1**. Do questions **4** and **5** on a new sheet of paper.

2 $\begin{bmatrix} ^-1 & 0 \\ 0 & ^-1 \end{bmatrix}$ **3** $\begin{bmatrix} 0 & 1 \\ ^-1 & 0 \end{bmatrix}$ **4** $\begin{bmatrix} 0{\cdot}7 & 0{\cdot}7 \\ ^-0{\cdot}7 & 0{\cdot}7 \end{bmatrix}$

5 $\begin{bmatrix} 0{\cdot}7 & ^-0{\cdot}7 \\ 0{\cdot}7 & 0{\cdot}7 \end{bmatrix}$

Matrices that shear

Exercise 16.5

Some matrices **shear** a shape. You will see exactly what a shear is when you have done this exercise.

1 **a** On squared paper mark the numbers 0 to 3 on the vertical axis, and 0 to 12 on the horizontal axis.

b Mark the points $A(2, 0)$, $B(2, 2)$, $C(6, 2)$, $D(6, 0)$ and join them to make a rectangle.

c Multiply the matrices of the four points by the matrix $\begin{bmatrix} 1 & 3 \\ 0 & 1 \end{bmatrix}$

d Mark in the four new points with the letters A', B', C', D'.

e You will see that the rectangle has been **sheared** into a parallelogram.

2 Repeat question **1** with the points (3, 0), (3, 3), (5, 3), (5, 0) and the matrix $\begin{bmatrix} 1 & 2 \\ 0 & 1 \end{bmatrix}$

3 Repeat question **1** with the points $A(2, 0)$, $B(2, 2)$, $C(4, 2)$, $D(4, 0)$ and the matrix $\begin{bmatrix} 1 & 1 \\ 0 & 1 \end{bmatrix}$ and then with the matrix $\begin{bmatrix} 1 & ^-1 \\ 0 & 1 \end{bmatrix}$

Exercise R 18

Part 1. Finding probabilities

Four boys, Peter, John, Simon and Ron, want to go on a school journey, but there are only two places left. The two who are to go will be chosen by having their names drawn from a hat. The number of possible pairs is 6.

Peter	Peter	Peter	John	John	Simon
John	Simon	Ron	Simon	Ron	Ron

Find the following probabilities (either as a fraction or a decimal).

1 That Peter is chosen

2 That Ron is chosen.

3 That Ron and Simon are chosen.

4 That Ron and John are chosen.

5 That Peter is not chosen.

6 That Ron is chosen, but Simon is not.

7 That neither Simon, nor Peter are chosen.

8 That Simon is chosen, but Peter is not.

Four cards marked with the numbers 2, 4, 6, 8 are placed in a box. One of the cards is taken out, and then a second card.

9 Write down all the possible pairs that can be taken out. (There are 12.) Find the following probabilities (either as a fraction or as a decimal).

10 That the number 24 is drawn.

11 That either 24 or 42 is drawn.

12 That 98 is drawn.

13 That 46 is not drawn.

14 That a number greater than 50 is drawn.

15 That a number less than 40 is drawn.

16 That a number less than 100 is drawn.

17 That a number between 40 and 60 is drawn.

Three cards marked with the letters T, O, R are placed in a box. One of the cards is taken out, then the second, then the third.

18 Write down all the possible groups of three that can be taken out.

Write down the following probabilities (either as a fraction or a decimal).

19 That the three cards form the word ROT.

20 That the three cards form the word TOR.

21 That the three cards form a word starting with the letter O.

22 That the three cards form a word ending with the letter T.

23 That the three cards have the letter O immediately before the letter R.

24 That the letter T is in the middle.

25 That the letter T is not in the middle.

The answers to the next questions can be given either as fractions or decimals.

26 Eight boys took an examination and 6 passed. What is the probability of a boy passing the examination?

27 Ten light bulbs were tested and 2 failed. What is the probability of a light bulb failing the test?

28 Forty TV sets were examined. Eight were found to be wrongly adjusted. Find the probability that a TV set will be found to be wrongly adjusted.

Part 2. Using probabilities

In these examples you may find the answer using either the fractional or decimal value for the probability.

1 The probability of a light bulb failing a test is $\frac{1}{4}$ or 0·25. How many would you expect to fail if the number tested was

a 4, **b** 8, **c** 12, **d** 16, **e** 20, **f** 64, **g** 440?

2 A TV rental firm has found that the chance of one of its sets breaking down in a year is $\frac{1}{10}$. How many sets would the firm expect to break down if the number of sets it rents out is

a 10, **b** 20, **c** 30, **d** 50, **e** 100, **f** 500, **g** 670?

3 The probability of a car passing a test of its lights is $\frac{4}{5}$ or 0·8. How many would you expect to pass if the number tested was

a 5, **b** 10, **c** 15, **d** 20, **e** 25, **f** 50, **g** 130?

4 A car manufacturer found that $\frac{2}{3}$ or 0·67 of the cars he produced were passed first time by the quality control inspectors. How many cars would be passed the first time if the number of cars he produced was

a 3, **b** 9, **c** 15, **d** 21, **e** 60, **f** 600, **g** 9000?

5 What is the probability that a car he produces will not pass the test in question **4**? How many cars would fail to pass the test the first time if the numbers inspected were

a 6, **b** 12, **c** 18, **d** 24, **e** 300, **f** 450, **g** 720?

In the next problems you have first to work out a probability, and then use the probability to answer some more questions. You may work out the probability as a fraction or a decimal.

6 A school entered 10 pupils for a new examination. Eight passed.

a What is the probability that a pupil passes the examination?

How many would you expect to pass the next year if the number entered was **b** 5, **c** 15, **d** 20, **e** 25, **f** 30, **g** 150?

7 A car hire firm ran 16 new cars for one year. Six of them developed serious faults in the year.

a What is the probability that one of their new cars develops a fault during the first year?

How many would you expect to develop a fault in the first year if the number of new cars was
b 8, **c** 24, **d** 32, **e** 40, **f** 70, **g** 101?

8 A car insurance firm found that on average out of each 35 motorists insured with them, 20 made a claim during one year.

a What is the probability of a motorist making a claim during one year?

How many claims can the firm expect each year if the number of motorists they insure is
b 42, **c** 56, **d** 70, **e** 200, **f** 2000, **g** 30 000?

unit 17 More on probability and gambling

Insurance

If you own a car you will insure it against accidents, theft, etc. You might have to pay £30 a year. You probably will not have an accident, but if you do the insurance company might have to pay you a large amount for repairs, perhaps several hundred pounds. If a lot of its customers have accidents an insurance company could lose a lot of money. To make sure this does not happen it does some calculations which use the probability of one of its customers having an accident.

Example

An insurance company has 100 customers. The probability of a customer having an accident in one year is $\frac{1}{10}$. The average cost of an accident is £200. It charges its customers £25 each per year.

a How many customers will the firm expect to have accidents?

b How much will they have to pay out?

c How much will it collect from its customers?

d How much profit, or loss, will they make?

a Most likely number of accidents $= \frac{1}{10} \times 100 = 10$.

b They will have to pay out $10 \times £200 = £2000$.

c They will collect $100 \times £25 = £2500$.

d Their profit will be $£2500 - £2000 = £500$.

Out of this profit they will have to pay wages, rent, etc. They could have a bad year with a lot of accidents, and so lose a lot of money. But the calculation shows that, on average, they should expect to make a profit of £500 per year.

Exercise 17.1

Do the above example using the following figures.

	Probability of accident	Number of customers	Average cost of accident	Yearly charge for each customer
1	$\frac{1}{10}$	200	£200	£25
2	$\frac{1}{10}$	400	£300	£40
3	$\frac{1}{5}$	50	£400	£50
4	$\frac{1}{5}$	400	£50	£20
5	$\frac{2}{5}$	200	£300	£100
6	$\frac{3}{10}$	1000	£150	£50

7 How much will they have to charge in question **1** to make a profit of £10 from each customer?

8 How much will they have to charge in question **3** to make a profit of £15 from each customer?

9 How much must they charge in question **2** to make a total profit of £2000?

10 How much must they charge in question **4** to make a total profit of £2000?

11 How much must they charge in question **6** to make a total profit of £10 000?

Gambling

Example

0	3	0
2	8	2
0	3	0

At a fairground a customer pays 3p a throw to throw a dart at this board. If he hits one of the numbered squares he will be paid that amount in prize money.

a What will his average winnings be in 9 throws?

b What are his average winnings per throw?

c What is his average profit, or loss, per throw?

Assume that he always hits one of the squares, and that he is equally likely to hit any square.

a Average winnings in 9 throws $= 0+3+0+2+8+2+0+3+0 = 18$p

b Average winnings per throw $= 18 \div 9 = 2$p

c His loss per throw $= 3\text{p} - 2\text{p} = 1$p

This means that every time a customer throws a dart the owner of this game will on average get 1p. If he gets a large number of customers who are good at darts he will of course get less or even make a loss.

Exercise 17.2

Do the above example using the following boards. The cost of one throw is given under each board.

1 **a**

1	1	1
2	8	2
1	1	1

3p

b

0	8	0
8	0	8
0	3	0

3p

c

10	0	10
0	10	0
0	6	0

5p

d

1	2	3
4	5	6
7	8	9

6p

e

14	3	8
16	4	10
18	6	2

8p

f

6	8	4
8	7	6
0	6	9

5p

2 How much would the owner have to charge per throw for each of the above boards to make a profit of 2p per throw?

3 Two dice are thrown and the score is got by taking away the scores on each one. For example, scores of 6 and 2 will give a final score of 4. Copy and complete this table which shows all the possible scores.

NUMBER ON SECOND DIE

NUMBER ON FIRST DIE	1	2	3	4	5	6
1	0	1	2			
2	1	0	1			
3	2	1	0			
4						
5						
6						

Find the probabilities of the following scores.

a 1, **b** 2, **c** 3, **d** 4, **e** 5, **f** 6, **g** 7, **h** 0

i What is the most likely score?

j What is the least likely score?

k What is the average score per throw?

revision 19 Bar charts and averages

Exercise R 19

1 Find the averages of the following:

a 3, 4, 7, 8, 5, 8, 7

b 2, 5, 5, 6, 6, 7, 7, 7, 9

c 7, 9, 11, 5, 7, 3

d £8, £2, £6, £4, £7, £8, £0

e £10, £12, £8, £10, £6, £4, £3, £8, £2

f £13, £14, £16, £14, £18

g 0, 0, 0, 0, 0, 0, 8, 16

h £1·43, £1·48, £1·46, £1·43, £1·42, £1·39, £1·40

i £1·34, £3·57, £5·76, £3·26, £3·51

j 4·6, 4·7, 3·8, 7·9, 3·9, 2·8

k 3, 7, 4, 6, 2, 5, 7, 8

l 3, 4, 5, 6, 7, 8, 9, 10, 11, 12

2 The bar chart shows the numbers of boys away in a class during one week.

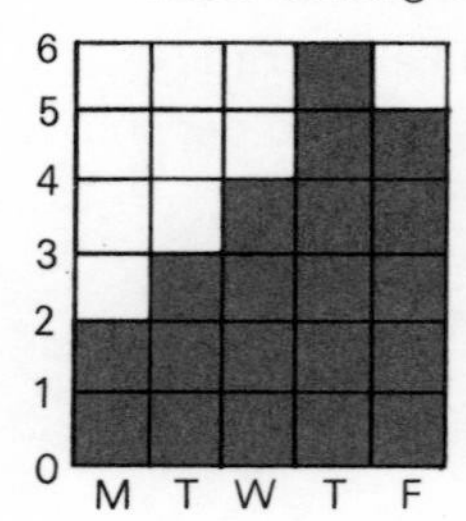

a Copy and complete this table.

Day	Mon	Tues	Wed	Thurs	Fri
Number away	2				

b What was the total number of absences for the week?

c What was the average number of boys absent each day?

3 The table shows the number of boys away during the next week.

Day	Mon	Tues	Wed	Thurs	Fri
Number away	5	4	3	7	6

a Draw a bar chart showing the number of boys away during this week.

b What was the total number of absences for the week?

c What was the average number of boys absent each day?

d Was the class attendance better during the first or second week?

4 The graph shows the average number of hours of sunshine per day in a sea side town.

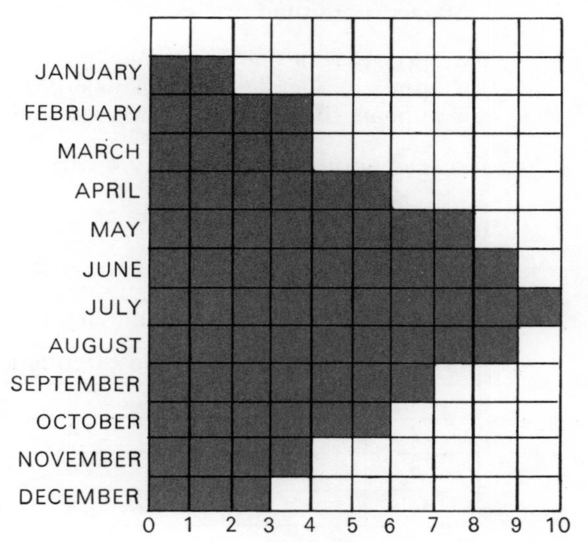

a Copy and complete this table.

Month	J	F	M	A	M	J	J	A	S	O	N	D
Hours of sunshine per day					8							

b Which is probably the best month in which to take a holiday in this town?

c Why is the word "probably" used in question **b**?

d Will a holiday taken in July always have more hours of sunshine per day than a holiday taken in April?

e Add up the numbers in the table in question **a**.

f What is the average number of hours of sunshine *per day* for the whole year?

g What is the average number of hours of sunshine *per week* during December?

unit 18 More on averages and pie charts

Suppose we have to find the average of £4, £4, £4, £4, £7, £7, £8, £8, £8. You will see that each of the numbers appears more than once.

£4 appears four times. We say that its **frequency** is 4.
£7 appears twice. We say that its **frequency** is 2.
£8 appears three times. We say that its **frequency** is 3.

To save adding them all up we can say: 4 lots of £4=£16
2 lots of £7=£14
3 lots of £8=£24
Add them up £54

There are 9 numbers, so average=£54÷9=£6

We usually set it out in a table like this –

Value	Frequency	Value × frequency
£4	4	£16
£7	2	£14
£8	3	£24
Totals	9	£54

Average=£54÷9=£6

Exercise 18.1

1 Find the average of these amounts of money by copying and completing the table below: £2, £2, £5, £5, £5, £5, £4, £4, £4.

Value	Frequency	Value × frequency
£2		
£4	3	£12
£5		
Totals		

2 Find the average of these amounts of money using these frequency tables.

a

Value	Frequency	Value × frequency
£2	2	
£3	5	
£5	1	
Totals		

b

Value	Frequency	Value × frequency
£5	4	
£6	1	
£7	0	
£8	2	
Totals		

3 Find the average of the following, using frequency tables.

a £5, £5, £5, £6, £8, £8, £8, £9, £9

b £2, £2, £2, £3, £3, £3, £4, £4

c 4m, 4m, 5m, 5m, 6m, 6m, 6m

d £10, £10, £10, £10, £11, £11, £11, £12, £12, £12, £12, £12, £12, £13, £13, £13

e 1, 1, 1, 1, 1, 1, 2, 2, 2, 3, 3, 4, 4, 4, 4, 5, 5, 6, 6, 6, 7, 7, 8, 8, 8

f 5, 5, 5, 5, 6, 6, 6, 7, 7, 8, 8, 8, 9, 9, 10, 10, 10, 11, 11, 11, 11

4 Two dice are thrown and their scores are shown on the frequency table.

Score	Frequency
2	0
3	1
4	2
5	4
6	6
7	7
8	4
9	5
10	2
11	0
12	2

a How many times were the dice thrown?

b What was the total score on all the throws?

c What was the average score per throw?

d Draw a bar chart showing the results given in the frequency table.

Averages of grouped numbers

Sometimes it is necessary to find the averages of numbers that we do not know exactly. If you look at the table below, which shows the earnings of 100 men in a factory, you can see that for example 20 men earn a wage which is between £32 and £34 (or more exactly between £32.00$\frac{1}{2}$ and £34), but we do not know exactly how much each man earns.

To do a problem like this we say that on average the 20 men earn £33 each, in the same way we say that the men in the £30 to £32 group earn on average £31 each. If we do the same thing for the other groups we get the second table, and then we can find the average in the usual way.

Wage	Frequency
Over £30 and up to £32	10
Over £32 and up to £34	20
Over £34 and up to £36	30
Over £36 and up to £38	30
Over £38 and up to £40	10

Wage	"Middle wage"	F	Middle wage × F
£30 to £32	£31	10	£310
£32 to £34	£33	20	£660
£34 to £36	£35	30	£1050
£36 to £38	£37	30	£1110
£38 to £40	£39	10	£390
Totals		100	£3520

Average = £3520 ÷ 100 = £35·20

We do not of course know what the true average is, but an average worked out this way is usually very close to the true value.

Exercise 18.2

1 Find the average of these wages using a "middle wage" table like the one above.

Wage	Frequency
Over £30 and up to £32	4
Over £32 and up to £34	1
Over £34 and up to £36	0
Over £36 and up to £38	2

2 Write down the frequency table for the following bar chart, and then find the average using the "middle wage" method.

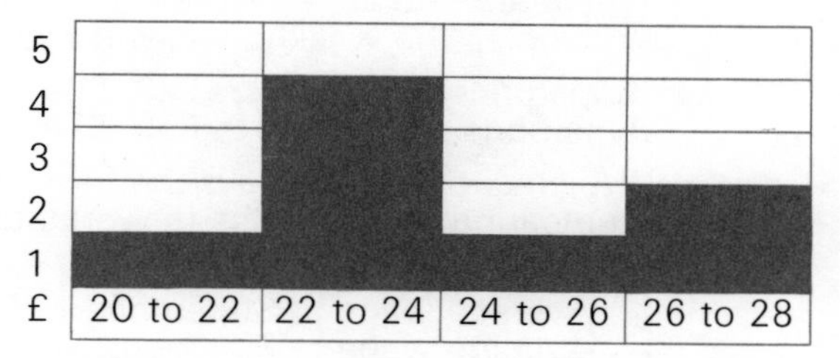

3 Find the averages of these sets of examination marks, and draw a bar chart for each set.

a

Mark	Frequency
1 to 20	1
21 to 40	2
41 to 60	3
61 to 80	1
81 to 100	2

b

Mark	Frequency
1 to 20	0
21 to 40	3
41 to 60	2
61 to 80	1
81 to 100	2

c

Mark	Frequency
1 to 10	0
11 to 20	2
21 to 30	4
31 to 40	7
41 to 50	11
51 to 60	12
61 to 70	14
71 to 80	8
81 to 90	5
91 to 100	1

d

Mark	Frequency
1 to 10	2
11 to 20	6
21 to 30	12
31 to 40	15
41 to 50	26
51 to 60	43
61 to 70	21
71 to 80	7
81 to 90	2
91 to 100	0

The "middle marks" you need to use are 10·5, 30·5, etc, for **a** and **b**, and 5·5, 15·5, etc, for **c** and **d**.

You can use 10, 30, etc, and 5, 15, etc instead, which will give averages slightly smaller than the correct value. You can then get the correct value by adding 0·5 to each average. Can you see why?

Another kind of average

There are two supermarkets in a town. Each one employs 9 people. The weekly wages they pay are shown below.

Superprice	£21, £22, £23, £24, £25, £26, £27, £28, £29
Finefoods	£18, £18, £19, £19, £19, £20, £20, £20, £81

Which supermarket pays the best wages? One way of comparing the wages is to work out the two averages. The averages are given below.

Average wage at Superprice = £25.
Average wage at Finefoods = £26.

It seems as though Finefoods pay the best wage. But do they?
The lowest wage at Superprice is £21, and *all* of the people but one at Finefoods get paid less than this. Unless you are the lucky one earning the £81 it is obviously better to work at Superprice in spite of the fact that they pay a lower average wage.
You can see that averages are sometimes not a very good way of comparing.

Another kind of average is called the **median**. It is the number in the middle when they are placed in order of size. You will see that:

The **median wage** at Superprice is £25.

The **median wage** at Finefoods is £19.

Since Superprice pay the best wage to most of the people working for them, you can see that the **median** is the best way of comparing the wages.

The average we have been using up to now is sometimes called the **mean**.

If there is no middle number we usually take the median as the number half way between the two middle numbers. For example the median of 2, 2, 5, 7, 8, 9 is taken to be 6.

Exercise 18.3

Work out the **mean** (average), and **median** of these sets of numbers.

1 £3, £5, £6, £7, £10, £11, £14

2 £6, £6, £7, £9, £11, £12, £12, £13, £14

3 20, 21, 15, 18, 20, 14. (Write them in order of size to get the median.)

4 £7, £11, £89, £5, £9, £13, £6

5 17, 95, 18, 17, 12, 9, 15, 9

6 £30, £33, £7, £6, £7, £8, £9, £9, £8

7 **a** Compare the median and the mean in questions **1**, **2**, and **3**. What do you notice?

b Compare the median and the mean in questions **4**, **5**, and **6**. What do you notice?

c What can you say about a set of numbers where the mean and the median are about the same?

d What can you say about a set of numbers where there is a large difference between the mean and the median?

e What can you say about sets of numbers where it is better to use the median rather than the mean?

Exercise 18.4

1 The pie chart shows how a man spends his weekly wage of £16.

a How many slices are there in the chart?

b How much does each slice stand for?

How much does he spend on:

c Fares?

d Entertainments?

e Food?

f Clothes?

g Rent?

h Savings?

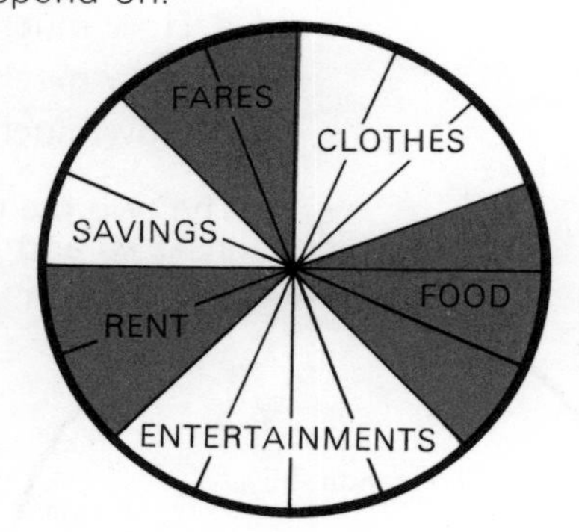

2 The pie chart shows how a family spend their weekly income of £32.

a How many slices are there in the chart?

b How much does each slice stand for?

How much do they spend on:

c Rent?

d Clothes?

e Savings?

f Entertainment?

g Food?

h Fares?

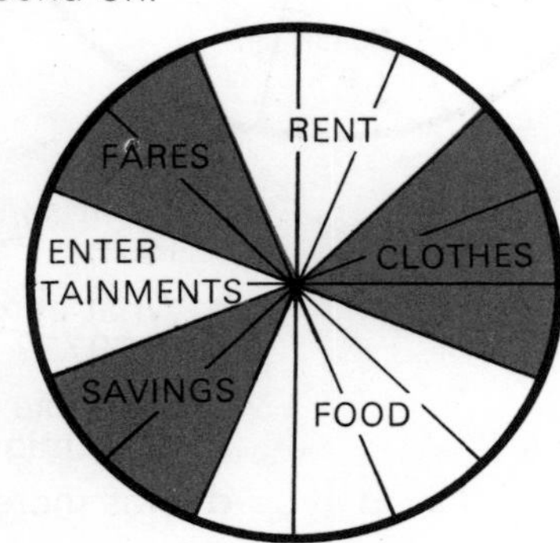

3 The pie chart shows how much each member of the Evans family earns. John earns £15.

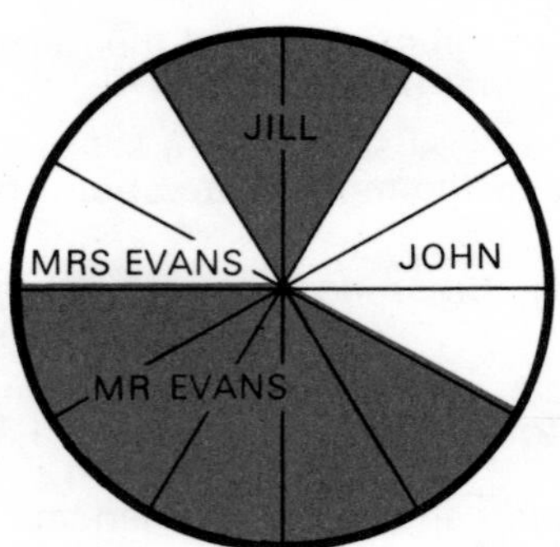

a How many slices represents John's earnings?

b How much does one slice stand for?

c How much does Mrs Evans earn?

d How much does Mr Evans earn?

e How much does Jill earn?

f How much does the whole family earn?

4 The two pie charts show how a Council spent the rates in 1970 and 1973.

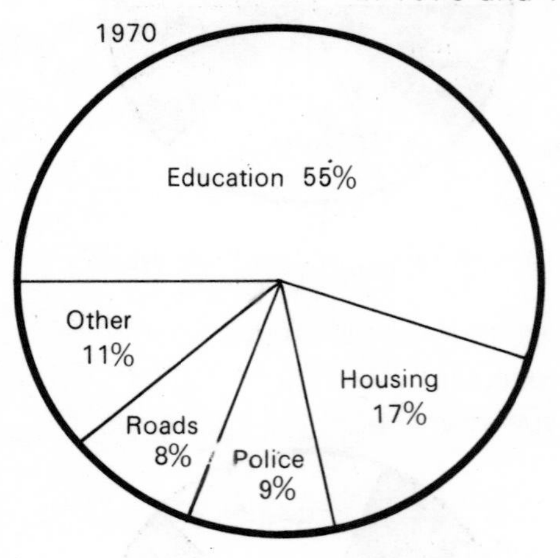

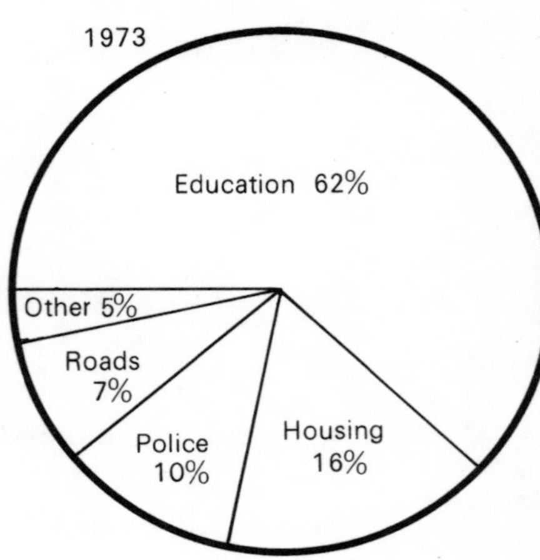

a Did it spend more than half the rates on education in 1970?

b What percentage did it spend on roads and housing in 1973?

c If the pie chart for 1973 were larger than for 1970, what might this tell you?

d Was more spent on education in 1973 than in 1970?

e Was a bigger share spent on housing in 1973 than in 1970?

f Was more spent on housing in 1973 than in 1970?

g What percentage was spent on other things in 1973?

h What items had a bigger share spent on them in 1973?

i What items had a smaller share spent on them in 1973?

j Make a list of some of the other things that Councils spend the rates on.

For each of the next questions you will need to draw a circle divided into 20 equal parts like the one shown. (It can be copied, using tracing paper.) Each of the parts will represent 5%.

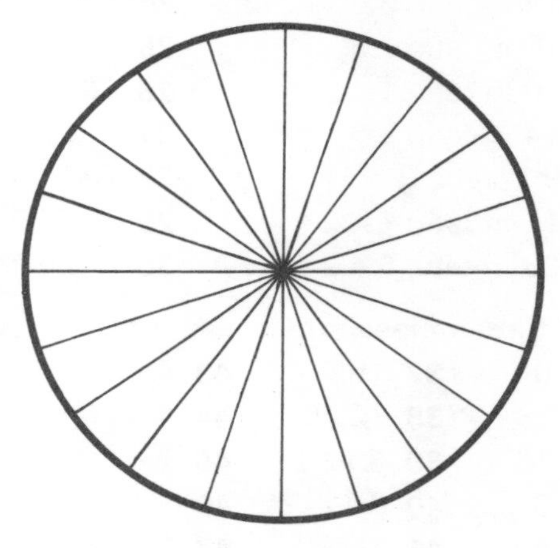

5 A man spends his income as follows: savings 15%, rent 35%, food 25%, fares 15%, clothes 5%, other 5%. Draw a pie chart to illustrate this.

6 The following figures show how a Council spends the rates: Education 40%, Housing 15%, Police 10%, Roads 5%, other 30%. Draw a pie chart to illustrate this.

7 The following figures show how Australian exports were divided up in 1969 (by value): Wool 25%, Meat 10%, Wheat 10%, Sugar 5%, Mining products 20%, other Raw Materials 15%, Manufactured products 15%. Draw a pie chart to illustrate this.

8 The following figures show how a Council spends the rates: Education 47%, Roads 14%, Police 11%, Housing 22%, Other 6%. Draw a pie chart to illustrate this. You will need to divide up some of the slices.

revision 20 Sharing and ratios

Exercise R 20

Divide these amounts in the ratio shown.

1 £20 in ratio 3:2
2 £6 in ratio 1:1
3 £12 in ratio 1:2
4 £8 in ratio 3:1
5 £9 in ratio 2:1
6 £21 in ratio 4:3
7 £15 in ratio 4:1
8 £12 in ratio 3:1
9 £14 in ratio 3:4
10 £10 in ratio 2:3
11 £10 in ratio 1:4
12 £14 in ratio 2:5
13 £9 in ratio 2:3:4
14 £11 in ratio 5:4:2
15 22p in ratio 5:4:2
16 6p in ratio 1:2:3
17 18p in ratio 2:3:4
18 £8 in ratio 2:3:3
19 16p in ratio 2:3:3
20 18p in ratio 1:2:3
21 £10 in ratio 3:4
22 £12 in ratio 2:3
23 £35 in ratio 3:5
24 £40 in ratio 5:4
25 43 kg in ratio 2:3:4
26 £34 in ratio 1:2:2
27 £26.34 in ratio 2:4:3
28 £24·56 in ratio 4:5:8
29 £14·75 in ratio 2:1:7
30 £0·98 in ratio 4:4:6:8

Write these ratios in their simplest form.

31 4:6
32 20:8
33 4:10
34 6:8
35 2:4
36 6:9
37 15:7
38 12:20
39 8:12
40 9:12
41 15:6
42 7:3
43 4:6:8
44 4:8:10
45 2:4:6
46 8:6:2
47 14:21:35
48 15:12:6
49 $2\frac{1}{2}:7\frac{1}{2}$
50 $5\frac{1}{4}:2\frac{1}{4}$
51 $2\frac{2}{3}:\frac{2}{3}$
52 0·8:0·2
53 1·4:2·1
54 0·2:0·3:0·6

55 Two men start a business. They agree to share the profits or losses in the ratio 2:3.

a How should they share a profit of £10?

b How should they share a profit of £35?

c How should they share a profit of £45?

d How should they share a loss of £15?

56 A certain type of concrete is made of cement powder and sand in the ratio 3:5. What weight of each would be needed to make the following weights of dry mix? (That is the weight before water is added.)

a 8 kg b 24 kg c 32 kg d 48 kg e 800 kg
f 16 tonnes g 144 kg

57 A man and his wife invest 12p and 16p in a pools entry.

a Write the ratio 12:16 in its simplest form.

How would they share out winnings of b £7 c £21
d £35 e £7700?

58 90 kg of alloy contains 20 kg copper and 70 kg zinc.

a Write the ratio 20:70 in its simplest form.

What weights of copper and zinc would be needed for the following weights of the alloy: **b** 81 kg **c** 54 kg **d** 45 kg **e** 720 tonnes **f** 50 kg?

59 A man and his wife occupy 120 square metres of floor space in their house. They use another 60 square metres as offices for the man's business, and another 40 square metres they rent out to the wife's brother for offices.

a Write the ratio 120:60:40 in its simplest form.

How should they apportion the following expenses between the man and his wife, the man's business, and the brother's business?

b Rates £220 **c** Mortgage interest £146

d Heating and light £214·45?

The first picture shows one of the ways that square tiles can be laid on a floor, the second picture shows one of the ways that wood blocks can be laid, and the third picture shows the way that bricks are usually placed when building a wall.

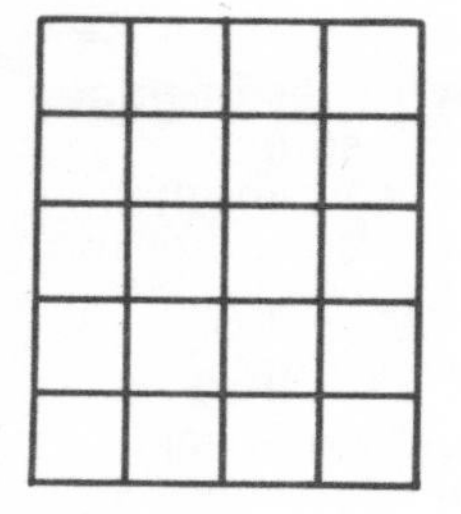

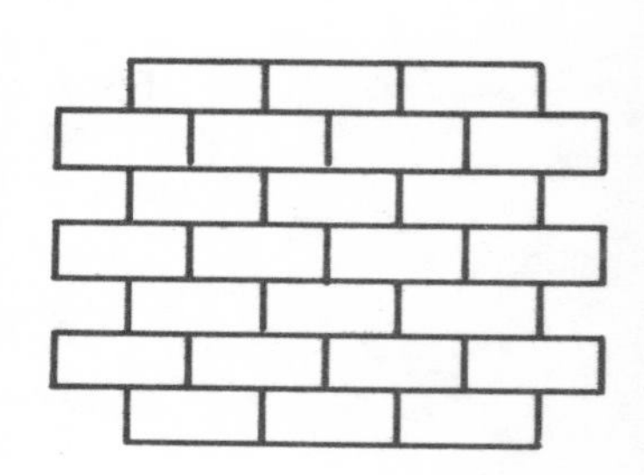

Regular arrangements of shapes like this are often called **tessellations**. Here are some more examples of tessellations.

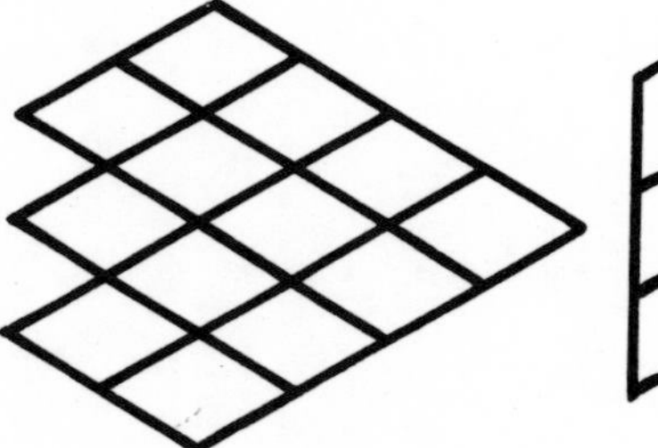

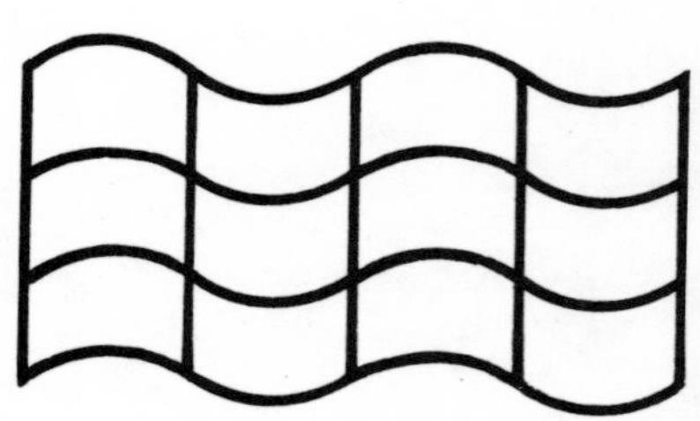

Not all shapes can be made into tessellations. Here are some that cannot. If you try to make tessellations with them you will find that there will be spaces between them.

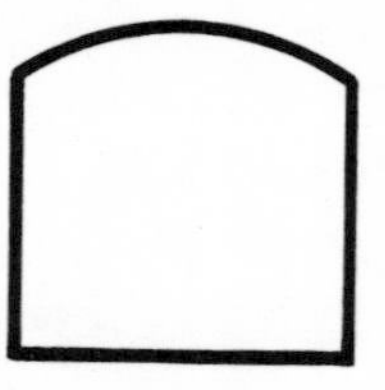

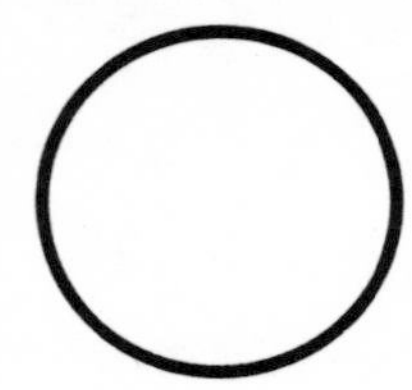

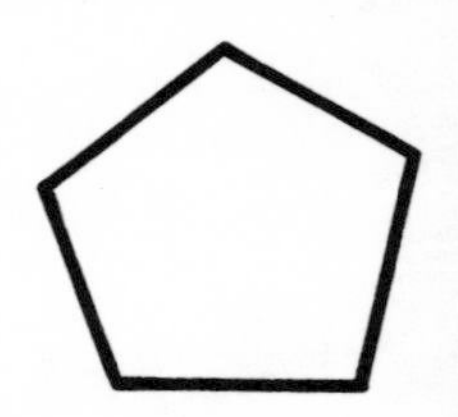

Exercise 19.1

Draw tessellations on squared paper with these shapes. The first one has been started to show you what to do. There may be more than one way for some.

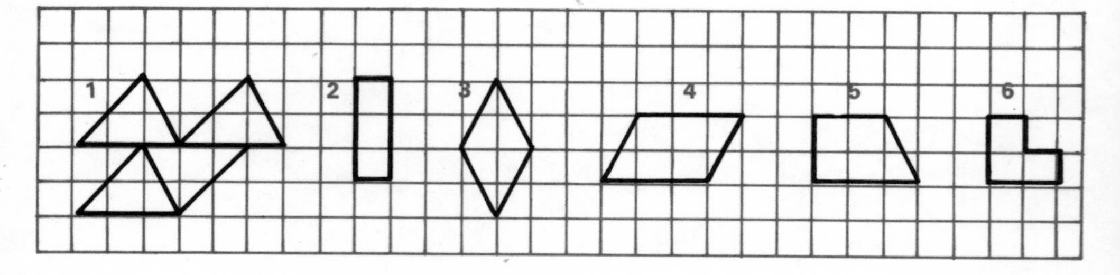

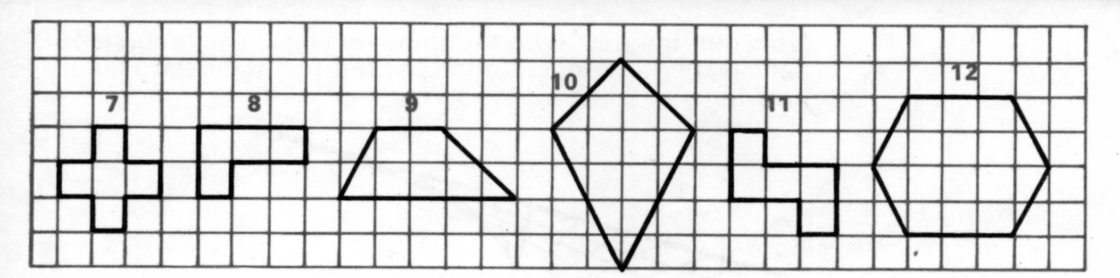

In Exercise 1 all of the shapes could be made into tessellations. In the next exercise some of the shapes cannot be made into tessellations.

Exercise 19.2

Decide which of these shapes can be used to make tessellations, and draw the tessellations.

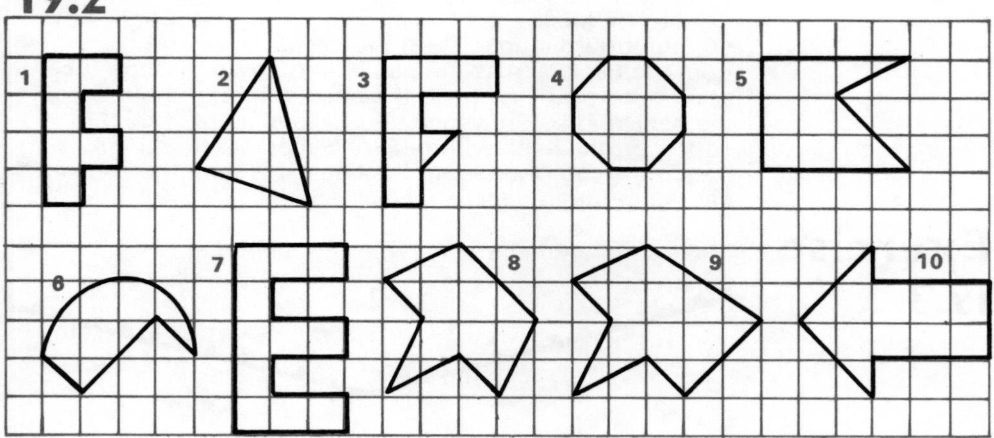

Curve stitching

Draw two lines starting from the same point and mark with dots a number of equal spaces on each line. Number the dots on one line in one direction, and on the other line in the other direction. Your drawing should now look something like this.

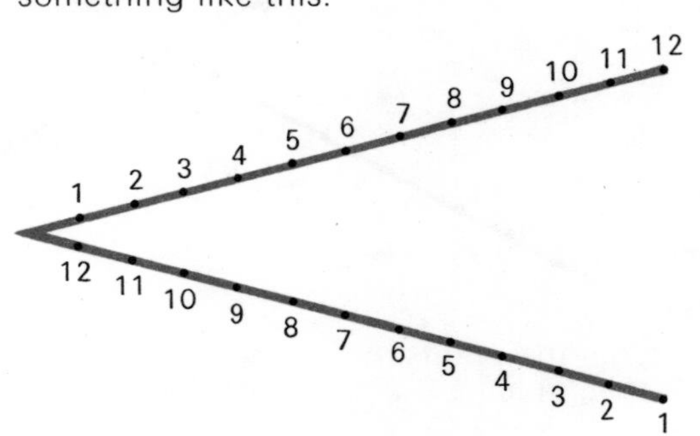

Now join the dots with the same numbers, using coloured pencils if you wish, and then rub out the numbers. Your drawing should now look something like this.

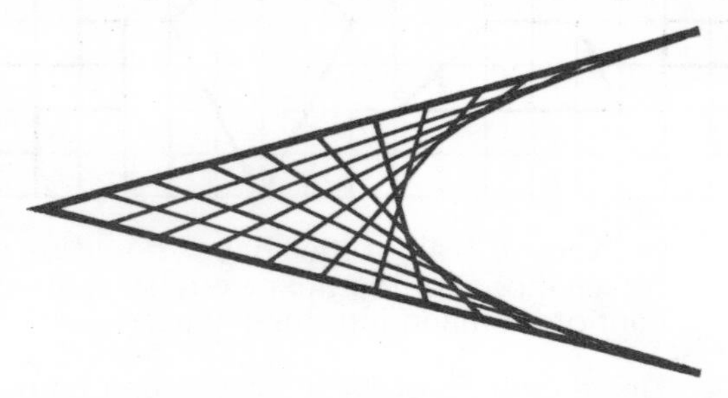

A more attractive way of doing this is to make holes where the dots are, and thread coloured cotton through using a needle. Tie a large knot in the cotton so that it does not pull through the hole. Push the needle through hole 1 from the back, across the paper, and down through the second hole 1. Now push the needle through hole 2 from the back and so on. When you have finished it fix the cotton on the back with glue or Sellotape. If you are using cotton instead of drawing the lines, it is better to use stiff card rather than paper.

Exercise 19.3

Make some more patterns with the following.

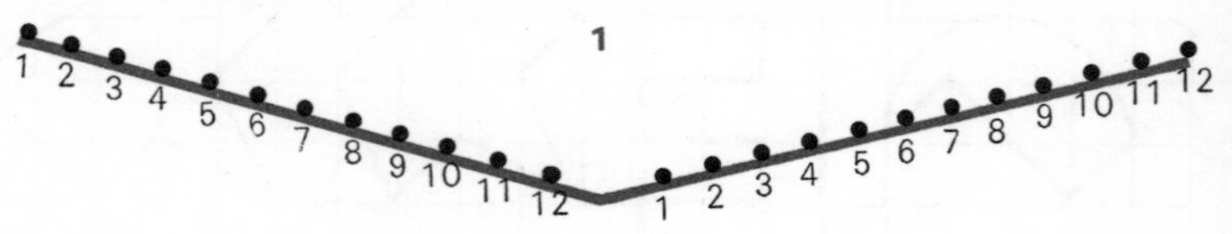

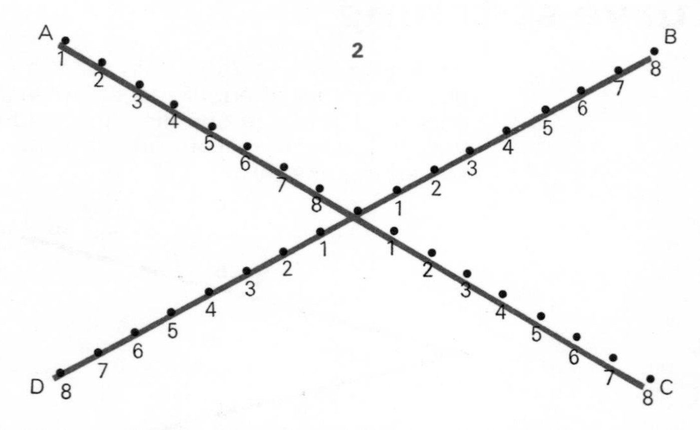

Join the points in lines A and B, then in lines B and C, then in lines C and D, then in lines D and A.

3 Join the points in lines A and D, and then the points in lines B and C.

An even better pattern can now be got by joining the points in A to the points in B, and the points in C to the points in D, and by using a different colour. This time join 0 to 7, 1 to 6, 2 to 5, and so on.

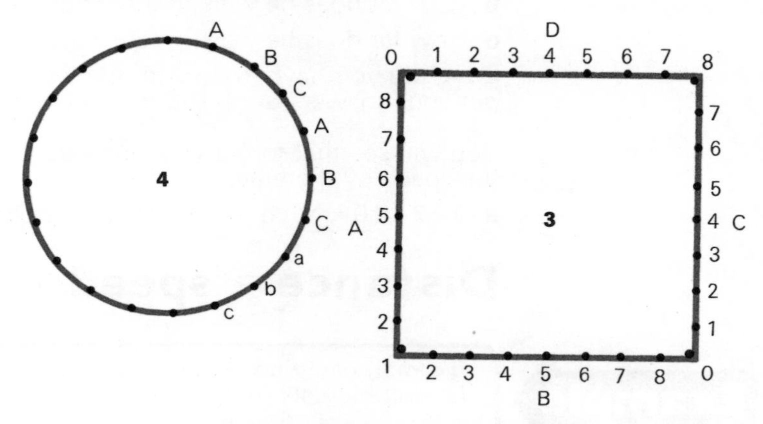

4 Join A to A, B to B, C to C, and so on all the way around the circle. Now using another colour join the A at the top of the circle to a, B to b, and C to c, and so on all the way around the circle.

5 Another way of doing question **4** is to join every point to every other point. For example 17 lines are drawn from A, 17 from B and so on.

6 There are many other beautiful shapes that can be drawn in this way, and if they are done carefully with coloured cotton on black, white or coloured card they can be framed and hung on the wall. Some have been started for you below. Try to make up some of your own.

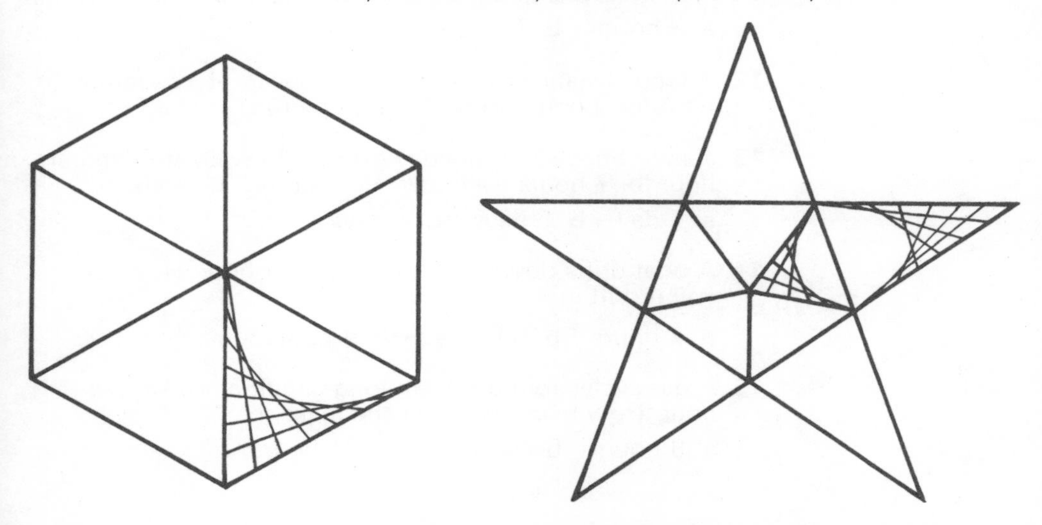

Finding the distance

If you are told that a man walks at 5 km per hour this means that in every hour he walks 5 km.

a How far does he walk in two hours?

b How far does he walk in three hours?

c How far does he walk in four hours?

What did you have to do with the time and the speed to get your answers. Add, subtract, multiply, or divide?

You will see that to get your answers you had to multiply the speed by the time.

a $5 \times 2 = \underline{10 \text{ km}}$ **b** $5 \times 3 = \underline{15 \text{ km}}$ **c** $5 \times 4 = \underline{20 \text{ km}}$

Distance = speed × time

Example

If a man can walk at a speed of 6 km per hour, how far will he walk in 7 hours?

Distance $= 6 \times 7 = \underline{42 \text{ km}}$.

Exercise 20.1

Find the distance travelled in each of the following examples.

	1	2	3	4	5	6	7	8	9	10
Speed in km per hour	4	9	7	12	9	20	17	25	16	85
Time in hours	3	2	4	6	7	3	5	7	3	9

11 A train travels at 90 km per hour. How far will it travel in
a 3 hours **b** 5 hours **c** $2\frac{1}{2}$ hours?

12 A long-distance runner runs at a speed of 12 km per hour for 3 hours. How far does he run?

13 A man on walking holiday expects to walk at 6 km per hour for 7 hours each day. How far will he walk in
a 1 day **b** 5 days **c** 8 days?

14 A boat drifts down river at $1\frac{1}{2}$ km per hour. How far will it drift in
a 2 hours **b** 6 hours **c** 9 hours?

15 A girl cycles round a 1 km long circuit at 18 km per hour. How many laps will she make in
a 3 hours **b** $5\frac{1}{2}$ hours **c** $7\frac{1}{3}$ hours?

16 A man cycles at 14 km per hour for $1\frac{1}{2}$ hours and then gets a puncture. He then pushes his bike for $\frac{3}{4}$ hour at 4 km per hour to the nearest village and catches a bus on which he travels for 20 min. at 30 km per hour. How far did he travel altogether?

Finding the time

A man can walk at 5 km per hour.

a How long will he take to walk 10 km?

b How long will he take to walk 15 km?

c How long will he take to walk 20 km?

What did you have to do with the distance and the speed to find the time – add, subtract, multiply, or divide?

You will see that to get your answers you had to divide the distance by the speed.

a $10 \div 5 = \underline{2 \text{ hours}}$ **b** $15 \div 5 = \underline{3 \text{ hours}}$ **c** $20 \div 5 = \underline{4 \text{ hours}}$

Time = distance ÷ speed

A teacher wants to take some pupils for a walk over a distance of 12 km. If they walk at 4 km per hour how long will it take?

$$\text{Time} = 12 \div 4 = \underline{3 \text{ hours}}$$

There is also another way of doing this problem. We can use the formula on page 172 and write down

$$\text{Distance} = \text{Speed} \times \text{Time}$$

We can then put in what we know and get

$$12 = 4 \times \text{Time}$$

This is an equation and we can write:

$$4t = 12$$

$$\Rightarrow t = \frac{12}{4}$$

$$\Rightarrow \underline{t = 3 \text{ hours}}$$

Exercise 20.2

Find the time taken in each of the following examples.

	1	2	3	4	5	6	7	8	9	10
Distance in km	10	8	20	36	48	60	100	50	72	144
Speed in km per hr	2	4	5	9	8	5	10	2	9	8

11 A man walks 12 km at 3 km per hour. How long does he take?

12 A man on horseback rides at 10 km per hour. How long will he take to travel

a 5 km **b** 20 km **c** 35 km **d** 2 km?

13 An airliner cruises at 750 km per hour. How long does it take to fly

a 1500 km **b** 7500 km **c** 9000 km?

14 A rally driver drives at 90 km per hour. How long does he take to finish the course, if it is

a 270 km **b** 360 km **c** 540 km long?

15 A bullet travels at 1000 m per second. How long does it take to reach the target if the distance is

a 500 m **b** 750 m?

16 A snail crawls at $12\frac{1}{2}$ cm per min. How long does it take to crawl

a 100 cm **b** 450 cm **c** $2\frac{1}{2}$ m?

Finding the speed

a A man walks 12 km in 2 hours. What is his speed?

b A man walks 12 km in 3 hours. What is his speed?

c A man walks 12 km in 4 hours. What is his speed?

What did you do with the distance and the time to get the speed – add, subtract, multiply or divide?
You will see that to get your answers you had to divide the distance by the time.

a $12 \div 2 =$ <u>6 km per hour</u> **b** $12 \div 3 =$ <u>4 km per hour</u>
c $12 \div 4 =$ <u>3 km per hour</u>

Speed = distance ÷ time

A car goes 80 km in 2 hours. What is its speed?

Speed $= 80 \div 2 =$ <u>40 km per hour</u>

There is another way of doing this problem. We can use the formula on page 172 and write down

Distance = Speed × Time

We can then put in what we know and get

$80 = \text{Speed} \times 2$

This is an equation and we can write

$2s = 80$

$\Rightarrow s = 80 \div 2$

$\Rightarrow \underline{s = 40 \text{ km per hour}}$

Example

A man walks 17 km in three hours. What is his speed?

Speed $= 17 \div 3 = \underline{5{\cdot}66 \text{ km per hour}}$

Exercise 20.3

Find the speed in each of the following examples.

	1	2	3	4	5	6	7	8	9	10
Distance in km	10	15	24	48	48	49	84	100	50	120
Time in hours	2	3	4	6	4	7	7	5	10	6

11 A man walks 15 km in 3 hours, what is his speed?

12 A man on horseback leaves his home at 8.30 and arrives 24 km away at 10.30. **a** How long does he take? **b** What is his speed?

13 A man cycles 65 km in 5 hours. What is his speed?

14 A racing car completes a $2\frac{1}{2}$ km circuit in 1 minute. At what speed is it travelling? Give the answer in km per hr.

15 A bus begins its journey at 9 a.m. and reaches its destination 17 km away at 11 a.m. What is the speed of the bus?

16 A man has to drive to a place 210 km distant. He starts out at 11 a.m. and arrives at 2 p.m. What is his speed?

17 An athlete runs 3000 m in 10 minutes. What is his speed **a** in metres per second, and **b** km per hour?

18 A girl walks 14 km in three hours. What is her speed?

19 A man drives 400 km in 3 hours. What is his speed?

20 A woman cycles a distance of 5 km to the shops in 16 minutes. What is her speed? Give the answer in km per hr.

More problems on speed, time and distance (mixed examples)

Example

A man walks 13 km in 2 hours 23 minutes. What is his speed?
The best way of doing this problem is to change the time to 143 minutes.

$$\text{Speed} = 13 \div 143 \text{ km per minute.}$$

This is the number of km he walks in one minute.
In one hour he walks 60 times as far.

$$\text{Speed} = 60 \times 13 \div 143 = 780 \div 143$$
$$= \underline{5{\cdot}45 \text{ km per hour}}$$

Example

How long will it take to drive 135 km at 42 km per hour? Answer in hours and minutes.

$$\text{Time} = 135 \div 42 = \underline{3{\cdot}21 \text{ hours}}$$

We now have to change the decimal part of the hour to minutes. To change hours to minutes we multiply by 60.

$$0.21 \text{ hour} = 0{\cdot}21 \times 60 = \underline{12{\cdot}6 \text{ minutes}}$$

Answer: 3 hours 13 minutes (to the nearest minute).

Exercise 20.4

1 A man drives 14 287 km in 13 days.

a How far does he average in one day?

b If he drives 7 hours a day, what is his average speed?

c How far would he travel in 17 days?

d How long would it take him to go 10 362 km, if he drove at the same speed for 6 hours a day?

2 A barge travels at 7 km per hour for 11 hours each day.

a The barge starts at 9 a.m. each day, and leaves Birmingham on Monday, arriving in London at 1 p.m. on Wednesday. What distance has it travelled?

b The barge departs again at 4 p.m. for Oxford, a distance of 91 km. When does it arrive? (Assume that the barge is at rest between 8 p.m. and 9 a.m.)

3 a A cruise liner steams 576 km in 24 hours. What is its speed?

b The ship is 1032 km from its next port of call. How long will it take to arrive there?

c The ship spends 9 days 13 hours actually sailing during the cruise. How far does it travel?

4 **a** An airliner flies from Marseille to New York, a distance of 7100 km, in $7\frac{1}{2}$ hours. What is its speed?

b It then flies on to Houston at the same speed in $2\frac{1}{2}$ hours. What is the distance from New York to Houston?

c From Houston the plane flies on to Los Angeles, a distance of 2200 km. How long does it take?

5 **a** A girl cycles 5 km in 12 minutes. What is her speed?

b If she continued cycling at the same speed for a further 33 minutes, how far would she then have gone?

unit 21 Proportion

Your friend has bought 3 pencils. They cost him 12p. You want to buy 7 of the same sort of pencil. How much will they cost?

One way is to find how much one cost. We can set out the working like this.

3 pencils cost 12p

1 pencil costs $\frac{12}{3} = 4p$

7 pencils cost $7 \times 4p = \underline{28p}$

Examples like this are usually called **proportion.**

When doing examples like this you should always write the first line so as to get your answer at the end of the line. In the example above the answer is in pence, so we write as the first line:

3 pencils cost 12p

not 12p buys 3 pencils.

Example

If you work for a man in the school holidays for 4 days and he pays you £8, how much should he pay you if you work for 5 days?

One way of doing this is to first find out how much you are paid for one day.
You can see this is $£8 \div 4 = £2$.
For 5 days' work he should pay you $5 \times £2 = \underline{£10}$.

The usual way of setting this out is like this:

For 4 days' work I get paid £8.

For 1 day's work I get paid $\frac{£8}{4} = £2$.

For 5 days' work I get paid $5 \times £2 = \underline{£10}$.

Exercise 21.1

Copy and complete these examples.

1 If for 6 days' work you get paid £12, how much should you be paid for 8 days?

For 6 days' work I get paid £12.

For 1 day's work I get paid $\frac{£12}{6} = £2$.

For 8 days' work I get paid $8 \times £2 =$

2 If 5 pencils cost 15p, find the cost of 9.

5 pencils cost 15p.

1 pencil costs $\frac{15p}{5} =$

9 pencils cost

3 If I can buy 16 fireworks for 8p, how many can I buy for 12p?

8p buys 16 fireworks.

1p buys   fireworks.

12p buys   fireworks.

4 A boy works for 3 days and is paid £6. How much will he be paid for 6 days?

5 A boy works for 3 days and is paid £9. How much will he be paid for 5 days?

6 If 4 pencils cost 16p, how much will 5 cost?

7 If 5 pencils cost 25p, how much will 8 cost?

8 If 4 apples cost 12p, how much will 7 cost?

9 If 5 lemons cost 10p, how much will 8 cost?

10 The fare for a bus journey of 6 km is 18p. What will a journey of 7 km cost if the fare is charged at the same rate?

11 The fare for a bus journey of 8 km is 24p. What will a journey of 5 km cost?

12 A factory can produce 42 cars in 7 days. How many cars will be produced in 6 days?

13 A craftsman can make 24 chairs in 2 weeks. How many will he make in 8 weeks?

14 A recipe states that at a certain temperature a joint of meat 6 cm thick takes an hour to cook. If cooking time is proportional to the thickness of the meat, how long will a joint 10 cm thick take to cook? (Change hours to minutes.)

15 A car travels 120 km in 2 hours. If it continues at the same speed

a How far will it go in 8 hours?

b How long will a journey 300 km take?

16 200 cc liquid weedkiller is required for an area of 100 sq m. How much would be required for a lawn 560 sq m?

17 A fertilizer has a spreading power of 1 kg per 250 sq m. How much would be required for a field 3500 sq m?

18 A factory hand can tighten 75 bolts in 15 minutes. How many will he tighten in an 8-hour day?

19 A train journey of 17 km costs 7p. If there is no reduction for long-distance travel, how much will a journey of 864 km cost?

20 A factory can produce 126 refrigerators in 14 days. Given continuous production, how many could be produced in 1 year (365 days)?

All the examples you have been doing come out exactly. If you think that an example will not come out exactly there is a slightly different method that you can use.

Example

If I get paid £7 for 3 days' work, how much will I get paid for 5 days' work?

For 3 days' work I get paid £7.

For 1 day's work I get paid $\frac{£7}{3}$.

For 5 days' work I get paid

$5 \times \frac{7}{3} = \frac{5 \times 7}{3} = \frac{35}{3} = \underline{£11 \cdot 66}$

$$\begin{array}{r} 11\cdot66 \\ 3\overline{)35\cdot00} \\ 3 \\ \hline 20 \\ 18 \\ \hline 20 \\ 18 \\ \hline \end{array}$$

The only difference here is that the second line is not worked out, but is left as $\frac{7}{3}$ pounds.

If the problem does come out exactly, you can often cancel in the last line.

Example

If 7 pencils can be bought for 3p, how many can be bought for 15p?

3p buys 7 pencils.

1p buys $\frac{7}{3}$ pencils.

15p buys $15 \times \frac{7}{3} = \frac{\overset{5}{\cancel{15}} \times 7}{\underset{1}{\cancel{3}}} = \underline{35 \text{ pencils.}}$

Exercise 21.2

Copy and complete these examples.

1 If I get paid £7 for 3 days' work, how much will I get paid for 4 days' work?

For 3 days' work I get paid £7.

For 1 day's work I get paid $\frac{£7}{3}$.

For 4 days' work I get paid $4 \times \frac{7}{3} = \frac{£28}{3} = \underline{£?}$

2 If I get paid £9 for 4 days' work, how much will I get paid for 3 days' work.

For 4 days' work I get paid £9.

For 1 day's work I get paid £?.

For 3 days' work I get paid £?.

3 If 5 pencils can be bought for 4p, how many can be bought for 16p?

4p buys 5 pencils.

1p buys ? pencils.

16p buys ? pencils.

4 If I get paid £5 for 2 days' work, how much will I get paid for 5 days' work?

5 If I get paid £12 for 5 days' work, how much will I get paid for 3 days' work?

6 If 8 pencils can be bought for 3p, how many can be bought for 12p?

7 If 10 kg potatoes cost 25p, what will 7 kg cost?

8 If 4 pencils cost 7p, what will 10 pencils cost?

9 The shadow of a tree is 3 m long, and that of a flag-pole is 2 m long. Three hours later the shadow of the tree is 4·5 m long. How long is the shadow of the flag-pole?

10 Four litres of paint are required to paint one wall of a shed 7 m long. How much paint is required for the other wall of the shed, 9 m long, if the height is the same for both walls?

11 Three spools of cotton are required for stitching 8 sheets. How much cotton is required for 20 sheets?

12 A factory hand can print a pattern on 29 balloons in 6 minutes. How many balloons will she print in 1 hour?

13 A car is travelling at a speed of 70 km per hour.

a How far will it go in 40 minutes?

b How long will a journey of 20 km take?

14 A woman earns £25 for a 38-hour week. How much will she earn if she works part-time for 20 hours a week at the same rate?

15 A Council can build 5 houses of a certain type for £22 500. How many can it build for £40 500?

Inverse proportions

If 2 boys take 6 hours to dig a large garden, how long will it take 6 boys?

2 boys take 6 hours.

1 boy takes $\frac{6}{2} = 3$ hours.

6 boys take $6 \times 3 = \underline{18 \text{ hours}}$.

Look at this answer. There is something wrong. If 2 boys take 6 hours, then 6 boys should take less time, not more. They should not take 18 hours. Can you find the mistake?

The correct way of doing the problem is this:

2 boys take 6 hours.

1 boy takes $2 \times 6 = 12$ hours.
(One boy takes longer than 2 boys, so we multiply.)

6 boys take $\frac{12}{6} = 2$ hours.

(6 boys take less time than 1 boy, so we divide.)

Examples like this are usually called **inverse proportion**.

Example

If I walk at 3 km per hour, it will take me 4 hours to do a walk. How long will it take me if I walk at 4 km per hour?

If I walk at 3 km per hour it takes me 4 hours.

If I walk at 1 km per hour it will take me $3 \times 4 = 12$ hours.
(It takes longer since I am walking more slowly.)

If I walk at 4 km per hour it will take $\frac{12}{4} = \underline{3 \text{ hours}}$

(It takes less time since I am walking faster.)

Exercise 21.3

Copy and complete these examples.

1 If 3 boys take 6 hours to dig a garden, how long will it take 9 boys?

3 boys take 6 hours.

1 boy will take $3 \times 6 = 18$ hours.

9 boys will take $\frac{18}{9} =$ $\underline{\qquad \text{ hours}}$

2 If 4 boys take 6 hours to dig a garden, how long will it take 3 boys?

4 boys take 6 hours.

1 boy will take hours.

3 boys will take hours

3 If I walk at 6 km per hour, it take me 4 hours to do a walk. How long will it take me if I walk at 8 km per hour?

If I walk at 6 km per hour it takes me 4 hours.

If I walk at 1 km per hour it takes me hours.

If I walk at 8 km per hour it will take me hours

4 If 2 boys take 4 hours to dig a garden, how long will 8 boys take?

5 If 1 boy takes 4 hours to dig a garden, how long will 2 boys take?

6 If I walk at 4 km per hour, it will take me 5 hours to do a walk. How long will it take if I walk at 5 km per hour?

7 If I walk at 6 km per hour, it will take me 5 hours to do a walk. How long will it take if I walk at 5 km per hour?

8 If I walk up a mountain at 2 km per hour, it will take me 3 hours to reach the top. How long will it take to come down, if I walk at 6 km per hour?

9 A boy takes 2 hours to get to a football match, cycling at 7 km per hour against the wind. Coming home the wind has changed, and he only takes 1 hour. What is his average speed on the return journey?

10 A barge travels upstream at 8 km per hour, taking 5 hours to reach its port. It returns downstream at 20 km per hour. How long will it take?

11 A fishing trawler with a crew of 8 takes on board enough food and water to last 9 days. If the boat sails without 2 of the men, how long will the stores last?

12 A family took 1 hour to drive from London to Brighton (87 km) for the day. It took them 3 hours slow driving in a queue coming back. What was their average speed on the return journey?

13 An expedition of 8 members in Alaska loads up enough food for 18 days. If they are joined by an Eskimo just as they start, how long will the food now last? (Assume that the Eskimo has the same rations as the rest of the party.)

14 It takes 8 men 15 weeks to make an aeroplane engine. For a rush job for export, 3 extra men are taken on. How long will the job take?

15 A man giving a party buys enough beer for 4 glasses each if 14 people come. Four more people arrive unexpectedly. How many glasses each will there now be?

16 A van driver finds he needs 39 l of petrol to drive from London to St Ives to pick up some goods, giving him 12 km per litre. Returning with the load, he gets only 10 km per l. How much petrol will he need?

Exercise 21.4

In the following examples you will need to decide whether to divide or multiply in the second line of the working.

1 Three climbers drove to North Wales for a weekend and agreed to share expenses. They were joined by a fourth for the journey home. If the three climbers each paid 72p for petrol on the way there, how much would each of the four pay on the return journey?

2 It usually takes 4 men 18 days to pick an apple crop in an orchard. How long would it take 6 men?

3 Six men can pick the apples from 48 trees in one day. How many trees can 9 men pick in 1 day?

4 If 4 men assemble 64 electric motors in 1 day, how many would 5 assemble?

5 A farmer usually employs 4 men at 75p per hour for haymaking. One year 5 boys apply for the work. How much does he pay them per hour if he keeps his wages bill the same?

6 114 people can each ride for one hour per week at a riding school that has 6 horses. How many people can ride every week when the riding school buys another horse?

7 A herd of 20 cows produces an average of 75 l of milk per day. If a farmer buys 5 more cows of the same breed, what will be his average milk yield per day?

8 A farmer can milk 18 cows in 1 hour. With the same number of milking machines, how long will milking take if he buys 10 more cows?

9 A type setter is given a typed manuscript of a novel consisting of 150 pages of 300 words a page. He is told to reduce it to 130 pages. How many words a page must he set?

10 If it takes 8 men 22 hours to paint a building, how long will it take if 2 men are off sick?

11 Twenty lorries were needed to bring in 150 tonnes of gravel to make a road. How many extra lorry loads will be required if 55 tonnes are required to extend the road?

12 A farmer finds that milking takes 85 minutes with 8 milking machines. How long will it take if one is out of order?

13 What is the cost of 70 kg of scrap metal if 6 kg cost £10·50?

14 At a hairdresser's 9 women work for 35 hours a week each. If one is away ill, and all the clients arrive, how much longer will each of the women have to work?

15 In an experiment it is found that 23 cc of acid will produce 5·6 litres of gas.

a How much gas will 33 cc of acid produce?

b How much acid will be needed to produce 7·5 litres of gas?

Exercise R 21

Write each of these as a percentage.

1a 8 out of 100. **b** 20 out of 100.
c 9 out of 100. **d** $8\frac{1}{2}$ out of 100

2a 4 out of 50 **b** 7 out of 50
c 10 out of 50 **d** 23 out of 50

3a 18 out of 200 **b** 40 out of 200
c 100 out of 200 **d** 160 out of 200

4a 3 out of 25 **b** 9 out of 25
c 10 out of 25 **d** 23 out of 25

5a 7 out of 20 **b** 4 out of 20
c 11 out of 20 **d** 19 out of 20

6a 9 out of 10 **b** 60 out of 300
c 4 out of 5 **d** 80 out of 500

7a 18 out of 24 **b** 9 out of 36
c 45 out of 60 **d** 32 out of 40

8a 7 out of 9 **b** 5 out of 7
c 5 out of 6 **d** 4 out of 9

9a 11 out of 13 **b** 17 out of 55
c 19 out of 73 **d** 201 out of 234

Find the following percentages:

		a	b	c	d	e	f	g
10	40% of	100	200	300	400	50	25	20
11	80% of	100	200	300	400	50	25	20
12	20% of	100	200	300	400	50	25	10
13	8% of	100	50	200	25	75	300	400
14	9% of	100	50	20	10	25	200	300
15	7% of	24	25	48	32	45	12	16
16	29% of	24	25	48	32	45	12	16
17	23% of	13	17	34	47	39	23	69
18	$2\frac{1}{2}$% of	40	80	60	90	200	24	43
19	$2\frac{3}{4}$% of	40	80	60	90	200	24	43

20 30% of a class passed an examination. What percentage failed?

21 47% of the pupils in a school are girls. What percentage are boys?

22 300 pupils took an examination. 240 passed and 60 failed.

a What percentage passed?

b What percentage failed?

23 50 boys entered an examination. 42 passed in English and 38 in Mathematics.

a What percentage passed in English?

b What percentage passed in Mathematics?

24 In form 5A, 15 out of 20 passed a swimming test, and in form 5B, 20 out of 25 passed.

a What percentage passed in 5A?

b What percentage passed in 5B?

c Which form did best?

25 A hotel adds 20% (Service Charge and Value Added Tax) on to every bill. How much would be added to each of the following bills?

a £100 **b** £50 **c** £200 **d** £20 **e** £25 **f** £30
g £45

26 An alloy is made of 30% copper and 70% iron. What weight of each would there be in the following weights of alloy?

a 100 kg **b** 50 kg **c** 25 kg **d** 60 kg **e** 42 kg
f 57 kg **g** 235 kg

unit 22 Borrowing and lending money

Simple interest

If you borrow money from a bank you have to pay back more than you borrow. This extra money is called **interest**.

If you borrow £100 for one year you might have to pay interest of 9%. This means that you will have to pay back an extra £9 for every £100 that you borrow. In this case you would have to pay back £109.
The rate of interest is usually given for one year.

Example

If you borrow some money at 8% interest for one year:

a How much interest will you pay if you borrow £100?

b How much interest will you pay if you borrow £200?

c How much interest will you pay if you borrow £50?

d How much interest will you pay if you borrow £25?

a On £100 you will obviously pay £8.

b On £200 you will pay £16. (Twice as much as in **a**)

c On £50 you will pay £4. (Half as much as in **a**)

d On £25 you will pay £2. (One quarter as much as in **a**)

Example

A man lends £300 for one year at 7% interest. How much interest will he be paid?

£100 will pay him £7.

£300 will pay him £21. (Three times as much.)

Exercise 22.1

How much interest will you have to pay if you borrow the following amounts for one year at the interest rates shown?

		a	b	c	d	e
1	£200 at	2%	7%	5%	8%	10%
2	£300 at	3%	4%	6%	9%	11%
3	£50 at	4%	8%	6%	7%	15%
4	£25 at	8%	4%	16%	12%	10%
5	£20 at	5%	10%	15%	20%	25%
6	£10 at	10%	20%	30%	5%	15%

7 A man borrows £200 at 8% interest for 1 year.

a How much interest does he have to pay?

b How much does he have to pay back altogether?

c If he makes his payments in four equal instalments, how much is each payment?

8 A man borrows £400 at 10% interest for four years.

a How much interest does he have to pay for one year?

b How much interest does he have to pay for four years?

c How much does he have to pay back altogether?

d If he makes his payments in four equal instalments how much is each payment?

9 A man buys a house with a mortgage of £9000. He pays interest of 9% per year.

a How much interest does he have to pay on £1000?

b How much interest does he have to pay on £9000?

10 A woman wins £200 000 on the pools. She invests this at 5% per year.

a How much interest would she get on £200?

b How much interest would she get on £200 000?

c How much per week would she get in interest?

11 A man buys a house with a mortgage of £8000. He pays interest of 8% per year.

a How much interest does he have to pay on £800?

b How much interest does he have to pay on £8000?

c If the interest rate was increased to 9%, how much extra interest would he pay every year?

12 A man borrows £200 at 5% per year. At the end of one year the interest is added to the £200.

a How much does he owe at the end of the first year?

b At the end of the second year interest is calculated on what he owes at the end of the first year. How much is this interest?

c What is the total including interest he owes at the end of the second year?

More difficult examples

A man lends £72 for one year at 8%. How much interest will he be paid?

The method we were using in the first part of this Unit will not work very well in this case, and another method is needed which will work for all problems of this type.

$$8\% \text{ of } £72 \text{ means } \frac{8}{100} \times £72 = 0{\cdot}08 \times £72 = \underline{£5{\cdot}76.}$$

Example

A woman borrows £130 for one year at 9%. What will the interest be?

$0{\cdot}09 \times 130 = \underline{£11{\cdot}70}$

To deal with the most common fractional percentages remember that $\frac{1}{2} = 0{\cdot}5$, $\frac{1}{4} = 0{\cdot}25$, $\frac{3}{4} = 0{\cdot}75$, $\frac{1}{3} = 0{\cdot}33$, $\frac{2}{3} = 0{\cdot}67$.

For example $4\frac{1}{2}\% = 4{\cdot}5\%$, so we would multiply by $0{\cdot}045$.

$17\frac{2}{3}\% = 17{\cdot}67\%$, so we would multiply by $0{\cdot}1767$.

If we had to find say $5\frac{2}{7}\%$ we would divide 2 by 7 to get $0{\cdot}28$, so $5\frac{2}{7}\% = 5{\cdot}28\%$, and we would multiply by $0{\cdot}0528$.

Example

A man invests £450 in a bank at $9\frac{3}{4}\%$. How much interest would he get every year?

$9\frac{3}{4} = 9{\cdot}75$, so Interest $= 0{\cdot}0975 \times 350 = £34{\cdot}125$
$= \underline{£34{\cdot}12\frac{1}{2}}$

Exercise 22.2

How much interest will you have to pay if you borrow the following amounts for one year at the interest rates shown?

		a	b	c	d	e	f	g
1	£8	5%	6%	7%	8%	9%	10%	20%
2	£9	7%	8%	9%	10%	11%	12%	13%
3	£40	7%	8%	9%	10%	15%	17%	21%
4	£500	8%	9%	$5\frac{1}{2}\%$	$6\frac{1}{2}\%$	$7\frac{1}{2}\%$	$8\frac{1}{4}\%$	$9\frac{3}{4}\%$
5	£357	$5\frac{1}{3}\%$	$7\frac{2}{3}\%$	$8\frac{5}{8}\%$	$6\frac{3}{4}\%$	$17\frac{3}{4}\%$	$4\frac{3}{7}\%$	$\frac{3}{4}\%$

6 A man borrows £240 at 6% interest for one year.

a How much interest does he have to pay?

b How much does he have to pay back altogether?

c If he makes his payments in 12 equal instalments, how much is each payment?

7 Repeat question **6** with £270 borrowed at 9%.

8 Repeat question **6** with £340 borrowed at 12%.

9 Repeat question **6** with £8000 borrowed at $13\frac{1}{2}\%$.

What interest are you being charged?

It can be shown that if you borrow P pounds for one year, and have to pay back interest of I pounds, the percentage interest rate you are being charged is given by:

$$r = \frac{100 \times I}{P}$$

Example

I borrow £300 for one year and have to pay interest of £21. What rate of interest am I being charged?

$$r = \frac{100 \times 21}{300} = \underline{7\%}$$

(If you are not sure how to cancel fractions like this you should practice with the examples on page 00.)

Example

I borrow £300 for one year and have to pay interest of £22. What rate of interest am I being charged?

$$r = \frac{100 \times 22}{300} = \frac{22}{3} = \underline{7\tfrac{1}{3}\%}$$

Exercise 22.3

In each of the following examples find the percentage rate of interest charged.

1

	a	b	c	d	e	f
Amount borrowed	£200	£300	£400	£500	£200	£600
Interest paid	£10	£18	£32	£35	£18	£30

	g	h	i	j	k	l
Amount borrowed	£800	£700	£800	£500	£400	£300
Interest paid	£56	£42	£64	£45	£20	£30

2

	a	b	c	d	e	f
Amount borrowed	£250	£350	£450	£550	£650	£750
Interest paid	£15	£14	£27	£44	£65	£60

	g	h	i	j	k	l
Amount borrowed	£850	£250	£450	£650	£250	£50
Interest paid	£51	£20	£36	£26	£15	£1

3

	a	b	c	d	e
Amount borrowed	£240	£460	£76	£400	£500
Interest paid	£16·80	£41·40	£3·80	£18	£37·50

	f	g	h	i
Amount borrowed	£56	£78	£560	£578
Interest paid	£1·96	£4·29	£40·60	£19·27

4 Each example below gives the amount borrowed, and the total paid back at the end of one year. In each example find the interest paid, and then the rate of interest being charged.

	a	b	c	d	e	f
Amount borrowed	£200	£300	£400	£600	£250	£350
Amount paid back	£210	£318	£428	£654	£260	£392

	g	h	i	j
Amount borrowed	£750	£780	£690	£1737
Amount paid back	£855	£838·50	£721·05	£1973·45

How much do you need to invest?

A man retires and wishes to get £450 interest from his savings every year. He can invest his savings at 9%. How much does he need to have saved?

The answer to problems like this is given by the formula

$$P = \frac{100 \times I}{r}$$

where I is the interest he wants and r is the rate of interest available. Using this for the example above we get

$$P = \frac{100 \times 450}{9} = \underline{£5000}$$

Exercise 22.4

In each of the following examples find the amount that would have to be invested to produce the interest shown at the rate of interest paid.

1

	a	b	c	e	e	f
Interest paid	£400	£600	£800	£540	£90	£250
Rate of interest	8%	6%	4%	9%	9%	5%

	g	h	i	j	k	l	m
Interest paid	£490	£63	£48	£65	£69	£70	£30
Rate of interest	7%	7%	12%	13%	15%	10%	12%

2

	a	b	c	d	e
Interest paid	£120	£980	£85	£80	£340
Rate of interest	20%	14%	17%	$9\frac{1}{2}$%	8·5%

	f	g	h	i	j
Interest paid	£450	£1550	£450	£675	£3450
Rate of interest	$6\frac{1}{4}$%	$7\frac{3}{4}$%	17%	$12\frac{1}{2}$%	$7\frac{2}{3}$%

Compound interest

If you borrow £100 for two years at 5% per year, how much has to be paid back at the end of two years?

Interest for one year =£5.
Interest for two years=£10.
Amount to be paid back=£100+£10=£110

This answer is not quite correct, because during the second year you owe £105, not £100, and the interest during the second year has to be worked out on £105. It can be shown that the amount to be paid back=£110·25.

You can see that there is not much difference between the two answers, but for large amounts of money borrowed over, say, ten years the difference could be quite large. Interest calculated like this is called **compound interest**, and is best worked out using a compound interest table.

Example

£5 is borrowed for 4 years at 7%. How much has to be paid back at the end of 4 years?

The table tells us that £1 at 7% for 4 years becomes £1·31.

At the end of 4 years the amount to be paid back $= 5 \times £1 \cdot 31 = \underline{£6 \cdot 55}$.

	7%	8%	9%	10%	11%	12%
1 year	1·07	1·08	1·09	1·10	1·11	1·12
2 years	1·14	1·17	1·19	1·21	1·23	1·25
3 years	1·23	1·26	1·30	1·33	1·37	1·40
4 years	1·31	1·36	1·41	1·46	1·52	1·57
5 years	1·40	1·45	1·54	1·61	1·69	1·76

Exercise 22.5

In each of the examples below the amount borrowed, the rate of interest, and the number of years it is borrowed for are given. Find out how much has to be paid back.

1

	a	b	c	d	e	f	g	h
Amount borrowed	£8	£5	£4	£9	£3	£8	£3	£7
Number of years	4	1	3	2	5	1	3	4
Rate of interest	9%	11%	8%	7%	10%	12%	10%	8%

	i	j	k	l	m	n	o
Amount borrowed	£9	£7	£21	£13	£25	£17	£35
Number of years	5	2	5	3	2	4	1
Rate of interest	7%	9%	12%	7%	8%	10%	11%

2

	a	b	c	d	e	f
Amount borrowed	£200	£300	£400	£5000	£6000	£450
Number of years	4	5	3	2	1	3
Rate of interest	9%	8%	10%	7%	11%	12%

3 In examples 1 and 2 work out the interest paid in each case. 1a is done for you.
Interest paid = Amount paid − Amount borrowed
= £11.28 − £8.00 = £3.28.

revision 22 Prime numbers, rectangular numbers and factors

Exercise R 22

With each of the following numbers state whether it is a prime number or a rectangular number. If it is a rectangular number show all the ways it can be arranged as a rectangle.

	a	b	c	d	e	f	g	h	i
1	3	4	9	14	5	12	20	8	17
2	30	10	15	24	19	21	23	27	33
3	31	26	29	32	28	40	38	41	44
4	81	93	91	107	60	100	67	144	113

Write each of these numbers as a product of prime numbers, and also write down the set of their factors.

	a	b	c	d	e	f	g	h	i
5	9	6	4	16	20	21	24	28	25
6	22	36	32	33	45	72	35	30	34
7	40	48	70	64	76	100	108	144	156

8 Write down the prime numbers between 80 and 150.

Which of these numbers are primes? If a number is not a prime, write down the smallest number that goes into it.

	a	b	c	d	e	f	g	h	i
9	101	109	113	129	139	151	209	173	213
10	217	181	237	233	391	529	643	1373	1447

unit 23 Everyday arithmetic

In this Unit you will learn how to do some of the problems that you will have to do after you have left school.

Multiplying money with half pennies

How do we do sums like this. $8 \times 7\frac{1}{2}$p? Two ways are shown below.

1 $8 \times 7\text{p} = 56\text{p}$
$8 \times \frac{1}{2}\text{p} = 4\text{p}$
Total 60p

2 We can also say that $7\frac{1}{2}\text{p} = 7{\cdot}5\text{p}$

$$\begin{array}{r} 7{\cdot}5 \\ 8\times \\ \hline 60{\cdot}0\text{p} \\ \hline \end{array}$$

Example

Here is another example: $7 \times 7\frac{1}{2}$p

$7 \times 7\text{p} = 49\text{p}$
$7 \times \frac{1}{2}\text{p} = 3\frac{1}{2}\text{p}$
Total $52\frac{1}{2}$p

$$\begin{array}{r} 7{\cdot}5 \\ 7\times \\ \hline 52{\cdot}5 \\ \hline \end{array}$$ which equals $52\frac{1}{2}$p

Exercise 23.1

1a $4 \times 3\frac{1}{2}$p **b** $6 \times 4\frac{1}{2}$p **c** $2 \times 9\frac{1}{2}$p
d $8 \times 6\frac{1}{2}$p **e** $6 \times 7\frac{1}{2}$p **f** $8 \times 5\frac{1}{2}$p

2a $3 \times 3\frac{1}{2}$p **b** $5 \times 4\frac{1}{2}$p **c** $3 \times 9\frac{1}{2}$p
d $7 \times 6\frac{1}{2}$p **e** $5 \times 7\frac{1}{2}$p **f** $7 \times 5\frac{1}{2}$p

3a $4 \times 13\frac{1}{2}$p **b** $6 \times 14\frac{1}{2}$p **c** $2 \times 45\frac{1}{2}$p
d $8 \times 36\frac{1}{2}$p **e** $6 \times 67\frac{1}{2}$p **f** $8 \times 95\frac{1}{2}$p

4a $3 \times 13\frac{1}{2}$p **b** $5 \times 14\frac{1}{2}$p **c** $3 \times 45\frac{1}{2}$p
d $7 \times 36\frac{1}{2}$p **e** $5 \times 67\frac{1}{2}$p **f** $7 \times 95\frac{1}{2}$p

In the next examples you may find it helps to write £$1{\cdot}23\frac{1}{2}$ as £1·235, and so on.

5a $4 \times$ £$1{\cdot}23\frac{1}{2}$ **b** $6 \times$ £$2{\cdot}54\frac{1}{2}$ **c** $8 \times$ £$3{\cdot}45\frac{1}{2}$
d $5 \times$ £$2{\cdot}39\frac{1}{2}$ **e** $7 \times$ £$5{\cdot}43\frac{1}{2}$ **f** $9 \times$ £$5{\cdot}13\frac{1}{2}$

6a $11 \times 23\frac{1}{2}$p **b** $23 \times 34\frac{1}{2}$p **c** $43 \times 72\frac{1}{2}$p
d $64 \times 56\frac{1}{2}$p **e** $68 \times 45\frac{1}{2}$p **f** $79 \times 57\frac{1}{2}$p

7a $11 \times$ £$1{\cdot}23\frac{1}{2}$ **b** $23 \times$ £$2{\cdot}54\frac{1}{2}$ **c** $43 \times$ £$3{\cdot}45\frac{1}{2}$
d $64 \times$ £$2{\cdot}39\frac{1}{2}$ **e** $68 \times$ £$5{\cdot}43\frac{1}{2}$ **f** $79 \times$ £$5{\cdot}13\frac{1}{2}$

Checking bills

When a factory or a shop sells something it often prepares a bill which looks like this:

Remember that 6 kg sugar @ $9\frac{1}{2}$p means that each kg costs $9\frac{1}{2}$p, etc.

Van Groceries: 134 High Rd.
To: Mrs. Smith, 6 Bryant Rd.

6 kg sugar @ $3\frac{1}{2}$p	0·57
12 eggs @ 3p	0·36
2 kg butter @ 34p	0·68
	£1·61

Exercise 23.2

Copy and complete these bills.

1 6 kg sugar @ 9p
12 eggs at 4p
2 kg butter @ 36p
Total

2 7 kg sugar @ 8p
8 eggs @ 5p
3 kg butter @ 32p
Total

3 5 kg sugar @ 7p
18 eggs @ 3p
5 kg butter @ 31p
Total

4 8 2p stamps
12 3p stamps
14 4p stamps
Total

5 7 3p stamps
9 4p stamps
25 5p stamps
Total

6 6 rolls of wallpaper @ 39p
2 packets of paste @ 17p
3 tins of paint @ 37p
Total

The following examples will give you practice in doing bills.

7 $4 \times 3\frac{1}{2}$p
$6 \times 5\frac{1}{2}$p
7×9p
Total

8 7×13p
$5 \times 6\frac{1}{2}$p
$9 \times 3\frac{1}{2}$p
Total

9 $17 \times 4\frac{1}{2}$p
16×9p
$21 \times 2\frac{1}{2}$p
$20 \times 9\frac{1}{2}$p
Total

10 6×53p
$8 \times 12\frac{1}{2}$p
9×53p
$12 \times 7\frac{1}{2}$p
Total

11 5×45p
$6 \times 17\frac{1}{2}$p
$23 \times 20\frac{1}{2}$p
47×34p
Total

12 $4 \times £2{\cdot}34$
$5 \times £4{\cdot}09$
$7 \times £3{\cdot}89$
$2 \times £12{\cdot}23$
Total

13 $7 \times £4{\cdot}37$
$8 \times £4{\cdot}94$
$4 \times £12{\cdot}56$
$6 \times £5{\cdot}89$
Total

14 $7 \times £2{\cdot}36\frac{1}{2}$
$9 \times £1{\cdot}08\frac{1}{2}$
$5 \times £7{\cdot}11\frac{1}{2}$
$4 \times £4{\cdot}00\frac{1}{2}$
Total

15 $5 \times £4{\cdot}78$
$6 \times £23{\cdot}42$
$17 \times £3{\cdot}42\frac{1}{2}$
$56 \times £2{\cdot}71\frac{1}{2}$
Total

Which is the best buy?

Example

Which is the better buy? 7 kg of potatoes costing 28p, or 4 kg costing 20p?

7 kg costs 28p
So 1 kg costs 28p ÷ 7 = 4p

4 kg costs 20p
So 1 kg costs 20p ÷ 4 = 5p

7 kg is the better buy.

Exercise 23.3

In each of the following examples work out which is the better buy.

1 Grapefruit at 4p each, or 3 for 15p.
2 Grapefruit at 4 for 16p, or 7 for 21p.
3 2 kg potatoes for 12p, or 3 kg for 24p.
4 10 kg potatoes for 30p, or 4 kg for 16p.
5 A 1 kg joint of beef costing £1·20, or a 2 kg joint of the same quality costing £3·30.
6 A 2 kg joint of beef costing £2·20, or a 3 kg joint of the same quality costing £3·30.
7 $\frac{1}{2}$ kg apples at 8p, or 1 kg at 14p.
8 $\frac{1}{2}$ kg apples at 9p, or 2 kg at 38p.
9 A 3 kg turkey costing £1·80, or a 10 kg turkey costing £5.
10 A 4 kg turkey costing £2·00, or a 5 kg turkey costing £2·75.
11 A packet of flour weighing $1\frac{1}{2}$ kg and costing 21p, or a $\frac{1}{2}$ kg packet costing 9p.
12 A packet of flour weighing $2\frac{1}{2}$ kg and costing 35p, or a $1\frac{1}{2}$ kg packet costing 20p.

Is it cheaper to buy or to make?

A child's dressing gown can be knitted from 22 balls of wool costing 9p per ball. A similar dressing gown can be bought for £2·35. Is it cheaper to buy or to make, and by how much?

Cost of wool $= 22 \times 9\text{p} = £1{\cdot}98$.

It is cheaper to make, and the amount saved $= £2{\cdot}35 - £1{\cdot}98 = \underline{£0{\cdot}37}$.

It might take quite a time to knit the dressing gown, and of course you might not think it worth while to spend all this time just to save 37p.

Exercise 23.4

In each of the following examples work out whether it is cheaper to make or to buy, and work out how much is saved.

1 A child's dressing gown knitted from 26 balls of wool costing 9p a ball, or a similar one bought at £3·20.
2 A pair of gloves knitted from 2 balls of wool at 13p a ball, or a similar pair bought for 49p.
3 A pair of gloves knitted from 3 balls of wool at 19p a ball, or a similar pair bought for 52p.
4 A pair of heavy woollen socks knitted from 4 balls of wool costing 15p per ball, or a similar pair costing 97p?

5 Do **1** again and include the cost of needles at 11p, and a pattern at 13p.

6 Do **2** again and include the cost of needles at 9p, and a pattern at 11p.

7 Do **3** again and include the cost of needles at 10p, and a pattern at 9p.

8 Do **4** again and include the cost of needles at 12p, and a pattern at $8\frac{1}{2}$p.

9 A cotton dress made from 2 m material at 73p per m, 1 reel of cotton costing 8p, 1 zip costing 14p, 1 m bias binding costing 9p; or a similar dress costing £5.

10 A party dress made from 3 m of nylon at £1·20 per m, 1 reel cotton costing 8p, one zip costing 23p, 2 m bias binding costing 9p per m; or a similar dress costing £3·80.

11 A long dress made from 5 m of silk at £4·32 per m, 2 reels of cotton at 9p each, one zip costing 32p, 4 m bias binding costing 8p per m; or a similar dress costing £25·50.

12 Do **9** again and include the cost of the dress being made by a dressmaker, £1·50.

13 Do **10** again and include the cost of the dress being made by a dressmaker, £2·30.

14 Do **11** again and include the cost of the dress being made by a dressmaker, £4·50.

15 A pair of curtains ready-made in a shop costs £10. Made by a dressmaker who charges £0·70, curtains of the same material and size require 6 m at £1·20 per m, 2 m bias binding at 10p per m, 2 m curtain tape at 15p per m, 2 reels of cotton at 8p per reel.

Exercise 23.5

Use this list of prices in this exercise. (1 kg=1000 g, 1 litre=1000 cc)

6 eggs cost 24p.
1 kg flour costs 20p.
1 kg sugar costs 10p.
100 g chocolate cake covering costs 8p.
250 g butter costs 15p.
1 litre milk costs 10p.
100 g crystallized cherries costs 8p.
1 kg bitter oranges costs 20p.
1 lemon costs 3p.
100 g yeast costs 15p.

1 Copy and complete the following. Keep your answers, because you will need them later on in the exercise.

1 egg costs
2 eggs cost
3 eggs cost
4 eggs cost
125 g butter costs
25 g butter costs
50 g butter costs
100 g butter costs
200 g butter costs

100 g flour costs
50 g flour costs
200 g flour costs
250 g flour costs
500 g flour costs
750 g flour costs
300 g flour costs
800 g flour costs

100 cc milk costs
50 cc milk costs
150 cc milk costs
100 g sugar costs
50 g sugar costs
150 g sugar costs
200 g sugar costs
250 g sugar costs

In each of the following examples find out whether it is cheaper to make or to buy, and work out how much is saved. Take the cost of gas or electricity used for cooking to be 2p in each example.

2 Chocolate cake requiring 3 eggs, 100 g flour, 150 g sugar, 50 g chocolate cake covering. Shop price of cake is 30p.

3 A cherry slab cake requiring 250 g butter, 250 g sugar, 4 eggs, 500 g flour, 150 cc milk, 200 g crystallized cherries. Shop price 58p.

4 5 kg marmalade requiring $1\frac{1}{2}$ kg oranges, 3 kg sugar, 2 lemons. Shop price 15p per kg.

5 Albert biscuits in a cake shop cost 9p for 12. If they are made at home you will need 125 g butter, 150 g sugar, 2 eggs, 300 g flour. This will make 48.

6 A small loaf of bread costs 15p. Two small loaves the same size can be made from 10 g yeast, 750 g flour, 25 g butter.

7 A bottle of lemon squash costs 16p. Two bottles can be made from 3 lemons and $1\frac{1}{2}$ kg of sugar.

8 1 kg sugar and 250 g butter will produce 1100 g butterscotch. 100 g costs 8p.

9 A large loaf of bread costs 28p. It can be made at home with 10 g yeast, 800 g flour, 50 g butter.

10 Shortbread shapes cost 1p each. Made at home they require 250 g flour, 250 g butter, 100 g sugar. This will produce 36 shapes.

Hire purchase

You might want to buy a record player which costs £20 (this is often called the **cash price**) but you do not have enough money. Another way of buying it is on **hire purchase**. In this case you might be able to buy it by paying a **deposit** of £5 and 6 monthly instalments of £3. When you buy something on hire purchase you are being lent some of the cost of what you are buying, so you will have to pay interest on what you are borrowing. It will cost you more to buy on hire purchase. This is called the **hire purchase** price. You can work out how much extra you are paying as follows.

Deposit =	£5
Monthly payments 6 × £3 =	£18 +
Hire purchase price =	£23
Take away cash price	£20 −
Extra payment =	£3

It is up to you to decide whether to pay the extra £3 and have your record player straight away, or to wait until you have saved up enough money.

In some of the examples in the next exercise you have to multiply by 12, 18, 24, or 36. If you are not very good at multiplying by these numbers you can use the following method.
To multiply by 12, multiply by 6, and then multiply the answer by 2.
To multiply by 18, multiply by 6, and then multiply the answer by 3.
To multiply by 24, multiply by 6, and then multiply the answer by 4.
To multiply by 36, multiply by 6, and then multiply the answer by 6.

Exercise 23.6

In each of the following examples find the hire purchase price, and then how much extra you are paying.

	1	2	3	4	5	6	7	8	9
Cash price	£18	£24	£30	£36	£42	£48	£32	£43	£160
Deposit	£2	£3	£3	£4	£5	£7	£4	£5	£20
Monthly payment	£3	£4	£5	£6	£7	£8	£3	£4	£15
Number of payments	6	6	6	6	6	6	12	12	12

	10	11	12	13	14	15	16	17
Cash price	£29	£43	£152	£314	£141	£225	£767	£7·60
Deposit	£5	£7	£20	£32	£21	£29	£123	£0·90
Monthly payment	£2	£3	£7	£13	£5	£8	£24	£1·23
Number of payments	18	18	24	24	36	36	36	6

	18	19	20	21	22
Cash price	£24·10	£90·10	£165·90	£62·50	£187·53
Deposit	£3	£9	£15·50	£6·50	£19·55
Monthly payment	£4·65	£7·64	£9·32	£2·71	£5·54
Number of payments	6	12	18	24	36

You may find the following useful in the next examples.
To multiply by 26, multiply by 13 then by 2.
To multiply by 52, multiply by 13 then by 4.
To multiply by 78, multiply by 13 then by 6.
To multiply by 104, multiply by 13 then by 8.
To multiply by 156, multiply by 13 then by 12 (or by 3 and 4).

	23	24	25	26	27
Cash price	£2·08	£3·39	£19·63	£40·46	£90·21
Deposit	£0·50	£0·43	£2·00	£4·50	£9·70
Weekly payment	£0·07	£0·13	£0·42	£0·86	£1·23
Number of payments	26	26	52	52	78

	28	29	30	31	32
Cash price	£109·84	£256·85	£858·50	£819·54	£2904·99
Deposit	£12	£31	£91	£93	£327
Weekly payment	£1·56	£2·79	£8·73	£5·96½	£21·32½
Number of payments	78	104	104	156	156

Paying the rates

Much of the cost of running schools, repairing roads, collecting rubbish, etc. has to be paid for by your local Council. They collect the money for this from the owners of the factories, shops and houses in their area. The money they collect is called the **rates**.
Each building is given a **rateable value**. This depends on the size and value of the building. A house might have a rateable value of between £100 and £200.
When the Council have worked out how much they need to spend in the next year they work out the rate. It might be 50p in the pound. This means that if you have a house with a rateable value of £200 you will have to pay 200 lots of 50p.

You will have to pay 200 × 50 = 10 000p = £100

Example

A man owns a small garage with a rateable value of £24. The rates are 80p in the £. How much will he have to pay each year in rates?

He will have to pay 24 × 80 = 1920p = £19·20

Exercise 23.7

Find our how much has to be paid each year in the following examples.

	1	2	3	4	5
Rateable value	£40	£50	£100	£200	£300
Rate in £	50p	60p	70p	90p	80p

	6	7	8	9	10
Rateable value	£40	£60	£70	£200	£300
Rate in £	45p	75p	83p	120p	110p

	11	12	13	14	15
Rateable value	£74	£63	£84	£120	£240
Rate in £	30p	60p	70p	80p	90p

	16	17	18	19	20
Rateable value	£75	£65	£84	£123	£235
Rate in £	43p	73p	86p	130p	116p

Paying for gas and electricity

There are two main ways of paying for gas and electricity. You can either pay in advance, using a slot-meter, or you can have a bill sent, usually every three months.

The bill is made up of two parts:

1 The standing charge. This will always be the same and does not depend on how much electricity or gas you use.

2 The charge for the amount of electricity or gas that you use.

Gas is sold by the **therm**. A gas cooker uses roughly 2 therms a week.
Electricity is sold by the **unit**. A 1-kilowatt fire uses 1 unit every hour.

If you use a large amount of gas or electricity for heating you can often get it cheaper than the usual price. Your electricity or gas showroom will be able to tell you about this.

Example

What will be the total of a three-monthly gas bill for 9 therms at 12p per therm? The standing charge is £0·67.

Cost of gas 9 × 12p	1·08
Standing charge	0·67
Total	£1·75

Exercise 23.8

Work out the total of each of these gas bills.

	1	2	3	4
Number of therms used	8	9	6	7
Cost of one therm	11p	12p	13p	14p
Standing charge	£0·76	£0·94	£0·45	£0·73

	5	6	7	8
Number of therms used	8	11	21	24
Cost of one therm	15p	12p	14p	16p
Standing charge	£1·05	£1·07	£0·99	£0·42

Work out the total of each of these electricity bills.

	9	10	11	12
Number of units used	50	60	70	100
Cost of one unit	1p	1p	2p	1·1p
Standing charge	£1·23	£0·97	£1·02	£0·84

	13	14	15	16
Number of units used	200	400	93	456
Cost of one unit	1·3p	0·8p	0·9p	1·2p
Standing charge	£1·04	£1·45	£2·17	£0·90

When the bill comes it will of course tell you how much you have to pay, but you might want to work out how much you have to pay before the bill comes.

To do this you need to know what the meter read when the last bill was sent to you. This will be on the bill. You also need to read the meter. You can work out how many units you have used by taking the previous reading from the present one.

Example

The previous reading (on the last bill) was 1132. The present reading is 1236. The cost per unit is 0·9p, and the standing charge is £1·29. What will the bill be?

Number of units used = 1236 − 1132 = 104

Cost of electricity = 104 × 0·9 = 93·6p = £0·936
Standing charge = £1·29
Total = £2·226

You will get a bill for either £2·22 or £2·23.

Exercise 23.9

In each of these examples work out the number of units used, and then the total for each bill.

	1	2	3	4	5
Present reading	2356	4375	3985	5953	4828
Previous reading	2256	4275	3935	5753	4628
Cost of one unit	1p	1p	2p	0·9p	0·7p
Standing charge	£1·32	£0·67	£1·06	£0·95	£1·43

	6	7	8	9	10
Present reading	48643	48375	59200	19385	38394
Previous reading	48493	48125	58876	18804	37415
Cost of one unit	1·1p	1·2p	1·3p	1·2p	1·4p
Standing charge	£1·43	£0·98	£0·45	£1·04	£0·99

Working out your gas bill from the meter is more difficult, because the meter tells you the number of cubic feet of gas that you have used and you have to change this to therms. The formula for doing this is –

$$\text{Number of therms} = \frac{\text{Calorific value} \times \text{Number of cubic ft}}{100\,000}$$

Your last gas bill will tell you what the calorific value of your gas is.

Example

How many therms in 30 000 cubic ft? Calorific value = 1000.

$$\text{Number of therms} = \frac{1000 \times 30\,000}{100\,000} = \underline{300}$$

Exercise 23.10

In each of the following examples find:

a The number of therms used.

b The total of each gas bill.

	1	2	3	4	5
Number of cubic ft	4000	5000	6000	7000	4000
Calorific value	1000	1000	500	500	500
Cost of one therm	9p	8p	10p	9p	11p
Standing charge	£0·90	£0·80	£0·95	£1·23	£1·23

	6	7	8	9
Number of cubic ft	5000	6000	8000	7000
Calorific value	450	1100	450	1200
Cost of one therm	13p	14p	15p	11p
Standing charge	£1·09	£0·78	£1·20	£0·78

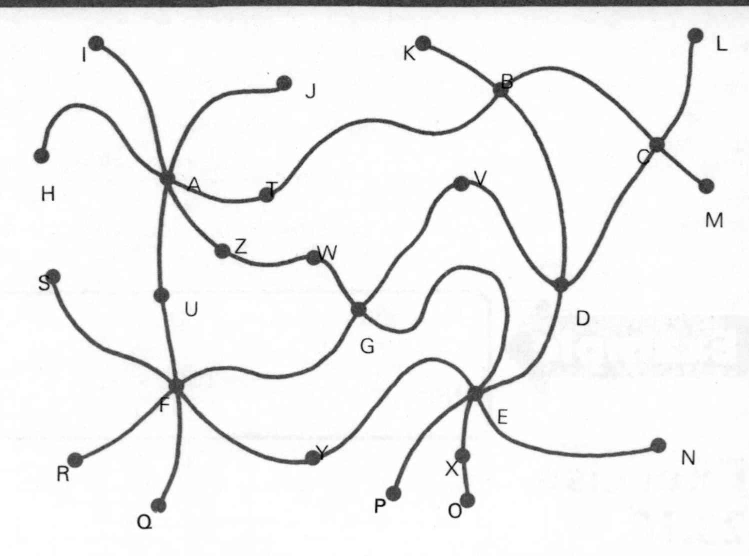

The first diagram shows a railway map. The towns are marked by letters. If you wanted to go from I to K where would you need to change? It is not very easy to get the answer quickly, but if the map is drawn as in the second diagram we can now see that to get from I to K we need to change at E, or we could change at A and B, or at G and D. We should have to look at the train times to see which of the three ways was the best.

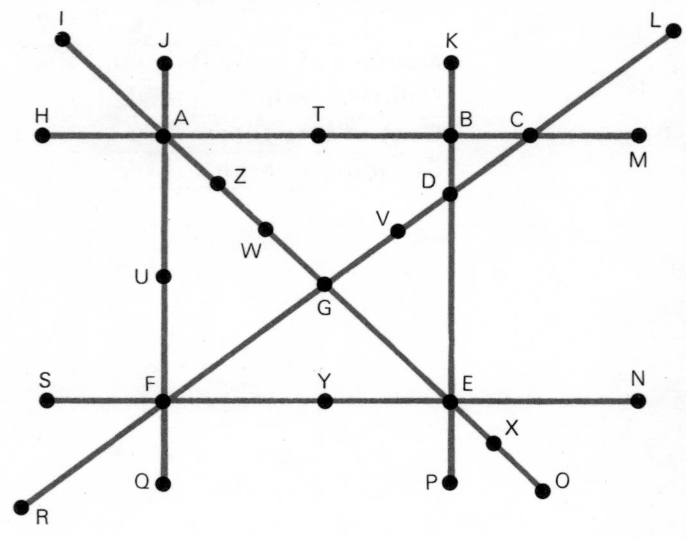

The second map is of course not a proper map, as all the distances have been changed; but it still shows which railway lines go through which towns. Two maps like this are called **topologically equivalent**.

To decide whether two shapes or maps are topologically equivalent, imagine they are both printed on rubber sheets and then decide whether you can stretch one into the other.

Here are some pairs of shapes that are topologically equivalent.

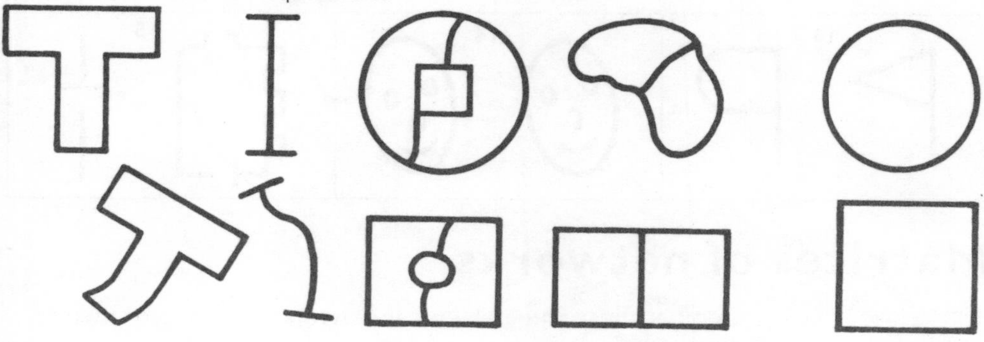

Here are some pairs of shapes that are not topologically equivalent.

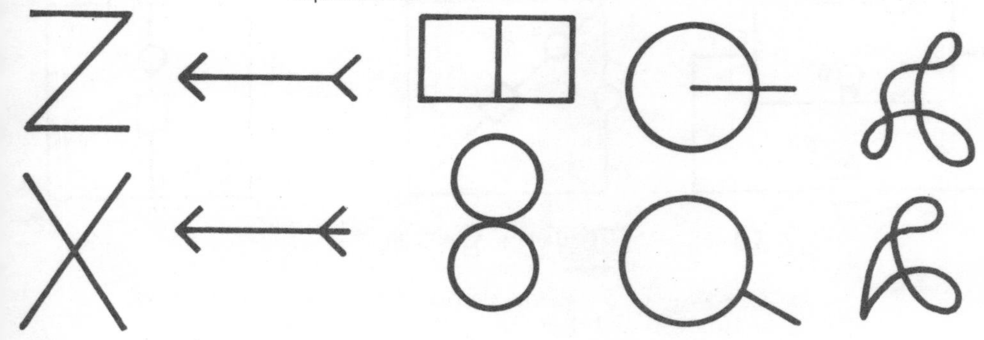

Exercise 24.1

Which of these pairs of shapes are topologically equivalent?

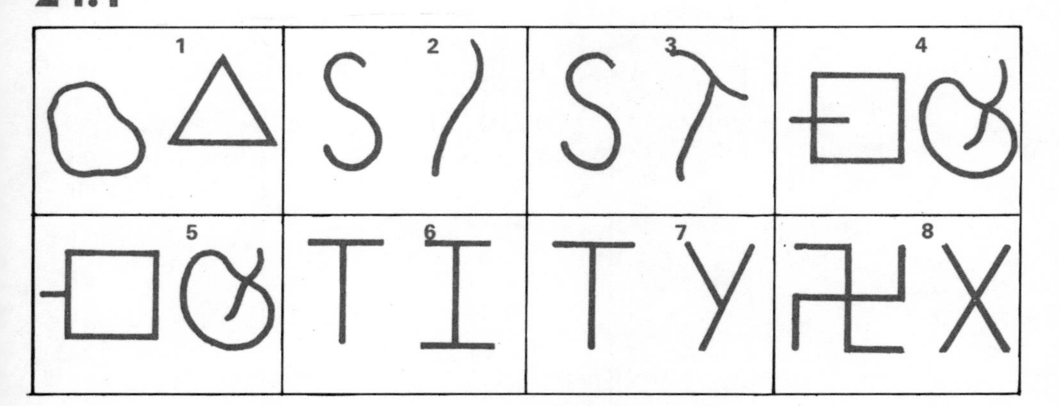

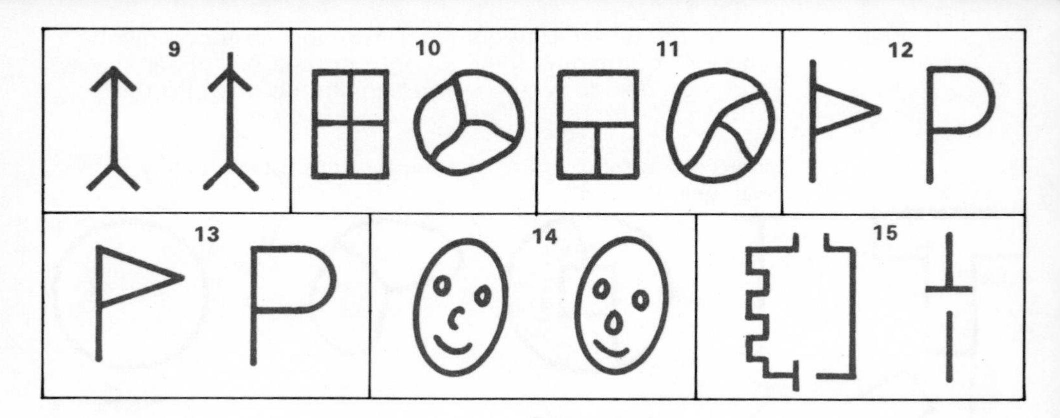

Matrices of networks

Look at these three electrical networks. Are any of them the same?

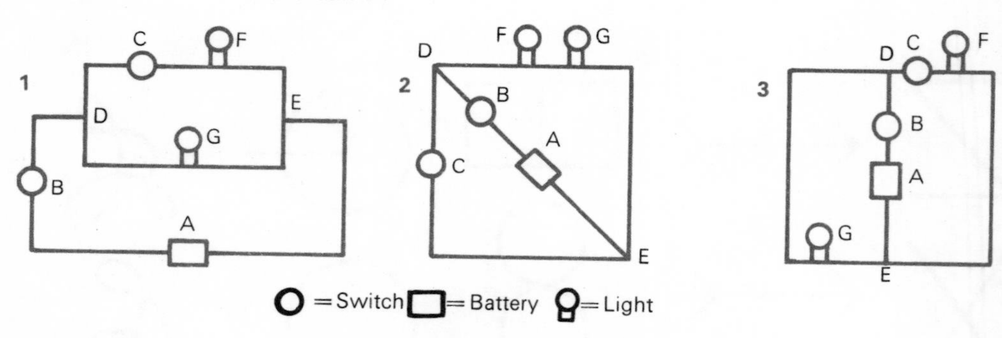

One way of answering a question like this is to draw a matrix for each network.
We will draw a matrix for the first network.
There is a wire from A to B so we mark in a 1.
There is no wire from A to C so we mark in a 0.
There is no wire from A to D so we mark in a 0.
And so on.

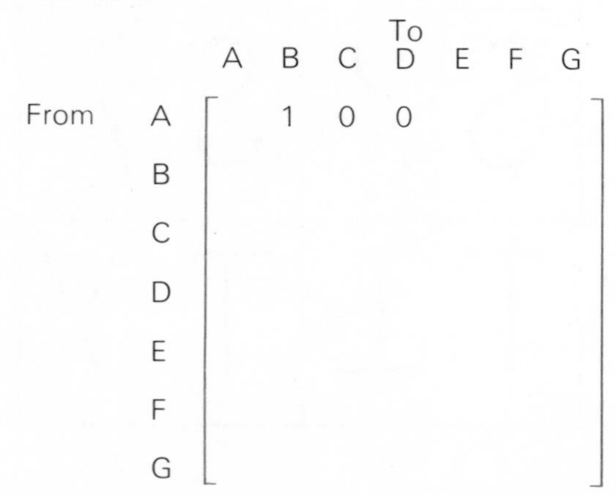

From \ To	A	B	C	D	E	F	G
A		1	0	0			
B							
C							
D							
E							
F							
G							

If we complete this matrix and do the same for the other networks we get the matrices shown below. (Because there are no wires from A to A, or B to B, and so on, we put in a 0.)

1

To

From	A	B	C	D	E	F	G
A	0	1	0	0	1	0	0
B	1	0	0	1	0	0	0
C	0	0	0	1	0	1	0
D	0	1	1	0	0	0	1
E	1	0	0	0	0	1	1
F	0	0	1	0	1	0	0
G	0	0	0	1	1	0	0

2

To

From	A	B	C	D	E	F	G
A	0	1	0	0	1	0	0
B	1	0	0	1	0	0	0
C	0	0	0	1	1	0	0
D	0	1	1	0	0	1	0
E	1	0	1	0	0	0	1
F	0	0	0	1	0	0	1
G	0	0	0	0	1	1	0

3

To

From	A	B	C	D	E	F	G
A	0	1	0	0	1	0	0
B	1	0	0	1	0	0	0
C	0	0	0	1	0	1	0
D	0	1	1	0	0	0	1
E	1	0	0	0	0	1	1
F	0	0	1	0	1	0	0
G	0	0	0	1	1	0	0

Look carefully at the matrices. You should see that 1 and 3 are the same, but 2 is different. This means that networks 1 and 3 are the same.

What will happen in 1 and 3 if we turn on switch B (leaving switch C off)?
What will happen in 2 if we turn on switch B (leaving switch C off)?

Exercise 24.2

Write the matrix for each of the following networks. Which of the networks are the same (topologically equivalent)?

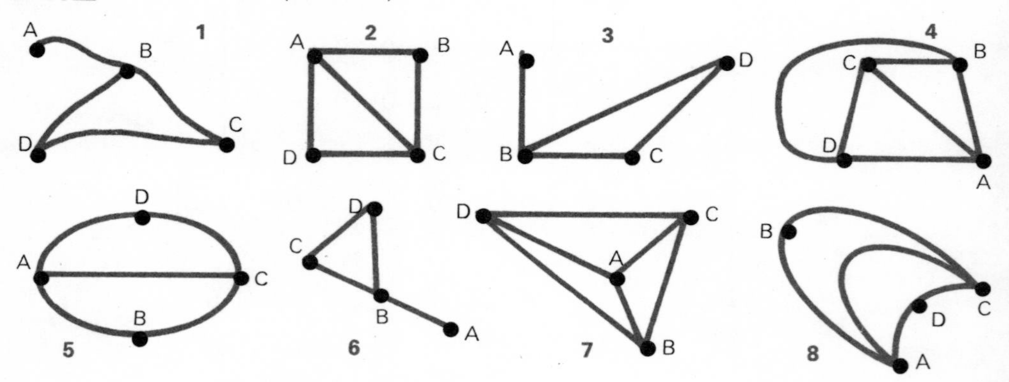

Look at this network and its matrix. Notice how we show in the matrix the three lines from A to B, the line from A to A, and the two lines from C to C.

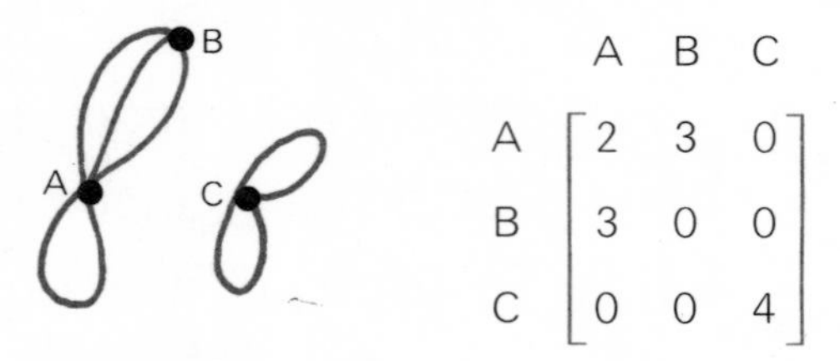

$$\begin{array}{c} \\ A \\ B \\ C \end{array} \begin{array}{c} \begin{array}{ccc} A & B & C \end{array} \\ \begin{bmatrix} 2 & 3 & 0 \\ 3 & 0 & 0 \\ 0 & 0 & 4 \end{bmatrix} \end{array}$$

There are 2 ways to go from A to A because we can go around the loop either way.
There are 4 ways to go from C to C because we can go around each loop either way.

Write the matrix for each of the following networks. Which of the networks are topologically equivalent?

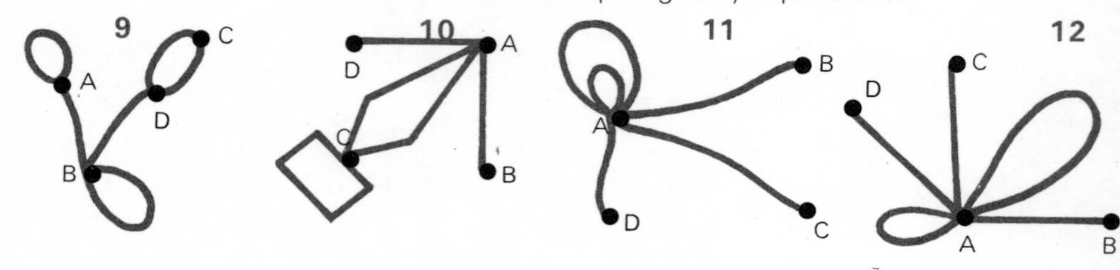

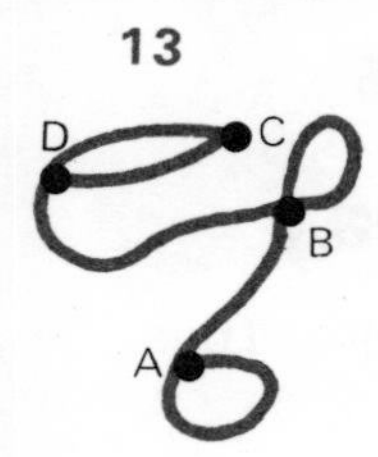

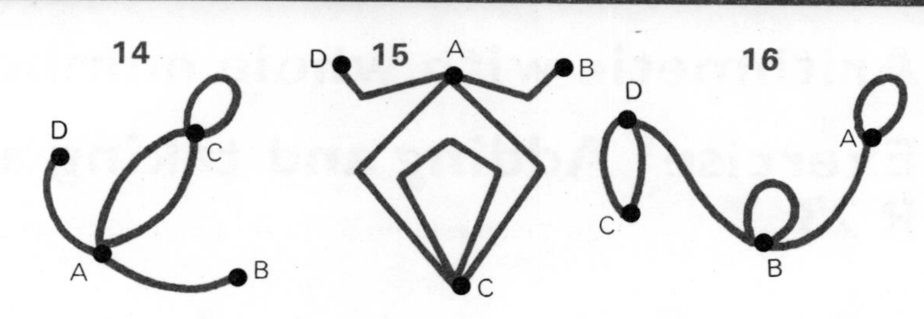

17 Two switches, a battery, a light, and two junctions where wires can be joined are fixed to a board. Copy the diagram and draw in the wires shown in the matrix.

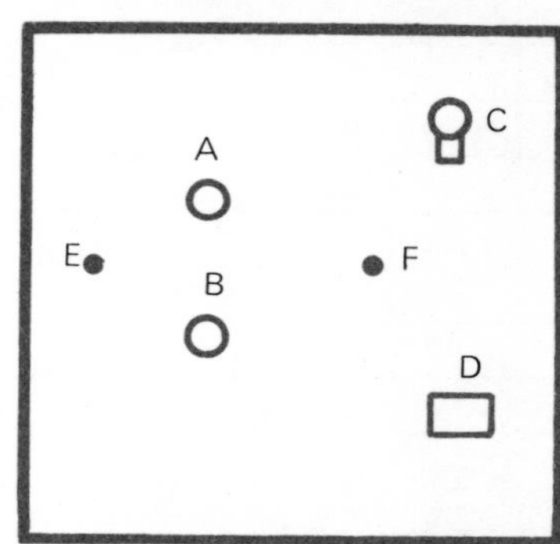

	A	B	C	D	E	F
A	0	0	0	0	0	0
B	0	0	1	1	0	0
C	0	1	0	1	0	0
D	0	1	1	0	0	0
E	0	0	0	0	0	0
F	0	0	0	0	0	0

a What happens when switch A is turned on?

b What happens when switch B is turned on?

c What happens when switches A and B are both turned on?

Repeat question **17** with the following matrices.

18

	A	B	C	D	E	F
A	0	1	1	0	0	0
B	1	0	0	1	0	0
C	1	0	0	1	0	0
D	0	1	1	0	0	0
E	0	0	0	0	0	0
F	0	0	0	0	0	0

19

	A	B	C	D	E	F
A	0	0	0	0	1	1
B	0	0	0	0	1	1
C	0	0	0	1	0	1
D	0	0	1	0	1	0
E	1	1	0	1	0	0
F	1	1	1	0	0	0

20

	A	B	C	D	E	F
A	0	0	1	0	1	0
B	0	0	0	0	1	1
C	1	0	0	0	1	1
D	0	0	0	0	1	1
E	1	1	1	1	0	0
F	0	1	1	1	0	0

revision 23 Arithmetic with whole numbers

Arithmetic with whole numbers

Exercise R 23 Adding and taking away

Do these adding sums. For example sum number **1b** is 4+5=

	a	b	c	d	e	f	g	h	i	j
	3	5	7	12	24	51	84	193	246	719
1 4										
2 7										
3 9										
4 15										
5 23										
6 57										
7 93										
8 302										
9 744										
10 368										

Do numbers **1** to **10** as taking away sums. Always take the smaller number from the larger. For example, sum number **3d** is 12−9=

Multiplying and dividing

Do these multiplication sums. Always put the smallest number underneath. For example, sum number **4b** is

$$\begin{array}{r} 23 \\ \underline{8\times} \end{array}$$

	a	b	c	d	e	f	g	h	i	j
	14	23	77	45	89	92	97	147	532	987
1 2										
2 6										
3 7										
4 8										
5 9										
6 14										
7 38										
8 42										
9 67										
10 891										

Do these division sums. If there is a remainder write it down as well.

1a 24÷4 **b** 16÷4 **c** 40÷8 **d** 35÷5
e 63÷7 **f** 54÷6 **g** 72÷8 **h** 48÷8
i 45÷5 **j** 50÷5

2a 64÷8 **b** 32÷4 **c** 18÷3 **d** 70÷7
e 15÷5 **f** 45÷9 **g** 21÷7 **h** 42÷6
i 25÷5 **j** 36÷6

3a 105÷5 **b** 124÷4 **c** 46÷2 **d** 129÷3
e 203÷7 **f** 312÷6 **g** 352÷8 **h** 837÷9
i 711÷9 **j** 672÷8

4a 26÷4 **b** 19÷4 **c** 43÷8 **d** 37÷5
e 65÷7 **f** 107÷5 **g** 127÷4 **h** 131÷3
i 208÷7 **j** 315÷6

5a 357÷21 **b** 742÷14 **c** 782÷17 **d** 374÷22
e 483÷23 **f** 512÷32 **g** 2397÷47 **h** 897÷69
i 3403÷83 **j** 4108÷79

6a 1551÷47 **b** 4624÷68 **c** 400÷25 **d** 858÷78
e 5022÷62 **f** 1462÷43 **g** 2456÷45 **h** 2784÷97
i 3497÷34 **j** 3498÷32

7a 456÷123 **b** 568÷121 **c** 4565÷23 **d** 2356÷37
e 9876÷34 **f** 8472÷73 **g** 3975÷19 **h** 2960÷28
i 3965÷53 **j** 9418÷82

Arithmetic with decimals

Exercise R 24 Adding and taking away decimals

Do these adding sums. For example, sum number **2c** is

8·0
5·3+
___

	a	b	c	d	e	f	g	h	i
	1·4	2·9	8	15·4	0·7	0·54	0·04	0·543	17·32
1 1·8									
2 5·3									
3 5									
4 14·3									
5 0·9									
6 0·46									
7 0·03									
8 0·738									
9 17·3									
10 43·82									

Do numbers **1** to **10** as take away sums. Always write the smaller number at the bottom.

Multiplying and dividing decimals by whole numbers

Do these multiplying sums. For example, sum number **3a** is 7 × 1·4 =

	a 1·4	b 2·3	c 0·7	d 0·4	e 3·25	f 5·62	g 24·9	h 45·8	i 13·74	j 16·753
1 2										
2 5										
3 7										
4 9										
5 13										
6 32										
7 63										
8 79										
9 83										
10 257										

Do these dividing sums.

1a 2·2 ÷ 2 **b** 6·3 ÷ 3 **c** 9·6 ÷ 3 **d** 4·8 ÷ 2
e 8·4 ÷ 4 **f** 1·6 ÷ 2 **g** 3·5 ÷ 7

2a 4·9 ÷ 7 **b** 3·2 ÷ 8 **c** 1·8 ÷ 6 **d** 1·5 ÷ 3
e 4·5 ÷ 5 **f** 2·1 ÷ 3 **g** 4·2 ÷ 7

3a 10·4 ÷ 8 **b** 9·8 ÷ 7 **c** 16·2 ÷ 9 **d** 22·5 ÷ 9
e 19·2 ÷ 8 **f** 28·8 ÷ 9 **g** 23·2 ÷ 8

4a 2·43 ÷ 9 **b** 2·16 ÷ 8 **c** 5·1 ÷ 7 **d** 3·6 ÷ 8
e 0·21 ÷ 6 **f** 0·83 ÷ 8 **g** 1·7 ÷ 3

5a 0·48 ÷ 5 **b** 2·2 ÷ 3 **c** 45 ÷ 7 **d** 8·5 ÷ 9
e 51 ÷ 3 **f** 0·92 ÷ 4 **g** 7·8 ÷ 6

6a 23·4 ÷ 13 **b** 27·6 ÷ 11 **c** 25·3 ÷ 21 **d** 7·54 ÷ 22
e 60·3 ÷ 23 **f** 2·46 ÷ 17 **g** 0·654 ÷ 22

7a 0·83 ÷ 32 **b** 0·84 ÷ 42 **c** 9·53 ÷ 46 **d** 97·4 ÷ 63
e 6·49 ÷ 18 **f** 9·23 ÷ 73 **g** 0·937 ÷ 91

Multiplying and dividing by decimals

Do the following multiplying sums. For example, sum number **2b** is 0·6 × 0·5 =

	a 0·4	b 0·5	c 0·8	d 2·1	e 3·2	f 5·3	g 0·75	h 0·06	i 2·563	j 0·0263
1 0·3										
2 0·6										
3 0·9										
4 1·2										
5 3·4										
6 0·64										
7 1·23										
8 8·32										
9 0·341										
10 26·8										

Do these dividing sums.

		a	b	c	d	e
1	Divide 4·8	by 0·2	0·3	0·4	0·06	0·08
2	Divide 7·2	by 0·3	0·4	0·6	0·03	0·04
3	Divide 1·44	by 0·4	0·04	0·8	0·03	0·003
4	Divide 0·72	by 0·3	0·02	0·004	0·006	0·08
5	Divide 3·6	by 1·2	2·4	0·24	0·12	0·036
6	Divide 0·264	by 1·1	0·12	1·2	0·11	0·011
7	Divide 39	by 1·5	0·15	1·3	0·13	0·013
8	Divide 38	by 1·7	0·16	0·023	20·1	0·0021
9	Divide 23·74	by 2·3	0·45	0·023	2·04	3·06
10	Divide 674	by 0·34	0·032	1·22	10·5	24·9

Multiplying and dividing decimals by 10, 100 and 1000

Do these multiplication and division sums. For example, **2c** is 6·42 × 1000 =

	a ×10	b ×100	c ×1000	d ÷10	e ÷100	f ÷1000
1 3·5						
2 6·42						
3 9						
4 0·8						
5 27·3						
6 84						
7 0·62						
8 47·39						

Fractions

Exercise R 25

Simplifying fractions

Simplify (cancel) these fractions.

	a	b	c	d	e	f	g	h	i
1	$\frac{2}{4}$	$\frac{2}{8}$	$\frac{3}{6}$	$\frac{3}{9}$	$\frac{2}{10}$	$\frac{2}{6}$	$\frac{6}{8}$	$\frac{6}{9}$	$\frac{4}{8}$
2	$\frac{3}{12}$	$\frac{4}{10}$	$\frac{6}{18}$	$\frac{2}{12}$	$\frac{12}{16}$	$\frac{5}{10}$	$\frac{8}{12}$	$\frac{12}{18}$	$\frac{14}{21}$
3	$\frac{6}{12}$	$\frac{18}{24}$	$\frac{21}{28}$	$\frac{6}{16}$	$\frac{24}{30}$	$\frac{30}{50}$	$\frac{9}{24}$	$\frac{28}{56}$	$\frac{25}{30}$

Changing fractions to decimals

Change these fractions to decimals.

	a	b	c	d	e	f	g	h	i
1	$\frac{3}{4}$	$\frac{1}{4}$	$\frac{1}{2}$	$\frac{2}{3}$	$\frac{1}{5}$	$\frac{1}{3}$	$\frac{3}{5}$	$\frac{1}{6}$	$\frac{4}{5}$
2	$\frac{3}{7}$	$\frac{2}{5}$	$\frac{5}{6}$	$\frac{1}{7}$	$\frac{3}{8}$	$\frac{4}{6}$	$\frac{5}{8}$	$\frac{6}{7}$	$\frac{2}{7}$
3	$\frac{6}{8}$	$\frac{9}{4}$	$\frac{5}{9}$	$\frac{7}{11}$	$\frac{1}{9}$	$\frac{9}{17}$	$\frac{11}{5}$	$\frac{4}{9}$	$\frac{7}{13}$
4	$\frac{8}{21}$	$\frac{32}{47}$	$\frac{73}{74}$	$\frac{16}{7}$	$\frac{1}{62}$	$\frac{9}{11}$	$\frac{65}{82}$	$\frac{163}{42}$	$\frac{13}{7}$

Finding fractions of numbers

		a	b	c	d	e
1	Find $\frac{1}{4}$ of	16	12	8	28	24
2	Find $\frac{3}{5}$ of	15	10	40	35	50
3	Find $\frac{3}{7}$ of	21	14	28	35	42
4	Find $\frac{5}{9}$ of	18	20	25	41	36
5	Find $\frac{7}{9}$ of	47	83	39	42	29
6	Find $\frac{3}{8}$ of	£1·20	£1·52	£4·38	£0·83	£24·46
7	Find $\frac{4}{7}$ of	£3·15	£4·69	£0·68	£34·89	£247
8	Find $\frac{10}{21}$ of	£54·76	£37·85	£7·98	£24·73	£0·95

unit 25 More arithmetic

In this Unit you can do the examples using logarithms, slide rules, or calculating machines if you know how to.

Exercise 25.1

1a $1{\cdot}36 \times 5{\cdot}47$ **b** $2{\cdot}94 \times 3{\cdot}19$ **c** $3{\cdot}95 \times 2{\cdot}27$
d $4{\cdot}89 \times 1{\cdot}74$ **e** $1{\cdot}56 \times 3{\cdot}42$

2a $4{\cdot}7 \times 2{\cdot}11$ **b** $6{\cdot}32 \times 1{\cdot}3$ **c** $9 \times 1{\cdot}08$
d $4 \times 2{\cdot}37$ **e** $8 \times 1{\cdot}21$

3a $2{\cdot}9 \times 3{\cdot}24$ **b** $7{\cdot}23 \times 1{\cdot}29$ **c** $4{\cdot}5 \times 2{\cdot}1$
d $1{\cdot}08 \times 3{\cdot}09$ **e** $1{\cdot}08 \times 1{\cdot}09$

4a $2{\cdot}3 \times 2{\cdot}3$ **b** $6{\cdot}03 \times 1{\cdot}2$ **c** $5{\cdot}36 \times 1{\cdot}67$
d $7 \times 1{\cdot}29$ **e** $3{\cdot}75 \times 2{\cdot}31$

5a $3{\cdot}67 \times 2{\cdot}71$ **b** $3{\cdot}16 \times 3{\cdot}14$ **c** $9 \times 1{\cdot}11$
d $5{\cdot}72 \times 1{\cdot}20$ **e** $2{\cdot}23 \times 4$

6a $9{\cdot}12 \times 1{\cdot}03$ **b** $3{\cdot}45 \times 2{\cdot}78$ **c** $1{\cdot}14 \times 1{\cdot}65$
d $1{\cdot}73 \times 1{\cdot}73$ **e** $3{\cdot}12 \times 3{\cdot}12$

Exercise 25.2

1a $5{\cdot}01 \div 2{\cdot}51$ **b** $5{\cdot}73 \div 1{\cdot}67$ **c** $5{\cdot}73 \div 3{\cdot}17$
d $9{\cdot}83 \div 3$ **e** $8{\cdot}93 \div 4{\cdot}71$

2a $9 \div 5{\cdot}51$ **b** $6{\cdot}58 \div 3{\cdot}35$ **c** $9{\cdot}99 \div 3{\cdot}33$
d $4{\cdot}97 \div 1{\cdot}23$ **e** $1{\cdot}09 \div 1{\cdot}03$

3a $5{\cdot}99 \div 4{\cdot}63$ **b** $7{\cdot}94 \div 3{\cdot}16$ **c** $4 \div 3{\cdot}73$
d $7{\cdot}51 \div 5{\cdot}8$ **e** $8 \div 2{\cdot}35$

4a $9{\cdot}00 \div 6{\cdot}91$ **b** $8{\cdot}13 \div 4{\cdot}56$ **c** $7 \div 6{\cdot}01$
d $3{\cdot}17 \div 1{\cdot}31$ **e** $9{\cdot}23 \div 4{\cdot}65$

5a $4{\cdot}99 \div 1{\cdot}63$ **b** $7{\cdot}53 \div 1{\cdot}04$ **c** $9{\cdot}98 \div 3{\cdot}59$
d $8{\cdot}11 \div 4{\cdot}04$ **e** $9{\cdot}23 \div 7$

6a $7{\cdot}13 \div 3{\cdot}26$ **b** $6{\cdot}31 \div 5{\cdot}01$ **c** $5{\cdot}48 \div 1{\cdot}11$
d $3{\cdot}17 \div 2{\cdot}97$ **e** $9 \div 7{\cdot}23$

Exercise 25.3

1 Find the area of a room 2·2 m by 4·17 m.

2 Find the area of a room 3·74 m by 2·18 m.

3 Find the area of a room 2·64 m by 2·43 m.

4 How many planks of wood 1·43 m long can be cut from a piece 8·58 m long?

5 How many planks of wood 2·31 m long can be cut from a piece 9·24 m long?

6 How many planks of wood 1·87 m long can be cut from a piece 7·48 m long?

7 Find the cost of 3·25 m dress material at £1·81 per m.

8 Find the number of lengths of fencing 1·3 m long required to fence a length of 9·10 m.

9 How many books at £1·36 each can be bought for £8·16?

10 Find the cost of 2·75 m of dress material at £2·57 per m.

11 Find the area of a farm 1·61 km wide by 4·81 km long.

12 The area of a room is 7·04 sq m and its length is 1·76 m. What is its width?

13 Find the cost of 3·75 kg plastic at £1·27 per kg.

14 How many books at £1·07 each can be bought for £9·63?

15 Find the cost of 3·73 kg prime steak at £2·44 per kg.

16 The area of a room is 8·63 sq m and its length is 2·45 m. What is its width?

17 How many complete planks of wood each 1·55 m long can be cut from a piece 8 m long?

18 How much change will be left from £10 after buying 2·35 kg salmon at £3·66 per kg?

19 There are 9·37 m dress material left on a roll. How many complete lengths each 1·5 m can be cut from it?

20 There is 9·31 m wire left on a reel. How many complete lengths each 1·82 m long can be cut from it?

Exercise 25.4

	a	b	c	d
1	35·6 × 48·8	46·7 × 23·8	56·3 ÷ 13·8	46·8 ÷ 34·8
2	245 × 1·56	67·3 × 3·98	24·8 ÷ 4·98	39·5 ÷ 3·86
3	67·4 × 2·98	56 × 6·78	259 ÷ 3·87	298 ÷ 23·9
4	23·9 × 85·7	58 × 96	35·6 ÷ 23·9	87 ÷ 45
5	346 × 3·97	146 × 89·7	79 ÷ 4·37	45 600 ÷ 876
6	13 × 67·8	9·35 × 89 700	345 ÷ 2·26	56 700 ÷ 56 400
7	134 × 65	6790 × 1·13	567 ÷ 238	564 ÷ 2·36

Exercise 25.5

1 Find the area of a room 5·75 m by 10·7 m.

2 Find the area of a room 12·8 m by 17·4 m.

3 How many lengths of wire 1·55 m long can be cut from 27·9 m?

4 How many lengths of wire 15·6 m long can be cut from 234 m?

5 Find the volume of a room 3·31 m by 3·46 m by 4·16 m.

6 Find the volume of a room 10·4 m by 13·6 m and 3·17 m high.

7 17 people hire a coach. The cost is £28. How much will each have to pay?

8 23 people hire a coach for £58·30. How much will each have to pay?

9 A pools win of £22 500 is to be divided between 14 people. How much will each get?

10 The area of a rectangular field is 73 100 sq m. If one side is 220 m, what is the length of the other side?

11 A school orders 73 books at £0·67 per book. What is the total cost of the books? (Change to pence before you multiply, and then change the answer back to £.)

12 136 000 tonnes of earth are to be removed to make a road through a hill. How many truck loads will be required if a truck can take 16·3 tonnes?

13 A room is 3·27 m by 4·92 m.

a What is the area of the room?

b How much will it cost to carpet at £1·89 per sq m?

14 A man earns £1·17 per hour.

a How much will he earn in a week of 41·5 hours?

b How much will he earn in 49 weeks?

15 A canal is to be dug. 83 m long, 7·75 m wide, and 2·25 m deep.

a Find the volume of earth to be removed.

b If each cubic m weighs 2·3 tonnes, what weight of earth has to be removed?

16 A patch of ground is to be planted with seed, one box of which will plant 14 sq m. The ground is 16·8 m by 25·9 m.

a What is the area of the ground?

b How many boxes of seed will be required?

17 A room 2·63 m by 3·48 m is to be covered with carpet costing £2·11 per sq m.

a Find the area of the room.

b Find the cost of the carpet.

18 A bathroom 360 cm by 240 cm is to be covered with tiles. Each tile is 30 cm square.

a Will the tiles fit the room exactly?

b What is the area of the room?

c What is the area of one tile?

d How many tiles will be needed?

19 A book has an average of 13 words per line, and an average of 38 lines on a page.

a What is the average number of words per page?

b About how many words will there be in the book if it has 179 pages?

20 A wall 23·5 m long and 1·7 m high has to be painted. Each tin of paint covers 11 sq m and costs 39p.

a What is the area of the wall?

b How many tins of paint must be bought?

c How much will this cost?

Exercise 25.6

1a $245 \times 0{\cdot}529$ **b** $6750 \times 0{\cdot}624$ **c** $356 \times 0{\cdot}0463$
d $34{\cdot}7 \times 0{\cdot}692$ **e** $5760 \times 0{\cdot}00\,927$

2a $73\,900 \times 0{\cdot}593$ **b** $98\,600 \times 0{\cdot}0836$ **c** $8360 \times 0{\cdot}0803$
d $62\,900 \times 0{\cdot}000\,32$ **e** $60\,400 \times 0{\cdot}00\,501$

3a $0{\cdot}0735 \times 23{\cdot}7$ **b** $0{\cdot}000\ 534 \times 246$ **c** $0{\cdot}00\ 812 \times 762$
d $2460 \times 0{\cdot}000\ 324$ **e** $0{\cdot}00\ 374 \times 56{\cdot}9$

4a $3{\cdot}67 \times 0{\cdot}637$ **b** $0{\cdot}284 \times 0{\cdot}826$ **c** $0{\cdot}647 \times 0{\cdot}068$
d $0{\cdot}0386 \times 0{\cdot}0063$ **e** $0{\cdot}746 \times 0{\cdot}00\ 853$

5a $45{\cdot}6 \div 456$ **b** $456 \div 6450$ **c** $56{\cdot}8 \div 3860$
d $4{\cdot}06 \div 45{\cdot}8$ **e** $6{\cdot}78 \div 765$

6a $3{\cdot}67 \div 0{\cdot}965$ **b** $46{\cdot}7 \div 0{\cdot}987$ **c** $3{\cdot}08 \div 0{\cdot}0827$
d $45{\cdot}9 \div 0{\cdot}00\ 714$ **e** $456 \div 0{\cdot}0076$

7a $0{\cdot}937 \div 475$ **b** $0{\cdot}0285 \div 67{\cdot}9$ **c** $0{\cdot}0534 \div 8760$
d $0{\cdot}728 \div 938$ **e** $0{\cdot}00\ 843 \div 56{\cdot}9$

8a $0{\cdot}286 \div 0{\cdot}054$ **b** $0{\cdot}937 \div 0{\cdot}00\ 845$ **c** $0{\cdot}0746 \div 0{\cdot}00\ 584$
d $0{\cdot}0546 \div 0{\cdot}0076$ **e** $0{\cdot}00\ 734 \div 0{\cdot}000\ 765$

9a $0{\cdot}0145 \div 0{\cdot}735$ **b** $0{\cdot}00\ 628 \div 0{\cdot}654$ **c** $0{\cdot}0045 \div 0{\cdot}0325$
d $0{\cdot}0043 \div 0{\cdot}017$ **e** $0{\cdot}00\ 078 \div 0{\cdot}00\ 404$

Exercise 25.7

1 Find the area of a room 7·8 m by 0·67 m.

2 What is the cost of 6·65 m of cloth at £0·98 per metre?

3 22·5 m of cloth cost £18·90. What was the cost per metre?

4 What is the area of an airfield which is 1·52 km long and 0·037 km wide?

5 A pools win of £55 has to be shared between 73 people. How much will each person get?

6 What is the area of a rectangular field 0·032 km by 0·145 km?

7 A factory needs to cut 2670 pieces of wire, each 0·43 m long, off a roll of wire. How long will the roll need to be?

8 0·73 tonne of fertilizer has to be put into 45 bags. How much will each bag contain?

9 How many bags, each holding 0·33 kg, can be filled from 74·5 kg?

10 One litre of gas weighs 0·017 grams. Find the weight of 7·75 litres.

11 2·7 litres of a liquid weigh 0·975 kg. Find the weight of one litre.

12 25 cc of gas weigh 0·735 g. Find the weight of 1 cc.

13 Find the cost of 0·75 m of lace edging at £0·89 per m.

14 0·45 kg of fertilizer costs £0·78. What is the cost per kg?

15 How many flour bags, each holding 0·0032 tonne, can be filled from 0·735 tonne?

16 The distance around a metal bar is 0·00 365 m. What length of wire will be required to put 78 turns around the bar?

17 How many cups of tea can be bought for £0·78 if one cup costs £0·05½ (£0·055)?

18 Find the cost of 0·65 sq m of wood if 1 sq m costs £0·63.

19 A shop sells cartons of salad for £0·27. Each carton weighs 0·35 kg. What is the price per kg being charged by the shop?

20 25 m of metal strip costs £13·50.

a Find the cost of one metre.

b What will be the cost of 0·85 m?

Exercise 25.8

1a $1{\cdot}24 \times 5{\cdot}67 \times 8{\cdot}86$ **b** $34{\cdot}6 \times 2{\cdot}45 \div 4{\cdot}78$
c $1{\cdot}5^2$ **d** $\sqrt{4{\cdot}56}$

2a $23{\cdot}8 \times 78 \times 4{\cdot}89$ **b** $2{\cdot}78 \times 24{\cdot}3 \div 58{\cdot}2$
c $34{\cdot}6^2$ **d** $\sqrt{13{\cdot}8}$

3a $456 \times 56{\cdot}7 \times 8{\cdot}56$ **b** $56 \times 57 \div 45$
c $2{\cdot}45^3$ **d** $\sqrt[3]{475}$

4a $34{\cdot}4 \times 34{\cdot}7 \times 458$ **b** $24 \times 46 \times 78 \div 56{\cdot}8$
c $45{\cdot}7^3$ **d** $\sqrt[3]{56{\cdot}9}$

5a $3{\cdot}14 \times 4{\cdot}5 \times 4{\cdot}5$ **b** $46{\cdot}7 \div (3{\cdot}45 \times 3{\cdot}51)$
c $3{\cdot}45^5$ **d** $\sqrt{23{\cdot}6 \times 15{\cdot}4}$

6a $45{\cdot}6 \times 3{\cdot}56 \times 0{\cdot}0567$ **b** $23{\cdot}6 \times 4{\cdot}56 \div 0{\cdot}862$
c $0{\cdot}46^2$ **d** $\sqrt{0{\cdot}0478}$

7a $567 \times 0{\cdot}729 \times 0{\cdot}827$ **b** $35{\cdot}8 \times 0{\cdot}00\ 765 \div 2{\cdot}67$
c $0{\cdot}0567^2$ **d** $\sqrt{0{\cdot}456}$

8a $0{\cdot}678 \times 0{\cdot}567 \times 0{\cdot}483$ **b** $0{\cdot}78 \times 0{\cdot}67 \div 0{\cdot}564$
c $0{\cdot}45^3$ **d** $\sqrt[3]{0{\cdot}234}$

9a $3{\cdot}14 \times 0{\cdot}675 \times 0{\cdot}675$ **b** $0{\cdot}567 \times 896 \div 0{\cdot}348$
c $0{\cdot}9901^3$ **d** $\sqrt[3]{0{\cdot}0234}$

10a $3{\cdot}14 \times 2{\cdot}45 \times 0{\cdot}756$ **b** $0{\cdot}467 \div (7{\cdot}56 \times 0{\cdot}123)$
c $0{\cdot}95^6$ **d** $\sqrt[3]{0{\cdot}00\ 234}$

11a $\sqrt{2{\cdot}34 \times 0{\cdot}045 \times 46}$ **b** $\sqrt{45{\cdot}6 \div 37{\cdot}2}$
c $\sqrt[3]{0{\cdot}067 \div 0{\cdot}0134}$ **d** $\sqrt{56 \div 65}$

12a $3{\cdot}14 \times 4{\cdot}5^2 \times 5{\cdot}75$ **b** $1{\cdot}33 \times 3{\cdot}14 \times 4{\cdot}5^3$
c $4 \times 3{\cdot}14 \times 5{\cdot}65^2$ **d** $\sqrt{1 \div 7{\cdot}56}$

page 222/Table of squares and square roots

x	x^2	$\sqrt{x}$
1	1	1·00
2	4	1·41
3	9	1·73
4	16	2·00
5	25	2·24
6	36	2·45
7	49	2·65
8	64	2·83
9	81	3·00
10	100	3·16
11	121	3·32
12	144	3·46
13	169	3·61
14	196	3·74
15	225	3·87
16	256	4·00
17	289	4·12
18	324	4·24
19	361	4·36
20	400	4·47
21	441	4·58
22	484	4·69
23	529	4·80
24	576	4·90
25	625	5·00
26	676	5·10
27	729	5·20
28	784	5·29
29	841	5·39
30	900	5·48
31	961	5·57
32	1024	5·66
33	1089	5·74
34	1156	5·83
35	1225	5·92
36	1296	6·00
37	1369	6·08
38	1444	6·16

x	x^2	$\sqrt{x}$
39	1521	6·24
40	1600	6·32
41	1681	6·40
42	1764	6·48
43	1849	6·56
44	1936	6·63
45	2025	6·71
46	2116	6·78
47	2209	6·86
48	2304	6·93
49	2401	7·00
50	2500	7·07
51	2601	7·14
52	2704	7·21
53	2809	7·28
54	2916	7·35
55	3025	7·42
56	3136	7·48
57	3249	7·55
58	3364	7·62
59	3481	7·68
60	3600	7·75
61	3721	7·81
62	3844	7·87
63	3969	7·94
64	4096	8·00
65	4225	8·06
66	4356	8·12
67	4489	8·19
68	4624	8·25
69	4761	8·31
70	4900	8·37
71	5041	8·43
72	5184	8·49
73	5329	8·54
74	5476	8·60
75	5625	8·66
76	5776	8·72

x	x^2	$\sqrt{x}$
77	5929	8·77
78	6084	8·83
79	6241	8·89
80	6400	8·94
81	6561	9·00
82	6724	9·06
83	6889	9·11
84	7056	9·17
85	7225	9·22
86	7396	9·27
87	7569	9·33
88	7744	9·38
89	7921	9·43
90	8100	9·49
91	8281	9·54
92	8464	9·59
93	8649	9·64
94	8836	9·70
95	9025	9·75
96	9216	9·80
97	9409	9·85
98	9604	9·90
99	9801	9·95
100	10000	10·0

A	Sine of *A*	Tangent of *A*
0	0·00	0·00
1	0·02	0·02
2	0·04	0·04
3	0·05	0·05
4	0·07	0·07
5	0·09	0·09
6	0·11	0·11
7	0·12	0·12
8	0·14	0·14
9	0·16	0·16
10	0·17	0·18
11	0·19	0·19
12	0·21	0·21
13	0·23	0·23
14	0·24	0·25
15	0·26	0·27
16	0·28	0·29
17	0·29	0·31
18	0·31	0·33
19	0·33	0·34
20	0·34	0·36
21	0·36	0·38
22	0·38	0·40
23	0·39	0·42
24	0·41	0·45
25	0·42	0·47
26	0·44	0·49
27	0·45	0·51
28	0·47	0·53
29	0·49	0·55
30	0·50	0·58
31	0·52	0·60
32	0·53	0·63
33	0·55	0·65
34	0·56	0·68
35	0·57	0·70
36	0·59	0·73
37	0·60	0·75
38	0·62	0·78
39	0·63	0·81
40	0·64	0·84
41	0·66	0·87
42	0·67	0·90
43	0·68	0·93
44	0·70	0·97
45	0·71	1·00
46	0·72	1·04
47	0·73	1·07
48	0·74	1·11
49	0·76	1·15
50	0·77	1·19
51	0·78	1·23
52	0·79	1·28
53	0·80	1·33
54	0·81	1·38
55	0·82	1·43
56	0·83	1·48
57	0·84	1·54
58	0·85	1·60
59	0·86	1·66
60	0·87	1·73
61	0·88	1·80
62	0·88	1·88
63	0·89	1·96
64	0·90	2·05
65	0·91	2·14
66	0·91	2·25
67	0·92	2·36
68	0·93	2·48
69	0·93	2·61
70	0·94	2·75
71	0·95	2·90
72	0·95	3·08
73	0·96	3·27
74	0·96	3·49
75	0·97	3·73
76	0·97	4·01
77	0·97	4·33
78	0·98	4·70
79	0·98	5·14
80	0·99	5·67
81	0·99	6·31
82	0·99	7·12
83	0·99	8·14
84	1·00	9·51
85	1·00	11·4
86	1·00	14·3
87	1·00	19·1
88	1·00	28·6
89	1·00	57·3
90	1·00	—

Exercise R1

Teachers will find it useful to have a copy of Book 1 available to assist them in preparing lessons which involve the work in the revision exercises.

1 (4, 3)
2 (5, 2)
3 (3, 1)
4 (2, 2)
5 (1, 4)
6 (3, 4)
7 ($5\frac{1}{2}$, 4)
8 (2, 3)
9 (5, 1)
10 (1, 1)
11 (5, 0)
12 (1, 0)
13 (0, 2)
14 (2, $4\frac{1}{2}$)
15 (3, $^{-}3$)
16 (2, $^{-}2$), ($^{-}3$, 1)
17 (1, $^{-}3$)
18 ($^{-}4$, 4)
19 ($^{-}1$, 3)
20 ($^{-}2\frac{1}{2}$, $4\frac{1}{2}$)
21 ($^{-}2$, $3\frac{1}{2}$)
22 (4, 0), (4, $^{-}2$), (2, $^{-}5$), ($^{-}4$, 2), (0, $^{-}2$)
23 (4, $^{-}5$), ($^{-}2$, 2), ($^{-}4$, 0), ($^{-}6$, $^{-}2$)
24 (5, $^{-}2$), ($^{-}6$, 3), ($^{-}4$, $^{-}3$), ($^{-}4$, $^{-}5$)
25 ($^{-}4$, $^{-}2$)
26 ($^{-}2$, $^{-}3$)
27 ($^{-}5\frac{1}{2}$, 0)
28 ($^{-}5\frac{1}{2}$, $^{-}3$)
29 ($^{-}2\frac{1}{2}$, $^{-}3\frac{1}{2}$)
30 ($^{-}2$, $^{-}2$)

The usual method is by taking back bearings, but if the boys walk at say 5km per hour they will arrive at B after about 1hr 10mins.

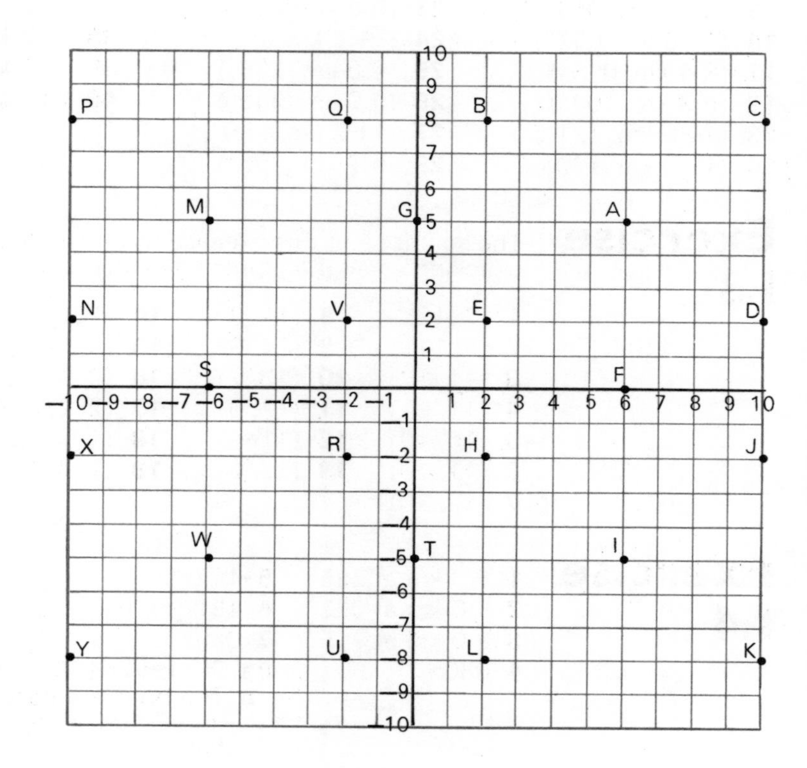

Exercise 1.1

Answers one degree above or below the answers given are acceptable.

A	012°	**F**	051°	**K**	090°	**Q**	150°	**V**	252°
B	024°	**G**	057°	**L**	102°	**R**	166°	**W**	288°
C	033°	**H**	065°	**M**	111°	**S**	199°	**X**	308°
D	039°	**I**	073°	**O**	120°	**T**	219°	**Y**	328°
E	042°	**J**	082°	**P**	135°	**U**	238°	**Z**	345°

Exercise 1.2

Numbers 1 to 10 give bearings between 0° and 90°, 11 to 20 bearings between 90° and 180°, 21 to 31 bearings between 180° and 270°, 32 to 40 bearings between 270° and 360°.

Answers one degree, or one tenth of a km, above or below the answers given are acceptable.

1 (7·8 km, 082°)
2 (9·7 km, 081°)
3 (5·3 km, 048°)
4 (7·0 km, 059°)
5 (8·7 km, 066°)
6 (6·0 km, 018°)
7 (6·9 km, 033°)
8 (8·2 km, 046°)
9 (11·3 km, 059°)
10 (11·8 km, 052°)
11 (8·4 km, 098°)
12 (9·3 km, 109°)
13 (10·0 km, 121°)
14 (11·0 km, 130°)
15 (6·6 km, 100°)
16 (7·6 km, 114°)
17 (8·5 km, 128°)
18 (9·6 km, 136°)
19 (5·7 km, 121°)
20 (5·3 km, 173°)
21 (5·9 km, 198°)
22 (7·7 km, 223°)
23 (9·6 km, 231°)
24 (11·6 km, 239°)
25 (6·8 km, 238°)
26 (9·0 km, 245°)
27 (11·2 km, 250°)
28 (5·7 km, 256°)
29 (8·0 km, 259°)
30 (10·4 km, 262°)
31 (8·3 km, 269°)
32 (8·4 km, 275°)
33 (9·0 km, 290°)
34 (9·9 km, 303°)
35 (4·9 km, 280°)
36 (5·9 km, 301°)
37 (7·2 km, 318°)
38 (4·2 km, 320°)
39 (6·0 km, 336°)
40 (3·1 km, 356°)

Exercise 1.3

The solutions of this exercise have been chosen for easy checking as all the points have integer co-ordinates.

1 (7, 12)
2 (9, 11)
3 (11, 12)
4 (11, 11)
5 (10, 10)
6 (11, 10)
7 (12, 10)
8 (11, 9)
9 (12, 9)
10 (13, 7)
11 (12, 5)
12 (11, 4)
13 (12, 2)
14 (10, 1)
15 (8, 1)
16 (3, 1)
17 (2, 2)
18 (1, 2)
19 (1, 4)
20 (1, 7)
21 (1, 8)
22 (0, 10)
23 (2, 10)
24 (3, 12)
25 (5, 12)

Exercise 1.4

1 6·0 km at 295°
2 5·0 km at 022°
3 9·0 km at 115°
4 080°
5 162°
6 151°
7 089°
8a 9·7 km at 108°
b The boat will pass over the second lot of hidden rocks and then hit the coast.

Exercise 2.1

The first three exercises deal only with positive numbers. The next three exercises repeat the work using negative numbers. If pupils need practice in adding negative numbers before starting Exercises 2.4 to 2.6, this will be found in Exercise 6.1

1 $\begin{pmatrix}2\\2\end{pmatrix}$ **6** $\begin{pmatrix}3\\3\end{pmatrix}$ **11** $\begin{pmatrix}2\\5\end{pmatrix}$ **16** $\begin{pmatrix}5\\0\end{pmatrix}$

2 $\begin{pmatrix}5\\2\end{pmatrix}$ **7** $\begin{pmatrix}5\\3\end{pmatrix}$ **12** $\begin{pmatrix}2\\2\end{pmatrix}$ **17** $\begin{pmatrix}3\\2\end{pmatrix}$

3 $\begin{pmatrix}2\\5\end{pmatrix}$ **8** $\begin{pmatrix}8\\3\end{pmatrix}$ **13** $\begin{pmatrix}2\\0\end{pmatrix}$ **18** $\begin{pmatrix}3\\5\end{pmatrix}$

4 $\begin{pmatrix}5\\5\end{pmatrix}$ **9** $\begin{pmatrix}3\\2\end{pmatrix}$ **14** $\begin{pmatrix}5\\0\end{pmatrix}$ **19** $\begin{pmatrix}0\\3\end{pmatrix}$

5 $\begin{pmatrix}10\\2\end{pmatrix}$ **10** $\begin{pmatrix}3\\5\end{pmatrix}$ **15** $\begin{pmatrix}0\\5\end{pmatrix}$ **20** $\begin{pmatrix}0\\0\end{pmatrix}$

21 to **31**

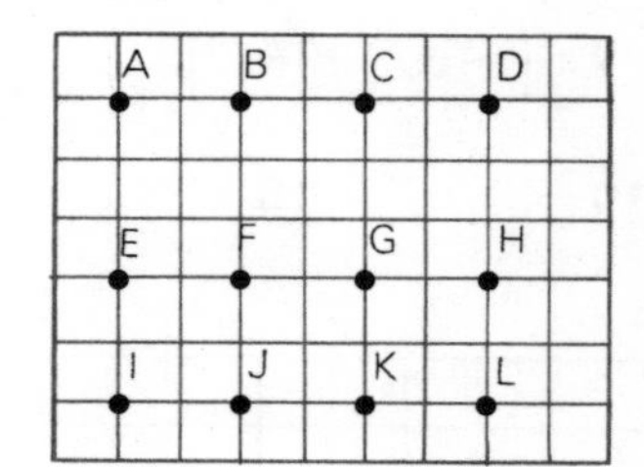

Exercise 2.2

1 $\begin{pmatrix}2\\3\end{pmatrix} + \begin{pmatrix}4\\2\end{pmatrix}$

2 $\begin{pmatrix}3\\2\end{pmatrix} + \begin{pmatrix}2\\3\end{pmatrix}$

3 $\begin{pmatrix}2\\2\end{pmatrix} + \begin{pmatrix}3\\1\end{pmatrix} + \begin{pmatrix}1\\2\end{pmatrix}$

4 $\begin{pmatrix}3\\1\end{pmatrix} + \begin{pmatrix}3\\2\end{pmatrix} + \begin{pmatrix}1\\2\end{pmatrix}$

5 $\begin{pmatrix}3\\0\end{pmatrix} + \begin{pmatrix}0\\3\end{pmatrix}$

6 $\begin{pmatrix}0\\4\end{pmatrix} + \begin{pmatrix}4\\0\end{pmatrix}$

7 $\begin{pmatrix}3\\0\end{pmatrix} + \begin{pmatrix}3\\2\end{pmatrix} + \begin{pmatrix}0\\2\end{pmatrix}$

8 $\begin{pmatrix}2\\2\end{pmatrix} + \begin{pmatrix}2\\0\end{pmatrix} + \begin{pmatrix}2\\2\end{pmatrix} + \begin{pmatrix}2\\1\end{pmatrix}$

Exercise 2.3

1 $\begin{pmatrix}5\\3\end{pmatrix}$ **3** $\begin{pmatrix}6\\5\end{pmatrix}$ **5** $\begin{pmatrix}6\\5\end{pmatrix}$ **7** $\begin{pmatrix}3\\4\end{pmatrix}$ **9** $\begin{pmatrix}7\\5\end{pmatrix}$

2 $\begin{pmatrix}4\\5\end{pmatrix}$ **4** $\begin{pmatrix}7\\5\end{pmatrix}$ **6** $\begin{pmatrix}5\\2\end{pmatrix}$ **8** $\begin{pmatrix}4\\4\end{pmatrix}$

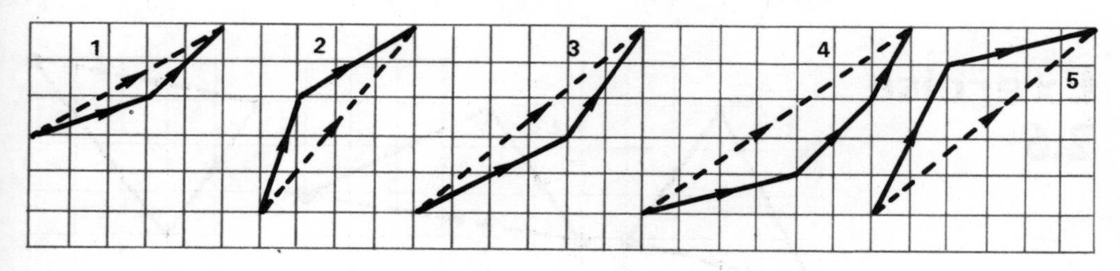

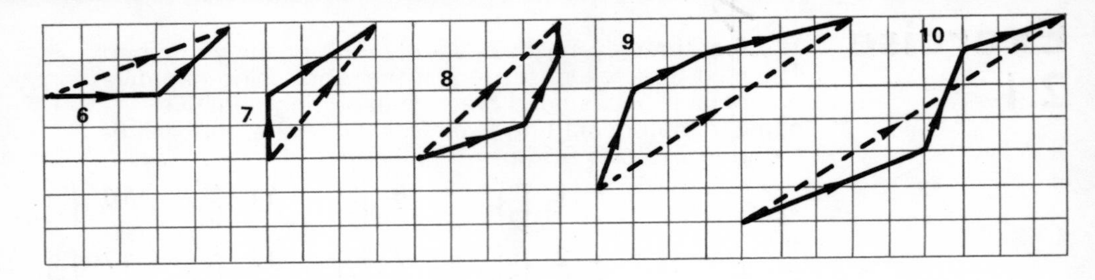

Exercise 2.4

1 $\begin{pmatrix} 1 \\ -2 \end{pmatrix}$

2 $\begin{pmatrix} 3 \\ -3 \end{pmatrix}$

3 $\begin{pmatrix} 6 \\ -2 \end{pmatrix}$

4 $\begin{pmatrix} -5 \\ 3 \end{pmatrix}$

5 $\begin{pmatrix} -4 \\ 5 \end{pmatrix}$

6 $\begin{pmatrix} -3 \\ 2 \end{pmatrix}$

7 $\begin{pmatrix} -3 \\ -1 \end{pmatrix}$

8 $\begin{pmatrix} -5 \\ -2 \end{pmatrix}$

9 $\begin{pmatrix} -4 \\ -4 \end{pmatrix}$

10 $\begin{pmatrix} 2 \\ 0 \end{pmatrix}$

11 $\begin{pmatrix} 5 \\ -1 \end{pmatrix}$

12 $\begin{pmatrix} -1 \\ -3 \end{pmatrix}$

13 $\begin{pmatrix} -5 \\ 0 \end{pmatrix}$

14 $\begin{pmatrix} -1 \\ 5 \end{pmatrix}$

15 $\begin{pmatrix} -2 \\ 0 \end{pmatrix}$

16 $\begin{pmatrix} 5 \\ -4 \end{pmatrix}$

17 $\begin{pmatrix} 0 \\ -4 \end{pmatrix}$

18 $\begin{pmatrix} -2 \\ 3 \end{pmatrix}$

19 $\begin{pmatrix} 1 \\ -5 \end{pmatrix}$

20 $\begin{pmatrix} 5 \\ 0 \end{pmatrix}$

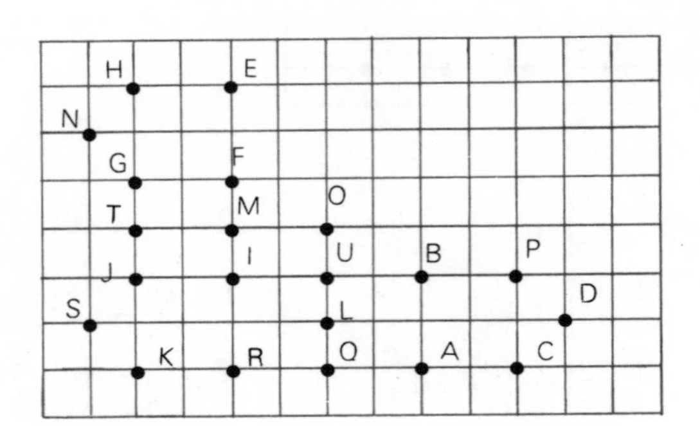

Exercise 2.5

1 $\begin{pmatrix} 4 \\ -2 \end{pmatrix} + \begin{pmatrix} 3 \\ 3 \end{pmatrix}$

2 $\begin{pmatrix} 3 \\ -2 \end{pmatrix} + \begin{pmatrix} 1 \\ -4 \end{pmatrix}$

3 $\begin{pmatrix} -1 \\ 3 \end{pmatrix} + \begin{pmatrix} 3 \\ 2 \end{pmatrix}$

4 $\begin{pmatrix} -1 \\ -4 \end{pmatrix} + \begin{pmatrix} 3 \\ -2 \end{pmatrix}$

5 $\begin{pmatrix} 3 \\ 1 \end{pmatrix} + \begin{pmatrix} -4 \\ 3 \end{pmatrix}$

6 $\begin{pmatrix} 3 \\ 0 \end{pmatrix} + \begin{pmatrix} 2 \\ -3 \end{pmatrix}$

7 $\begin{pmatrix} 2 \\ 3 \end{pmatrix} + \begin{pmatrix} 2 \\ -5 \end{pmatrix}$

8 $\begin{pmatrix} 3 \\ 2 \end{pmatrix} + \begin{pmatrix} 2 \\ -3 \end{pmatrix} + \begin{pmatrix} 1 \\ 3 \end{pmatrix}$

9 $\begin{pmatrix} 4 \\ -2 \end{pmatrix} + \begin{pmatrix} 5 \\ 3 \end{pmatrix} + \begin{pmatrix} -1 \\ -3 \end{pmatrix}$

Exercise 2.6

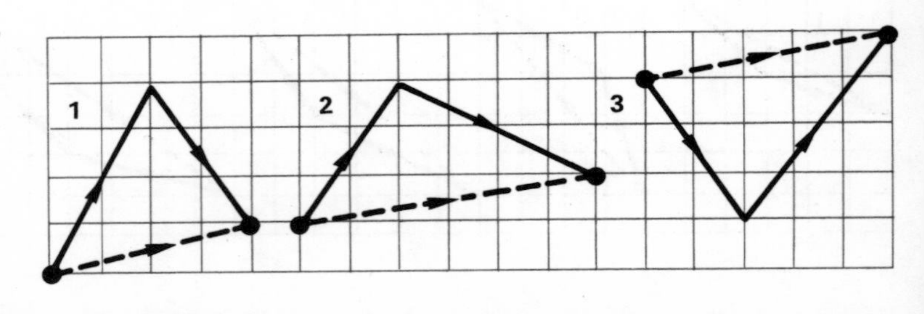

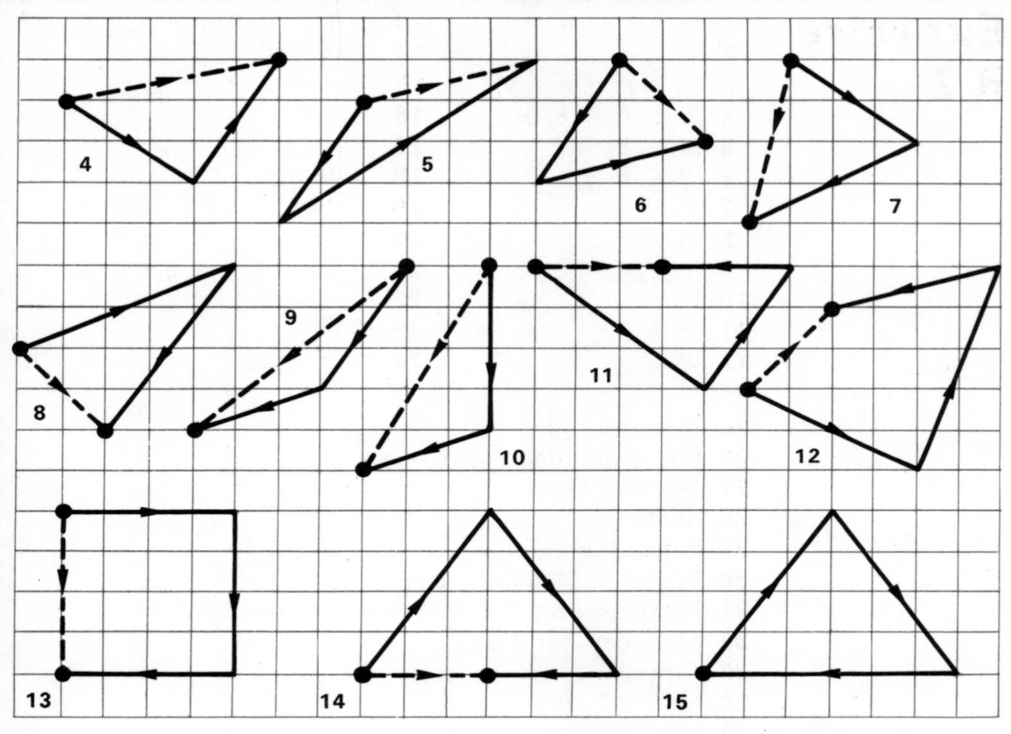

1 $\begin{pmatrix}4\\1\end{pmatrix}$ **4** $\begin{pmatrix}5\\1\end{pmatrix}$ **7** $\begin{pmatrix}-1\\-4\end{pmatrix}$ **10** $\begin{pmatrix}-3\\-5\end{pmatrix}$ **13** $\begin{pmatrix}0\\-4\end{pmatrix}$

2 $\begin{pmatrix}6\\1\end{pmatrix}$ **5** $\begin{pmatrix}4\\1\end{pmatrix}$ **8** $\begin{pmatrix}2\\-2\end{pmatrix}$ **11** $\begin{pmatrix}3\\0\end{pmatrix}$ **14** $\begin{pmatrix}3\\0\end{pmatrix}$

3 $\begin{pmatrix}5\\1\end{pmatrix}$ **6** $\begin{pmatrix}2\\-2\end{pmatrix}$ **9** $\begin{pmatrix}-5\\-4\end{pmatrix}$ **12** $\begin{pmatrix}2\\2\end{pmatrix}$

Exercise 2.7

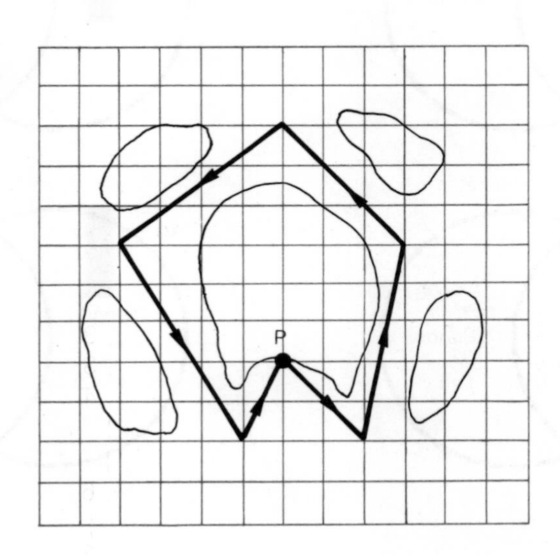

Exercise R 2

1 $\{a, b, c, d, e, f\}$
2 $\{a, b, c, d\}$
3 $\{a, b, c, d, e, f, g\}$
4 $\{3, 5, 6, 7, 8, 9\}$
5 $\{4, 6, 8\}$
6 $\{3, 4, 5, 6\}$
7 $\{p, r\}$
8 $\{q, r\}$
9 $\{r, t\}$
10 $\{\ \}$
11 $\{3, 5, 7\}$
12 $\{8\}$
13 $\{2, 3, 4, 5, 6, 7, 8, 9, 11\}$
14 $\{2, 3, 4, 6, 8\}$
15 $\{3, 4, 5, 7, 9, 10, 11\}$
16 $\{3\}$
17 $\{4\}$
18 $\{4, 6, 8\}$
19 $\{\ \}$
20 $\{2, 3, 4, 6, 7, 8, 10\}$
21 $\{3, 4, 5, 6, 7, 8, 9, 10, 11\}$
22 $\{2, 3, 4, 6, 8\}$
23 $\{3, 7\}$
24 $\{3, 4, 7, 10\}$

25 Roy, Ron, Alan
26 Roy, Ron, Alan, Peter, John
27 Michael, Simon
28 Michael, Simon, Peter, John
29 Peter, John
30 Roy, Ron, Alan, Michael, Simon
31 People, News, Globe, Times, Graphic, Sun
32 Mr Green, Mr Brown
33 Times, Graphic
34 Mr Smith. 6
35 15p
36 48p

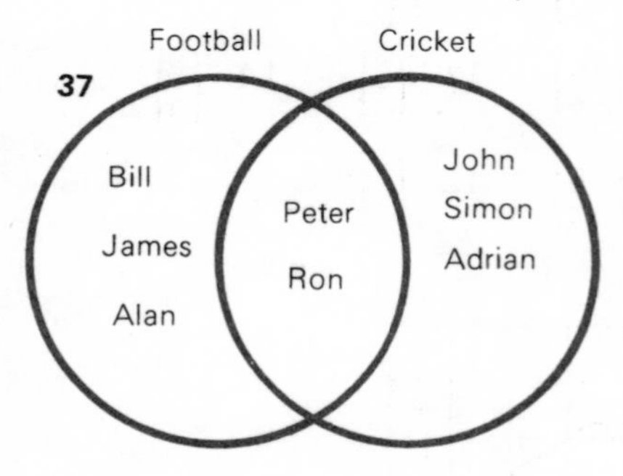

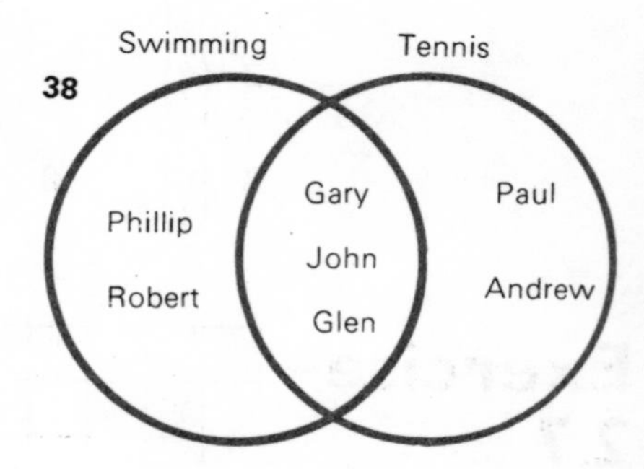

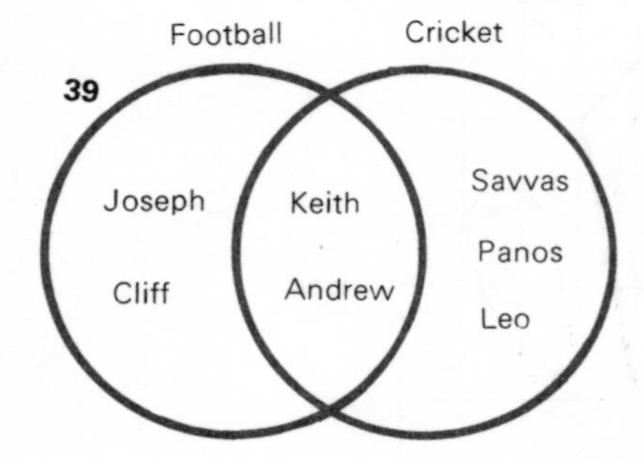

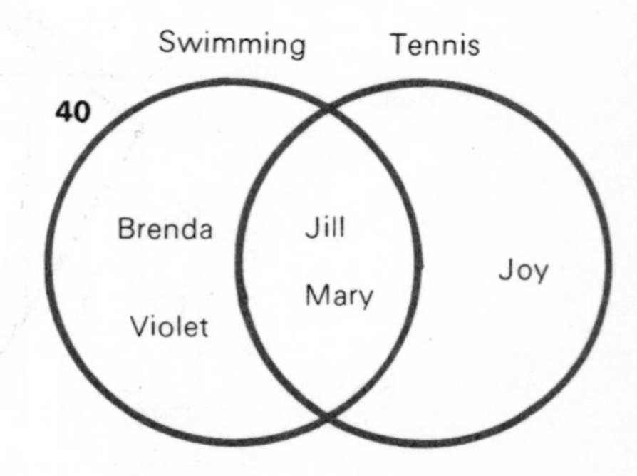

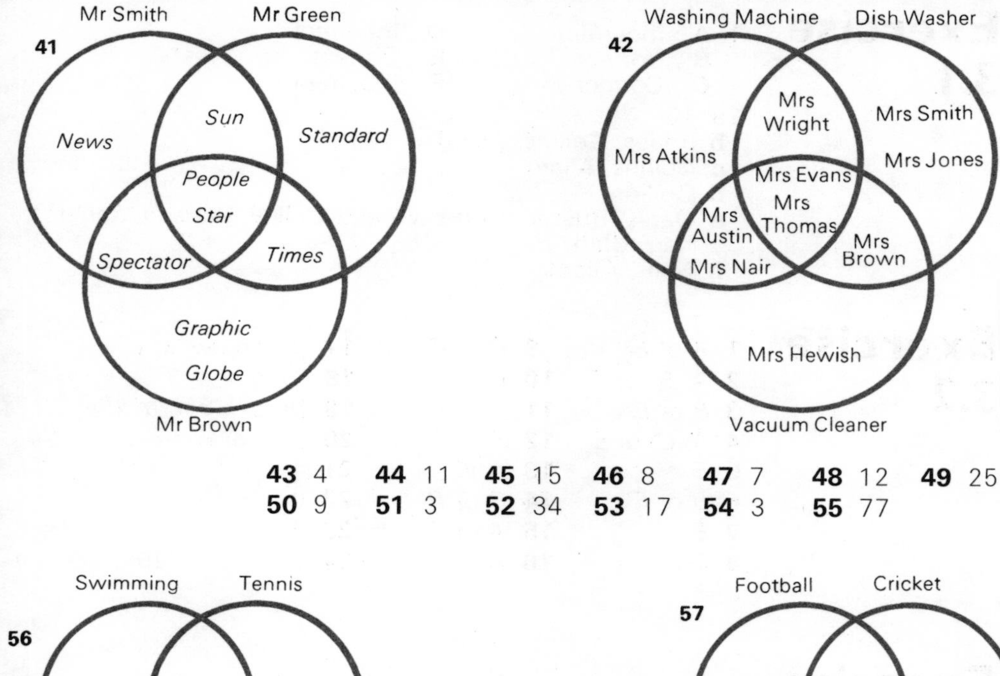

43 4 **44** 11 **45** 15 **46** 8 **47** 7 **48** 12 **49** 25
50 9 **51** 3 **52** 34 **53** 17 **54** 3 **55** 77

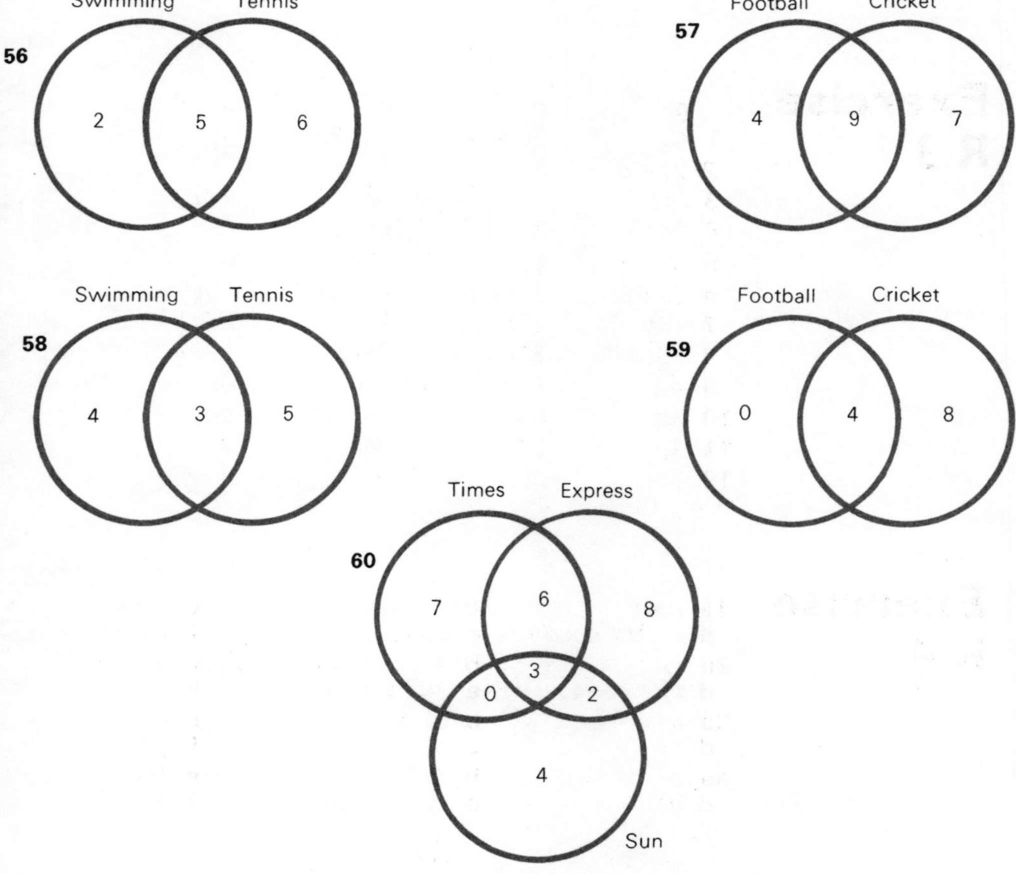

Exercise 3.1

1 A Incorrect
B Correct
C Correct
D Incorrect
E Correct
F Incorrect

2a 16
b Jones, Collins, Allard
c Collins, Allard
d Collins
e Oates, Russell, Curtis, Weeks, Colley, Watson, Jones, Kerr, Clark
f Mills, Allard

Exercise 3.2

1 A or E
2 A, B, or E
3 B or E
4 B, C, or E
5 E
6 B or E
7 E
8 E
9 B or C
10 C
11 C or A
12 A
13 A, B, or C
14 B or C
15 A or C
16 A
17 $\{t, u, v, w, x, y\}$
18 $\{p, q, x, y\}$
19 $\{p, q, r, s, t, u, x, y\}$
20 $\{\ \}$ or ϕ
21 4
22 6
23 2
24 10
25 $\{\ \}$ or ϕ

Exercise R 3

	a	b	c	d
1	$4a$	$2p+2q$	$2l+2m$	$6z$
2	$2c+d$	$3c+2d$	N	$15f$
3	PQ	CD	$6g$	$4AB$
4	$2d$	N	$6s$	0
5	N	$11z$	$10t$	N
6	$7s+3t$	$3m+3n$	$w+5x$	$10y$
7	$12b$	$20t$	$9pq$	PRQ
8	$28LM$	$21hm$	$35ab$	$12mn$
9	$4w$	$4a$	$5p$	$3z$
10	$3b$	$6u$	$4y$	$4q$
11	7	24	N	7
12	$3bc$	N	$3h$	4

Exercise R 4

1a $b \times b$ **b** $c \times c \times c$ **c** $a \times a$
d $d \times d \times d \times d \times d$ **e** $e \times e \times e$ **f** $v \times v \times v \times v$

2a $t \times t$ **b** $w \times w \times w \times w \times w$ **c** q
d $t \times t \times t \times t \times t \times t$ **e** $z \times z \times z \times z$ **f** s

3a a^3 **b** c^4 **c** j^3
d n^2 **e** h^5 **f** y^3

4a b^4 **b** A^5 **c** D^2
d e^3 **e** t^2 **f** P^6

5a 9	**b** 3	**c** 27
d 16	**e** 8	**f** 4
6a 2	**b** 81	**c** 5
d 25	**e** 32	**f** 125
7a 5	**b** 25	**c** 65
d 61	**e** 9	**f** 5
8a 33	**b** 15	**c** 32
d 141	**e** 6	**f** 29
9a 25	**b** 64	**c** 4
d 1	**e** 8	**f** 1
10a 512	**b** 5	**c** 89
d 68	**e** 21	**f** 24
11a 2	**b** 256	**c** 14
d 92	**e** 25	**f** 8
12a L^2	**b** S^2	
13c a^3	**d** b^3	

Exercise 4.1

1 a^6	**4** d^4	**7** g^4	**10** h^5	**13** m^{12}
2 b^7	**5** e^8	**8** g^4	**11** j^{15}	**14** n^6
3 c^9	**6** f^8	**9** h^5	**12** k^{21}	**15** p^{33}

Exercise 4.2

The pupil should realise that $x^0 = 1$.

1 a^3	**4** d^4	**7** g^3	**10** m^1 or m	**13** 1 or q^0
2 b^4	**5** e^3	**8** h^5	**11** n^1 or n	**14** 1 or r^0
3 c^4	**6** f^3	**9** k^4	**12** 1 or p^0	**15** 1 or r^0

Exercise 4.3

1 $10a^7$	**8** $7a^6b^4$	**15** $\frac{6}{5a^3}$
2 $18a^5$	**9** $3a^3$	**16** $2a^4b^1$ or $2a^4b$
3 $20a^8$	**10** $8a^2$	**17** $6a^3b^4$
4 $21a^5$	**11** $6a^2$	**18** $\frac{4d^3}{e^5}$
5 $6a^5b^4$	**12** $4b^1$ or $4b$	**19** $\frac{4f^1}{g^7}$ or $\frac{4f}{g^7}$
6 $20a^7b^7$	**13** $7a^3$	**20** $\frac{5}{a^2b^3}$
7 $14a^7b^{10}$	**14** $\frac{12a^5}{5}$	

Exercise 4.4

1 $3d^3$	**6** $9f^3$	**11** N	**16** $12ab^6$
2 $4f^2$	**7** N	**12** N	**17** $10ab^2$
3 N	**8** N	**13** $7ab^2$	**18** N
4 $2d^2$	**9** $14w^4$	**14** $6a^2b^2$	**19** $4a^2+4b^2$
5 N	**10** $8g^2$	**15** N	**20** 0

Exercise 4.5

	a	b	c
1	a^2	12	$12a^2$
2	s^2	8	$8s^2$
3	d^2	5	$5d^2$
4	t^2	14	$14t^2$
5	c^2	6	$6c^2$
10	L^3	10	$10L^3$
11	d^3	36	$36d^3$
12	c^3	15	$15c^3$

6 $8a^2$
7 $15b^2$
8 $3OL^2$
9 $2D^2$

Exercise R 5

1 $D=4S$
2 $D=3L$
3 $D=2L+2W$
4 $D=2a+b$
5 $D=2S+2T$
6 $D=b+2s+t$
7 $D=2p+2q+4r$
8 $D=4a$
9 $D=h+2s$
10 $W=8S+4L$
11 $W=4x+4y+4z$
12 $W=4L+4S$
13 $W=6E+3L$
14 $B=N-G$, $N=B+G$
15 $T=A+B$, $A=T-B$
16 $G=J+B$, $J=G-B$
17 $T=S+2A$, $S=T-2A$
18 $T=BM$, $B=T\div M$
19 $C=T\div P$, $T=CP$
20 $C=Q\div 10$, $Q=10C$
21a $T=M+W$ **b** $W=T-M$ **c** $M=T-W$
22a $T=B+S$ **b** $B=T-S$ **c** $S=T-B$
23a $T=BC$ **b** $C=T\div B$ **c** $B=T\div C$
24 $P=AC+BL$
25 $BC+DE$
26 $BC+2BC=3BC$

Exercise 5.1

A suitable treatment for less able pupils would be Exercise 5.1, and the examples in the first part of Exercise 5.2.

	a	b	c	d	e
1	27	15	30	10	16
2	56	9	44	32	28
3	80	30	30	44	16
4	8	36	56	40	18
5	35	21	20	32	24
6	18	12	20	55	42
7	5	3	8	3	5·5
8	2·5	10	20	18	16
9	17·5	17·5	22·5	10·5	9
10	49	6	40	16	63
11	18	60	20	36	0
12	5	5	2	3	5·5
13	20	8	24	14	24·5
14	10·5	10·5	100	40	15
15	120	60	39	100	83·7

Exercise 5.2

1a 14 cm **b** 18 cm **c** 28 cm
d 38 cm **e** 44 cm **f** 58 cm

2a 27 cm **b** 39 cm **c** 54 cm
d 63 cm **e** 93 cm

3a 40 sq m **b** 54 sq m **c** 72 sq m
d 100 sq m

4a 36 m **b** 48 m **c** 52 m
d 60 m

5a 20 sq m **b** 24 sq m **c** 56 sq m
d 31·5 sq m **e** 58·5 sq m

6a 48 sq cm **b** 60 sq cm **c** 117 sq cm
d 135 sq cm **e** 240 sq cm

7a 3024 g **b** 3780 g **c** 2016 g
d 3136 g **e** 16 380 g **f** 18 928 g

8a 28·26 cm **b** 40·82 cm **c** 56·52 cm
d 65·94 cm **e** 97·34 cm

9a 50·24 sq cm **b** 62·80 sq cm **c** 122·48 sq cm
d 141·30 sq cm **e** 251·20 sq cm

Exercise 5.3

1a $4p+4q$ **b** $5a+5b$
c $2x+2y$ **d** $7m+7n$

2a $3a-3b$ **b** $6t-6s$
c $4f-4e$ **d** $5p-5q$

3a $3a+3b+3c$ **b** $2r+2s+2t$
c $7x+7y-7z$ **d** $5b-5d+5c$

4a b^2+bd **b** $rs-rt$
c $ef+e^2$ **d** $ap+aq$

5a $7x+14$ **b** $3d+9$
c $2a+6b$ **d** $4a-12$

6a $2v-4w$ b $2a+4b-6c$
c $4s-4t$ d $12+6f$

7a $2a^2+16ab$ b $6va+12vb+18vc$
c $abd-2abe$ d $5z^2+25zx$

8a $6a^2+3ab$ b $10tc-25tn$
c $3act+7acr$ d $4w-4we-12wf$

Exercise 5.4

	a	b	c	d
1	$3(a+b)$	$6(p+q)$	$4(x+y)$	$5(m+n)$
2	$7(x-y)$	$2(s-t)$	$8(w-v)$	$3(t-s)$
3	$4(a+b+c)$	$9(p+q+r)$	$4(d+e-f)$	$5(a+b+c)$
4	$a(b+d)$	$s(t+r)$	$d(e+d)$	$a(a-e)$
5	$5(a+1)$	$5(a+2)$	$5(a+2b)$	$3(x+4)$
6	$x(x+3)$	$2(x+2y+4z)$	$x(x-4)$	Not possible
7	$2a(b+2)$	$6a(1-2b)$	$2a(b+c)$	$n(3-8m)$
8	$y(x+y)$	Not possible	Not possible	$3(s-2t+r)$
9	Not possible	$2(a+2b+3c)$	$a(b-4a)$	Not possible
10	$a(1-b+c)$	Not possible	Not possible	Not possible

Exercise 5.5

1 $D=2(a+b)$
2 $D=2(p+q)$
3 $D=2(x+y)$
4 $D=2(c+d)$
5 $D=2(e+f)$
6 $D=2(s+t)$
7 $D=2(f+g+h)$
8 $D=2(L+2W)$
9 $D=2(3a+b)$
10 $D=2(d+2c)$
11 $D=4(h+l+w)$
12 $D=4(l+2s)$
13 $D=2(2a+b+c)$
14 $D=4(a+b)$
15 $T=5(B+G)$
16 $T=25(P-N)$
17 $T=F(M+W)$
18 $T=P(Y-Z)$
19 $T=9(d+b)$
20 $T=m(X+Y)$
21 $D=2(L+W)$
22 $T=P(A-B)$
23 a $B+G$ b $T=7(B+G)$
24 a $6M$ b $6W$ c $T=6(M+W)$
25 a AC b BC c $D=C(A+B)$
26 a KL b JL c $M=L(K-J)$
27 a $P+Q+R$ b $T=S(P+Q+R)$
28 a $H+G$ b $P+T$, $C=(H+G)(P+T)$
29 a $P-N$ b $C+M$ c $T=(P-N)(C+M)$
30 a $P-R$ b $C+T$ c $Y=(P-R)(C+T)$
31 a NA b NB c $T=N(A+B)$
32 a $W-2S$ b $L-2S$ c $A=(W-2S)(L-2S)$
33 a $12r$ b $6p$ c $T=6(2r+p)$
34 a $6B$ b $12P$ c $8R$ d $T=2(3B+6P+4R)$
35 $4(L+W+H)$

Exercise R 6

Numbers 1 to 4 are either sequences where each term is obtained by adding or subtracting a fixed number from the previous term. e.g. 2(a) and 4(c); or where each term is obtained by multiplying or dividing the previous term by a fixed number. e.g. 1(b) and 3(d).
Many of the remainder can be done by looking at the pattern in the differences between the terms. For example 6(c) gives

4 7 13 22 34
3 6 9 12

which can be continued to give

4 7 13 22 34 49 67
3 6 9 12 15 18

	a	b	c	d
1	11, 13	32, 64	35, 40	81, 243
2	36, 41	48, 96	28, 35	1000, 10 000
3	6, 3	31, 38	8, 5	2, 1
4	44, 41	324, 972	6·8, 6·1	1·5, 0·5
5	17, 23	36, 46	15, 21	35, 44
6	30, 42	13, 7	49, 67	3, 2
7	77, 107	29, 18	10, 13	150·1, 210·1
8	⁻21, ⁻46	102, 137	4, 4	11, 5

9 **a** $5 \times 5 \times 5$ **b** $180 = 5 \times 6 \times 6$ **c** 245 **d** 320
e 605 **f** 4805

Exercise R 7

	a	b	c	d	e	f	g
1	⁻2	⁻2	2	⁻4	⁻6	⁻2	⁻2
2	⁻1	0	4	⁻1	⁻4	⁻8	⁻18
3	4	⁻5	0	⁻12	5	⁻7	⁻10
4	1	⁻9	⁻1	3	⁻13	⁻3	9
5	⁻3	⁻15	5	⁻1	⁻17	⁻23	⁻5

6 £8 overdraft **7** £4 **8** £2 overdraft **9** £2
10 £15 overdraft

The first five exercises are aimed at giving practice in the manipulation of negative numbers, whereas Exercise 6.6 gives examples on the interpretation of negative answers. For pupils who will not need to be able to manipulate negative numbers, or who will find this too difficult, a suitable treatment of negative numbers would be Exercise R7 followed by Exercise 6.6

Exercise 6.1

	a	b	c	d	e	f
1	3	6	3	0	⁻1	⁻3
2	⁻5	⁻3	⁻7	⁻3	5	6

3 3	3	−1	−6	−3	−5
4 −9	−10	−11	−10	−10	−9
5 3	−7	−9	−1	−28	−8

6 £30 **7** £9 overdraft **8** £50 overdraft **9** £0 **10** £0

Exercise 6.2

	a	b	c	d	e	f
1	17	14	10	10	10	12
2	5	3	4	6	5	0
3	−3	−6	−7	−5	−2	−6
4	16	10	21	5	−18	−20
5	−4	−2	−4	20	−3	−3

Exercise 6.3

	a	b	c	d	e	f
1	−12	6	−24	−15	−21	14
2	14	−18	24	−2	24	30
3	−30	8	−8	8	18	0
4	−24	24	−24	16	16	36
5	−90	−32	60	1	−27	−8

Exercise 6.4

	a	b	c	d	e	f
1	−8	−3	6	7	−6	−6
2	−6	−7	−8	−7	−3	7
3	8	3	6	8	8	3
4	−6	−2	−6	−10	10	−6
5	−2	−15	−12	−24	2·666	−1

Exercise 6.5

	a	b	c	d	e	f
1	16	−4	−20	−21	−6	−15
2	35	1	−1	−3	6	8
3	3	−3	14	−5	4	−3
4	4	−8	17	−3	8	−24
5	105	74	−24	−8	−23	−14

Exercise 6.6

1 **a** 24 m per sec
b 10 m per sec
c 0 m per sec
d It is at the top of its flight.
e −20 m per sec
f It is now falling at 20 m per sec.

2 **a** 105 m
b 120 m
c 125 m
d 120 m
e 105 m
f It is rising, and then falling.
g 0 m
h It has reached the ground again.
i −55 m
j It has fallen 55 m below the edge of the cliff.

3 **a** 5 m **b** 10 m **c** 15 m **d** 40 tonnes
e 0 m **f** $^{-}$3m **g** The counterweight is 3 m the other side of C. **h** 5 m the other side of C.

Exercise R 8

	a	b	c	d	e
1	36	20	7	0	81
2	13	3	1	9	10
3	9	16	6	12	13
4	5	1	−5	2	7
5	38	11	−17	4	31
6	0	5	14	−4	3
7	36	8	2	4	0
8	108	32	18	14	90
9	3	8·5	1·5	5·5	7
10	2·75	6·5	$1\frac{1}{9}$=1·111	7·5	3·333
11	£30	£60	£40	£50	£210
12	£22	£42	£62	£102	£182
13	18 sq m	36 sq m	60 sq m	36 sq m	231 sq m
14	36 sq m	40 sq m	126 sq m	286 sq m	—
15	216 g	540 g	5400 g	315 g	2835 g

Exercise R 9

1 **a** 20°C **b** 17°C **c** 8 a.m. **d** 12 a.m. and 6 p.m. **e** 21°C **f** 15°C

2 **a** 200 km **b** 100 km **c** 0 km **d** 1 p.m. and 11 p.m. **e** 200 km and 200 km **f** At rest **g** At rest **h** 50 km per hour **i** 12 a.m. and 2 p.m., 100 km per hr

3 **a** 7·5 cm **b** 22·5 cm **c** 5 inches **d** 7 inches

4 **a** £1·50 **b** £7·50 **c** 12 m **d** 28 m **e** £4·12$\frac{1}{2}$ **f** £6·37$\frac{1}{2}$ **g** 26·7 m

0	8	16	24	32
0	3	6	9	12

Exercise 7.1 gives practice in drawing and reading a straight line graph given the equation of the line. Exercise 7.2 repeats this with curved lines. Exercise 7.3 contains more difficult examples on straight line graphs.

Exercise 7.1

1

0	2	4	6	8
0	8	16	24	32

a 12 cm **b** 5 cm **c** 28 cm **d** $3\frac{1}{2}$ cm **e** 12 cm
f $4\frac{1}{2}$ cm

2

0	1	2	3	4	5	6	7
0	5	10	15	20	25	30	35

a $12\frac{1}{2}$ km **b** $17\frac{1}{2}$ km **c** $1\frac{1}{2}$ hr **d** $4\frac{1}{2}$ hr **e** $27\frac{1}{2}$ km
f $6\frac{1}{2}$ hr

3

0	2	4	6	8	10
0	5	10	15	20	25

a $12\frac{1}{2}$ cm **b** $17\frac{1}{2}$ cm **c** 3·2 inches **d** 4·8 inches
e $7\frac{1}{2}$ cm **f** 7·2 inches

4

0	2	4	6	8
0	6·2	12·4	18·6	24·8

a 15·5 cm **b** 9·3 cm **c** 4·84 cm **d** 4·19 cm
e 21·7 cm **f** 5·16 cm

0	2	4	6	8
0	6·28	12·56	18·84	25·12

a 15·7 cm **b** 9·42 cm **c** 4·78 cm **d** 4·14 cm
e 22·0 cm **f** 5·10 cm

5

0	2	4	6	8	10
0	10	20	30	40	50

a 4·4 **b** 3·6 **c** 8·6

Exercise 7.2

1

0	1	2	3	4	5	6	7	8	9	10
0	1	4	9	16	25	36	49	64	81	100

a 20·25 sq cm **b** 56·25 sq cm **c** 42·25 sq cm
d 6·71 cm **e** 8·37 cm **f** 5·48 cm **g** 72·25 sq cm
h 6·32 cm

2

0	1	2	3	4	5
0	3	12	27	48	75

a 6·75 sq cm **b** 18·75 sq cm **c** 60·75 sq cm
d 4·09 cm **e** 4·47 cm **f** 5·32 cm

3

0	1	2	3	4	5
0	3·1	12·4	27·9	49·6	77·5

0	1	2	3	4	5
0	3·14	12·56	28·26	50·24	78·50

a 7·00 sq cm **b** 19·4 sq cm **c** 62·8 sq cm
d 4·01 cm **e** 4·40 cm **f** 5·23 cm

a 7·07 sq cm **b** 19·6 sq cm **c** 63·6 sq cm
d 3·99 cm **e** 4·37 cm **f** 5·21 cm

4

0	1	2	3	4	5	6
0	5	20	45	80	125	180

a 31·25 m **b** 101·25 m **c** 5·48 sec **d** 4·47 sec
e 61·25 m

5

0	1	4	9	16	25	36	49	64	81	100
0	1	2	3	4	5	6	7	8	9	10

a 4·47 **b** 5·48 **c** 8·37 **d** 5·92 **e** 7·42 **f** 8·66
g 6·24 **h** 7·28

6

2	3	4	6	9	12
18	12	9	6	4	3

a 2·4 **b** 3·27 **c** 5·14 **d** 7·20 **e** 4·5 **f** 3·60

Exercise 7.3

1

0	2	4	6	8	10
10	20	30	40	50	60

a £25 **b** £55 **c** £47·50 **d** 1t **e** 3t **f** 4·6t

2

0	2	4	6	8	10
4	10	16	22	28	34

a £13 **b** £31 **c** £26·50 **d** 3·67t **e** 7·00t
f 9·67t

3

0	2	4	6	8	10
7	15	23	31	39	47

a £19 **b** £43 **c** £37 **d** 2t **e** $4\frac{1}{2}t$ **f** $6\frac{1}{2}t$

4

0	400	800	1200
4	5	6	7

a £5·50 £5·75 **b** 200, 1080

5

0	400	800	1200
0	4	8	12

a £2, £7 **b** 300, 850 **c** 534 **d** 1p
e The cost per leaflet depends on the number bought.
f Offset, 0·65p.

Exercise R 10

1 6 sq cm, 10 cm
2 20 sq cm, 18 cm
3 63 sq cm, 32 cm
4 161 sq cm, 60 cm
5 24 sq m, 22 m
6 26 sq m, 26 m
7 48 sq m, 38 m
8 56 sq m, 40 m
9 3 sq cm
10 10 sq cm
11 15 sq cm
12 21 sq cm
13 31·5 sq cm
14 32·5 sq cm
15 36 sq cm
16 50 sq cm

Costs of carpeting for numbers **5–8** above:
5 £72, £104·88 **6** £78, £113·62 **7** £144, £209·76
8 £168, £244·72

Exercise 8.1

1 20 sq cm
2 18 sq cm
3 15 sq cm
4 35 sq cm
5 16 sq cm
6 40 sq cm
7 16·5 sq cm
8 27·5 sq cm
9 58·5 sq cm
10 85·5 sq cm
11 59·6 sq cm
12 46·25 sq cm

Exercise 8.2

1 8 sq cm
2 15 sq cm
3 35 sq cm
4 60 sq cm
5 77 sq cm
6 117 sq cm
7 19·2 sq cm
8 22·72 sq cm
9 7 cm, 56 sq cm
10 5 cm, 30 sq cm
11 11 cm, 99 sq cm

Exercise 8.3

1 **a** 12 sq m
b 32 sq m
c 44 sq m
2 **a** 30 sq m
b 80 sq m
c 110 sq m
3 **a** 18 sq m
b 40 sq m
c 58 sq m
4 **a** 32 sq m
b 60 sq m
c 92 sq m
5 **a** 10 sq cm
b 12 sq cm
c 22 sq cm
6 **a** 24·5 sq cm
b 45·5 sq cm
c 70 sq cm
7 **a** 10 sq cm
b 18 sq cm
c 28 sq cm
8 **a** 5 tins
b £3·75
9 **a** 13 tins
b £9·75
10 **a** 7 tins
b £5·25
11 **a** 11 tins
c £8·25

Exercise R11

Questions 1 to 12. All answers in metres.

1 3, 3·1, 3·14
2 6, 6·2, 6·28
3 9, 9·3, 9·42
4 12, 12·4, 12·56
5 15, 15·5, 15·70
6 18, 18·6, 18·84
7 21, 21·7, 21·98
8 24, 24·8, 25·12
9 27, 27·9, 28·26
10 30, 31, 31·4
11 39, 40·3, 40·82
12 87, 89·9, 91·06

Questions 13 to 24. All answers in square metres.

13 3, 3·1, 3·14
14 12, 12·4, 12·56
15 27, 27·9, 28·26
16 48, 49·6, 50·24
17 75, 77·5, 78·5
18 108, 111·6, 113·04
19 147, 151·9, 153·86
20 192, 198·4, 200·96
21 243, 251·1, 254·34
22 300, 310, 314·00
23 432, 446·4, 452·16
24 6075, 6277·5, 6358·50

25 **a** 2 m
b 6 m, 6·2 m, 6·28 m
c £4·50, £4·65, £4·71
d 1 m
e 3 sq m, 3·1 sq m, 3·14 sq m
f £12, £12·40, £12·56

26 **a** 4 m
b 12 m, 12·4 m, 12·56 m
c £9, £9·30, £9·42
d 2 m
e 12 sq m, 12·4 sq m, 12·56 sq m
f £48, £49·60, £50·24

27 **a** 6 m
b 18 m, 18·6 m, 18·84 m
c £13·50, £13·95, £14·13
d 3 m
e 27 sq m, 27·9 sq m, 28·26 sq m
f £108, £111·60, £113·04

28 **a** 10 m
b 30 m, 31 m, 31·4 m
c £22·5, £23·25, £23·55
d 5 m
e 75 sq m, 77·5 sq m, 78·5 sq m
f £300, £310, £314

29 **a** 26 m
b 78 m, 80·6 m, 81·64 m
c £58·50, £60·45, £61·23
d 13 m
e 507 sq m, 523·9 sq m, 530·66 sq m
f £2028, £2095·6, £2122·64

Exercise R 12

1 24 cubic cm, 52 sq cm
2 120 cubic cm, 140 sq cm
3 90 cubic m, 126 sq m
4 504 cubic m, 382 sq m
5 210 cubic cm, 214 sq cm
6 336 cubic m, 292 sq m
7 350 cubic cm, 310 sq cm
8 196 cubic m, 217 sq m

Exercise 9.1

1 12 cubic m
2 45 cubic m
3 135 cubic m
4 200 cubic m
5 63 cubic m
6 180 cubic m
7 324 cubic m
8 640 cubic m
9 480 cubic m
10 105 cubic m
11 80 cubic m
12 224 cubic m
13 220 cubic m
14 168 cubic m
15 396 cubic m

Exercise 9.2

1 72, 74·4, 75·36 cubic m
2 135, 139·5, 141·3 cubic m
3 288, 297·6, 301·44 cubic m
4 300, 310, 314 cubic m
5 540, 558, 565·2 cubic m
6 1701, 1757·7, 1780·38 cubic m
7 525, 542·5, 549·5 cubic cm
8 756, 781·2, 791·28 cubic cm
9 735, 759·5, 769·3 cubic cm
10 1944, 2008·8, 2034·72 cubic cm
11 4719, 4876·3, 4939·22 cubic cm

Exercise 9.3

1 40 cubic m
2 28 cubic m
3 56 cubic m
4 46·66 cubic m
5 93·33 cubic m
6 93·33 cubic m
7 12, 12·4, 12·56 cubic m
8 36, 37·2, 37·68 cubic m
9 96, 99·2, 100·48 cubic m
10 45, 46·5, 47·1 cubic m
11 20, 20·66, 20·93 cubic m
12 112, 115·7, 117·2 cubic m
13 128, 132·2, 133·9 cubic m
14 250, 258·3, 261·6 cubic m

Exercise 9.4

1 108, 111·6, 113·04 cubic m
2 864, 892·8, 904·32 cubic m
3 2916, 3013·2, 3052·08 cubic m
4 32, 33·06, 33·49 cubic m
5 256, 264·5, 267·9 cubic m
6 4000, 4133, 4186 cubic m
7 1372, 1417, 1436 cubic m
8 2048, 2116, 2143 cubic m

Exercise 9.5

A
1 96, 99·2, 100·4 sq m
2 144, 148·8, 150·72 sq m
3 240, 248, 251·2 sq m
4 270, 279, 282·6 sq m
5 396, 409·2, 414·48 sq m
6 864, 892·8, 904·32 sq m
7 360, 372, 376·8 sq cm
8 468, 483·6, 489·84 sq cm
9 504, 520·8, 527·52 sq cm
10 918, 948·6, 960·84 sq cm
11 1584, 1636·8, 1657·92 sq cm

B
1 108, 111·6, 113·04 sq m
2 432, 446·4, 452·1 sq m
3 972, 1004·4, 1017·36 sq m
4 48, 49·6, 50·24 sq m
5 192, 198·4, 200·96 sq m
6 1200, 1240, 1256 sq m
7 588, 607·6, 615·44 sq m
8 768, 793·6, 803·84 sq m

Exercise 9.6

1 36, 37·2, 37·68 sq m
2 60, 62, 62·8 sq m
3 105, 108·5, 109·9 sq m
4 144, 148·8, 150·72 sq m

From Ex 9.3. Answers to **b** are in sq m.

	7	**8**	**9**	**10**
a	3·61 m	9·22 m	7·21 m	5·83 m
b	21·66	55·32	86·52	52·47
	22·38	57·16	89·40	54·21
	22·67	57·90	90·55	54·91
	11	**12**	**13**	**14**
a	5·39 m	8·06 m	8·94 m	11·2 m
b	32·34	96·72	107·2	168·0
	33·41	99·94	110·8	173·6
	33·84	101·2	112·2	175·8

Exercise 9.7

1 **a** 136 sq cm **b** 400 cm **c** 54 400 cubic cm
d 435 200 g **e** 435·2 kg
2 **a** 840 sq cm **b** 4 200 000 cubic cm **c** 25 200 000 g
d 25 200 kg
3 **a** 72 sq m **b** 5760 kg **c** 48 cubic m **d** 96*t*
a 74·4 sq m **b** 5952 kg **c** 49·6 cubic m **d** 99·2*t*
a 75·36 sq m **b** 6028 kg **c** 50·24 cubic m
d 100·48*t*

4 **a** 180 sq m **b** 20 tins **c** £14·60
a 186 sq m **b** 21 tins **c** £15·33
a 188·4 sq m **b** 21 tins **c** £15·33
5 **a** 16 cubic m **b** 96*t* **c** 4·47 m **d** 26·82 sq m
e 1877 kg
a 16·53 cubic m **b** 99·18*t* **c** 4·47 m
d 27·71 sq m **e** 1939 kg
a 16·74 cubic m **b** 100·44*t* **c** 4·47 m
d 28·07 sq m **e** 1964 kg
6 **a** 5324 cubic m **b** 10 648*t* **c** 1452 sq m
d 182 tins **e** £223·86
a 5501 cubic m **b** 11 002*t* **c** 1500 sq m
d 188 tins **e** £231·24
a 5572 cubic m **b** 11 144*t* **c** 1519 sq m
d 190 tins **e** £233·70
7 **a** 221 sq cm **b** 4862 cubic cm

Exercise R 13

	a	b	c	d	e	f
1	30_4	22_5	20_6	15_7	14_8	13_9
2	101_3	22_4	20_5	14_6	13_7	12_8
3	100_4	31_5	24_6	22_7	20_8	17_9
4	2211_3	1030_4	301_5	204_6	136_7	114_8
5	3312_4	1441_5	1050_6	501_7	366_8	303_9
6	8	17	9	12	9	5
7	9	7	11	15	13	10
8	14	26	27	15	25	14
9	71	31	46	24	31	69
10	137	303	98	131	1650	2427
11	7	8	5	6	9	11
12	10	13	12	14	15	16
13	27	23	21	33	87	119
14	1110	1101	10001	10111	11011	11111
15	100001	101011	110000	110101	111110	1000000
16	1000001	1000100	1001010	1011001	1101100	11110101

Exercise R 14

	a	b	c	d	e
1	$a=9$	$b=5$	$c=1$	$d=11$	$e=11$
2	$q=20$	$r=19$	$s=18$	$t=21$	$u=32$
3	$a=6$	$b=9$	$c=7$	$d=12$	$e=12$
4	$q=27$	$r=31$	$s=37$	$t=47$	$u=55$
5	$a=4$	$b=9$	$c=3$	$d=9$	$e=7$

6	$q=7$	$r=9$	$s=10$	$t=8$	$u=6$
7	$a=6$	$b=6$	$c=24$	$d=20$	$e=10$
8	$q=100$	$r=99$	$s=120$	$t=90$	$u=104$
9	$a=7$	$u=4$	$n=7$	$r=3$	$t=2$
10	$b=3$	$m=5$	$p=3$	$r=3$	$h=11$
11	$a=1{\cdot}5$	$b=5{\cdot}4$	$c=4{\cdot}166$	$d=2{\cdot}857$	$e=3{\cdot}875$
12	$p=2{\cdot}5$	$q=4{\cdot}666$	$r=5{\cdot}2$	$s=2{\cdot}375$	$t=4{\cdot}777$
13	$v=5{\cdot}4$	$w=5{\cdot}571$	$x=11{\cdot}75$	$y=7{\cdot}625$	$z=0{\cdot}6153$
14	$a=-3$	$b=-5$	$c=18{\cdot}5$	$d=1{\cdot}5$	$e=-1$

15 3 m
16 7 m
17 9 m
18 8.75 m

19 6 kg, 12 kg, 30 kg
20 4 kg, 8 kg, 16 kg
21 4 kg, 12 kg, 20 kg
22 40 g, 60 g, 100 g
23 200 g, 300 g, 400 g

24 20 km per hr
25 50 km per hr
26 50 km per hr
27 80 km per hr
28 69·4 km per hr
29 44·53 km per hr

The answers in Exercise 10.2 are all whole numbers. Exercise 10.3 can be omitted if necessary. The work in Exercise 10.5 is necessary for some of the work in Units 11 and 13.

Exercise 10.1

	a	b	c	d	e
1	289	529	4225	3481	784
2	313	2554	867	1111	89
3	1730	121	1331	169	28 561

Exercise 10.2

	a	b	c	d
1	4	3	1	5
2	9	2	7	8
3	11	10	6	0
4	17	13	12	16
5	23	20	15	14
6	30	21	18	19
7	53	58	76	86

Exercises 10.3 & 10.4

	a	b	c	d	e	f
1	2·24	2·65	2·83	3·32	4·12	4·24
2	4·80	5·20	5·48	5·83	6·08	6·48
3	6·71	7·87	8·54	9·43	9·59	9·90

Exercise 10.5

Two solutions are given in question 5. The first is obtained by correcting up to the nearest whole number before taking the square root, the second is correct to three significant figures.

1 **a** $z=4$ **b** $s=8$ **c** $r=9$
d $d=5$ **e** $f=4$

2 **a** $y=3{\cdot}61$ **b** $b=5{\cdot}39$ **c** $x=7{\cdot}48$
d $m=8{\cdot}19$ **e** $n=9{\cdot}64$

3 **a** $a=3$ **b** $b=4$ **c** $c=2$
d $d=5$ **e** $c=7$

4 **a** $p=3{\cdot}16$ **b** $q=2{\cdot}45$ **c** $r=2{\cdot}83$
d $s=4{\cdot}47$ **e** $t=3{\cdot}32$

5 **a** $x=2{\cdot}24$, $2{\cdot}28$ **b** $y=3$, $2{\cdot}92$ **c** $z=2{\cdot}83$, $2{\cdot}89$
d $w=3{\cdot}46$, $3{\cdot}45$ **e** $v=3{\cdot}32$, $3{\cdot}33$

6 **a** $a=4$ **b** $b=6$ **c** $c=8$
d $d=7$ **e** $e=3$

7 **a** $p=4$ **b** $q=5$ **c** $r=6$
d $s=7$ **e** $t=10$

8 **a** $v=4{\cdot}90$ **b** $w=4{\cdot}36$ **c** $x=7{\cdot}14$
d $y=8{\cdot}06$ **e** $z=6{\cdot}86$

A suitable treatment of the work in this unit for the less able pupils is to do Exercises 11.1(a) to (k), and Exercise 11.2 (Uusing $\pi=3$) 1(a) to 1(f), 2(a) to 2(c).

Exercise 11.1

Numbers (a) to (k) have whole number answers.

1 **a** 3 m **b** 4 m **c** 5 m **d** 6 m **e** 9 m **f** 8 m
g 9 m **h** 21 m **i** 11 m **j** 12 m **k** 11 m **l** 4·666 m
m 2·5 m **n** 6·5 m **o** 5·5 m **p** 2 m **q** 2·355 m

2 5·5 m
3 4·285 m
4 5·208 m

In the answers following three values are given because of the three values of π that can be used.

Exercise 11.2

If 3 is used as the value for π the answers to 1(a) to 1(f) and 2(a) to 2(c) are whole numbers.

1 **a** 2 m, 1·935 m, 1·91 m
b 5 m, 4·838 m, 4·777 m
c 6 m, 5·806 m, 5·732 m
d 7 m, 6·774 m, 6·687 m
e 8 m, 7·741 m, 7·643 m
f 9 m, 8·709 m, 8·598 m
g 2·333 m, 2·258 m, 2·229 m
h 3·666 m, 3·548 m, 3·503 m
i 3·333 m, 3·225 m, 3·184 m

2 a 11 cm, 10·64 cm, 10·50 cm
b 14 cm, 13·54 cm, 13·37 cm
c 24 cm, 23·22 cm, 22·92 cm
d 26·66 cm, 25·80 cm, 25·47 cm
e 83·33 cm, 80·64 cm, 79·61 cm
3 a 240 m **b** 80 m, 77·41 m, 76·43 m
4 a 270 m **b** 90 m, 87·09 m, 85·98 m
5 a 260 m **b** 86·66 m, 83·87 m, 82·80 m
6 a 200 m **b** 66·66 m, 64·51 m, 63·69 m

Exercise 11.3

Practice in solving the equations used in this Exercise will be found in Exercise 10.5. Using $\pi=3$ numbers 1 to 10 give whole number answers.

1 2 m, 1·97 m, 1·95 m
2 3 m, 2·95 m, 2·93 m
3 5 m, 4·92 m, 4·89 m
4 6 m, 5·90 m, 5·87 m
5 8 m, 7·87 m, 7·82 m
6 7 m, 6·88 m, 6·84 m
7 9 m, 8·85 m, 8·80 m
8 10 m, 9·84 m, 9·77 m
9 11 m, 10·8 m, 10·8 m
10 12 m, 11·8 m, 11·7 m
11 3·16 cm, 3·11 cm, 3·09 cm
12 3·74 cm, 3·67 cm, 3·66 cm
13 4·47 km, 4·40 km, 4·37 km
14 3·87 cm, 3·81 cm, 3·78 cm
15 4·87 cm, 4·79 cm, 4·74 cm
16 2·02 km, 1·99 km, 1·98 km

Exercise 11.4

Using $\pi=3$ numbers (a) to (j) give whole number answers.

a 3 m, 2·908 m, 2·866 m
b 2 m, 1·935 m, 1·910 m
c 4 m, 3·870 m, 3·821 m
d 3 m, 2·903 m, 2·866 m
e 2 m, 1·935 m, 1·910 m
f 5 m, 4·838 m, 4·777 m
g 6 m, 5·806 m, 5·732 m
h 2 m, 1·935 m, 1·910 m
i 3 m, 2·903 m, 2·866 m
j 6 m, 5·806 m, 5·732 m
k 3·24 m, 3·135 m, 3·095 m
l 2·48 m, 2·4 m, 2·370 m
m 6·563 m, 6·351 m, 6·270 m

Exercise 11.5

Using $\pi=3$ numbers (a) to (g) give whole number answers.

a 2 m, 1·97 m, 1·95 m
b 3 m, 2·95 m, 2·93 m
c 2 m, 1·97 m, 1·95 m
d 4 m, 3·94 m, 3·91 m
e 4 m, 3·94 m, 3·91 m
f 5 m, 4·92 m, 4·89 m
g 5 m, 4·92 m, 4·89 m
h 2·83 m, 2·78 m, 2·76 m
i 3·16 m, 3·11 m, 3·09 m
j 3·87 m, 3·81 m, 3·78 m
k 4·69 m, 4·62 m, 4·58 m
l 6·89 m, 6·78 m, 6·74 m

Exercise 12.5 is rather more difficult than the previous exercises. It shows that some equations in clock arithmetic have either two solutions or no solutions, rather than the one solution expected.

Exercise 12.1

1 5 p.m. **2** 4 p.m. **3** 3 p.m. **4** 3 p.m. **5** 2 a.m. **6** 2 a.m. **7** 3 a.m. **8** 9 a.m.

Exercise 12.2

1 9 a.m. **2** 9 a.m. **3** 10 a.m. **4** 6 a.m. **5** 11 p.m. **6** 7 p.m. **7** 4 p.m. **8** 6 p.m.

Exercise 12.3

1 9 a.m. **2** 8 p.m.

	3	**4**	**5**	**6**	**7**	**8**	**9**	**10**	**11**
Modulo 12	6	9	5	8	4	5	2	2	0
Modulo 24	18	21	17	20	16	17	2	2	0
Modulo 10	8	1	3	6	2	3	6	6	0

	12	**13**	**14**	**15**	**16**	**17**	**18**
Modulo 12	9	4	6	11	3	1	6
Modulo 24	21	4	6	11	3	13	6
Modulo 10	7	4	2	7	3	9	0

Exercise 12.4

1 9 **2** 3 **3** 4 **4** 6 **5** 0 **6** 4 **7** 6 **8** 0 **9** 4 **10** 2 **11** 1 **12** 6 **13** 4 **14** 0 **15** 7 **16** 6

Exercise 12.5

1

0 0 0 0 0 0
0 1 2 3 4 5
0 2 4 0 2 4
0 3 0 3 0 3
0 4 2 0 4 2
0 5 4 3 2 1

2 2
3 $a=3$
4 $b=4$
5 $c=2$ or 5
6 $d=1$ or 4
7 $e=1$ or 4
8 No solution
9 No solution

Exercise R 15

Questions 7 and 8 show the pupils that it is possible to enlarge figures either by multiplying the matrix of their co-ordinates by the scale factor, or by a geometrical construction. (To do it by construction make OA′=2OA etc in 7, and OQ′=3OQ etc in question 8)

1 Congruent
2 Congruent. They are mirror images.
3 Congruent. They are mirror images.
4 Not congruent

5 Not congruent

6 Congruent

7 **d** A′(6, 8), B′(2, 8), C′(2, 16)

e $\begin{bmatrix} 6 & 8 \\ 2 & 8 \\ 2 & 16 \end{bmatrix}$

8 **d** P′(6, 3), Q′(12, 3), R′(6, 9), S′(12, 9)

e $\begin{bmatrix} 6 & 3 \\ 12 & 3 \\ 6 & 9 \\ 12 & 9 \end{bmatrix}$

9 Scale factor = 2, $a = 14$ cm

10 Scale factor = 3, $b = 12$ cm, $c = 15$ cm

11 Scale factor = 5, $d = 5$ cm, $e = 5$ cm, $f = 15$ cm

12 Scale factor = 4, $g = 12$ cm, $h = 4$ cm, $i = 4$ cm

13 Scale factor = $\frac{1}{2}$, $j = 3$ cm, $k = 4$ cm

14 Scale factor = $\frac{1}{3}$, $l = 3$ m, $m = 8$ m, $n = 6$ m, $p = 4$ m

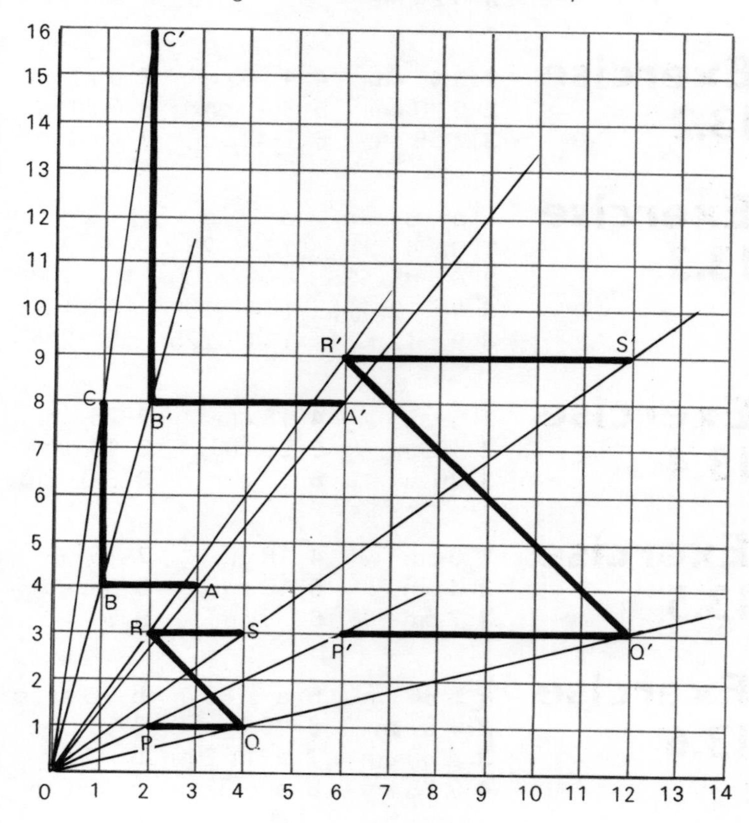

Exercise R 16

A=60°	H=145°	P=313°	W=45°	d=91°
B=63°	I=80°	Q=66°	X=70°	e=20°
C=30°	J=57°	R=108°	Y=67°	f=112°
D=53°	K=31°	S=72°	Z=70°	g=128°
E=67°	L=108°	T=72°	a=71°	h=75°
F=33°	M=298°	U=135°	b=20°	i=103°
G=120°	N=50°	V=45°	c=77°	j=68°

Exercises 13.1 and 13.2 involve the solution by scale drawings of problems usually done by Pythagoras. The less able pupils need only do this part of the Unit. Exercise 13.3 leads the pupil to discover the Rule of Pythagoras, and the rest of the Unit deals with calculations involving Pythagoras. Before Exercise 13.5 is done some pupils might need to revise the work in Exercise 10.5 numbers 6, 7, 8. Exercise 13.6 also provides additional work for the less able pupils to do by scale drawing.

Exercise 13.1

1 6·40 cm **5** 6·32 cm **8** 7·28 m **11** 4·47 cm
2 5·66 cm **6** 3·61 m **9** 4·24 m **12** 101 m
3 7·62 cm **7** 11·7 m **10** 10·4 m **13** 7·65 m
4 9·49 cm

Exercise 13.2

1 3·46 cm **4** 4·90 cm **7** 6·71 m **10** 4·47 m
2 5·20 cm **5** 3·32 cm **8** 6·24 m **11** 9·17 m
3 2·65 cm **6** 7·42 cm **9** 98·0 m **12** 86·2 m

Exercise 13.3

	a	b	c	d	e	f	g	h
1	3	4	9	16	25	5	25	The same
2	6	8	36	64	100	10	100	The same
3	5	12	25	144	169	13	169	The same
4	9	12	81	144	225	15	225	The same

Exercise 13.4

1 5 cm **4** 13 cm **7** 25 cm **10** 5·39 cm
2 10 cm **5** 25 cm **8** 26 cm **11** 6·32 cm
3 20 cm **6** 17 cm **9** 3·61 cm **12** 8·60 cm

Exercise 13.5

1 6 cm **4** 18 cm **7** 15 cm **10** 5·20 cm
2 4 cm **5** 10 cm **8** 9 cm **11** 6·32 cm
3 7 cm **6** 36 cm **9** 2·24 cm **12** 6·93 cm

Exercise 13.6

1 5·66 cm **5 a** 2·83 m **b** 10 m **c** 12·66 m **d** £4·05
2 7·81 m **6** 20·6 m, £2·27
3 4·58 cm **7 a** 11·2 m **b** 44·8 m **c** £4·03
4 7·28 km **8 a** 6·32 m **b** 4·90 m

This unit introduces the tangent and the sine. The cosine is not introduced as problems involving right angled triangles can be solved without its use. The second section of the Unit (page 126) is a repetition of the first part with the problems solved by calculation. The more able pupils could start here, while the less able pupils could first solve the problems by scale drawing (starting at the beginning of the Unit). With those pupils who are not capable of solving these problems by calculation the first part of the Unit together with Exercise 14.9 is sufficient. Although the tables in the book only give the sine and tangent of whole numbers of degrees, the answers are given to the nearest tenth of a degree in case the pupils have access to 3 figure tables.

Exercise 14.1

	1	2	3	4	5	6
A	26·5°	76·0°	36·9°	51·3°	36·9°	59·0°
B	63·5°	14·0°	53·1°	38·7°	53·1°	31·0°

7 16·7° **9** 11·3° **11** 66·8° **13** 14·0°
8 33·7° **10** 33·7° **12** 20·0° **14** 66·4°

Exercise 14.2

	1	2	3	4	5	6	7	8
A	36·9°	23·6°	48·6°	53·1°	60·0°	55·2°	68·0°	63·6°
B	53·1°	66·4°	41·4°	36·9°	30·0°	34·8°	22·0°	26·4°

9 14·5° **11** 53·1° **13** 14·5° **15** 51·0°
10 23·6° **12** 23·6° **14** 19·4° **16** 16·6°

Exercise 14.3

1 5·19 cm **5** 6·16 cm **9** 8·88 m **13** 301·6 m
2 4·76 cm **6** 6·42 cm **10** 84·0 m **14** 244·0 m
3 5·00 cm **7** 2·596 cm **11** 114·40 m **15** 460·0 km
4 4·524 cm **8** 2·120 cm **12** 76·80 m **16** 273·35 km

Exercise 14.4

	1	2	3	4
a	1·168 cm	1·875 cm	2·00 cm	4·092 cm
b	3·824 cm	4·635 cm	3·464 cm	4·386 cm
	5	**6**	**7**	**8**
a	2·650 cm	5·936 cm	5·346 cm	7·568 cm
b	4·240 cm	3·710 cm	2·724 cm	2·608 cm
	9	**10**	**11**	**12**
a	3·180 m	16·56 m	195·5 m	684·6 m
b	5·088 m	57·66 m	460·5 m	145·6 m
	13	**14**	**15**	
a	2·65 m	11·25 m	360·5 m	
b	4·24 m	48·70 m	599·9 m	

Exercise 14.5

Same answers as Exercise 1.

Exercise 14.6

Same answers as Exercise 2.

Exercise 14.7

First part has same answers as Exercise 3.

1 64°, 10·25 cm
2 47°, 7·49 cm
3 38°, 6·248 cm
4 27°, 4·590 cm
5 66°, 9·00 cm
6 51°, 11·07 cm
7 37°, 6·032 cm
8 12°, 1·491 cm

Exercise 14.8

Same answers as Exercise 4.

Exercise 14.9

1 A=38·7°, B=51·3°
2 A=51·5°, B=38·5°
3 A=39·2°, B=50·8°
4 a=0·9752 cm
5 2·524 cm
6 a=4·788 cm, b=5·511 cm
7 A=37·8°, B=52·2°
8 A=68·0°, B=22·0°
9 A=45·2°, B=44·8°
10 a=5·664 cm
11 a=13·40 cm
12 a=3·913 cm, b=5·193 cm

Exercise 15.1

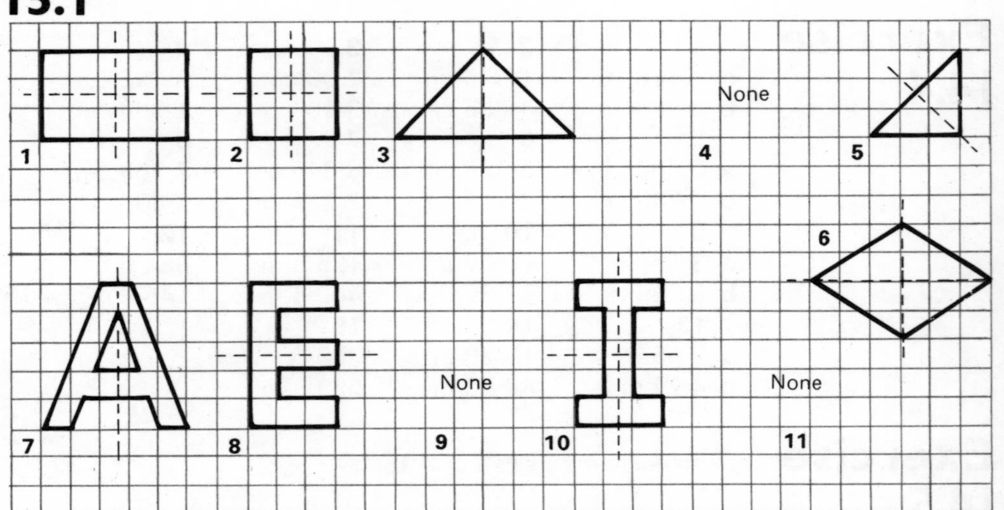

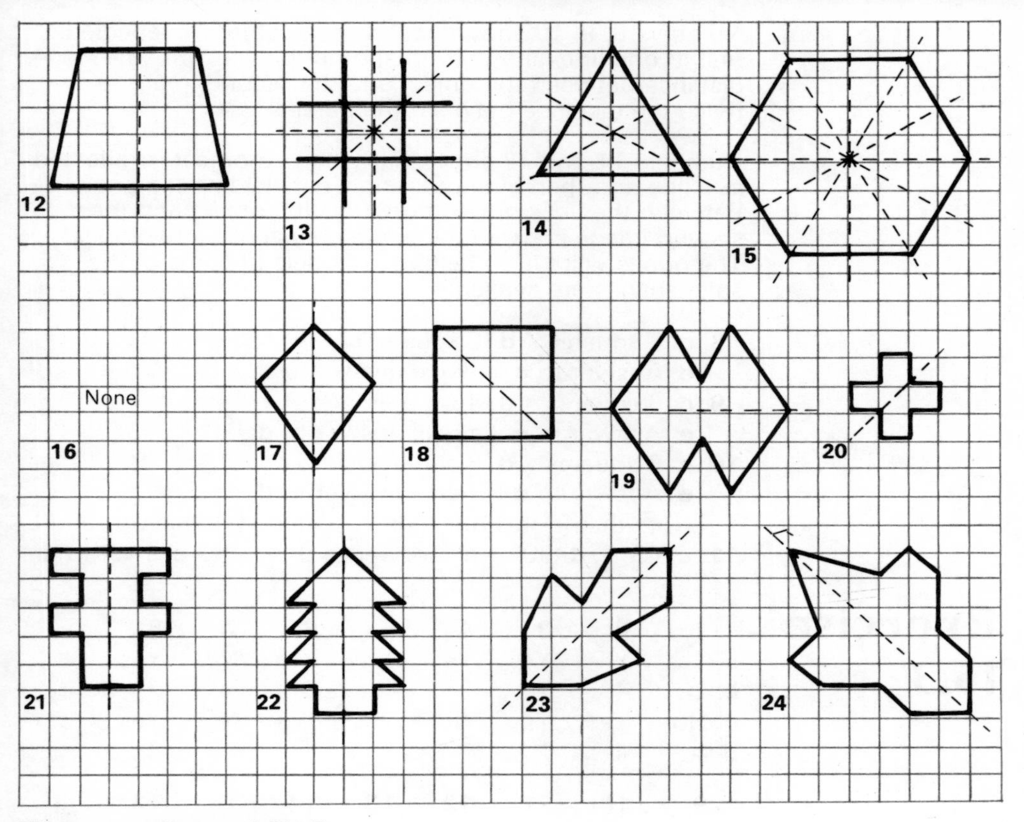

Exercise 15.2

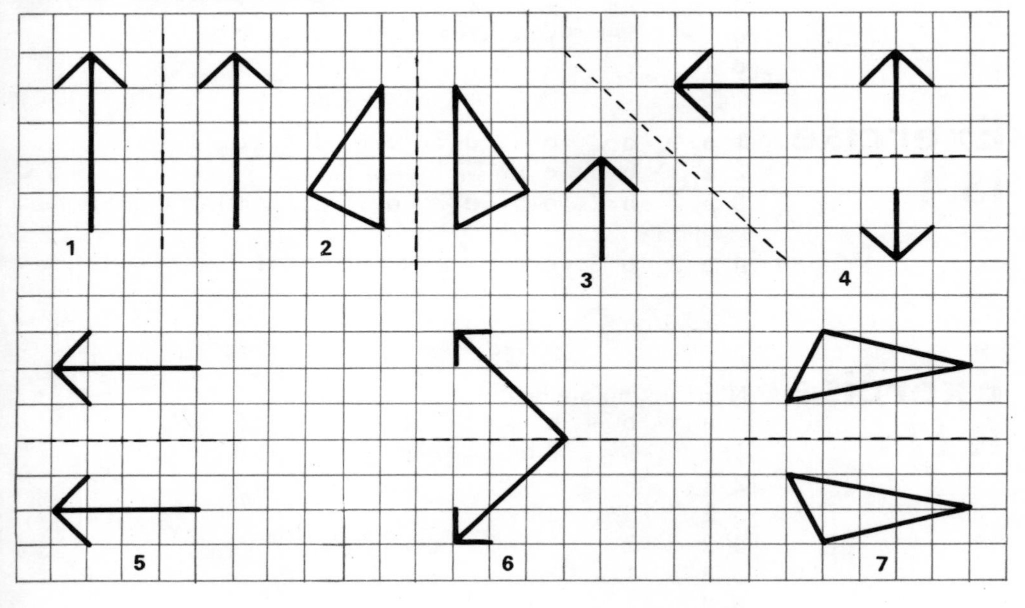

Numbers 8, 9, 10 show that when an object is reflected first in one line, and then in another parallel line, the distance between the object and the second image is twice the distance between the parallel lines.

Numbers 11 and 12 show that when an object is reflected in a line and then in another non-parallel line the angle between the lines drawn from 0 to the object and the second image is twice the angle between the two lines. The pupils will find it helpful if models of some of the solid shapes are available.

8 **c** 6 squares **d** 3 squares
e Answer to **c** is twice answer to **d**.
9 **c** 8 squares **d** 4 squares
e Answer to **c** is twice answer to **d**.
10 **c** 6 squares **d** 3 squares
e Answer to **c** is twice answer to **d**.
Note that in this question A_2 will be between L_1 and L_2.
11 **c** 45° **d** 90° **e** Answer to **d** is twice answer to **c**.

Exercise 15.3

	1	2	3	4	5	6	7	8
a	Yes	Yes	Yes	Yes	No	Yes	No	Yes
b	2	4	2	2	–	5	–	2
c	No	Yes	Yes	No	Yes	Yes	No	Yes
d	–	4	2	–	1	5	–	2

	9	10	11	12	13	14	15	16
a	No	Yes	Yes	Yes	Yes	Yes	Yes	No
b	–	2	4	2	4	8	2	–
c	Yes	Yes	Yes	Yes	Yes	Yes	No	Yes
d	1	2	4	2	4	8	–	1

Exercise 15.4

1 **a** 4 **b** 2 **c** 3 **d** 3 **e** 6 **f** 4 **g** 13
2 **a** 3 **b** 6 **c** 9
3 **a** 2 **b** 2 **c** 3 **d** 3 **e** Vertically down through the middle **f** 6 **g** 7
4 **a** 3 **b** 3 **c** A horizontal plane **d** 7

Exercise R 17

N means no answer.

	a	b	c	d	e	f
1	2	3	2 by 3	6	12	8
2	2	3	2 by 3	6	10	4
3	2	3	2 by 3	6	31	24
4	2	2	2 by 2	4	10	N
5	3	1	3 by 1	3	N	N
6	3	1	3 by 1	3	N	N

7 $\begin{bmatrix} 6 & 9 & 8 \\ 6 & 13 & 12 \end{bmatrix}$ **8** N **9** N **10** $\begin{bmatrix} 2 & 1 & 4 \\ 0 & 1 & 4 \end{bmatrix}$ **11** N

12 N **13** $\begin{bmatrix} 12 & 15 & 18 \\ 9 & 21 & 24 \end{bmatrix}$ **14** $\begin{bmatrix} 2 \\ 4 \\ 1 \end{bmatrix}$ **15** $\begin{bmatrix} 8 \\ 0 \\ 10 \end{bmatrix}$

16 $\begin{bmatrix} 8 & 10 & 12 \\ 6 & 14 & 16 \end{bmatrix}$ **17** $\begin{bmatrix} 6 & 12 & 6 \\ 9 & 18 & 12 \end{bmatrix}$ **18** $\begin{bmatrix} 14 & 22 & 18 \\ 15 & 32 & 28 \end{bmatrix}$

19 $\begin{bmatrix} 8 \\ 8 \\ 7 \end{bmatrix}$ **20** $\begin{bmatrix} ^-2 & ^-1 & ^-4 \\ 0 & ^-1 & ^-4 \end{bmatrix}$ **21** N **22** $\begin{bmatrix} 44 & 60 & 52 \\ 56 & 64 & 96 \end{bmatrix}$

23 $\begin{bmatrix} 8 & 16 \\ 6 & 4 \end{bmatrix}$ **24** N **25** $\begin{bmatrix} 2 & 4 \\ 1{\cdot}5 & 1 \end{bmatrix}$ **26** $\begin{bmatrix} ^-2 \\ ^-4 \\ ^-1 \end{bmatrix}$

27 **a** 6 **b** 8 **c** John (24 marks) **d** Brian and Victor (15 marks)

e $\begin{bmatrix} 15 & 17 & 7 & 9 \\ 9 & 19 & 9 & 11 \\ 13 & 13 & 11 & 9 \end{bmatrix}$ **f** $\begin{bmatrix} 8 & 6 & 4 & 5 \\ 4 & 9 & 6 & 5 \\ 5 & 6 & 4 & 6 \end{bmatrix}$ The marks for March

g $\begin{bmatrix} 70 & 80 & 30 & 40 \\ 50 & 90 & 40 & 50 \\ 60 & 70 & 50 & 50 \end{bmatrix}$ **h** Out of 100 **i** 50%
j Brian (41 marks)
k Out of thirty **l** 3

A suitable treatment for the less able pupils would be Exercise 16.1, numbers 1 to 5. The rest of the Unit deals with geometrical transformations associated with matrices. Enlargement 16.2, Reflection 16.3, Rotation 16.4, Shearing 16.5. It could be pointed out to the pupils that the area of a shape is not altered by shearing, and this leads to simple proofs for the areas of parallelograms and triangles.
It could also be pointed out that multiplication by the matrix

$$\begin{bmatrix} n & 0 \\ 0 & n \end{bmatrix}$$

gives enlargement with a scale factor n.

Exercise 16.1

1 $\begin{bmatrix} 117 \\ 52 \\ 94 \end{bmatrix}$ **2 a** $\begin{bmatrix} 58 \\ 46 \\ 52 \end{bmatrix}$ **b** $\begin{bmatrix} 35 \\ 23 \\ 52 \end{bmatrix}$ **c** $\begin{bmatrix} 62 \\ 31 \\ 33 \end{bmatrix}$ **d** $\begin{bmatrix} 57 \\ 68 \\ 30 \end{bmatrix}$

3 a $\begin{bmatrix} 26 \\ 29 \end{bmatrix}$ **b** $\begin{bmatrix} 45 \\ 43 \end{bmatrix}$ **c** $\begin{bmatrix} 41 \\ 39 \\ 19 \end{bmatrix}$ **4** $\begin{bmatrix} 76 \\ 54 \\ 26 \\ 88 \end{bmatrix}$

5 a $\begin{bmatrix} 48 \\ 42 \\ 51 \\ 65 \end{bmatrix}$ **b** $\begin{bmatrix} 88 \\ 48 \\ 72 \\ 116 \end{bmatrix}$ **c** $\begin{bmatrix} 82 \\ 102 \\ 24 \\ 58 \end{bmatrix}$ **d** $\begin{bmatrix} 45 \\ 70 \\ 90\frac{1}{2} \\ 77\frac{1}{2} \end{bmatrix}$

6 a $\begin{bmatrix} 1 \\ 2 \\ 3 \end{bmatrix}$ Leaves it unchanged. **b** $\begin{bmatrix} 3 \\ 2 \\ 1 \end{bmatrix}$ Turns it upside down.

c $\begin{bmatrix} 2 \\ 4 \\ 6 \end{bmatrix}$ Doubles each element of the matrix. **d** $\begin{bmatrix} 6 \\ 0 \\ 0 \end{bmatrix}$ Adds the elements together and places the result in the first element of the new matrix.

e $\begin{bmatrix} 15 \\ 10 \\ 5 \end{bmatrix}$ Turns the matrix upside down and multiplies each element by 5.

7 a $\begin{bmatrix} 10 & 5 \\ 6 & 3 \\ 6 & 3 \end{bmatrix}$ **b** $\begin{bmatrix} 200 \\ 100 \end{bmatrix}$ **c** $\begin{bmatrix} 2500 \\ 1500 \\ 1500 \end{bmatrix}$ Orange, Lemon, Lime

Exercise 16.2

1 A′(9, 9), B′(3, 9), C′(3, 18)
2 P′(12, 2), Q′(8, 2), R′(12, 8), S′(8, 8).
3 I′(16, 12), J′(8, 10), K′(6, 12), L′(8, 14)
The scale factors are 3, 2, and 2.

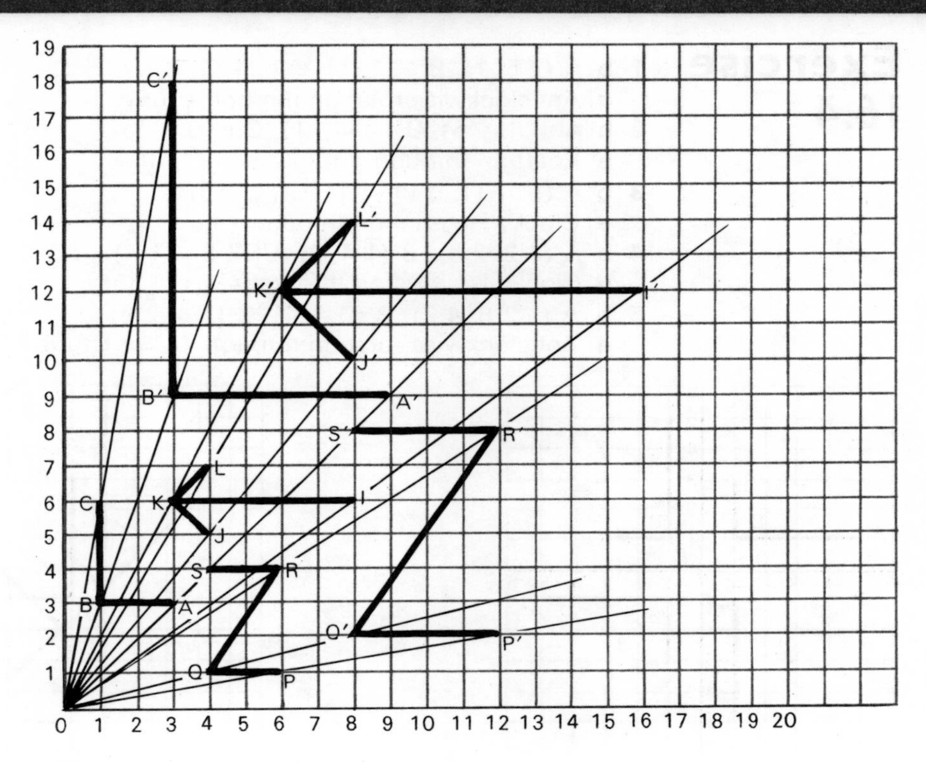

Exercise 16.3

1 A′(1, 2), B′(1, 4), C′(4, 5)

3 b A′($^-$5, $^-$1), B′($^-$1, $^-$1), C′($^-$2, $^-$3)
d Reflection in dotted line

4 b A′(1, $^-$5), B′(1, $^-$1), C′(3, $^-$2)
d Reflection in horizontal axis

5 b A′($^-$1, 5), B′($^-$1, 1), C′($^-$3, 2)
d Reflection in vertical axis

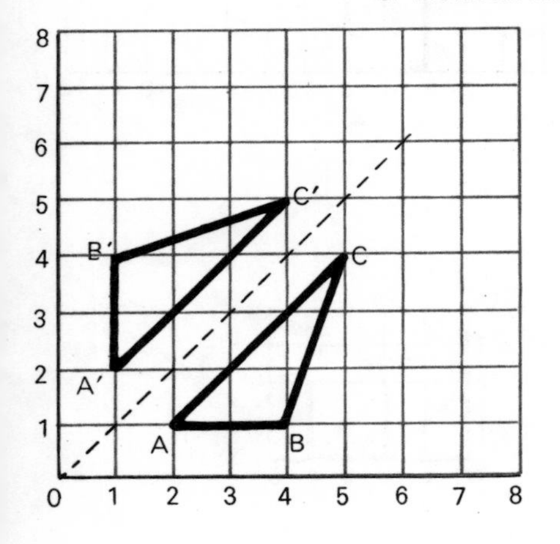

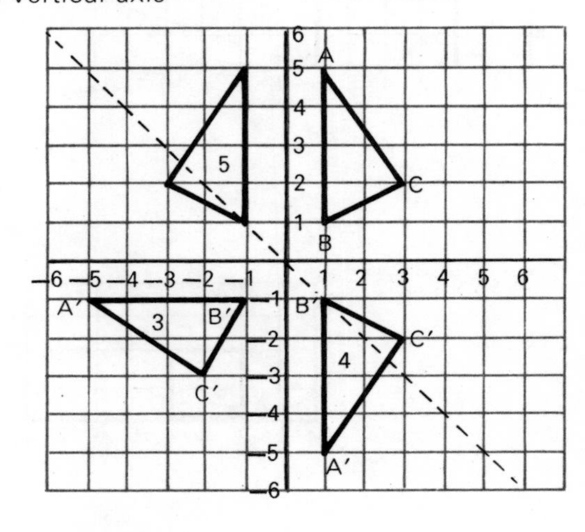

Exercise 16.4

1 b A′($^-5$, 1), B′($^-1$, 1), C′($^-1$, 3)
e Anticlockwise rotation through $\frac{1}{4}$ turn
2 b A′($^-1$, $^-5$), B′($^-1$, $^-1$), C′($^-3$, $^-1$)
e Rotation through $\frac{1}{2}$ turn
3 b A′(5, $^-1$), B′(1, $^-1$), C′(1, $^-3$)
e Clockwise rotation through $\frac{1}{4}$ turn
4 b A′(4·2, 2·8), B′(1·4, 0), C′(2·8, $^-1$·4)
e Clockwise rotation through 45° or $\frac{1}{8}$ turn.
5 b A′($^-2$·8, 4·2), B′(0, 1·4), C′(1·4, 2·8)
e Anticlockwise rotation through 45° or $\frac{1}{8}$ turn

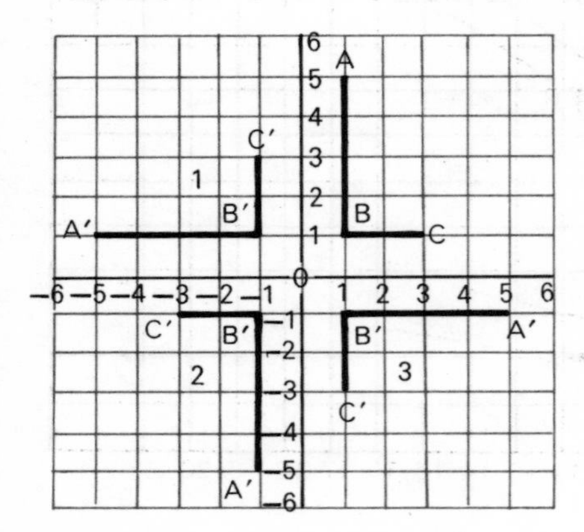

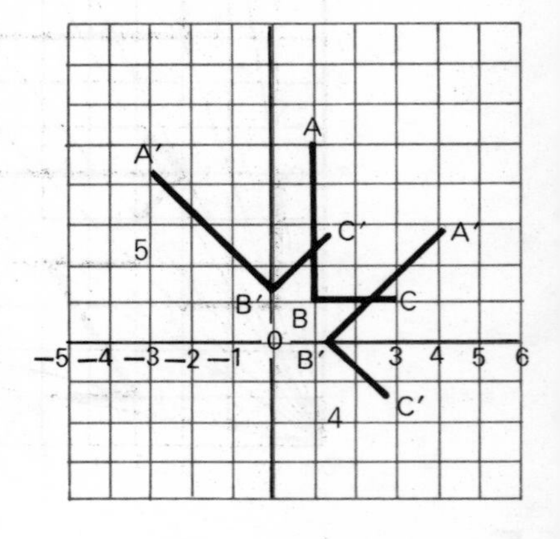

Exercise 16.5

1 c A′(2, 0), B′(8, 2), C′(12, 2), D′(6, 0)
2 c A′(3, 0), B′(9, 3), C′(11, 3), D′(5, 0)
3 c A′(2, 0), B′(4, 2), C′(6, 2), D′(4, 0)
c A′(2, 0), B′(0, 2), C′(2, 2), D′(4, 0)

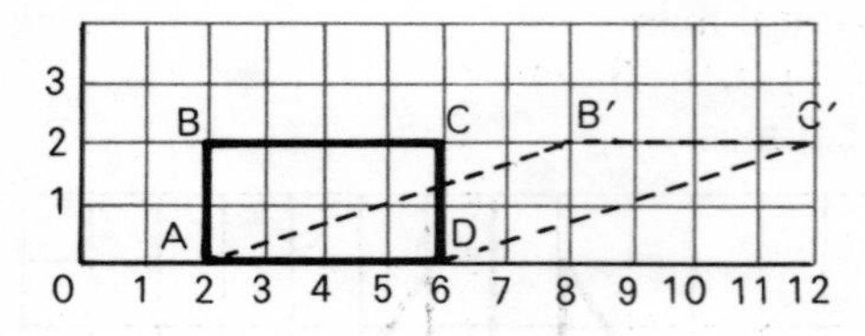

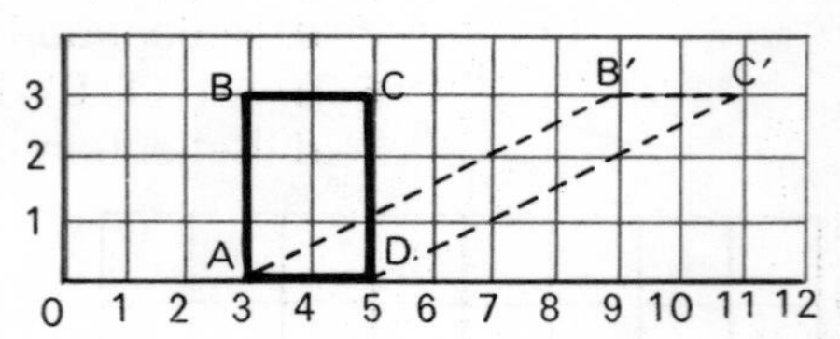

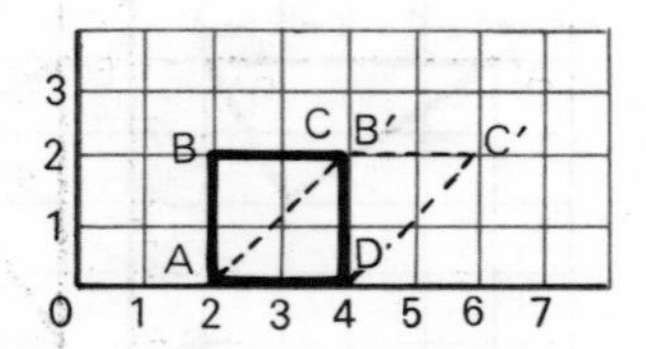

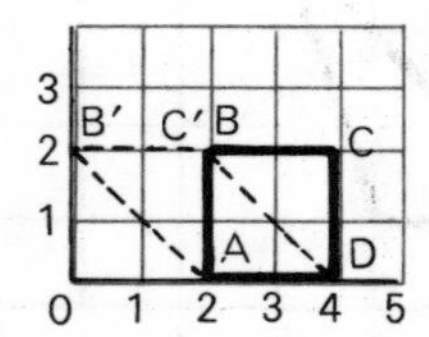

Exercise R 18

Part 1

Practice in cancelling (Exercise R25) might be needed before this exercise is done. If the probabilities are to be expressed as decimals then practice in changing fractions to decimals might be needed (Exercise R25).

1 $\frac{1}{2}$ or 0·5
2 $\frac{1}{2}$ or 0·5
3 $\frac{1}{6}$ or 0·1666
4 $\frac{1}{6}$ or 0·1666
5 $\frac{1}{2}$ or 0·5
6 $\frac{1}{3}$ or 0·3333
7 $\frac{1}{6}$ or 0·1666
8 $\frac{1}{3}$ or 0·333
9 Answer below
10 $\frac{1}{12}$ or 0·08333
11 $\frac{1}{6}$ or 0·1666
12 0
13 $\frac{11}{12}$ or 0·9166
14 $\frac{1}{2}$ or 0·5
15 $\frac{1}{4}$ or 0·25
16 1
17 $\frac{1}{4}$ or 0·25
18 Answer below
19 $\frac{1}{6}$ or 0·1666
20 $\frac{1}{6}$ or 0·1666
21 $\frac{1}{3}$ or 0·3333
22 $\frac{1}{3}$ or 0·3333
23 $\frac{1}{3}$ or 0·3333
24 $\frac{1}{3}$ or 0·3333
25 $\frac{2}{3}$ or 0·6666
26 $\frac{3}{4}$ or 0·75
27 $\frac{1}{5}$ or 0·2
28 $\frac{1}{5}$ or 0·2

Answer to **9** 24, 26, 28, 42, 46, 48, 62, 64, 68, 82, 84, 86
Answer to **18** ORT, OTR, ROT, RTO, TOR, TRO

Part 2

Practice in finding fractions of numbers (Exercise R25) might be needed before this is done, as well as practice in finding decimal parts of quantities (Exercise R24) if the probabilities are expressed as decimals.

	a	b	c	d	e	f	g
1	1	2	3	4	5	16	110
2	1	2	3	5	10	50	67
3	4	8	12	16	20	40	104
4	2	6	10	14	40	400	6000
5	2	4	6	8	100	150	240
6	$\frac{4}{5}$=0·8	4	12	16	20	24	120
7	$\frac{3}{8}$=0·375	3	9	12	15	26(26·25)	38(37·87)
8	$\frac{4}{7}$=0·5714	24	32	40	114(114·2)	1143(1142·8)	17 143

The two sections on insurance and gambling are independent of each other and they may be done in any order.

Exercise 17.1

	a	b	c	d
1	20	£4000	£5000	£1000 profit
2	40	£12 000	£16 000	£4000 profit
3	10	£4000	£2500	£1500 loss
4	80	£4000	£8000	£4000 profit
5	80	£24 000	£20 000	£4000 loss
6	300	£45 000	£50 000	£5000 profit

7 £30 **8** £95 **9** £35 **10** £15 **11** £55

Exercise 17.2

		a	b	c	d	e	f
1	(i)	18p	27p	36p	45p	81p	54p
	(ii)	2p	3p	4p	5p	9p	6p
	(iii)	1p loss	0p profit	1p loss	1p loss	1p profit	1p profit
2		4p	5p	6p	7p	11p	8p

3

0	1	2	3	4	5
1	0	1	2	3	4
2	1	0	1	2	3
3	2	1	0	1	2
4	3	2	1	0	1
5	4	3	2	1	0

a $\frac{10}{36}=\frac{5}{18}$ **b** $\frac{8}{36}=\frac{2}{9}$ **c** $\frac{6}{36}=\frac{1}{6}$ **d** $\frac{4}{36}=\frac{1}{9}$ **e** $\frac{2}{36}=\frac{1}{18}$
f $\frac{0}{36}=0$ **g** $\frac{0}{36}=0$ **h** $\frac{6}{36}=\frac{1}{6}$ **i** 1 **j** 5
k $70 \div 36 = 1\frac{17}{18} = 1{\cdot}944$

Exercise R 19

1 a 6 **b** 6 **c** 7 **d** £5 **e** £7 **f** £15 **g** 3
h £1·43 **i** £3·488 **j** 4·616 **k** 5·25 **l** 7·5
2 a 2, 3, 4, 6, 5 **b** 20 **c** 4
3 b 25 **c** 5 **d** First week
4 a 2, 4, 4, 6, 8, 9, 10, 9, 7, 6, 4, 3 **b** July
c The figures are averages taken over several years. The figure for a given month could be much better, or much worse, than the average.
d No **e** 72 **f** 6 hours **g** 21 hours

A suitable treatment of the work in this Unit for the less able pupils is Exercise 18.1 and Exercise 18.4.

Exercise 18.1

1 £4 **2 a** £3 **b** £6 **3 a** £7 **b** £2·875 **c** 5·142 m
d £11·50 **e** 4 **f** 8 **4 a** 33 **b** 237 **c** 7·181

Exercise 18.2

1 a £33 **2 a** £24
3 a 52·72 (52·22) **b** 55·5 (55·0)
c 56·125 (55·625) **d** 49·15 (48·65)

18.3

	1	2	3	4	5	6
MEAN	£8	£10	18	£20	24	£13
MEDIAN	£7	£11	19	£9	16	£8

7 a They are almost the same.
b They differ considerably from each other.
c They are more or less equally spread out.
d One or two of the numbers is much larger than any of the others.
e It is better to use the median where a few of the numbers are much larger than the others.

Exercise 18.4

1 a 16 **b** £1 **c** £2 **d** £4 **e** £3 **f** £3 **g** £2 **h** £2
2 a 16 **b** £2 **c** £6 **d** £6 **e** £4 **f** £4 **g** £8 **h** £4
3 a 3 **b** £5 **c** £10 **d** £25 **e** £10 **f** £60
4 a Yes **b** 23% **c** The Council spent more **d** Yes
e No **f** No **g** 5%
h Education, Police **i** Housing, Roads, Other things
j Libraries, Swimming pools, Parks, Welfare, Street lighting

The answers to **d** and **f** assume that the total expenditure is the same each year.

Equations can be used to solve these problems. For example to do number 1 we can say:

Let s = one share
Hence $3s+2s=20$
$5s=20$
$s=£4$ etc.

Pupils should be told to check that their answers add up to the original quantity being divided.

Exercise R 20

1 £12, £8
2 £3, £3
3 £4, £8
4 £6, £2
5 £6, £3
6 £12, £9
7 £12, £3
8 £9, £3
9 £6, £8
10 £4, £6
11 £2, £8
12 £4, £10
13 £2, £3, £4
14 £5, £4, £2
15 10p, 8p, 4p
16 1p, 2p, 3p
17 4p, 6p, 8p
18 £2, £3, £3

19 4p, 6p, 6p
20 3p, 6p, 9p
21 £4·285, £5·714
22 £4·80, £7·20
23 £13·125, £21·875
24 £22·222, £17·777
25 9·555 kg, 14·333 kg, 19·11 kg
26 £6·80, £13·60, £13·60
27 £5·853, £11·706, £8·779
28 £5·778, £7·223, £11·557
29 £2·95, £1·475, £10·325
30 17·8p, 17·8p, 26·7p, 35·6p
31 2:3
32 5:2
33 2:5
34 3:4
35 1:2
36 2:3
37 15:7
38 3:5
39 2:3
40 3:4
41 5:2
42 7:3
43 2:3:4
44 2:4:5
45 1:2:3
46 4:3:1
47 2:3:5
48 5:4:2
49 1:3
50 7:3
51 4:1
52 4:1
53 2:3
54 2:3:6
55 **a** £4, £6 **b** £14, £21 **c** £18, £27 **d** £6, £9
56 **a** 3 kg, 5 kg **b** 9 kg, 15 kg **c** 12 kg, 20 kg, **d** 18 kg, 30 kg **e** 300 kg, 500 kg **f** ·6*t*, 10*t* **g** 54 kg, 90 kg
57 **a** 3:4 **b** £3, £4 **c** £9, £12 **d** £15, £20 **e** £3300, £4400
58 **a** 2:7 **b** 18 kg, 63 kg **c** 12 kg, 42 kg **d** 10 kg, 35 kg **e** 160*t*, 560*t* **f** 11·11 kg, 38·88 kg
59 **a** 6:3:2 **b** £120, £60, £40 **c** £79·636, £39·818, £26·545 **d** £116·972, £58·486, £38·990

Exercise 19.1

Only one possible tessellation is shown for each shape. There may be others.

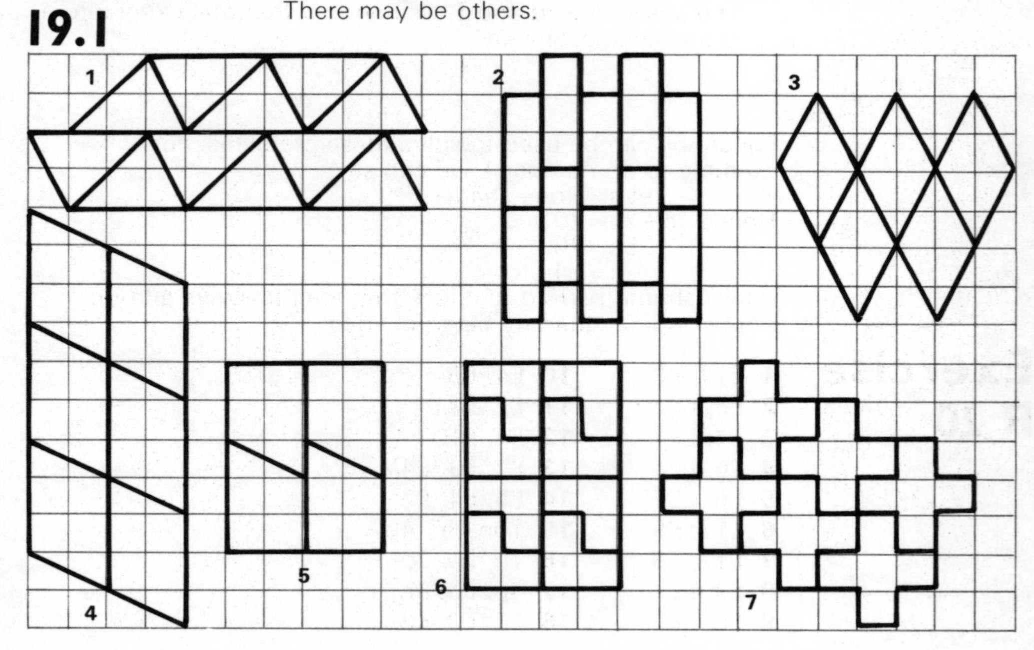

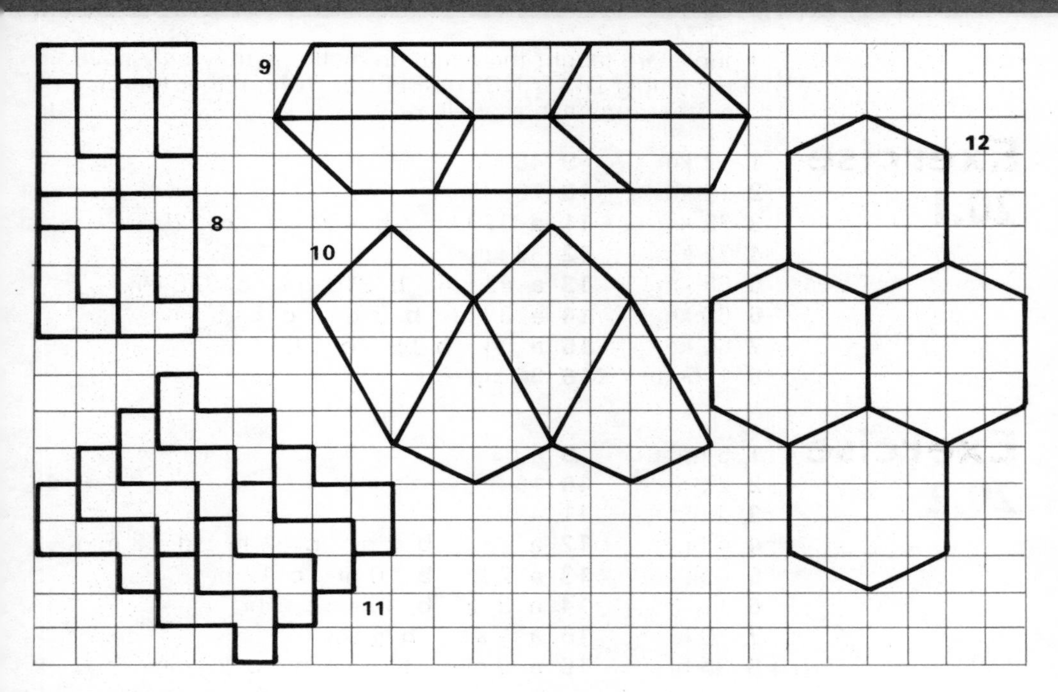

Exercise 19.2

Only one possible tessellation is shown for each shape.
Tessellations not possible with shapes 3, 4, 6, 9.

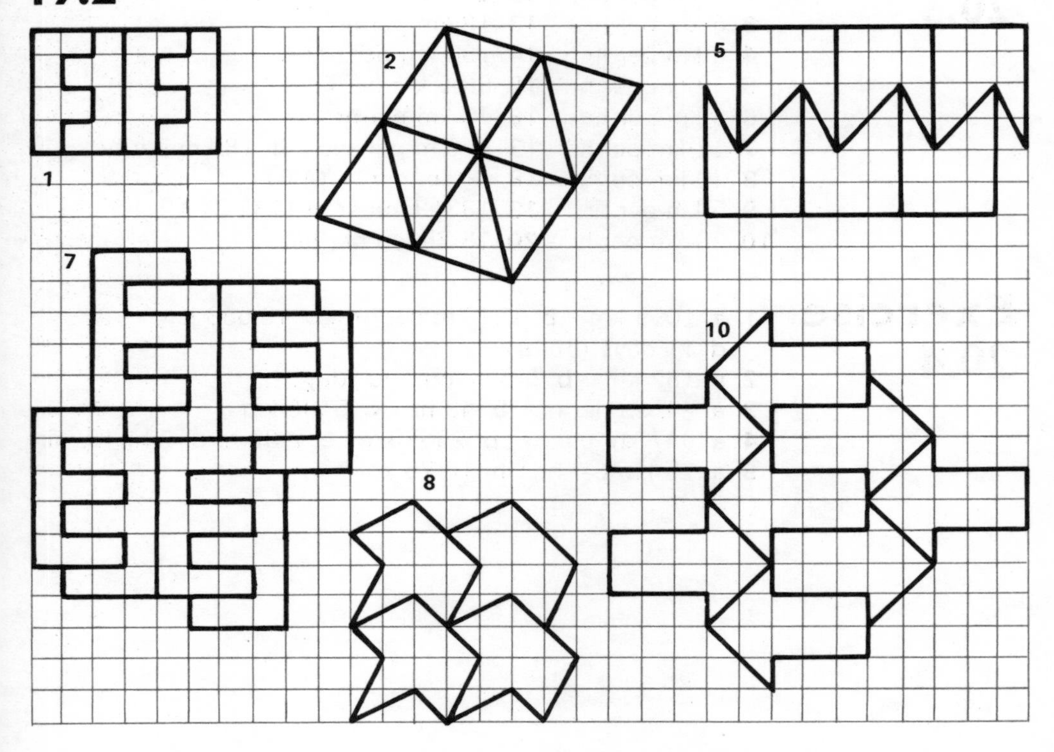

If pupils are taught the equation methods they only have to remember DISTANCE = SPEED × TIME, rather than the two other formulae as well.

Exercise 20.1

1 12 km
2 18 km
3 28 km
4 72 km
5 63 km
6 60 km
7 85 km
8 175 km
9 48 km
10 765 km
11 **a** 270 km **b** 450 km **c** 225 km
12 36 km
13 **a** 42 km **b** 210 km **c** 336 km
14 **a** 3 km **b** 9 km **c** 13·5 km
15 **a** 54 **b** 99 **c** 132
16 34 km

Exercise 20.2

1 5 hr
2 2 hr
3 4 hr
4 4 hr
5 6 hr
6 12 hr
7 10 hr
8 25 hr
9 8 hr
10 18 hr
11 4 hr
12 **a** $\frac{1}{2}$ hr **b** 2 hr **c** $3\frac{1}{2}$ hr **d** 12 min
13 **a** 2 hr **b** 10 hr **c** 12 hr
14 **a** 3 hr **b** 4 hr **c** 6 hr
15 **a** $\frac{1}{2}$ sec **b** $\frac{3}{4}$ sec
16 **a** 8 min **b** 36 min **c** 20 min

Exercise 20.3

1 5 km per hr
2 5 km per hr
3 6 km per hr
4 8 km per hr
5 12 km per hr
6 9 km per hr
7 12 km per hr
8 20 km per hr
9 5 km per hr
10 20 km per hr
11 5 km per hr
12 **a** 2 hr **b** 12 km per hr
13 13 km per hr
14 150 km per hr
15 88·5 km per hr
16 70 km per hr
17 **a** 5 m per sec **b** 18 km per hr
18 $4\frac{2}{3}$ km per hr (4·67)
19 $133\frac{1}{3}$ km per hr (133·3)
20 18·75 km per hr

Exercise 20.4

1 **a** 1099 km **b** 157 km per hr **c** 18 683 km
d 11 days (66 hr)
2 **a** 182 km **b** 6 p.m. on Thursday
3 **a** 24 km per hr **b** 43 hr **c** 5496 km
4 **a** 947 km per hr **b** 2367 km **c** 2·32 hr or 2 hr 19 min
5 **a** 25 km per hr **b** 18·75 km

For the less able pupils Exercises 21.1 and 21.3 only need be done.

Exercise 21.1

1 £16 **6** 20p **11** 15p **16** 1120 cc
2 27p **7** 40p **12** 36 **17** 14 kg
3 24 **8** 21p **13** 96 **18** 2400
4 £12 **9** 16p **14** 100 min **19** £35·576
5 £15 **10** 21p **15** **a** 480 km **b** 5 hr **20** 3285

Exercise 21.2

1 £9·333 **5** £7·20 **9** 3 m **13** **a** 46·66 km
2 £6·75 **6** 32 **10** 5·14 l **b** 17·14 min
3 20 **7** $17\frac{1}{2}$p **11** 7·5 **14** £13·157
4 £12·50 **8** $17\frac{1}{2}$p **12** 290 **15** 9

Exercise 21.3

1 2 hr **5** 2 hr **9** 14 km per hr **13** 16 days
2 8 hr **6** 4 hr **10** 2 hr
3 3 hr **7** 6 hr **11** 12 days
4 1 hr **8** 1 hr **12** 29 km per hr
14 About 11 weeks, assuming that the time decreases in proportion to the number of men employed
15 3·11 **16** 46·8 l

Exercise 21.4

1 54p
2 12
3 72
4 80
5 60p
6 133
7 93·75 l
8 93·33 min
9 346·1
10 29·33 hr = 29 hr 20 min
11 7·333 (8 loads)
12 97·14 min
13 £122·50
14 4·375 hr = 4 hr $22\frac{1}{2}$ min
15 **a** 8·034 l **b** 30·80 cc

Exercise R 21

With less able pupils many of the examples can be done by inspection rather than by calculation. e.g. 3 out of 50 is the same as 6 out of 100 which equals 6%.

	a	b	c	d
1	8%	20%	9%	$8\frac{1}{2}$%
2	8%	14%	20%	46%
3	9%	20%	50%	80%
4	12%	36%	40%	92%
5	35%	20%	55%	95%
6	90%	20%	80%	16%
7	75%	25%	75%	80%
8	77·77%	71·42%	83·33%	44·44%
9	84·61%	30·90%	26·02%	85·89%

	a	b	c	d	e	f	g
10	40	80	120	160	20	10	8
11	80	160	240	320	40	20	16
12	20	40	60	80	10	5	4
13	8	4	16	2	6	24	32
14	9	4·5	1·8	0·9	2·25	18	27
15	1·68	1·75	3·36	2·24	3·15	0·84	1·12
16	6·96	7·25	13·92	9·28	13·05	3·48	4·64
17	2·99	3·91	7·82	10·81	8·97	5·29	15·87
18	1	2	1·5	2·25	5	0·6	1·075
19	1·1	2·2	1·65	2·475	5·5	0·66	1·182

20 70% **21** 53% **22** 80%, 20% **23** **a** 84% **b** 76%

24 **a** 75% **b** 80% **c** 5B

25 **a** £20 **b** £10 **c** £40 **d** £4 **e** £5 **f** £6 **g** £9

26 **a** 30 kg, 70 kg **b** 15 kg, 35 kg **c** 7·5 kg, 17·5 kg
d 18 kg, 42 kg **e** 12·6 kg, 29·4 kg
f 17·1 kg, 39·9 kg **g** 70·5 kg, 164·5 kg

The less able pupils can do just Exercise 22.1 where many of the examples can be done by inspection. i.e. If £100 gives £8 interest, then £50 gives £4 interest. For the rest of the Unit pupils may need revision on multiplication of decimals (Exercise R24) and cancelling (Exercise R25).

Exercise 22.1

	a	b	c	d	e
1	£4	£14	£10	£16	£20
2	£9	£12	£18	£27	£33
3	£2	£4	£3	£3·50	£7·50
4	£2	£1	£4	£3	£2·50
5	£1	£2	£3	£4	£5
6	£1	£2	£3	£0·50	£1·50
7	£16	£216	£54	—	—
8	£40	£160	£560	£140	—
9	£90	£810	—	—	—
10	£10	£10 000	£192·307	—	—
11	£64	£640	£80	—	—
12	£214	£10·50	£220·50	—	—

Exercise 22.2

	a	b	c	d	e	f	g
1	£0·40	£0·48	£0·56	£0·64	£0·72	£0·80	£1·60
2	£0·63	£0·72	£0·81	£0·90	£0·99	£1·08	£1·17
3	£2·80	£3·20	£3·60	£4	£6	£6·80	£8·40
4	£40	£45	£27·50	£32·50	£37·50	£41·25	£48·75
5	£19·039	£27·369	£30·791	£24·097	£63·367	£15·809	£2·677
6	£14·40	£254·40	£21·20	—	—	—	—
7	£24·30	£294·30	£24·525	—	—	—	—
8	£40·80	£380·80	£31·733	—	—	—	—
9	£1080	£9080	£756·666	—	—	—	—

Exercise 22.3

	a	b	c	d	e	f	g	h	i	j	k	l
1	5%	6%	8%	7%	9%	5%	7%	6%	8%	9%	5%	10%
2	6%	4%	6%	8%	10%	8%	6%	8%	8%	4%	6%	2%
3	7%	9%	5%	$4\frac{1}{2}$%	$7\frac{1}{2}$%	$3\frac{1}{2}$%	$5\frac{1}{2}$%	$7\frac{1}{4}$%	3·333%	—	—	—
4	£10	£18	£28	£54	£10	£42	£105	£58·50	£31·05	£236·45	—	—
	5%	6%	7%	9%	4%	12%	14%	$7\frac{1}{2}$%	$4\frac{1}{2}$%	13·61%	—	—

Exercise 22.4

	a	b	c	d	e
1	£5000	£10 000	£20 000	£6000	£1000
2	£600	£7000	£500	£842·105	£4000
	f	**g**	**h**	**i**	**j**
1	£5000	£7000	£900	£400	£500
2	£7200	£20 000	£2647·05	£5400	£45 000
	k	**l**	**m**		
	£460	£700	£250		

Exercise 22.5

	a	b	c	d	e	f	g	h
1	£11·28	£5·55	£5·04	£10·26	£4·83	£8·96	£3·99	£9·52
2	£282	£435	£532	£5700	£6660	£630		
3	£3·28	£0·55	£1·04	£1·26	£1·83	£0·96	£0·99	£2·52
	£82	£135	£132	£750	£660	£180		

	i	j	k	l	m	n	o
1	£12·60	£8·33	£36·96	£15·99	£29·25	£24·82	£38·85
3	£3·60	£1·33	£15·96	£2·99	£4·25	£7·82	£3·85

Exercise R 22

	1	2	3
a	P	$2 \times 15 = 3 \times 10 = 5 \times 6$	P
b	2×2	2×5	2×13
c	3×3	3×5	P
d	2×7	$2 \times 12 = 3 \times 8 = 4 \times 6$	$2 \times 16 = 4 \times 8$
e	P	P	$2 \times 14 = 4 \times 7$
f	$2 \times 6 = 3 \times 4$	3×7	$2 \times 20 = 4 \times 10 = 5 \times 8$
g	$2 \times 10 = 4 \times 5$	P	2×19
h	2×4	3×9	P
i	P	3×11	$4 \times 11 = 2 \times 22$

4
a $3 \times 27 = 9 \times 9$
b 3×31
c 7×13
d P
e $2 \times 30 = 3 \times 20 = 4 \times 15 = 5 \times 12 = 6 \times 10$
f $2 \times 50 = 4 \times 25 = 5 \times 20 = 10 \times 10$
g P
h $2 \times 72 = 3 \times 48 = 4 \times 36 = 6 \times 24 = 8 \times 18 = 9 \times 16 = 12 \times 12$
i P

5 **a** 3×3 $\{1, 3, 9\}$ **b** 2×3 $\{1, 2, 3, 6\}$ **c** 2×2 $\{1, 2, 4\}$
d $2 \times 2 \times 2 \times 2$ $\{1, 2, 4, 8, 16\}$
e $2 \times 2 \times 5$ $\{1, 2, 4, 5, 10, 20\}$ **f** 3×7 $\{1, 3, 7, 21\}$
g $2 \times 2 \times 2 \times 3$ $\{1, 2, 3, 4, 6, 8, 12, 24\}$
h $2 \times 2 \times 7$ $\{1, 2, 4, 7, 14, 28\}$ **i** 5×5 $\{1, 5, 25\}$

6 **a** 2×11 $\{1, 2, 11, 22\}$
b $2 \times 2 \times 3 \times 3$ $\{1, 2, 3, 4, 6, 9, 12, 18, 36\}$
c $2 \times 2 \times 2 \times 2$ $\{1, 2, 4, 8, 16, 32\}$
d 3×11 $\{1, 3, 11, 33\}$
e $3 \times 3 \times 5$ $\{1, 3, 5, 9, 15, 45\}$
f $2 \times 2 \times 2 \times 3 \times 3$ $\{1, 2, 3, 4, 6, 8, 9, 12, 18, 24, 36, 72\}$
g 5×7 $\{1, 5, 7, 35\}$
h $2 \times 3 \times 5$ $\{1, 2, 3, 5, 6, 10, 15, 30\}$
i 2×17 $\{1, 2, 17, 34\}$

7 **a** $2 \times 2 \times 2 \times 5$ $\{1, 2, 4, 5, 8, 10, 20, 40\}$
b $2 \times 2 \times 2 \times 2 \times 3$ $\{1, 2, 3, 4, 6, 8, 12, 16, 24, 48\}$
c $2 \times 5 \times 7$ $\{1, 2, 5, 7, 10, 14, 35, 70\}$
d $2 \times 2 \times 2 \times 2 \times 2 \times 2$ $\{1, 2, 4, 8, 16, 32, 64\}$
e $2 \times 2 \times 19$ $\{1, 2, 4, 19, 38, 76\}$
f $2 \times 2 \times 5 \times 5$ $\{1, 2, 4, 5, 10, 20, 25, 50, 100\}$

g $2 \times 2 \times 3 \times 3 \times 3$ {1, 2, 3, 4, 6, 9, 12, 18, 27, 36, 54, 108}

h $2 \times 2 \times 2 \times 2 \times 3 \times 3$ {1, 2, 3, 4, 6, 8, 9, 12, 16, 18, 24, 36, 48, 72, 144}

i $2 \times 2 \times 3 \times 13$ {1, 2, 3, 4, 6, 12, 13, 26, 39, 52, 78, 156}

8 83, 89, 97, 101, 103, 107, 109, 113, 127, 131, 137, 139, 149

	a	b	c	d	e	f	g	h	i
9	P	P	P	3	P	P	11	P	3
10	7	P	3	P	17	23	P	P	P

Exercise 23.1

Each of the sections is self contained, and they may be done in any order.

	a	b	c	d	e	f
1	14p	27p	19p	52p	45p	44p
2	$10\frac{1}{2}$p	$22\frac{1}{2}$p	$28\frac{1}{2}$p	$45\frac{1}{2}$p	$37\frac{1}{2}$p	$38\frac{1}{2}$p
3	54p	87p	91p	£2·92	£4·05	£7·64
4	$40\frac{1}{2}$p	$72\frac{1}{2}$p	£1·$36\frac{1}{2}$	£2·$55\frac{1}{2}$	£3·$37\frac{1}{2}$	£6·$68\frac{1}{2}$
5	£4·94	£15·27	£27·64	£11·$97\frac{1}{2}$	£38·$04\frac{1}{2}$	£46·$21\frac{1}{2}$
6	£2·$58\frac{1}{2}$	£7·$93\frac{1}{2}$	£31·$17\frac{1}{2}$	£36·16	£30·94	£45·$42\frac{1}{2}$
7	£13·$58\frac{1}{2}$	£58·$53\frac{1}{2}$	£148·$56\frac{1}{2}$	£153·28	£369·58	£405·$66\frac{1}{2}$

Exercise 23.2

1
54p
48p
72p
£1·74

2
56p
40p
96p
£1·92

3
35p
54p
155p
£2·44

4
16p
36p
60p
£1·12

5
21p
36p
125p
£1·82

6
234p
34p
111p
£3·79

7
14p
33p
63p
£1·10

8
91p
$32\frac{1}{2}$p
$31\frac{1}{2}$p
£1·55

9
$76\frac{1}{2}$p
144p
$52\frac{1}{2}$p
190p
£14·63

10
318p
100p
477p
90p
£9·85

11
225p
105p
$471\frac{1}{2}$p
1598p
£23·$99\frac{1}{2}$

12
£9·36
£20·45
£27·23
£24·46
£81·50

13
£30·59
£39·52
£50·24
£35·34
£155·69

14
£16·$55\frac{1}{2}$
£9·$76\frac{1}{2}$
£35·$57\frac{1}{2}$
£16·02
£77·$91\frac{1}{2}$

15
£23·90
£140·52
£58·$22\frac{1}{2}$
£152·04
£374·$68\frac{1}{2}$

Exercise 23.3

1 4p each
2 7 for 21p
3 2 kg for 12p
4 10 kg for 30p
5 1 kg costing £1·20
6 Both the same value
7 1 kg at 14p
8 $\frac{1}{2}$ kg at 9p
9 10 kg costing £5
10 4 kg costing £2
11 $1\frac{1}{2}$ kg costing 21p
12 $1\frac{1}{2}$ kg costing 20p

Exercise 23.4

1 Make 86p
2 Make 23p
3 Buy 5p
4 Make 37p
5 Make 62p
6 Make 3p
7 Buy 24p
8 Make $16\frac{1}{2}$p
9 Make £3·23
10 Buy 29p
11 Make £3·08
12 Make £1·73
13 Buy £2·59
14 Buy £1·42
15 Make £1·44

Exercise 23.5

1

Eggs		*Butter*		*Flour*		*Milk*		*Sugar*	
1	4p	125 g	$7\frac{1}{2}$p	100 g	2p	100 cc	1p	100 g	1p
2	8p	25 g	$1\frac{1}{2}$p	50 g	1p	50 cc	$\frac{1}{2}$p	50 g	$\frac{1}{2}$p
3	12p	50 g	3p	200 g	4p	150 cc	$1\frac{1}{2}$p	150 g	$1\frac{1}{2}$p
4	16p	100 g	6p	250 g	5p			200 g	2p
		200 g	12p	500 g	10p			250 g	$2\frac{1}{2}$p
				750 g	15p				
				300 g	6p				
				800 g	16p				

2 Make $8\frac{1}{2}$p
3 Buy 5p
4 Make 7p on 5 kg
5 Make 11p on 48
6 Make 5p per loaf
7 Make 3p per bottle
8 Make 61p on 1100 g
9 Make $5\frac{1}{2}$p
10 Make 13p on 36

Exercise 23.6

	1	**2**	**3**	**4**	**5**	**6**	**7**	**8**	**9**	**10**	**11**
H.P. Price	£20	£27	£33	£40	£47	£55	£40	£53	£200	£41	£61
Extra Payment	£2	£3	£3	£4	£5	£7	£8	£10	£40	£12	£18

12	**13**	**14**	**15**	**16**	**17**	**18**	**19**	**20**
£188	£344	£201	£317	£987	£8·28	£30·90	£100·68	£183·26
£36	£30	£60	£92	£220	£0·68	£6·80	£10·58	£17·36

21	**22**	**23**	**24**	**25**	**26**	**27**	**28**
£71·54	£218·99	£2·32	£3·81	£23·84	£49·22	£105·64	£133·68
£9·04	£31·46	£0·24	£0·42	£4·21	£8·76	£15·43	£23·84

29	30	31	32
£321·16	£998·92	£1023·54	£3653·70
£64·31	£140·42	£204·00	£748·71

Exercise 23.7

1 £20	**6** £18	**11** £22·20	**16** £32·25
2 £30	**7** £45	**12** £37·80	**17** £47·45
3 £70	**8** £58·10	**13** £58·80	**18** £72·24
4 £180	**9** £240	**14** £96	**19** £159·90
5 £240	**10** £330	**15** £216	**20** £272·60

Exercise 23.8

1 £1·64	**5** £2·25	**9** £1·73	**13** £3·64
2 £2·02	**6** £2·39	**10** £1·57	**14** £4·65
3 £1·23	**7** £3·93	**11** £2·42	**15** £3·007
4 £1·71	**8** £4·26	**12** £1·94	**16** £6·372

Exercise 23.9

	1	2	3	4	5
Units	100	100	50	200	200
Bill	£2·32	£1·67	£2·06	£2·75	£2·83

	6	7	8	9	10
Units	150	250	324	581	979
Bill	£3·08	£3·98	£4·662	£8·012	£14·696

Exercise 23.10

	1	2	3	4	5	6	7	8	9
Therms	40	50	30	35	20	22·5	66	36	84
Bill	£4·50	£4·80	£3·95	£4·38	£3·43	£4·015	£10·02	£6·60	£10·02

Exercise 24.1

Less able pupils can do just Exercise 24.1

The following are topologically equivalent: 1, 2, 4, 7, 8, 11, 13, 15.
The following are not: 3, 5, 6, 9, 10, 12, 14.

Light G goes on. Lights F and G go on.

Exercise 24.2

1

	A	B	C	D
A	0	1	0	0
B	1	0	1	1
C	0	1	0	1
D	0	1	1	0

2

	A	B	C	D
A	0	1	1	1
B	1	0	1	0
C	1	1	0	1
D	1	0	1	0

3

	A	B	C	D
A	0	1	0	0
B	1	0	1	1
C	0	1	0	1
D	0	1	1	0

4

	A	B	C	D
A	0	1	1	1
B	1	0	1	1
C	1	1	0	1
D	1	1	1	0

5

	A	B	C	D
A	0	1	1	1
B	1	0	1	0
C	1	1	0	1
D	1	0	1	0

6

	A	B	C	D
A	0	1	0	0
B	1	0	1	1
C	0	1	0	1
D	0	1	1	0

7

	A	B	C	D
A	0	1	1	1
B	1	0	1	1
C	1	1	0	1
D	1	1	1	0

8

	A	B	C	D
A	0	1	1	1
B	1	0	1	0
C	1	1	0	1
D	1	0	1	0

1, 3, 6 are equivalent; 2, 5, 8 are equivalent; 4, 7 are equivalent.

9

	A	B	C	D
A	2	1	0	0
B	1	2	0	1
C	0	0	0	2
D	0	1	2	0

10

	A	B	C	D
A	0	1	2	1
B	1	0	0	0
C	2	0	2	0
D	1	0	0	0

11

	A	B	C	D
A	4	1	1	1
B	1	0	0	0
C	1	0	0	0
D	1	0	0	0

12

	A	B	C	D
A	4	1	1	1
B	1	0	0	0
C	1	0	0	0
D	1	0	0	0

13

	A	B	C	D
A	2	1	0	0
B	1	2	0	1
C	0	0	0	2
D	0	1	2	0

14

	A	B	C	D
A	0	1	2	1
B	1	0	0	0
C	2	0	2	0
D	1	0	0	0

15

	A	B	C	D
A	0	1	2	1
B	1	0	0	0
C	2	0	2	0
D	1	0	0	0

16

	A	B	C	D
A	2	1	0	0
B	1	2	0	1
C	0	0	0	2
D	0	1	2	0

9, 13, 16 are equivalent; 10, 14, 15 are equivalent; 11, 12 are equivalent.

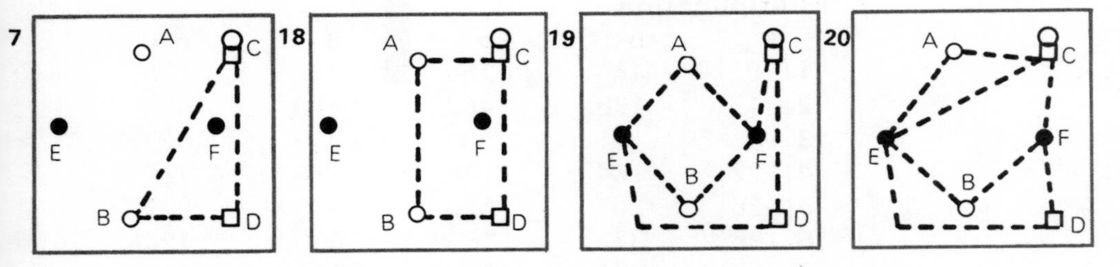

17 a Nothing **b** Light goes on **c** Light goes on.
18 a Nothing **b** Nothing **c** Light goes on.
19 a Light goes on **b** Light goes on **c** Light goes on.
20 a The light remains on. (The light will be on even if both switches are off.)
b The light will dim or go out, and the battery will short and will probably be damaged.
c Same answer as for **b**.

Exercise R 23

Addition

	a	b	c	d	e	f	g	h	i	j
1	7	9	11	16	28	55	88	197	250	723
2	10	12	14	19	31	58	91	200	253	726
3	12	14	16	21	33	60	93	202	255	728
4	18	20	22	27	39	66	99	208	261	734
5	26	28	30	35	47	74	107	216	269	742
6	60	62	64	69	81	108	141	250	303	776
7	96	98	100	105	117	144	177	286	339	812
8	305	307	309	314	326	353	386	495	548	1021
9	747	749	751	756	768	795	828	937	990	1463
10	371	373	375	380	392	419	452	561	614	1087

Subtraction

	a	b	c	d	e	f	g	h	i	j
1	1	1	3	8	20	47	80	189	242	715
2	4	2	0	5	17	44	77	186	239	712
3	6	4	2	3	15	42	75	184	237	710
4	12	10	8	3	9	36	69	178	231	704
5	20	18	16	11	1	28	61	170	223	696
6	54	52	50	45	33	6	27	136	189	662
7	90	88	86	81	69	42	9	100	153	626
8	299	297	295	290	278	251	218	109	56	417
9	741	739	737	732	720	693	660	551	498	25
10	365	363	361	356	344	317	284	175	122	351

Multiplication

	a	b	c	d	e
1	28	46	154	90	178
2	84	138	462	270	534
3	98	161	539	315	623
4	112	184	616	360	712
5	126	207	693	405	801
6	196	322	1078	630	1246
7	532	874	2926	1710	3382
8	588	966	3234	1890	3738
9	938	1541	5159	3015	5963
10	12 474	20 493	68 607	40 095	79 299

	f	g	h	i	j
1	184	194	294	1064	1974
2	552	582	882	3192	5922
3	644	679	1029	3724	6909
4	736	776	1176	4256	7896

	f	g	h	i	j
5	828	873	1323	4788	8883
6	1288	1358	2058	7448	13 818
7	3496	3686	5586	20 216	37 506
8	3864	4074	6174	22 344	41 454
9	6164	6499	9849	35 644	66 129
10	81 972	86 427	130 977	474 012	879 417

Division

	a	b	c	d	e
1	6	4	5	7	9
2	8	8	6	10	3
3	21	31	23	43	29
4	6 r 2	4 r 3	5 r 3	7 r 2	9 r 2
5	17	53	46	17	21
6	33	68	16	11	81
7	3 r 87	4 r 84	198 r 11	63 r 25	290 r 16

	f	g	h	i	j
1	9	9	6	9	10
2	5	3	7	5	6
3	52	44	93	79	84
4	21 r 2	31 r 3	43 r 2	29 r 5	52 r 3
5	16	51	13	41	52
6	34	54 r 26	28 r 68	102 r 29	109 r 10
7	116 r 4	209 r 4	105 r 20	74 r 43	114 r 70

Exercise R 24

Addition

	a	b	c	d	e	f	g	h	i
1	3·2	4·7	9·8	17·2	2·5	2·34	1·84	2·343	19·12
2	6·7	8·2	13·3	20·7	6	5·84	5·34	5·843	22·62
3	6·4	7·9	13	20·4	5·7	5·54	5·04	5·543	22·32
4	15·7	17·2	22·3	29·7	15	14·84	14·34	14·843	31·62
5	2·3	3·8	8·9	16·3	1·6	1·44	0·94	1·443	18·22
6	1·86	3·36	8·46	15·86	1·16	1	0·5	1·003	17·78
7	1·43	2·93	8·03	15·43	0·73	0·57	0·07	0·573	17·35
8	2·138	3·638	8·738	16·138	1·438	1·278	0·778	1·281	18·058
9	18·7	20·2	25·3	32·7	18	17·84	17·34	17·843	34·62
10	45·22	46·72	51·82	59·22	44·52	44·36	43·86	44·363	61·14

Subtraction

	a	b	c	d	e	f	g	h	i
1	0·4	1·1	6·2	13·6	1·1	1·26	1·76	1·257	15·52
2	3·9	2·4	2·7	10·1	4·6	4·76	5·26	4·757	12·02
3	3·6	2·1	3	10·4	4·3	4·46	4·96	4·457	12·32
4	12·9	11·4	6·3	1·1	13·6	13·76	14·26	13·757	3·02
5	0·5	2	7·1	14·5	0·2	0·36	0·86	0·357	16·42
6	0·94	2·44	7·54	14·94	0·24	0·08	0·42	0·083	16·86
7	1·37	2·87	7·97	15·37	0·67	0·51	0·01	0·513	17·29
8	0·662	2·162	7·262	14·662	0·038	0·198	0·698	0·195	16·582
9	15·9	14·4	9·3	1·9	16·6	16·76	17·26	16·757	0·02
10	42·42	40·92	35·82	28·42	43·12	43·28	43·78	43·277	26·5

Multiplication

	a	b	c	d	e
1	2·8	4·6	1·4	0·8	6·5
2	7	11·5	3·5	2	16·25
3	9·8	16·1	4·9	2·8	22·75
4	12·6	20·7	6·3	3·6	29·25
5	18·2	29·9	9·1	5·2	42·25
6	44·8	73·6	22·4	12·8	104
7	88·2	144·9	44·1	25·2	204·75
8	110·6	181·7	55·3	31·6	256·75
9	116·2	190·9	58·1	33·2	269·75
10	359·8	591·1	179·9	102·8	835·25

	f	g	h	i	j
1	11·24	49·8	91·6	27·48	33·506
2	28·1	124·5	229	68·7	83·765
3	39·34	174·3	320·6	96·18	117·271
4	50·58	224·1	412·2	123·66	150·777
5	73·06	323·7	595·4	178·62	217·789
6	179·84	796·8	1465·6	439·68	536·096
7	354·06	1568·7	2885·4	865·62	1055·439
8	443·98	1967·1	3618·2	1085·46	1323·487
9	466·46	2066·7	3801·4	1140·42	1390·499
10	1444·34	6399·3	11 770·6	3531·18	4305·521

Division

	a	b	c	d	e	f	g
1	1·1	2·1	3·2	2·4	2·1	0·8	0·5
2	0·7	0·4	0·3	0·5	0·9	0·7	0·6
3	1·3	1·4	1·8	2·5	2·4	3·2	2·9
4	0·27	0·27	0·7285	0·45	0·035	0·1037	0·5666
5	0·096	0·7333	6·428	0·9444	17	0·23	1·3
6	1·8	2·509	1·204	0·3427	2·621	0·1447	0·02 972
7	0·02 593	0·02	0·2071	1·546	0·3605	0·1264	0·01 029

Multiplication of decimals by decimals

	a	b	c	d	e
1	0·12	0·15	0·24	0·63	0·96
2	0·24	0·3	0·48	1·26	1·92
3	0·36	0·45	0·72	1·89	2·88
4	0·48	0·6	0·96	2·52	3·84
5	1·36	1·7	2·72	7·14	10·88
6	0·256	0·32	0·512	1·344	2·048
7	0·492	0·615	0·984	2·583	3·936
8	3·328	4·16	6·656	17·472	26·624
9	0·1364	0·1705	0·2728	0·7161	1·0912
10	10·72	13·4	21·44	56·28	85·76

	f	g	h	i	j
1	1·59	0·225	0·018	0·7689	0·007 89
2	3·18	0·45	0·036	1·5378	0·015 78
3	4·77	0·675	0·054	2·3067	0·023 67
4	6·36	0·9	0·072	3·0756	0·031 56
5	18·02	2·55	0·204	8·7142	0·089 42
6	3·392	0·48	0·0384	1·64 032	0·016 832
7	6·519	0·9225	0·0738	3·15 249	0·032 349
8	44·096	6·24	0·4992	21·32 416	0·218 816
9	1·8073	0·25 575	0·02 046	0·873 983	0·0089 683
10	142·04	20·1	1·608	68·6884	0·70 484

Division of decimals by decimals

	a	b	c	d	e
1	24	16	12	80	60
2	24	18	12	240	180
3	3·6	36	1·8	48	480
4	2·4	36	180	120	9
5	3	1·5	15	30	100
6	0·24	2·2	0·22	2·4	24
7	26	260	30	300	3000
8	22·35	237·5	1652	1·890	18 090
9	10·32	52·75	1032	11·63	7·758
10	1982	21 060	552·4	64·19	27·06

Multiplication and division by 10, 100, 1000

	a	b	c	d	e	f
1	35	350	3500	0·35	0·035	0·0035
2	64·2	642	6420	0·642	0·0642	0·006 42
3	90	900	9000	0·9	0·09	0·009
4	8	80	800	0·08	0·008	0·0008
5	273	2730	27 300	2·73	0·273	0·0273
6	840	8400	84 000	8·4	0·84	0·084
7	6·2	62	620	0·062	0·0062	0·000 62

	a	b	c	d	e	f
8	473·9	4739	47 390	4·739	0·4739	0·04 739

Exercise R 25

Simplifying fractions

	a	b	c	d	e	f	g	h	i
1	$\frac{1}{2}$	$\frac{1}{4}$	$\frac{1}{2}$	$\frac{1}{3}$	$\frac{1}{5}$	$\frac{1}{3}$	$\frac{3}{4}$	$\frac{2}{3}$	$\frac{1}{2}$
2	$\frac{1}{4}$	$\frac{2}{5}$	$\frac{1}{3}$	$\frac{1}{6}$	$\frac{3}{4}$	$\frac{1}{2}$	$\frac{2}{3}$	$\frac{2}{3}$	$\frac{2}{3}$
3	$\frac{1}{2}$	$\frac{3}{4}$	$\frac{3}{4}$	$\frac{3}{8}$	$\frac{4}{5}$	$\frac{3}{5}$	$\frac{3}{8}$	$\frac{1}{2}$	$\frac{5}{6}$

Changing fractions to decimals

	a	b	c	d	e
1	0·75	0·25	0·5	0·6666	0·2
2	0·4285	0·4	0·8333	0·1428	0·3750
3	0·75	2·25	0·5555	0·6363	0·1111
4	0·3809	0·6808	0·9864	2·285	0·016 12

f	g	h	i
0·3333	0·6	0·1666	0·8
0·6666	0·6250	0·8571	0·2857
0·5294	2·2	0·4444	0·5384
0·8181	0·7926	3·880	1·857

Finding fractions of numbers

	a	b	c	d	e
1	4	3	2	7	6
2	9	6	24	21	30
3	9	6	12	15	18
4	10	11·11	13·88	22·77	20
5	36·55	64·55	30·33	32·66	22·55
6	£0·45	£0·57	£1·642	£0·311	£9·172
7	£1·80	£2·68	£0·388	£19·937	£141·142
8	£26·076	£18·023	£3·80	£11·776	£0·452

The exercises in this Unit are arranged so that they can be used for the teaching of logarithms.

Exercises 25.1 25.2, and 25.3 involve only zero characteristics.

Exercises 25.4 and 25.5 involve only positive characteristics.

Exercises 25.6 to 25.8 involve negative characteristics.

Exercise 25.1

	a	b	c	d	e
1	7·439	9·379	8·967	8·509	5·335

	a	b	c	d	e
2	9·917	8·216	9·720	9·480	9·680
3	9·396	9·327	9·450	3·337	1·177
4	5·29	7·236	8·951	9·030	8·663
5	9·946	9·922	9·990	6·864	8·920
6	9·394	9·591	1·881	2·993	9·734

Exercise 25.2

	a	b	c	d	e
1	1·996	3·431	1·808	3·277	1·896
2	1·633	1·964	3·000	4·041	1·058
3	1·294	2·513	1·072	1·295	3·404
4	1·302	1·783	1·165	2·420	1·984
5	3·061	7·240	2·780	2·007	1·319
6	2·187	1·259	4·937	1·067	1·245

Exercise 25.3

1 9·174 sq m
2 8·153 sq m
3 6·415 sq m
4 6
5 4
6 4
7 £5·883
8 7
9 6
10 £7·068
11 7·744 sq km
12 4 m
13 £4·763
14 9
15 £9·10
16 3·522 m
17 5 (5·161)
18 £1·399
19 6 (6·247)
20 5 (5·115)

Exercise 25.4

	a	b	c	d
1	1737	1111	4·080	1·345
2	382·2	267·9	4·980	10·23
3	200·9	379·7	66·93	12·47
4	2048	5568	1·490	1·933
5	1374	13 100	18·08	52·05
6	881·4	838 700	152·7	1·005
7	8710	7673	2·382	239·0

Exercise 25.5

1 61·53 sq m
2 222·7 sq m
3 18
4 15
5 47·64 cubic m
6 448·4 cubic m
7 £1·647
8 £2·535
9 £1607
10 332·3 m
11 £48·91
12 8344
13 a 16·09 sq m **b** £30·41
14 a £48·56 **b** £2379
15 a 1447 cubic m **b** 3329 tonnes
16 a 435·1 sq m **b** 31·08
17 a 9·152 sq m **b** £19·31
18 a Yes. **b** 86 400 sq cm **c** 900 sq cm **d** 96
19 a 494 **b** 88 430
20 a 39·95 sq m **b** 4 (3·632) **c** £1·56

Exercise 25.6

	a	b	c	d	e
1	129·6	4212	16·48	24·01	53·40
2	43 820	8243	671·3	20·13	302·6
3	1·742	0·1314	6·187	0·7970	0·02 128
4	2·338	0·2346	0·04 400	0·0 002 432	0·006 363
5	0·1	0·07 070	0·01 472	0·08 865	0·008 863
6	3·803	47·32	37·24	6429	60 000
7	0·001 973	0·0 004 197	0·000 006 096	0·0 007 761	0·0 001 482
8	5·296	110·9	12·77	7·184	9·595
9	0·01 973	0·009 602	0·1385	0·02 529	0·1931

Exercise 25.7

1 5·226 sq m
2 £6·517
3 £0·84
4 0·05 624 sq km
5 £0·753
6 0·004 640 sq km
7 1148 m
8 0·01 622*t*
9 225 (225·8)
10 0·1318 g
11 0·3611 kg
12 0·0294 g
13 £0·6675
14 £1·733
15 229 (229·7)
16 0·2847 m
17 14 (14·18)
18 £0·4095
19 £0·7714
20 **a** £0·54 **b** £0·4590

Exercise 25.8

	a	b	c	d
1	62·29	17·73	2·25	2·135
2	9078	1·161	1197	3·715
3	221 300	70·93	14·71	7·802
4	546 700	1516	95 440	3·846
5	63·59	3·856	488·8	19·06
6	9·204	124·8	0·2116	0·2186
7	341·8	0·1026	0·003 215	0·6753
8	0·1857	0·9266	0·09 113	0·6162
9	1·431	1460	0·0 007 314	0·2860
10	5·816	0·5022	0·7351	0·1328
11	2·201	1·107	1·710	0·9282
12	365·6	380·6	400·9	0·3637

THE SCHOOL MATHEMATICS PROJECT

When the SMP was founded in 1961, its main objective was to devise radically new secondary-school mathematics courses (and corresponding GCE and CSE syllabuses) to reflect, more adequately than did the traditional syllabuses, the up-to-date nature and usages of mathematics.

This objective has now been realized. SMP *Books 1–5* form a five-year course to the O-level examination 'SMP Mathematics'. *Books 3T, 4* and *5* give a three-year course to the same O-level examination (the earlier *Books T* and *T4* being now regarded as obsolete). *Advanced Mathematics Books 1–4* cover the syllabus for the A-level examination 'SMP Mathematics' and five shorter texts cover the material of the various sections of the A-level examination 'SMP Further Mathematics'. Revisions of the first two books of *Advanced Mathematics* are available as *Revised Advanced Mathematics Books 1* and *2*. There are two books for 'SMP Additional Mathematics' at O-level. All the SMP GCE examinations are available to schools through any of the Examining Boards.

Books A–H, originally designed as a CSE course, cover broadly the same development of mathematics as do the first few books of the O-level series. Most CSE Boards offer appropriate examinations. In practice, this series is being used very widely in comprehensive schools, and its first seven books, followed by *Books X, Y* and *Z*, provide a course leading to the SMP O-level examination. An alternative treatment of the material in *SMP Books A, B, C* and *D* is available as *SMP Cards I* and *II*.

Teacher's Guides accompany all series of books.

The SMP has produced many other texts, and teachers are encouraged to obtain each year from the Cambridge University Press, Bentley House, 200 Euston Road, London, NW1 2DB, the full list of SMP books currently available. In the same way, help and advice may always be sought by teachers from the Director at the SMP Office, Westfield College, Hampstead, London, NW3 7ST, from which may also be obtained the annual Reports, details of forthcoming in-service training courses and so on.

The completion of this first ten years of work forms a firm base on which the SMP will continue to develop its research into the mathematical curriculum and is described in detail in Bryan Thwaites's *SMP: The First Ten Years*. The team of SMP writers, numbering some forty school and university mathematicians, is continually evaluating old work and preparing for new. At the same time, the effectiveness of the SMP's future work will depend, as it always has done, on obtaining reactions from a wide variety of teachers – and also from pupils – actively concerned in the class-room. Readers of the texts can therefore send their comments to the SMP, in the knowledge that they will be warmly welcomed.

1975

ACKNOWLEDGEMENTS

The principal authors, on whose contributions the SMP texts are largely based, are named in the annual reports. Many other authors have also provided original material, and still more have been directly involved in the revision of draft versions of chapters and books. The Project gratefully acknowledges the contributions which they and their schools have made.

This book – *Book Y* – has been written by

T. Easterbrook
D. Hale
E. W. Harper
Joyce Harris
B. Jefferson

and edited by Mary Tait.

The Project owes a great deal to its Secretaries Miss Jacqueline Sinfield and Miss Julie Baker, for their careful typing and assistance in connection with this book.

We would especially thank Professor J. V. Armitage and P. G. Bowie for the advice they have given on the fundamental mathematics of the course.

Some of the drawings at the chapter openings in this book are by Ken Vail.

We are grateful to the Oxford and Cambridge Schools Examination Board and the Southern Regional Examinations Board for permission to use questions from their examination papers, and to Frederick Parker Limited, P.O. Box 146, Leicester, for providing us with the photograph of the 'Little Giant' 150 litre/5T cement mixer and for allowing us to reproduce it.

We are very much indebted to the Cambridge University Press for their cooperation and help at all times.

THE SCHOOL MATHEMATICS PROJECT

Book Y

CAMBRIDGE UNIVERSITY PRESS

Preface

This is the second of three books designed for O-level candidates who have previously followed the *A–H* series of books. *X*, *Y* and *Z* follow on from *Book G* and cover the remainder of the course for the O-level examination in 'SMP Mathematics'. The books will also be found suitable for students following a one year revision course for O-level and for those who have previously taken the CSE examination.

Many of the topics introduced in *Books A–G* are extended in *X*, *Y* and *Z* and several new topics are also introduced. *Book Z* contains review chapters covering the complete O-level course.

Book Y commences with a graphical and algebraic approach to Rates of Change. This dual approach is also used in the chapters on Simultaneous Equations and Quadratic Functions. The chapter on Coordinates extends the work in 3D introduced in *Book X* and reviews polar coordinates. Algebraic structure is dealt with in some depth in the chapters on Looking for an Inverse, Sets of Numbers and Looking for Functions. The Slide Rule chapter introduces a method for testing for proportionality between ordered sets, and the method is applied in Looking for Functions.

The chapters on Plans and Elevations and Linear Programming form complete sections of the syllabus, and the Tangent chapter completes the work on trigonometry required for the O-level course.

The formula for the volume of a tetrahedron is established in the Mensuration chapter and this is used to deduce volumes of other solids.

Three interludes are included in the book. These are Incidence Matrices, Circle, and Units and Dimensions and they should be viewed as an integral part of the course.

Each of the three books is accompanied by a Teacher's Guide. The Teacher's Guides contain answers, teaching suggestions and ideas.

Contents

Contents

1 Rates of Change

1 The way things change

(*a*) When a bar of metal is heated, it expands, causing its length to increase. The arrow diagram in Figure 1 shows the results of an experiment in which the length of a bar of metal was measured at various temperatures.

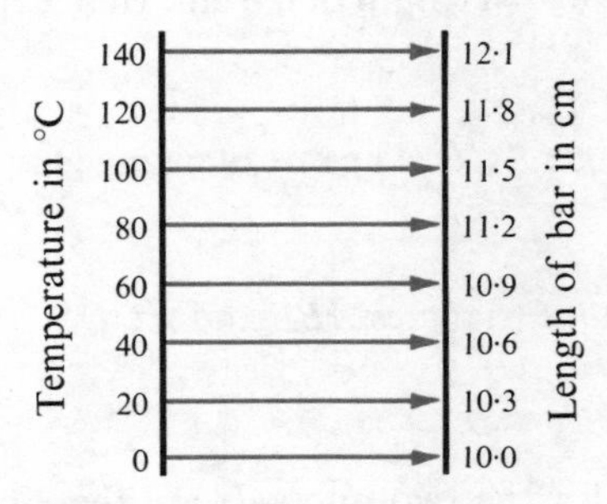

Fig. 1

Suppose that the metal bar at a temperature of 40 °C is heated until its temperature is 100 °C. Check from Figure 1 that the corresponding change in length is 0·9 cm.

We say that the *rate of change* of length with respect to temperature is

$$\frac{11{\cdot}5 - 10{\cdot}6}{100 - 40} = \frac{0{\cdot}9}{60} = 0{\cdot}015\,\text{cm/°C}.$$

Notice that a rate of change has units: in this case the units are cm/°C.

(*b*) Suppose that we heat the bar from 20 °C to 40 °C. What is the rate of change of length with respect to temperature for this interval?

Work out the rates of change for some more intervals of your own choice. What do you notice?

(*c*) If we graph the relation represented by Figure 1 we obtain the straight line graph shown in Figure 2.

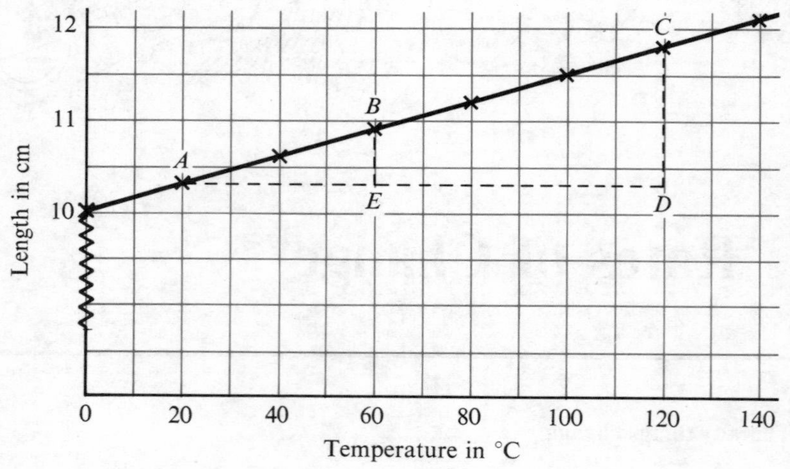

Fig. 2

Consider again the rate of change of length with respect to temperature.

In Figure 2, the distance AE represents a change of temperature of 40 °C and BE represents a change in length of 0·6 cm. Hence the rate of change for this interval is measured by $\dfrac{BE}{AE}$.

Now look at the triangle ACD. The rate of change for this interval is measured by $\dfrac{CD}{AD}$.

What can you say about triangles ABE and ACD?

Do you agree that $\dfrac{BE}{AE} = \dfrac{CD}{AD}$?

The rates of change for the two intervals are therefore equal. If you choose a different interval will you obtain the same rate of change?

(*d*) You should have found in (*b*) and (*c*) that the rate of change of length with respect to temperature is *constant* and equal to 0·015 cm/°C.

When the graph of a relation is a *straight line* then the rate of change of one quantity with respect to the other is constant.

Exercise A

1

x	0	2	4	6	8
y	1	7	13	19	25

Find the rate of change of y with respect to x for the intervals:
(i) $x = 0$ to $x = 6$; (ii) $x = 2$ to $x = 4$; (iii) $x = 4$ to $x = 8$.

2

x	$^-3$	0	3	6	9
y	$^-4$	1	6	11	16

Find the rate of change of y with respect to x for the intervals:
(i) $x = {}^-3$ to $x = 3$; (ii) $x = 0$ to $x = 6$; (iii) $x = 3$ to $x = 9$.

3 The distance a car has travelled from a starting point is recorded at various times and the results shown in the table. Find the rate of change of distance with respect to time for three intervals of your choice. Draw a graph of distance against time.

Time (s)	0	5	10	15	20	25	30
Distance (m)	0	$7\frac{1}{2}$	15	$22\frac{1}{2}$	30	$37\frac{1}{2}$	45

4 The temperature of oil being heated is measured at various times, and the results shown in the table.

Time (s)	0	10	20	30	40	50	60
Temperature (°C)	10	12·1	14·3	16	17·7	19·2	20

Find the rate of change of temperature with respect to time for three intervals. Comment on your results.

5 An engineering firm tests bolts to see how much they stretch under given loads. The table shows the results of a particular test.

Load (Newtons)	0	8	16	24	32
Length (mm)	500	502·5	505	507·5	510

Find the rate of change of length with respect to load for three intervals of your choice.

Draw a graph of the table to check that the rate of change is constant over the whole range.

6 (*a*)

x	0	1	2	3	4	5
y	1	3	5	7	9	11

Calculate the rate of change of y with respect to x, and draw a graph.

(*b*) Repeat (*a*) for the following table:

x	0	1	2	3	4	5
y	9·5	8	6·5	5	3·5	2

In what respect do the two graphs differ? How is this difference reflected in the rates of change?

2 Gradient

(*a*)

(*b*)

Fig. 3

(*a*) What do the road signs in Figure 3 tell you? Which hill is the steeper?

We say that the *gradient* of the hill represented in Figure 3(*a*) is 1 in 4, that is, the road climbs 1 m in every 4 m.

What would appear on a road sign for a road which climbs 100 m in 500 m?

Gradient measures the steepness of a road. It is the ratio of the vertical distance to the corresponding horizontal distance travelled in driving up the hill.

(*b*) In Figure 4 in going from A to B what is the vertical distance travelled? What is the horizontal distance travelled? Now write down the gradient and express it in its simplest form.

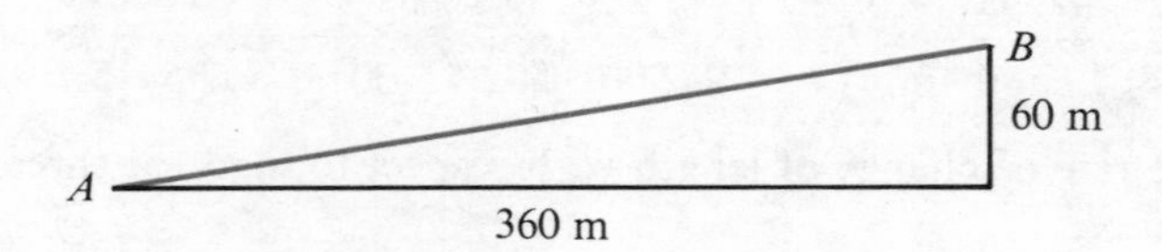

Fig. 4

(*c*) Look at the straight line graphs in Figure 5.

Suppose we wish to measure the gradient of the line in Figure 5(*a*).

What is the horizontal distance from A to B?

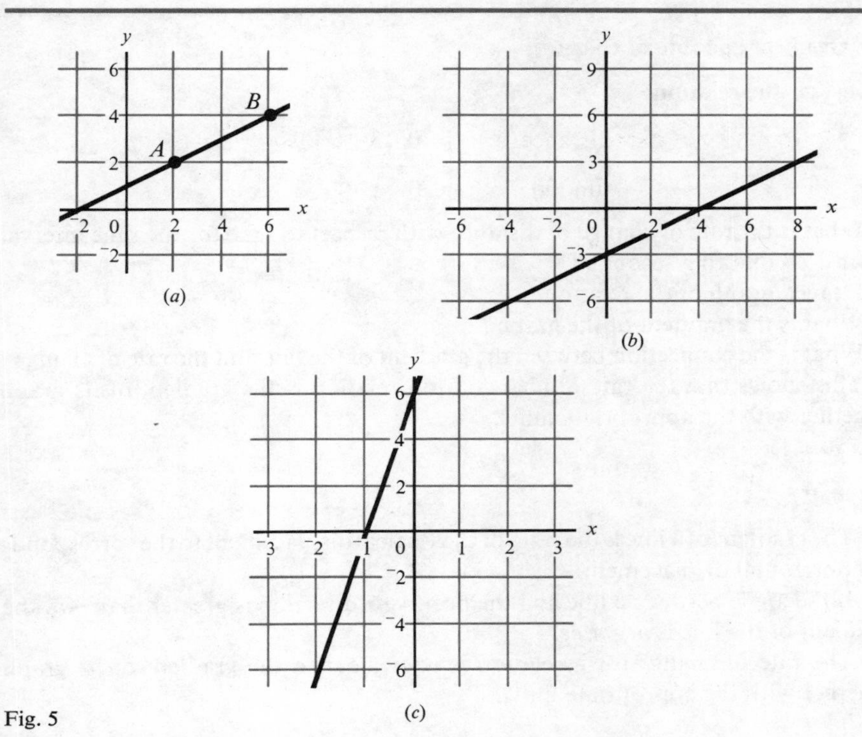

Fig. 5

What is the vertical distance from A to B?

Do you agree that the gradient of the line is 1 in 2?

The gradient of a line is generally written in fractional form and hence we would write the gradient of this line as $\frac{1}{2}$.

Now find the gradients of the lines in Figure 5(b) and (c) giving your answers in fractional form.

(d) Consider the graph shown in Figure 6. The horizontal distance from A to B is $^{-}2$. The vertical distance from A to B is 4.

Hence the gradient of the line is $\frac{4}{^{-}2} = {}^{-}2$.

In what way does the slope of the line in Figure 6 differ from those in Figure 5? How is this difference shown in the value of the gradient?

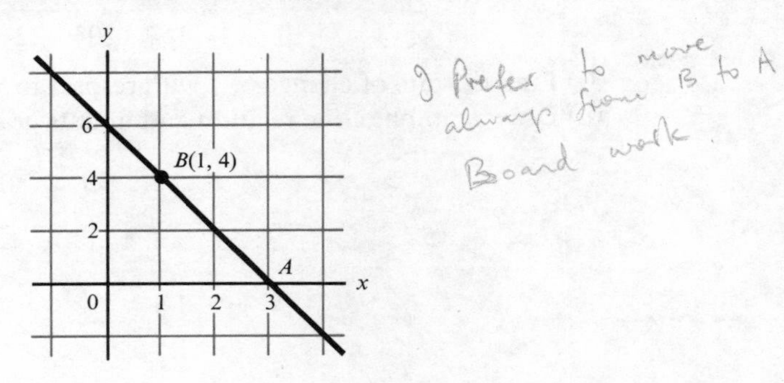

Fig. 6

3 Gradient and rate of change

Consider the relation:

distance (m)	0	3	6	9
time (s)	0	1	2	3

What is the rate of change of distance with respect to time for the time interval from 1 second to 3 seconds?

Draw a graph of the relation.

What is the gradient of the graph?

What is the connection between the gradient of the line and the rate of change?

This shows that the rate of change for a relation is the gradient of its graph together with the appropriate units.

Summary

1 The gradient of a line is the ratio of the vertical displacement to the corresponding horizontal displacement.
2 If the angle between a line and the positive x direction is greater than 90°, the gradient of the line is *negative*.
3 The rate of change for a relation is equivalent to the gradient of its graph together with the appropriate units.

Exercise B

1 Draw graphs showing examples of lines of gradients (*a*) 2; (*b*) $\frac{3}{5}$; (*c*) $^-\frac{1}{4}$; (*d*) 0.

2 Write down the gradients of the lines joining the following pairs of points and illustrate your answers with a sketch to check them:

(*a*) (3, 1) and ($^-1$, $^-2$); (*b*) (1, $^-1$) and (3, 1);
(*c*) ($^-4$, $^-3$) and ($^-2$, 5); (*d*) (4, 1) and (2, 5);
(*e*) (4, 1) and (0, 6); (*f*) (3, $^-2$) and ($^-2$, 3);
(*g*) (6, $^-5$) and (0, 0); (*h*) ($^-1$, $^-2$) and ($^-6$, 8);
(*i*) (4, 2) and (6, 2); (*j*) (4, 2) and (4, 8).

3

x	0	3	5	8	9
y	10	14	$16\frac{2}{3}$	$20\frac{2}{3}$	22

(i) Find the rate of change of y with respect to x.
(ii) Draw a graph of the relation and find its gradient.

4 Repeat Question 3 for each of the following tables:

(*a*)

x	$^-2$	0	1	3	6
y	6	5	$4\frac{1}{2}$	$3\frac{1}{2}$	2

(*b*)

x	$^-2$	$^-1$	1	5
y	7	$4\frac{2}{3}$	0	$^-9\frac{1}{3}$

5 Repeat Question 3 for each of the following tables:

(*a*)

length (cm)	10	25	30	40
mass (g)	7	$17\frac{1}{2}$	21	28

(*b*)

cost price (p)	10	15	20	25
selling price (p)	12	18	24	30

(*c*)

length (cm)	1	2	3	4
area (cm²)	0·5	2	4·5	8

4 Gradients of algebraic functions

(*a*) Consider the gradient of the graph of $f: x \rightarrow 2x + 5$ as shown in Figure 7.

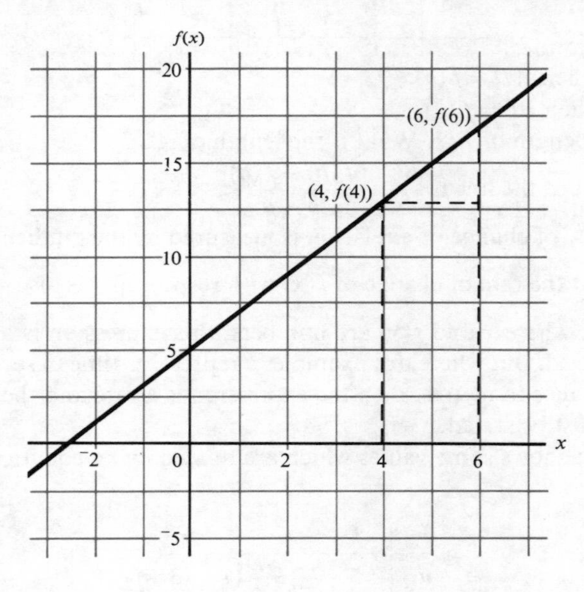

Fig. 7

Find $f(4)$ and $f(6)$.
Did you obtain 13 and 17?

Can you explain why the gradient of the line is $\dfrac{17-13}{6-4} = \dfrac{4}{2} = 2$?

Use this method to find the gradient of the graphs of the following functions:

(i) $f: x \to \frac{1}{2}x + 1$;
(ii) $f: x \to 3x - 2$;
(iii) $f: x \to {}^{-}2x + 3$.

What do you notice? Find the gradient of some more functions of the same form. Do they confirm your findings?

You should have found that a function of the form $f: x \to mx + c$ has gradient m. We say that m is *the coefficient of x in $mx + c$*. The gradients in (i), (ii), (iii) are $\frac{1}{2}$, 3 and $^{-}2$, respectively.

(*b*) Figure 8 shows the graph of the linear function $f: x \to f(x)$.

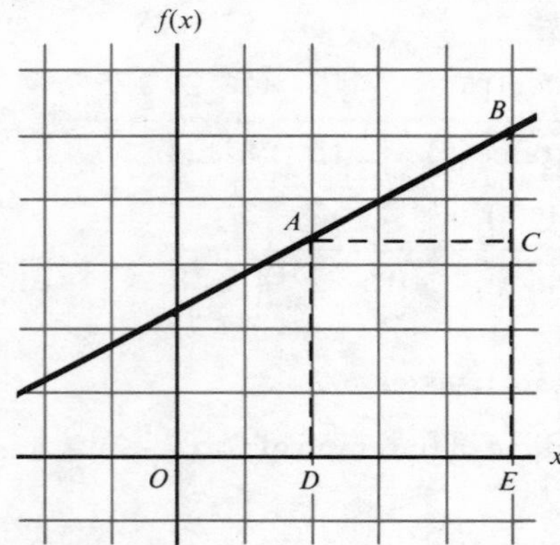

Fig. 8

If $OD = a$, then $AD = f(a)$.
If $OE = b$, then $BE = f(b)$.
What is the length of BC? What is the length of AC?

The gradient of the line is $\dfrac{BC}{AC} = \dfrac{f(b) - f(a)}{b - a}$.

Since the rate of change of a relation is measured by the gradient of its graph, we can say that the rate of change of $f(x)$ with respect to x is also $\dfrac{f(b) - f(a)}{b - a}$.

Notice that, where x and $f(x)$ are numbers, this expression is a ratio and no units are involved. But where, for example, x represents time in seconds and $f(x)$ represents distance in metres, then the expression is a rate and the units, metres per second, must be stated.

(*c*) The table shows some values which are related by an equation of the form $y = mx + c$.

x	1	2	3	4	5
y	5	7	9	11	13

What is the rate of change of y with respect to x?
Do you agree that the gradient of the graph of the relation is 2?
So we can now write $y = 2x + c$.
But when $x = 1$, $y = 5$ and therefore we can find the value of c.
Do you agree that $y = 2x + 3$?

Exercise C

1 Draw the graph that represents the function $f\colon x \to 3x + 1$, and find its gradient.

2 Calculate the gradient of the line $y = 1 + \frac{3}{4}x$.

3 Write down the gradients of the following functions:

(a) $x \to 3x + 4$; (b) $z \to 2 + 5z$;
(c) $x \to 3 - 8x$; (d) $t \to \frac{2}{3}t - 7$.

4 Write down the gradients of the following lines:

(a) $y = 4x - 3$; (b) $2y = 3x + 4$;
(c) $\frac{1}{3}y = 3 - \frac{1}{5}x$; (d) $\frac{3}{4}y = 7x - 1$.

5 Express y as a function of x of the form $y = mx + c$ for the following tables:

(a)

x	1	2	3	4
y	2	8	14	20

(b)

x	$^-5$	$^-3$	$^-1$	1
y	0	$^-4$	$^-8$	$^-12$

(c)

x	6	12	18	24
y	5	9	13	17

6 The following table shows the length of a piece of wire carrying various masses:

mass m (kg)	5	10	20	25
length l (cm)	51·5	53	56	57·5

Express l as a function of the form $l = am + b$. Use your expression to find (a) l when $m = 28$ and (b) the unstretched length of the wire (that is, the value of l when $m = 0$).

5 Rate of change at an instant

(a) A man drove from Manchester to Bristol, a distance of 255 km, in 3 hours. What was his average speed for the journey?

Do you think it likely that he travelled at exactly 85 km/hour for the whole of the journey?

(b) Here is a table giving more details of his journey:

time (hours)	0	0·4	0·68	1·18	1·68	2·18	2·37	3
distance (km)	0	20	50	100	150	200	220	255

The rate of change of distance with respect to time from 50 km to 200 km is $\dfrac{200 - 50}{2{\cdot}18 - 0{\cdot}68} = \dfrac{150}{1{\cdot}5} = 100$ km/hour. Notice that this is the average speed for the interval 50 km to 200 km.

Calculate the average speed for the following intervals:

(i) 150 km to 200 km; (ii) 0 km to 20 km;
(iii) 200 km to 255 km. Comment on your results.

(*c*) All the speeds calculated so far have been average speeds, but a speedometer measures speed at an instant.

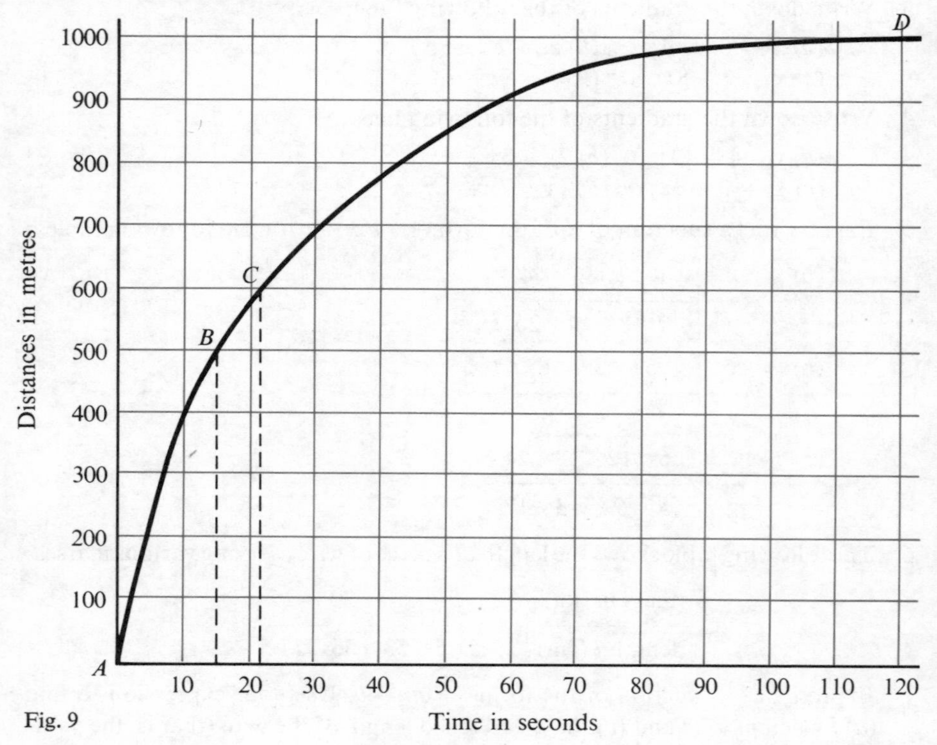

Fig. 9

Figure 9 shows a distance–time graph for a car which is slowing to rest over a distance of 1 km.

Calculate its average speed in metres per second over the whole interval from *A* to *D*.

Do you think this would be a good estimate of its speed as it passes the 500 m mark, point *B* on the graph?

The car was also timed from 500 m to 600 m, the interval *B* to *C* on the graph. Calculate its average speed over this interval.

Do you think this would be a better estimate of its speed at *B*?

How might we improve on our estimate of the speed at *B*?

In general, the shorter the distance over which the car is timed, the more accurate will be our estimate of its speed at *B*.

This is shown in Figure 10. The lines BC_1, BC_2, . . ., are chords all passing through *B*. As the point *C* gets closer to *B*, our idea that the curve possesses a direction at *B* suggests that there is exactly one line which 'just touches' the curve there and that the chords get closer to it in the sense of having the same direction

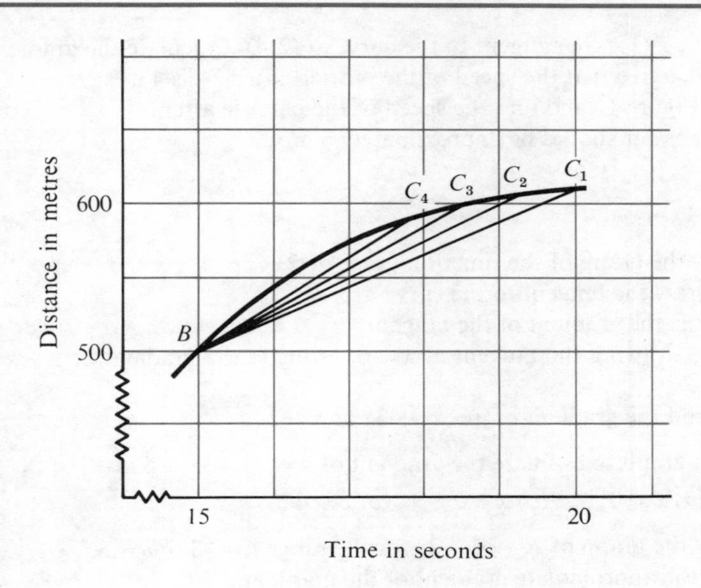

Fig. 10

at *B*. If there is such a line it is called the *tangent* at *B*. So the gradient of the chord *BC* approaches the gradient of the tangent as *C* approaches *B*.

We say that the speed at a given instant is the gradient of the tangent to the graph at that instant with appropriate units, that is the rate of change of the function at that instant.

To find the rate of change of distance with respect to time at a point we

(i) draw the tangent to the curve at that point;
(ii) calculate the gradient of the tangent;
(iii) state the units.

Figure 11 shows a distance time graph of a particle over an interval of 4 s.

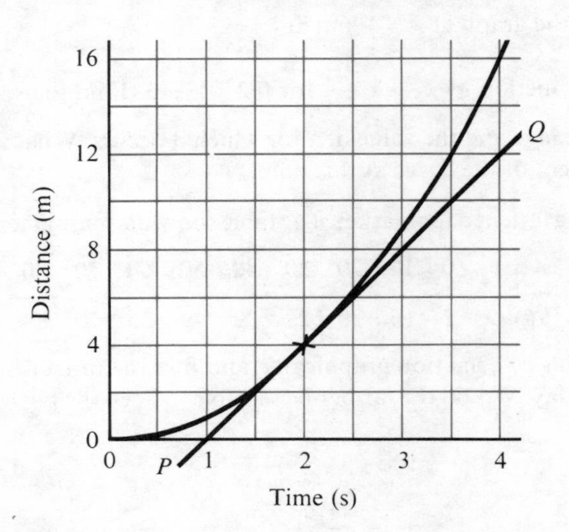

Fig. 11

The line PQ is the tangent to the curve at $(2,4)$. Calculate the gradient of PQ.
Do you agree that the speed of the particle after 2 s is 4 m/s?
Copy Figure 11 and find the speed of the particle after 3 s.
Your answer should be approximately 6 m/s.

Exercise D

1 Draw the graph of the function $y = x^2 + 2x$ for $^{-}1 \leqslant x \leqslant 3$.
(*a*) Draw the tangent to the curve at $x = 1$.
What is the gradient of the tangent?
(*b*) By drawing the tangent at $x = 0$, estimate the gradient of $y = x^2 + 2x$ at $x = 0$.
(*c*) Find the gradient of $y = x^2 + 2x$ at $x = 2$.

2 Use a graph to estimate the gradient of $y = x^2 - 3x + 5$ at
(*a*) $x = 0$; (*b*) $x = 2$; (*c*) $x = 0{\cdot}5$.

3 Draw the graph of $R = v^3 - 2v$ for the range $v = {}^{-}3$ to $v = 3$.
Find the approximate gradient of the graph at
(*a*) $v = 1$; (*b*) $v = {}^{-}2$; (*c*) $v = 0$.

4 A sledge slides down a slope and its distance travelled, s metres, at time t seconds is given by

$$s = 10t + 2t^2.$$

Draw a graph of s against t for $0 \leqslant t \leqslant 4$. Estimate the speed of the sledge at
(*a*) $t = 1$; (*b*) $t = 3$; (*c*) $t = 2{\cdot}4$.
What are the units of your answer?

5 Draw the curve $y = 2 - \frac{1}{x}$ for $\frac{1}{4} \leqslant x \leqslant 4$. Plot at least 8 points. Estimate the gradient of the graph at $x = \frac{1}{2}$ and at $x = 2$.

6 Graph the function $g: x \rightarrow x + \frac{1}{x}$ for $0{\cdot}2 \leqslant x \leqslant 3$. Find the smallest value of $g(x)$ in this range and the value of x for which it occurs. What is your estimate of the gradient of the curve at this point?

7 A pig is being fattened for market. The table shows its mass at ten-day intervals.

Day	0	10	20	30	40	50	60	70	80	90
Mass (kg)	2	16	20	28	39	49	56	70	84	101

Represent this information graphically, and find the rate of increase of mass on the 20th day. Would the farmer be sensible to keep the pig longer? Explain your answer.

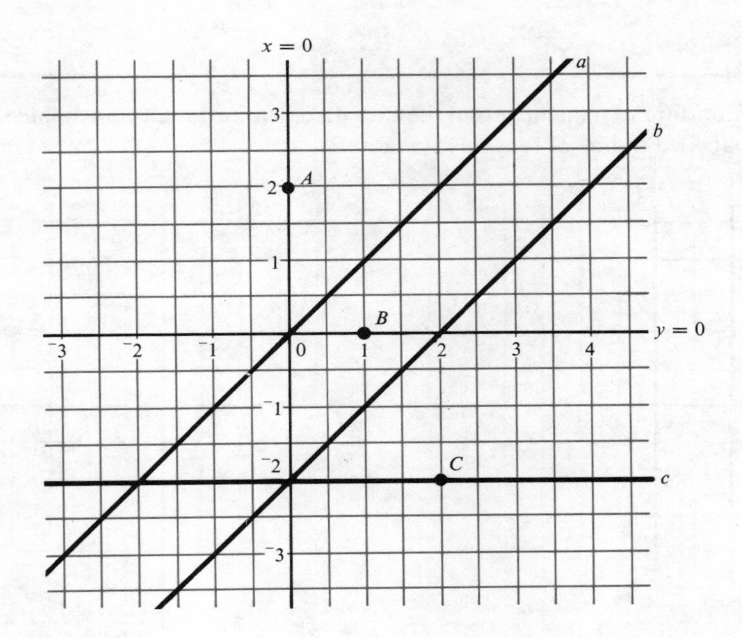

2 Coordinates

1 Coordinates in two dimensions

(*a*) A review

Make a copy of the diagram at the head of this chapter and use it to answer the following:

(i) What are the coordinates of A, B and C?

(ii) Draw the line through A, B, C. What is its gradient? If the equation of this line is $y = mx + c$, what are the values of m and c? (Check that the coordinates of A, B and C really do satisfy your equation.)

(iii) Write down the equations of the lines a, b and c.

(iv) Mark with a cross all points belonging to the set

$$\{(x, y)\colon\ y < x - 2;\ x, y \text{ are whole numbers}\}.$$

(v) What values of x and y satisfy both $y = {}^{-}2x + 2$ and $y = x - 2$? Check by substitution.

(*b*) Cartesian coordinates

The (x, y) system of coordinates for describing the position of a point was invented by the seventeenth century mathematician René Descartes. For this reason, they are generally called *Cartesian coordinates*.

It is important to remember that the axes used in a Cartesian coordinate system may be labelled in one of two ways (Figure 1).

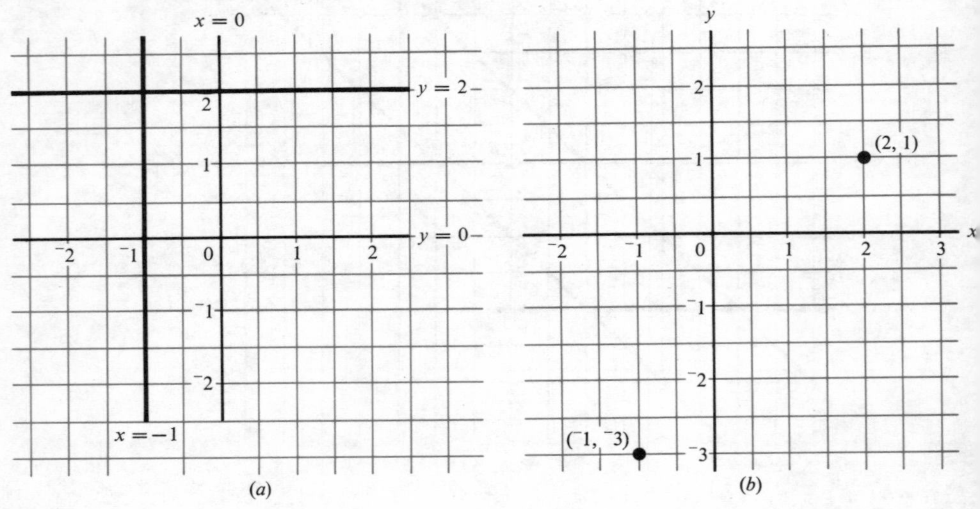

Fig. 1

Figure 1(a) reminds us that lines parallel to $y = 0$ have equations of the form $y =$ constant and lines parallel to $x = 0$ have equations of the form $x =$ constant. Figure 1(b) shows that the x and y coordinates of a point are read on the x and y axes.

(c) **Distance on a graph**

What are the coordinates of A and B in Figure 2? How can you obtain the length of AP from the x coordinates? How long is BP?

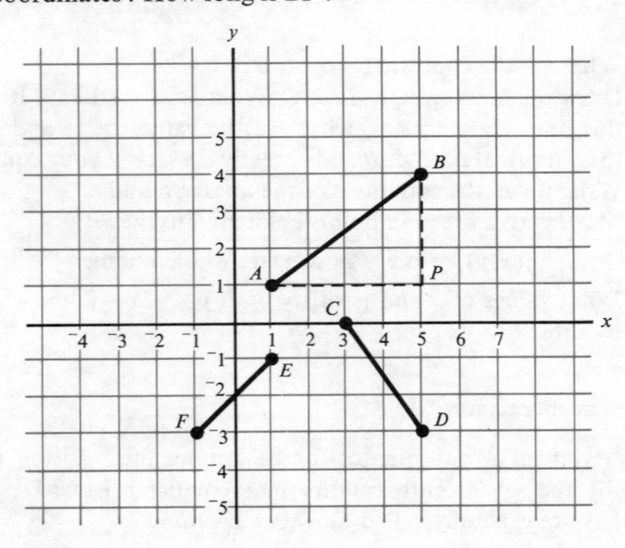

Fig. 2

Now use Pythagoras' rule to work out the distance from A to B. Copy Figure 2 and draw suitable right angled triangles to help you find the lengths CD and EF correct to 2 S.F.

Exercise A

1 Write down the equation of the line through $(0, 3)$ which has a gradient of $^{-}2$. Which of the following points lie on this line?

$$(1, 2), (1, 1), (2, ^{-}1), (^{-}1, 5), (3, ^{-}2).$$

2 Show on a graph the set of points $\{(x, y): y > 2x\}$.

3 What is the gradient of the line joining $(1, 2)$ and $(4, ^{-}1)$?

4 A is the point $(1, 0)$ and B is the point $(5, 0)$. Write down the coordinates of three points which belong to the set $\{P: PA = PB\}$.

5 Find the area of the region bounded by the x and y axes and the line $4x + 3y = 12$.

6 If $A = \{(x, y): x > 2\}$ and $B = \{(x, y): y < x\}$, show clearly on a diagram the set $A \cap B$.

7 A pentagon $ABCDE$ is symmetrical about the line $y = x$. A is $(3, 0)$, C is $(6, 6)$ and D is $(3, 6)$. What are the coordinates of B and E?

8 State a set of inequalities satisfied by the coordinates of all the points of the shaded region in Figure 3. (The boundary is included.)

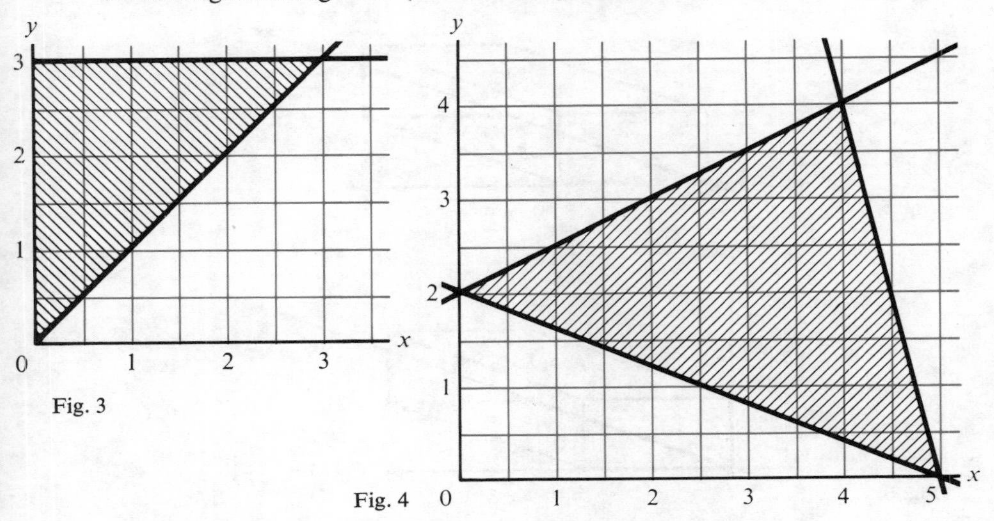

Fig. 3

Fig. 4

9 Two sides of the shaded triangle (Figure 4) have equations $2x + 5y = 10$ and $4x + y = 20$. What is the equation of the third side? Say whether the following points are inside the triangle, outside the triangle, or on its boundary.

$$(2, 2), (3, 2), (1, 4), (2\tfrac{1}{2}, 1), (1, 1), (3, 1).$$

10 What are the coordinates of the point $(3,2)$ after reflection in the x axis? Find the image of $(3,2)$ under the composite transformation – 'reflection in the x axis followed by a half-turn about the origin'.

11 If O is the origin and A, B are the points $(3,4)$, $(7,1)$ respectively, calculate the lengths OA, OB and AB. Show that the triangle OAB is right-angled.

12 Show that the points with coordinates $(7,10)$, $(^-5,^-6)$ and $(7,^-6)$ lie on a circle with its centre at $(1,2)$. What is the radius of this circle?

13 Check that $(0,2)$ belongs to the set of points $\{(x,y): x^2+y^2=4\}$ and find three other members of this set. Draw a graph to show the complete set.

14 P is the point $(2,0)$, Q is the point $(1,2)$ and P', Q' are their images under a transformation whose matrix is $\begin{pmatrix} 1 & 0 \\ 2 & 1 \end{pmatrix}$. Calculate the lengths PQ and $P'Q'$ and explain why the transformation cannot be a reflection or rotation.

2 Coordinates in three dimensions

(*a*) Fred in space

Fred, the mathematical fly, has landed at the centre of the floor, F_1 (Figure 5). His position on the floor is given by $(3,4)$. If Fred decides to land on a wall or ceiling, how can we describe his position? Fred's second landing place, F_2, is in the middle of a wall (Figure 6).

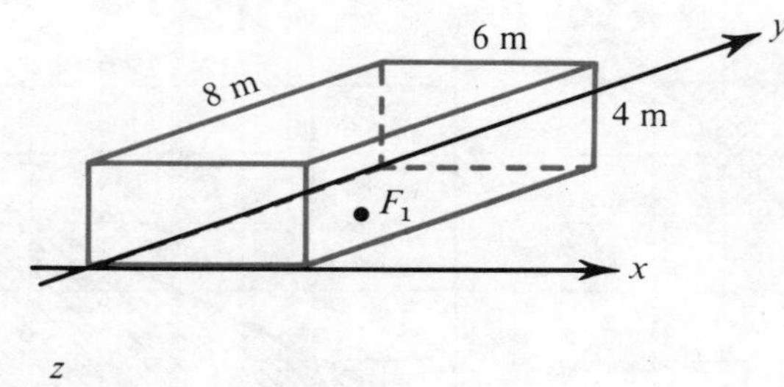

Fig. 5

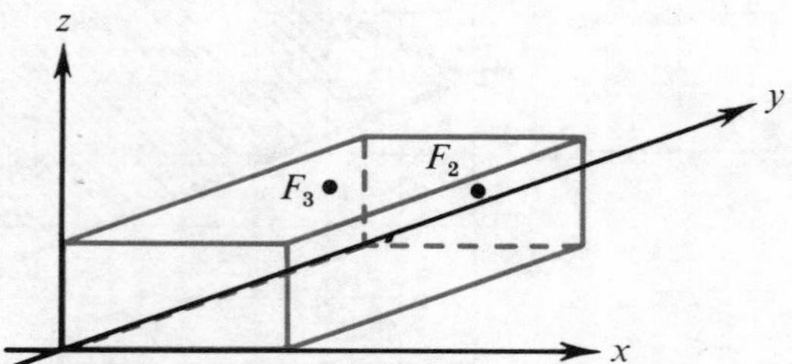

Fig. 6

Do you agree that his position is given by the three coordinates $(3,8,2)$? We can say that F_2 has x-coordinate 3, y-coordinate 8 and z-coordinate 2. What are the coordinates of Fred's third landing place in the centre of the ceiling (F_3 in Figure 6)?

We can now look again at F_1. Its position can be given by three coordinates. Do you agree that the z-coordinate of F_1 is 0? Write down all three coordinates of F_1.

In three dimensions, we establish three axes perpendicular to each other. Provided each axis has a scale, we can define the position of any point by giving three coordinates.

Make two cubes and label them as in Figure 7.

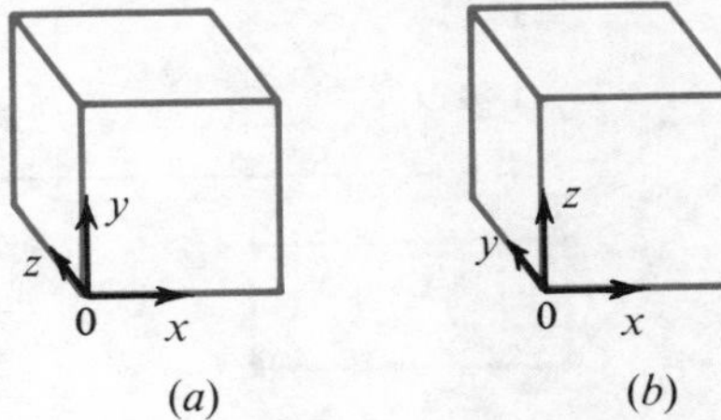

Fig. 7

Look at the set of axes shown in Figure 7(a). Is it possible to rotate them about O into the position shown in Figure 7(b)? Do you agree that the two sets of axes are not equivalent? They are related to one another as right-handed and left-handed gloves are. Figure 7(a) shows the left-handed set and Figure 7(b) the right-handed set. In practice, a right-handed set of axes is always used.

Exercise B

1 Write down the coordinates of the points A, B, C, D, E, F, G, H in Figure 8.

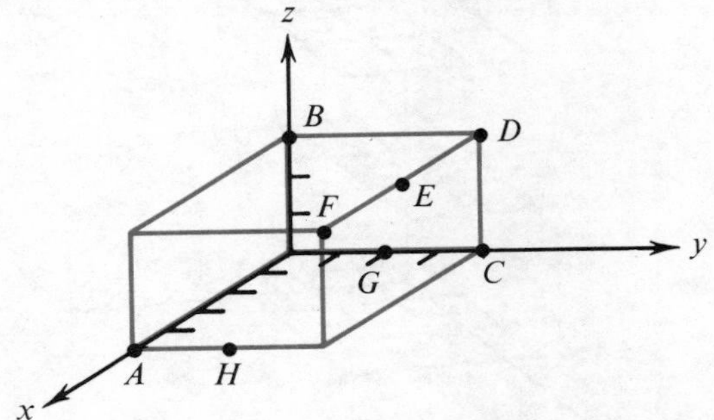

Fig. 8

2 Which sets of axes in Figure 9 are right-handed? (It will probably help to make and label some cubes.)

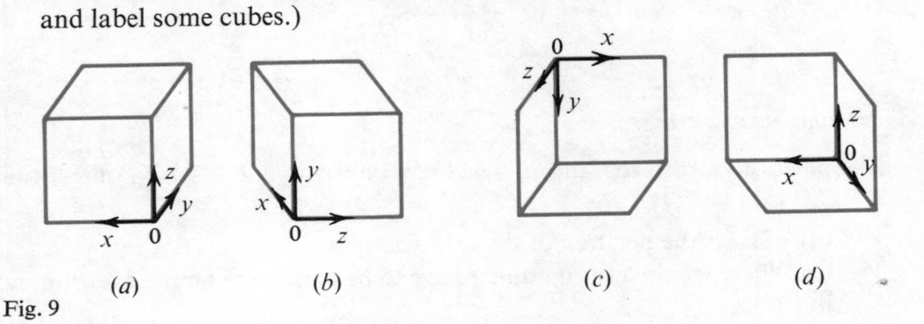

Fig. 9

3 *ABCDPQRS* is a cuboid (Figure 10). The origin is at its centre and the axes are parallel to its edges. If A is the point $(3, 2, 1)$, find the coordinates of the other seven vertices.

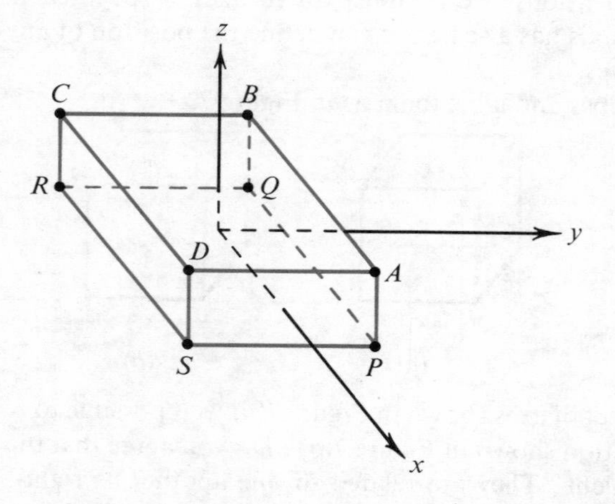

Fig. 10

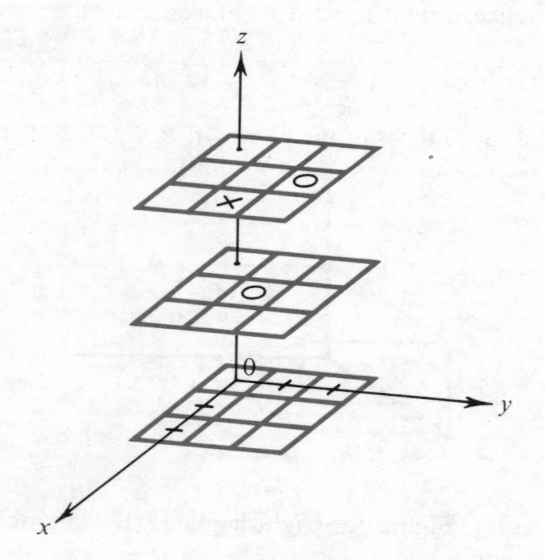

Fig. 11

4 *3D noughts and crosses*

The position of the cross in Figure 11 can be written $(2, 1, 2)$ and one of the noughts is at $(1, 1, 1)$.

(*a*) What is the position of the other nought?

(*b*) Where would a third nought have to be placed to complete a winning line?

5 In 3D noughts and crosses, where would crosses have to be placed to make winning lines if there were already crosses at the following pairs of points?

(*a*) (2,0,2), (1,0,2); (*b*) (0,0,0), (0,0,1);
(*c*) (0,2,2), (1,1,1); (*d*) (1,0,2), (1,2,2);
(*e*) (0,1,2), (2,1,0); (*f*) (1,1,1), (1,1,2).

(*b*) **Planes**

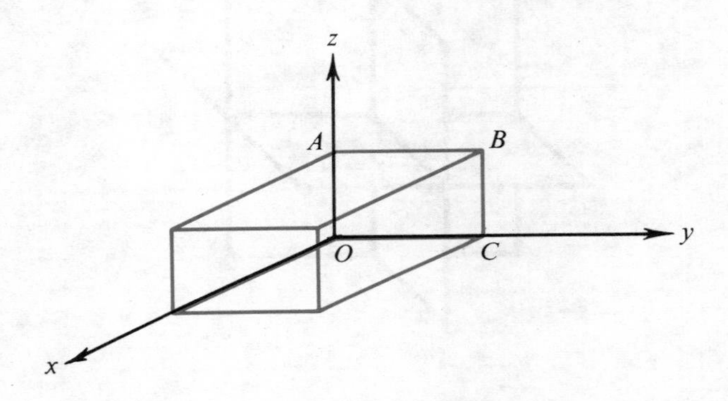

Fig. 12

In Figure 12,

O is the point (0,0,0),
A is the point (0,0,4),
B is the point (0,6,4),
C is the point (0,6,0).

These four points lie in a plane. Can you visualize this plane; it also contains the y and z axes? Write down the coordinates of two other points in the plane $OABC$.

Can you describe the set of points in space which satisfy the relation $x = 0$? Does your description agree with Figure 13?

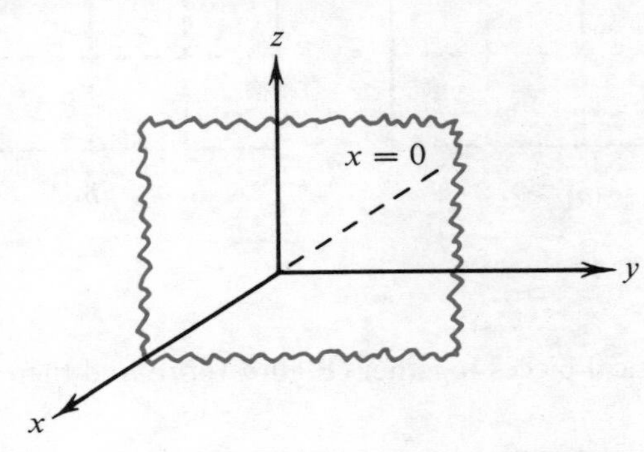

Fig. 13

$x = 0$ is the equation of a plane which contains the origin and is perpendicular to the x axis. Describe the planes $y = 0$ and $z = 0$ in a similar way.

(*c*) A model

In answering the questions of Exercise C, it will be useful to have a model of the three planes $x = 0$, $y = 0$ and $z = 0$ (Figure 14).

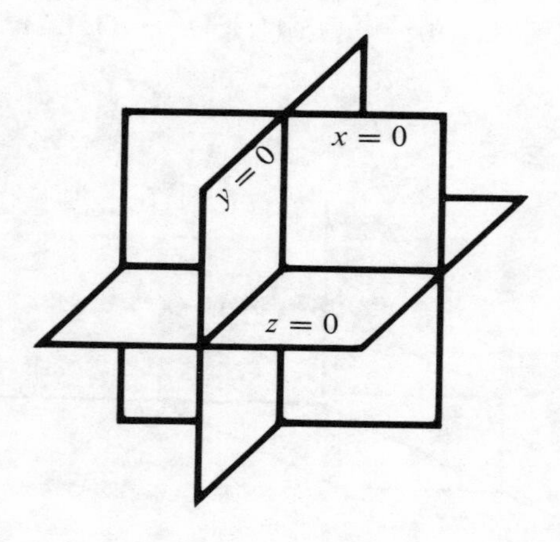

Fig. 14

To make the model you will need three squares (with sides about 15 cm long) of thin card. Two of these are cut as shown in Figure 15(*a*) and the third as in Figure 15(*b*).

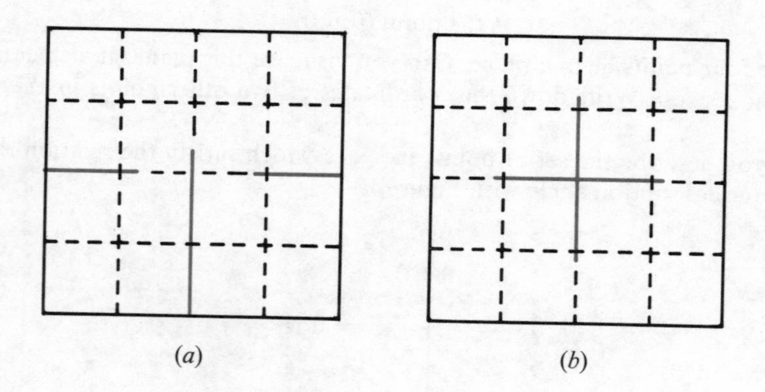

Fig. 15

Fit the two identical pieces together (Figure 16(*a*)) and then fold as shown in Figure 16(*b*).

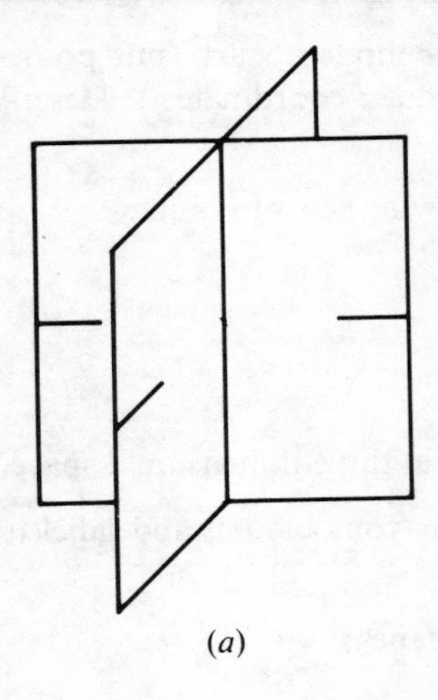

(a)

(b)

Fig. 16

The third square can now be placed half way down and when unfolded the model should appear as in Figure 14. Label the planes and the x, y and z axes.

Exercise C

1 In Figure 17, the shaded plane is parallel to the plane $z = 0$ and 3 units above it. Write down the equation of the shaded plane. What is the equation of a plane parallel to $z = 0$ and 3 units below it?

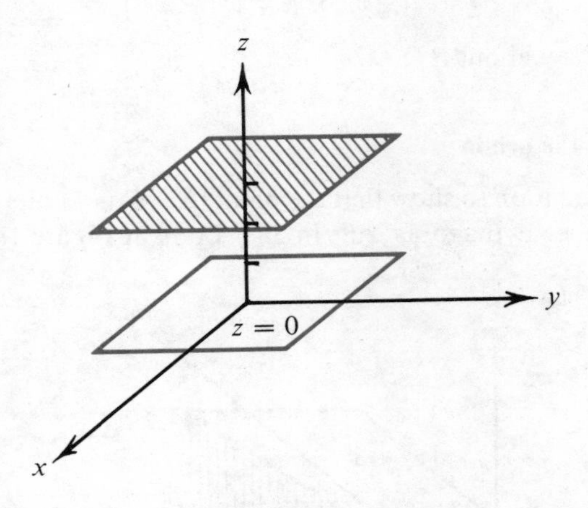

Fig. 17

2 Say which of these statements is true;

(a) the planes $x = 1$, $x = 5$ are parallel;
(b) the planes $x = 2$, $y = 3$ are perpendicular;
(c) the planes $x = 2$, $z = 3$ are perpendicular.

3 Put a suitable scale on the axes of your model. Mark four points on your model which have a y-coordinate, 0, and a z-coordinate, 1. Describe the set $\{(x,y,z): y=0, z=1\}$.

4 Draw and label on your model the following sets of points:

(*a*) $A=\{(x,y,z): y=0, z=2\}$;
(*b*) $B=\{(x,y,z): x=1, z=0\}$;
(*c*) $C=\{(x,y,z): x=0, z=2\}$;
(*d*) $D=\{(x,y,z): x=0, y=2\}$.

5 What does the equation $x=y$ represent in three-dimensional space?

6 Find $\{(x,y,z): x=y, z=0\}$. Draw it on your model, and label it with the letter E.

7 Find the angles between these pairs of planes:

(*a*) $x=0, y=0$;
(*b*) $x=0, y=x$;
(*c*) $y=0, y=x$;
(*d*) $z=0, y=x$;
(*e*) $x=3, y={}^{-}100$.

8 Give the coordinates of four points which satisfy the equation $x+y+z=2$. What do you think this equation represents?

If $$P=\{(x,y,z): x+y+z=2\}$$

and

$$Q=\{(x,y,z): x+y+z=3\},$$

what can you say about $P \cap Q$?

(*d*) Distance from the origin

Can you do a calculation to show that the point $(2,3,6)$ is 7 units from the origin?

Do you remember Pythagoras' rule in 3D? Look at Figure 18 for a reminder.

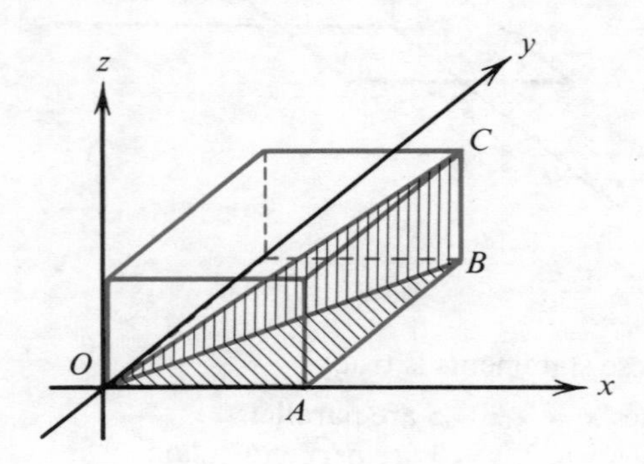

Fig. 18

$$OC^2 = OB^2 + BC^2 \quad \text{and} \quad OB^2 = OA^2 + AB^2$$

therefore

$$OC^2 = OA^2 + AB^2 + BC^2$$

or

$$OC = \sqrt{(OA^2 + AB^2 + BC^2)}.$$

With the axes shown, the lengths of OA, AB, BC are respectively the x, y, z coordinates of the point C. So if C is the point (x, y, z), its distance from the origin O is

$$\sqrt{(x^2 + y^2 + z^2)}.$$

The point (4, 5, 20) is also a whole number of units from the origin. Can you find this distance?

(*e*) Distance between two points

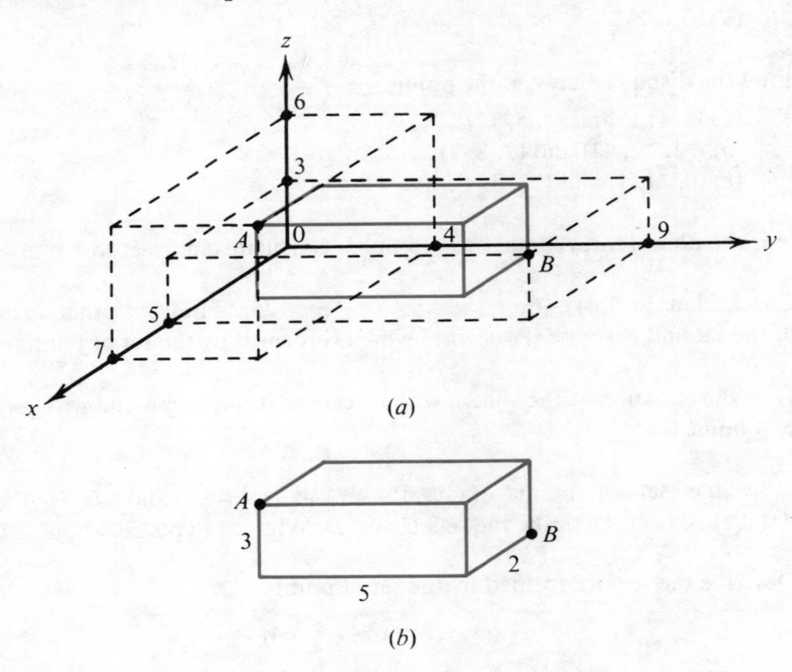

Fig. 19

In Figure 19(*a*), A is the point (7, 4, 6) and B is (5, 9, 3). Figure 19(*b*) shows a box with A and B at opposite corners and its edges parallel to the axes. How are the dimensions of this box obtained from the coordinates of A and B? Check by a calculation that $AB \approx 6{\cdot}2$.

The method used in this particular case will enable us to find the distance between any two points in space whose coordinates are known. We visualize a box with edges parallel to the axes and with the two points at opposite corners.

The dimensions of the box are obtained by taking the numerical difference between each pair of coordinates in turn. Finally, the 3D version of Pythagoras' rule is used to calculate the required distance. Special care is needed if some of the co-ordinates are negative numbers. For example,

$$\left.\begin{array}{l} P \text{ is } (^-2, 1, 5) \\ Q \text{ is } (^-3, ^-1, 2) \end{array}\right\} \Rightarrow \text{box dimensions are } 1, 2, 3$$

$$\Rightarrow PQ \approx 3{\cdot}7.$$

Exercise D

1 How far from the origin are the points:

(*a*) $(3, 4, 12)$;
(*b*) $(1, 0, 1)$;
(*c*) $(^-1, 4, 8)$;
(*d*) (a, b, c)?

2 Find the distance between the points:

(*a*) $(1, 1, 1)$ and $(4, 5, 13)$;
(*b*) $(^-1, ^-2, ^-3)$ and $(2, 2, 9)$;
(*c*) $(0, ^-1, 4)$ and $(^-2, 3, 4)$.

3 A is the point (a, a, a) and $OA = 15$ units. Calculate a correct to 1 D.P.

4 Check that $(0, 3, 4) \in \{(x, y, z): x^2 + y^2 + z^2 = 25\}$. Find six other members of the set and describe the surface which is formed by the set of points.

5 Give the equation of the sphere with its centre at the origin and with a radius of 4 units.

6 C is the set of points $\{(x, y, z): x^2 + y^2 + z^2 = 9\}$ and D is the set $\{(x, y, z): x = 4\}$. Describe the sets C and D. What can you say about $C \cap D$?

7 Describe the surface formed by the set of points

$$\{(x, y, z): x^2 + y^2 = 4\}.$$

(First consider what $x^2 + y^2 = 4$ represents in 2D.)

8 Find p if the points $(p, 1, 2)$, $(2, 3, 5)$, $(4, 7, 11)$ are in a straight line.

3 Polar coordinates

(*a*) A reminder

If you have a long memory, Figure 20 may be familiar. It is a drawing of a radar screen which appeared in *Book B*.

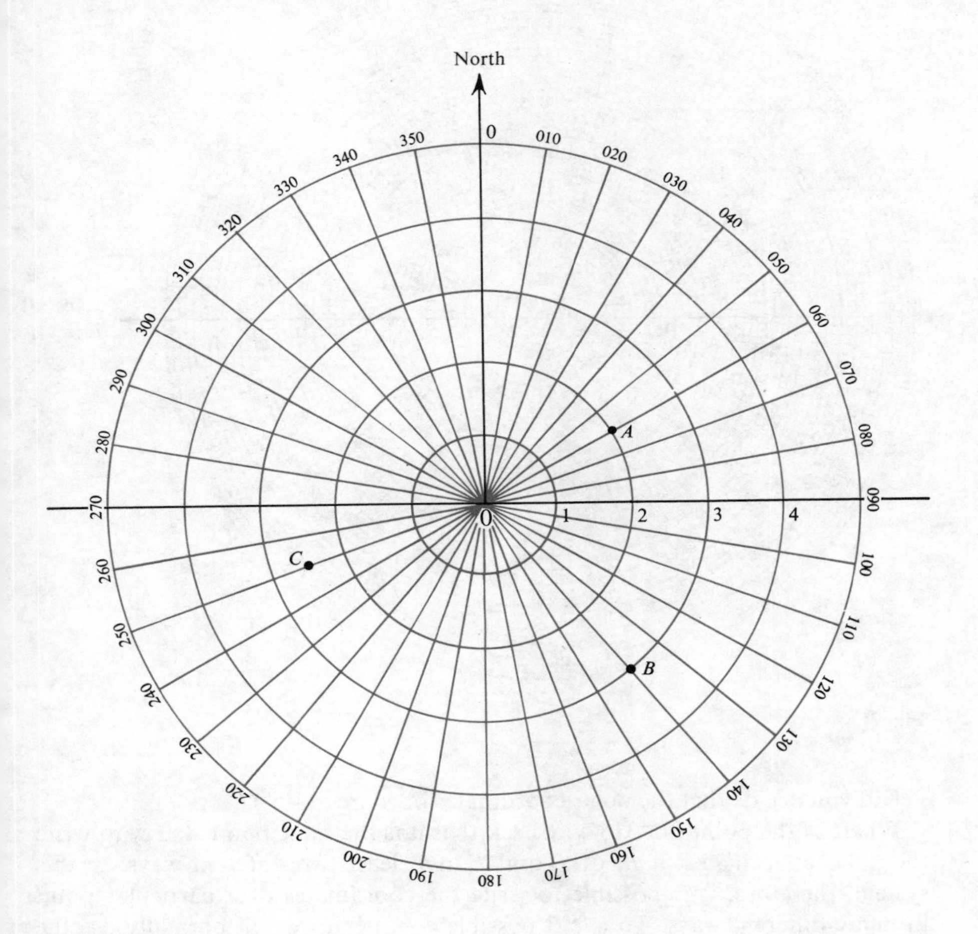

Fig. 20

The position of A is given by the two numbers 2 and 060°. We could refer to A as the point $(2, 060°)$ remembering that the first number is a *range* and the second a *bearing*. Write B and C in a similar way.

This provides another example of two numbers specifying the position of a point in a plane. The numbers are known as the *polar coordinates* of the point.

When bearing is not specifically involved, it is customary to use the convention 'anti-clockwise is positive' in measuring angles. The starting line for angle measure is sometimes referred to as the *initial direction* or central direction.

Figure 21 shows some points plotted on polar graph paper. What are their polar coordinates?

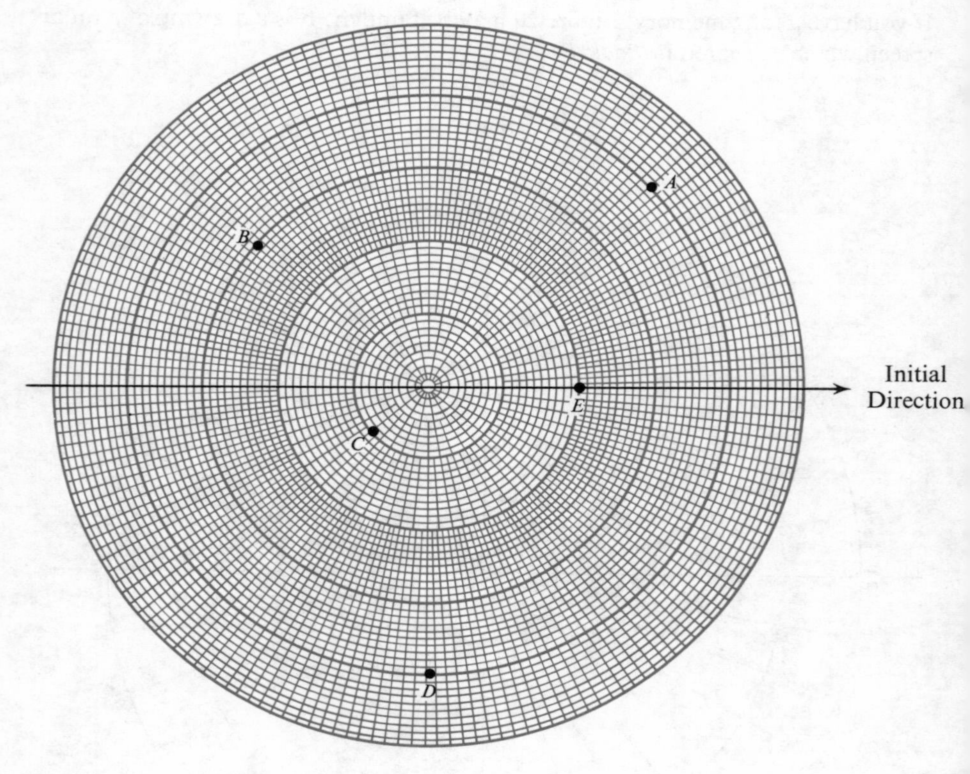

Fig. 21

Did you decide that the polar coordinates of A are $(4, 45°)$?

Where is the point $(4, {}^{-}315°)$? Check that it is the same point A. Try to write the polar coordinates of B, C, D and E in at least two different ways. In this system, therefore, it is possible to write the coordinates of a particular point in many different ways. To avoid possible confusion we will normally use the coordinates $(r, \alpha°)$ of a point which satisfy the following inequalities:

$$r \geqslant 0$$
$$0 \leqslant \alpha < 360.$$

(*b*) Polar and Cartesian

We now have two ways of giving the position of a point in a plane: polar coordinates and Cartesian coordinates.

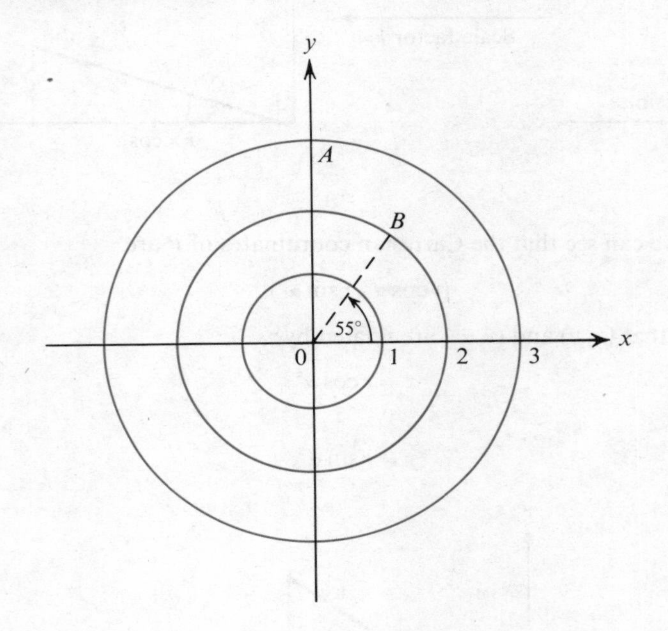

Fig. 22

How are the two types of coordinates related? In Figure 22, A has Cartesian coordinates $(0, 3)$ and polar coordinates $(3, 90°)$. In polar coordinates, B is $(2, 55°)$. What are its (x, y) coordinates? Figure 23 helps to answer this using trigonometry.

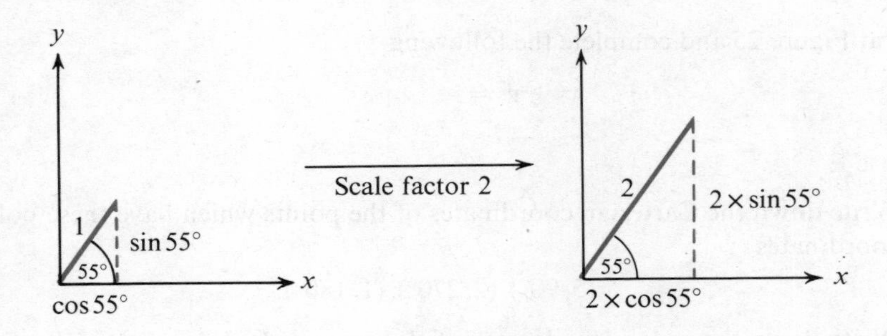

Fig. 23

The x coordinate of B is $2 \times \cos 55° \approx 1{\cdot}15$.

The y coordinate of B is $2 \times \sin 55° \approx 1{\cdot}64$. B has Cartesian coordinates $(1{\cdot}15, 1{\cdot}64)$.

Figure 24 shows a general point P *in the first quadrant*.

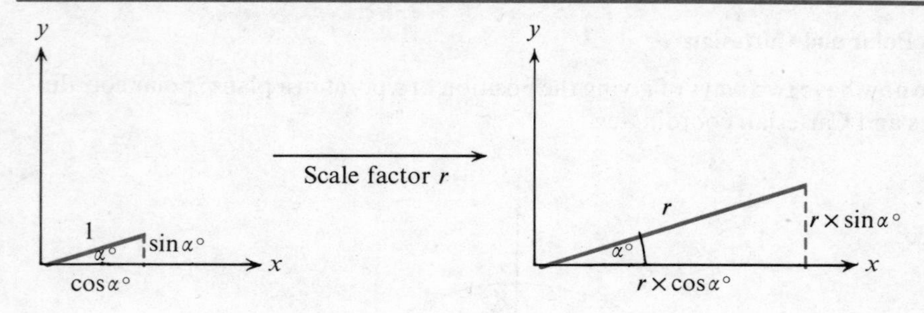

Fig. 24

From it we can see that the Cartesian coordinates of P are

$$(r\cos\alpha°, r\sin\alpha°).$$

This means that (x, y) and $(r, \alpha°)$ are related by:

$$x = r\cos\alpha°$$

and

$$y = r\sin\alpha°.$$

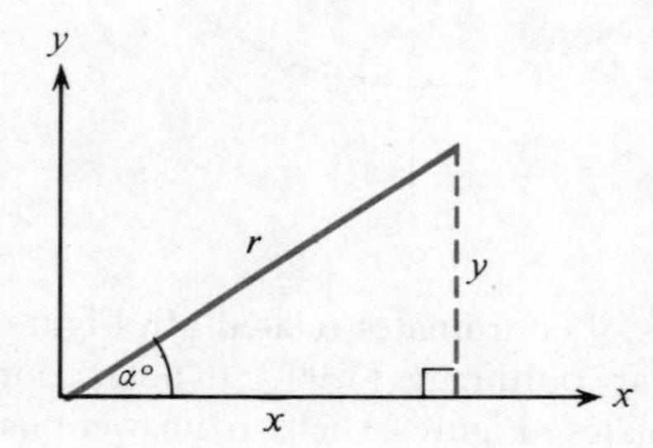

Fig. 25

Look at Figure 25 and complete the following:

$$x^2 + y^2 =$$

Exercise E

1 Write down the Cartesian coordinates of the points which have these polar coordinates:

$$(5, 90°), (2, 270°), (1, 180°).$$

2 Write down the polar coordinates of the points which have the following Cartesian coordinates:

$$(0, 2), (^-3, 0), (5, 0), (0, ^-1), (0, 0), (1, 1).$$

3 Describe the sets of points $\{(r, \alpha°)\colon\ r = 2\}$ and $\{(r, \alpha°)\colon\ r = 3\}$. If $A = \{(r, \alpha°)\colon\ 2 < r < 3\}$ and $B = \{(r, \alpha°)\colon\ 30 < \alpha < 60\}$, show shaded in separate diagrams the sets of points, A, B, $A \cup B$, $A \cap B$.

4 Calculate, correct to 2 S.F., the Cartesian coordinates of points which have polar coordinates:

$$(1, 20°), (10, 75°), (2{\cdot}5, 45°).$$

5 What are the polar coordinates of points which have Cartesian coordinates:

$$(1, 1), (3, 3), (3, 4), (4, 3), (1, 2)?$$

6 Use your tables to find $\sin 140°$ and $\cos 140°$ and then work out $5 \times \sin 140°$ and $5 \times \cos 140°$. The point $(5, 140°)$ is shown in Figure 26. What are its Cartesian coordinates?

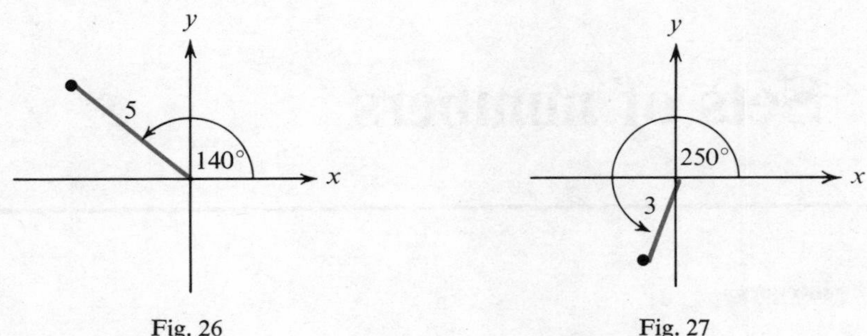

Fig. 26

Fig. 27

7 Work out $3 \times \sin 250°$ and $3 \times \cos 250°$. What are the Cartesian coordinates of the point $(3, 250°)$ shown in Figure 27?

8 Work out the Cartesian coordinates of the point with polar coordinates $(10, 330°)$.

9 We have seen that the equations $x = r\cos\alpha°$, $y = r\sin\alpha°$ relate the Cartesian and polar coordinates of any points in the first quadrant. Do you think that they apply to points in the other three quadrants?

10 Work out the polar coordinates of the points A, B, C in Figure 28.

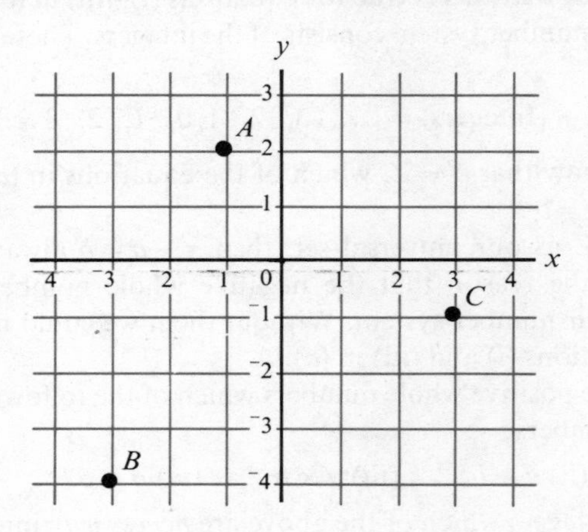

Fig. 28

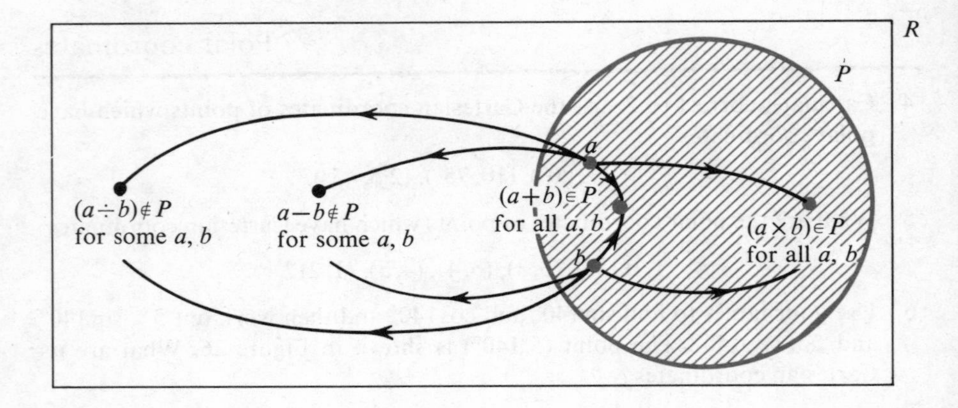

3 Sets of numbers

1 The integers

(*a*) Let us denote the set of positive whole numbers by P:

$$P = \{\text{Positive whole numbers}\} = \{^{+}1, ^{+}2, ^{+}3, \ldots\}.$$

Because these numbers behave like the counting numbers, we usually omit the upper positive and write, for example, $^{+}2$ as 2.

If a number system consists of only the positive whole numbers (i.e. if $\mathscr{E} = P$), which of the following equations have solutions:

(i) $x + 3 = 2$; (ii) $x + 2 = 3$; (iii) $x + 2 = 2$?

If the equation $x + a = b$ has a solution, what can you say about a and b?

(*b*) You should have found that $x + a = b$ has a positive whole number solution only if $b > a$. Check that this is true for equations (i)–(iii) in (*a*).

Now suppose a number system consists of the integers. These numbers form the set Z:

$$Z = \{\text{Integers}\} = \{\ldots, ^{-}3, ^{-}2, ^{-}1, 0, ^{+}1, ^{+}2, ^{+}3 \ldots\}.$$

Remembering now that $\mathscr{E} = Z$, which of the equations in (*a*) have solutions? What about $x + a = b$?

(*c*) If we take Z as our universal set, then $x + a = b$ always has a solution. This is precisely the reason that the negative whole numbers and zero were introduced into our number system. Without them we could not possibly solve, for example, equations (i) and (iii) in (*a*).

(*d*) If a and b are positive whole numbers which of the following are *necessarily* positive whole numbers:

(i) $a + b$; (ii) $a - b$; (iii) $a \times b$; (iv) $a \div b$?

If a and b are integers which of the above are *necessarily* integers?

(*e*) For each pair of positive whole numbers (a,b), $a+b$ is always a positive whole number. The same is true of $a \times b$, but not of $a - b$ and $a \div b$.

We say that P is *closed* under the operations of addition and multiplication. P is *not closed* under the operations of subtraction and division. The diagram at the head of the chapter attempts to represent the idea of closure for P.

Under which operations is the set of integers closed? Draw a diagram like the heading diagram to illustrate your answer.

Summary

1 $P = \{\text{Positive whole numbers}\} = \{^{+}1, {}^{+}2, {}^{+}3, \ldots\}$.
$Z = \{\text{Integers}\} = \{\ldots, {}^{-}3, {}^{-}2, {}^{-}1, 0, {}^{+}1, {}^{+}2, {}^{+}3, \ldots\}$.
We usually omit the upper positives and write the members of P as $1, 2, 3, \ldots$.

2 P is closed under addition and multiplication, i.e. if $a, b \in P$ then $a+b \in P$ for all a, b, and $a \times b \in P$ for all a, b.
Z is closed under addition, multiplication and subtraction.

Exercise A

1 Draw a Venn diagram to represent the relationship between P and Z.

2 If (*a*) $\mathscr{E} = Z$; (*b*) $\mathscr{E} = P$, which of the following equations have solutions?
(i) $x + 1 = 2$; (ii) $2x - 1 = 3$; (iii) $x + 1 = 2x + 1$;
(iv) $3x = 9$; (v) $\frac{x}{4} = 2$.

3 If (*a*) $\mathscr{E} = Z$; (*b*) $\mathscr{E} = P$, which of the following problems always have a solution?
(i) A boy buys x identical pens for n pence. How much did he pay for each?
(ii) Find the number of cm in p metres.
(iii) Solve the inequality $x + a \leqslant b$.
(iv) The temperature at 12.00 hours was n °C. If it fell by t deg C per hour what was the temperature at 15.00 hours?

4 If $\mathscr{E} = P$, when has the equation $ax = b$ a solution? What if $\mathscr{E} = Z$?

5 If the domain of the functions (i) $f\colon x \to x$; (ii) $g\colon x \to 2x$; (iii) $h\colon x \to x^2$; (iv) $m\colon x \to 2x - 1$ is P, what is the range of each function? What if the domain is Z?

6 State whether P and Z are closed or not closed under the operations $*$, $\oplus$, $\ominus$, @ defined below:
(i) $a * b = a \times 2b$; (ii) $a \oplus b = 2a + 3b$;
(iii) $a \ominus b = 2a - 3b$; (iv) $a \mathbin{@} b = \pm\sqrt{ab}$.

7 Graph the functions (i) $f\colon x \to x + 2$; (ii) $g\colon x \to 2 - x$ when the domains of f and g are (*a*) P; (*b*) Z. In each case describe the range of the functions.

2 The rational numbers

(*a*) Solve the equations (i) $4x = 2$; (ii) $3x = {}^{-}2$; (iii) $ax = b$ $(a \neq 0)$.

(*b*) The solution to (i) is $x = \frac{2}{4}$. You might have written $x = \frac{1}{2}$, which is equally correct. In fact, any one of the members of the set of equivalent fractions

$$\left\{\ldots, \frac{^{-}2}{^{-}4}, \frac{^{-}1}{^{-}2}, \frac{1}{2}, \frac{2}{4}, \ldots\right\}$$

could be given as the solution. It is usual, however, for us to write as the representative of such a set the fraction which is in its lowest terms – in this case $\frac{1}{2}$. $\frac{1}{2}$ is an example of a *rational* number. The rational numbers are the set Q of numbers of the form p/q where p and q are integers with no common divisor, and $q \neq 0$:

$$Q = \left\{\text{numbers}\,\frac{p}{q}\,\text{such that}, p, q \in Z\,\text{and have no common divisor, and}\,q \neq 0\right\}.$$

Which rational numbers represent the sets of equivalent fractions of which

(i) $\frac{10}{15}$; (ii) $\frac{^{-}17}{51}$; (iii) $\frac{10}{2}$

are members?

Would you say that $Z \subset Q$ is a true statement?

(*c*) The solution to $ax = b$ is $x = \frac{b}{a}$. Without the rational numbers as a further extension to our number system the equation would only have a solution if a was a factor of b. (Did you obtain this answer for Question 4, Exercise A?) This is one reason why the rational numbers are very important to us. Without them we would not be able to solve such simple problems as 'Divide 18 into 24 equal parts', i.e. 'Solve the equation $24x = 18$'.

(*d*) If a and b are rational which of the following are rational (if necessary after 'cancelling down'):

(i) $a + b$; (ii) $a - b$; (iii) $a \times b$; (iv) $a \div b$?

(*e*) Your answer to (*d*) should suggest to you an important property of the rational numbers; the set is closed under all four simple arithmetical operations, so long as we do not allow division by zero. This means that we can apply each operation and always obtain an answer which is rational.

Summary

1 $Q = \{\text{Rationals}\} = \{p/q$ such that p and q are integers with no common divisor, and $q \neq 0\}$.

2 Q is closed under addition, subtraction, multiplication and division (so long as we do not allow division by zero).

Exercise B

1 Draw a Venn diagram to represent the relationship between P, Z and Q.

2 If you were measuring (i) the diameter of a piston for a car engine; (ii) the length of a garden ready for turfing, to which particular set of numbers would your answer probably belong?

3 Which of the problems in Question 3, Exercise A always have a solution if $\mathscr{E} = Q$?

4 Write down the rational number which is representative of the following sets of equivalent fractions:

(i) $\left\{\ldots, \frac{^-1}{^-2}, \frac{1}{2}, \frac{2}{4}, \frac{3}{6}, \ldots\right\}$; (ii) $\left\{\ldots, \frac{^-3}{^-18}, \frac{^-2}{^-12}, \frac{^-1}{^-6}, \frac{1}{6}, \frac{2}{12}, \ldots\right\}$.

5 Which rational number is representative of the set of equivalent fractions of which

(i) $\frac{3}{9}$; (ii) $\frac{^-4}{20}$; (iii) $\frac{^-9}{^-27}$; (iv) $\frac{20}{20}$

are members?

6 Express the following as rational numbers – i.e., in the form $\frac{p}{q}$:

(i) 0·25; (ii) 0·0018; (iii) 2·5; (iv) 0·03.

7 Write down a rational number $\frac{p}{q}$ such that:

(i) $\frac{1}{12} < \frac{p}{q} < \frac{1}{13}$; (ii) $\frac{p}{q} + \frac{1}{2} = \frac{1}{7}$;

(iii) $2p = 4q$; (iv) $2p - 3q = 0$.

8 Write down the identity element for Q under the operations of (i) addition; (ii) subtraction; (iii) multiplication; (iv) division. (If an identity element does not exist, say so.)

3 More about the rationals

(*a*) Convert the rational numbers (i) $\frac{1}{2}$; (ii) $\frac{1}{3}$; (iii) $\frac{1}{8}$ to decimals.

(*b*) You will have used the method shown at the top of p. 34 to convert $\frac{4}{7}$ to a decimal.

We can see that $\frac{4}{7} = 0{\cdot}\dot{5}7142\dot{8}$ which is a *recurring* decimal. $\frac{1}{3} = 0{\cdot}\dot{3}$ is also a recurring decimal, but if we express $\frac{1}{2}$ and $\frac{1}{8}$ as decimals we obtain: $\frac{1}{2} = 0{\cdot}5$ and $\frac{1}{8} = 0{\cdot}125$. These are examples of *terminating* decimals.

(*c*) *All* rational numbers can be expressed as terminating or recurring decimals. The reason for this can be seen by studying the remainders [ringed in (*b*)] after each successive division. The remainders must always be less than the divisor (7 in this case). At some stage during the division one of the remainders will be

```
       0·5714285...
     7|4·0000000000
       3 5
        (5)0
         49
         (1)0
           7
          (3)0
           28
           (2)0
            14
            (6)0
             56
             (4)0
              35
              (5)0
               ⋮
```

repeated (in which case the decimal begins to recur), or the remainder will become zero (in which case the decimal terminates). Notice that when the divisor is 7, not more than seven successive divisions could be made without one of these two results being true, because the remainders can only be 0, 1, 2, 3, 4, 5, or 6.

(*d*) Now let us consider the reverse process of converting decimals to rational form.

To convert a terminating decimal to rational form we have only to express it as a fraction and then 'cancel down'. For example,

$$0{\cdot}204 = \frac{204}{1000} = \frac{51}{250}, \text{ which is rational.}$$

Convert the following to rational form:

(i) 0·6; (ii) 0·888; (iii) 0·41.

(*e*) Now consider the recurring decimal $0{\cdot}\dot{6}\dot{4}$.

To convert this to rational form we proceed as follows:

Let $D = 0{\cdot}64646464\ldots$.

Multiply by 100,

$100D = 64{\cdot}646464\ldots$.

Subtract the first equation from the second:

$99D = 64$.

Hence $D = \frac{64}{99}$, which is rational.

Now convert $0{\cdot}\dot{5}8\dot{2}$ to rational form (multiply by 1000).

What multiplier would you use to convert $0{\cdot}\dot{5}54\dot{2}$ to rational form?

Summary

1 Rational numbers, when expressed as decimals, either terminate or recur.

2 Terminating and recurring decimals can be expressed in rational form.

3 To convert $0{\cdot}\dot{a}_1 a_2 a_3 \ldots \dot{a}_n$ to rational form use the multiplier 10^n.

Exercise C

1 Convert the following to decimal form:

(i) $\frac{1}{2}$; (ii) $\frac{3}{8}$; (iii) $2\frac{1}{7}$; (iv) $\frac{13}{27}$.

2 What rationals are represented by:

(i) $0{\cdot}875$; (ii) $7{\cdot}5$; (iii) $6{\cdot}002$; (iv) $0{\cdot}\dot{7}$;
(v) $0{\cdot}\dot{8}8\dot{0}$; (vi) $1{\cdot}\dot{1}$; (vii) $3{\cdot}345$; (viii) $0{\cdot}\dot{4}24\dot{5}$?

3 If the denominator of a rational number is 2 or 5, what can you say about the decimal representation of the number? Can you find other examples of denominators for which the same is true? Can you generalize?

4 Divide 4 by the primes in turn up to 13. After how many places of decimals does each begin to recur? Try some more primes as divisors. Is there a general rule about the length of the recurring sequence?

5 Simplify (i) $\frac{1}{2}+\frac{1}{4}$; (ii) $3\frac{1}{4}+4\frac{1}{16}$; (iii) $2\frac{1}{8}-\frac{7}{16}$; (iv) $\frac{3}{7}-\frac{2}{21}$; (v) $\frac{13}{27}-\frac{4}{31}$; (vi) $\frac{3}{7}\times 1\frac{1}{8}$; (vii) $\frac{5}{8}\times 2\frac{1}{4}$; (viii) $\frac{3}{8}\div\frac{1}{4}$; (ix) $\frac{5}{7}\div 2\frac{1}{2}$.

6 Place the following in order of size, the smallest first: $\frac{1}{2}, \frac{3}{7}, \frac{5}{8}, \frac{4}{9}, \frac{6}{13}$.

7 If $a=\frac{1}{2}$, $b=\frac{1}{3}$, draw a number line and mark a and b on it. Calculate $\dfrac{a+b}{2}$ and mark this number on the number line. Calculate a number lying between (i) a and $\dfrac{a+b}{2}$; (ii) $\dfrac{a+b}{2}$ and b, and mark it on the number line.

8 $\dfrac{a}{b}$ and $\dfrac{p}{q}$ are two rational numbers, and $\dfrac{a}{b}<\dfrac{p}{q}$. Place these and $\dfrac{aq+pb}{2bq}$ in order of size, the smallest first. Explain why it is impossible to say that two numbers 'lie next to each other' on the number line.

4 The irrational numbers

(*a*) Consider the number 0·070070007000070000007...

If the sequence of 0's and 7's is continued in this way the decimal obviously does not terminate. Does it recur?

(*b*) Numbers which do not terminate or recur when expressed as decimals $\left(\text{i.e., numbers which cannot be expressed in the form } \dfrac{p}{q}\right)$ are called *irrational* numbers. Write down two more examples of irrational numbers. We will denote the set of irrational numbers by *I*.

(*c*) Two examples of irrational numbers which you will have used are π and $\sqrt{2}$. $\sqrt{2}$ is the solution to the equation $x^2 = 2$. $\sqrt{2}$ units is also the length of the hypotenuse of a right-angled isosceles triangle with its equal sides of length 1 unit (see Figure 1).

We can prove that $\sqrt{2}$ is irrational by using a powerful mathematical technique known as an 'indirect proof'. We first of all assume that $\sqrt{2}$ is a rational number and then show that our assumption leads us to contradict ourselves. This being so, our original assumption must have been wrong. Try to follow the argument below.

Assume $\sqrt{2}$ is a rational number $\frac{p}{q}$, where p and q are integers with *no common divisor*. Then

$$\frac{p^2}{q^2} = 2.$$

Hence $\quad p^2 = 2q^2.$

This tells us that p^2 must be an even number. Since p^2 is even then p must be even, and we can write

$$p = 2t.$$

Hence $\quad p^2 = 4t^2.$

However, since $\quad p^2 = 2q^2$

then $\quad 2q^2 = 4t^2$

so $\quad q^2 = 2t^2.$

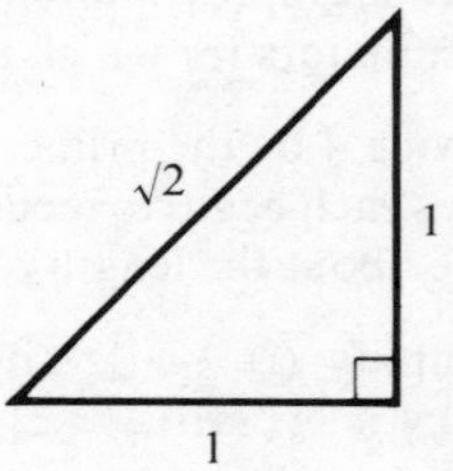

$\sqrt{2}$ 1 1

Fig. 1

This means that q is also an even number. Since both p and q have been shown to be even numbers they have a common factor 2. This contradicts our original assumption. Therefore $\sqrt{2}$ cannot be rational. By definition it must therefore be irrational.

Now try to prove $\sqrt{3}$ is irrational.

(*d*) The rational numbers and the irrational numbers taken together form the set of REAL NUMBERS, *R*. Can you think of a number which is not a real number?

Summary

1 The irrational numbers cannot be expressed in the form $\frac{p}{q}$, where p and q are integers.

2 {Rational numbers} ∪ {Irrational numbers} = {Real numbers}.

Exercise D

1 Draw a Venn diagram to show the sets of numbers described in this chapter. ($\mathscr{E}$ = {Real numbers}).

2 Simplify (i) Q'; (ii) I'; (iii) $R \cap Q$; (iv) $Z \cap Q$; (v) $Q' \cap I$.

3 Explain why $2 + \sqrt{2}$ is irrational.

4 If x is irrational is x^2 always irrational?

5 Multiply $(2 + \sqrt{2})$ by $(2 - \sqrt{2})$. If a and b are irrational is ab necessarily irrational?

6 Give an example to show that the irrationals are not closed under (i) addition; (ii) subtraction; (iii) division; (iv) multiplication.

7 Write down two irrational numbers whose (i) sum; (ii) product; (iii) difference is 2.

8 To which particular number sets do the solutions to the following belong? (i) $x^2 = 3$; (ii) $x^2 + 1 = 3$; (iii) $x^3 = 8$; (iv) $x^2 + x = 0$; (v) $x^2 = 8$.

5 Solution sets

(*a*) Figure 2 shows the set $A = \{^-1, 0, 1, 2, 3\}$ represented on the number line.

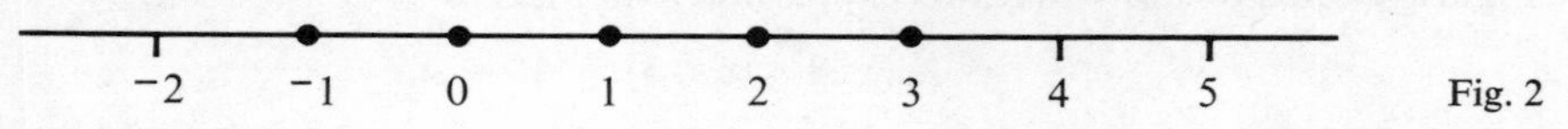

Fig. 2

Represent the sets

$$B = \{^-2, ^-1, 0, 1, 2\},$$
$$C = \{0, 1, 2\}$$

on the number line. List the members of $A \cap B \cap C$. What is $n(A \cap B \cap C)$?

(*b*) Suppose $\mathscr{E}$ = {Real numbers}. Check that the solution to the inequality $2x + 1 > 3$ is $x > 1$.

The solution to the inequality is an infinite set of numbers which we call the *solution set*. We can write this set as

$$\{x\colon x > 1\}$$

(i.e. the set of numbers x 'such that' x is greater than one).

Figure 3 shows the solution set represented on the number line.

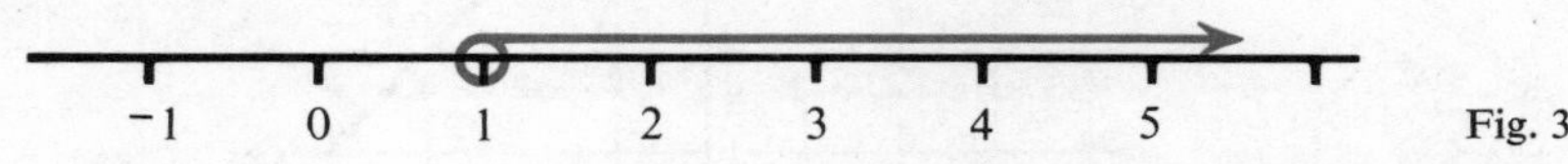

Fig. 3

Notice that the 'end point', 1, is ringed. This signifies that 1 is not, in fact, a member of the set. How do you think we would represent the solution set to the inequality $2x + 1 \geqslant 3$ on the number line (notice that the 'end point' *is* included in the set)?

(*c*) The solution set of the equation $2x + 3 = 0$ has only one member. Write the solution set in the form $\{x\colon x = \quad\}$.

(*d*) Solve the inequality $2x + 1 \leqslant 7$. Write the solution set in the form $\{x\colon x \leqslant \quad\}$, and represent it on the number line.

(*e*) We would solve the double inequality

$$3 < 2x + 1 \leqslant 7$$

in two stages:

(i) solve $$3 < 2x + 1,$$

and

(ii) solve $$2x + 1 \leqslant 7.$$

The solution sets to the two parts are respectively $\{x: x > 1\}$ and $\{x: x \leqslant 3\}$ (did you obtain this second answer in (*d*)?). Figure 4 shows the two sets represented on one number line. The numbers which satisfy the original inequality are members of each of these sets. The solution set is therefore

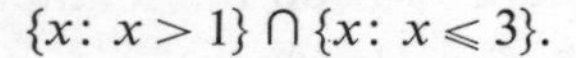

$$\{x: x > 1\} \cap \{x: x \leqslant 3\}.$$

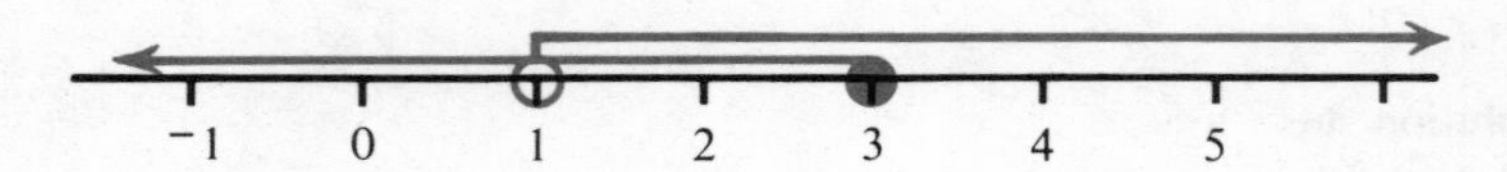

Fig. 4

Figure 4 helps to show that this expression simplifies to

$$\{x: 1 < x \leqslant 3\}.$$

(*f*) $\{(x,y): x + y = 3\}$ is a set of *points*. Figure 5 shows the set plotted on a graph. Copy the diagram and plot the points

$$\{(x,y): x = 2y\}.$$

Use your graph to help you simplify

$$\{(x,y): x + y = 3\} \cap \{(x,y): x = 2y\}.$$

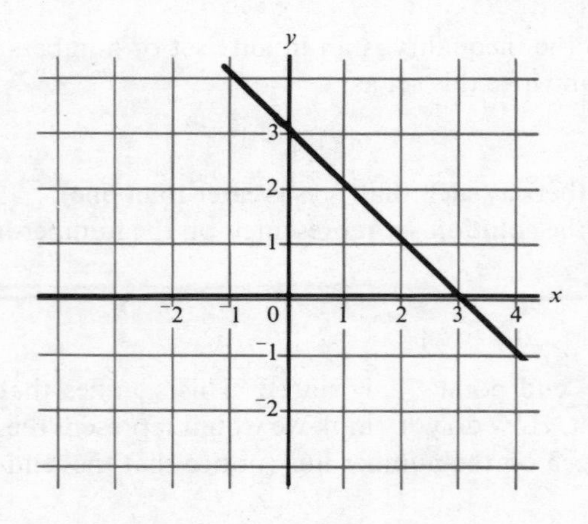

Fig. 5

Write the solution set to the simultaneous equations

$$\begin{cases} x + y = 3 \\ \quad x = 2y, \text{in the form} \{(x,y): x = \quad , y = \quad \}. \end{cases}$$

Exercise E

1 Represent the following sets on the number line in the cases when (*a*) $\mathscr{E} = \{\text{Reals}\}$; (*b*) $\mathscr{E} = \{\text{Integers}\}$.

(i) $A = \{x: x = 1\}$; (ii) $B = \{x: x \leqslant 1\}$; (iii) $C = \{x: 1 \leqslant x < 4\}$.

List the members of $A \cap B \cap C$ in each case.

2 $\mathscr{E} = \{\text{Reals}\}$. Represent the solution sets of the following in the form $\{x: x \quad \}$ and on the number line:

(i) $3x + 7 = 4$; (ii) $4 - x = 9$; (iii) $x - 8 > {}^{-}2$; (iv) $4x - 1 < 5$;
(v) $1 - x \leqslant x$; (vi) $x^2 = 4$; (vii) $x^2 + 3 = 4$.

3 $\mathscr{E} = \{\text{Integers}\}$. Represent the solution sets of the inequalities:

(i) ${}^{-}6 < 3x - 1 \leqslant 4$; (ii) $2 \leqslant 1 - x < 4$ on the number line.

Take $\mathscr{E} = \{Reals\}$ for the questions which follow, unless stated otherwise.

4 Plot the following sets on the number line:

(i) $\{A = x: x \leqslant 4\}$; (ii) $B = \{x: x > 5\}$; (iii) $C = \{x: x \geqslant 4\}$.

What is (*a*) $A \cap B$; (*b*) $A \cap C$; (*c*) $A \cup C$; (*d*) $B \cup C$?

5 Graph the sets of points:

$A = \{(x, y): 2x = y\}$; $B = \{(x, y): x = 6\}$.

What is $A \cap B$?

6 Draw a diagram to show the sets of points:

$A = \{(x, y): x > y\}$ and $B = \{(x, y): x + y \leqslant 6\}$.

(Use your own notation to denote whether a boundary line is included or is not included in the set.) Shade in the region $A \cap B$. Which of the following points are members of $A \cap B$: (i) (2,0); (ii) (3,3); (iii) (3,4); (iv) (0,2)?

7 B is a fixed point. Draw diagrams to represent the sets:

(i) $\{P: PB = 2 \text{ units}\}$; (ii) $\{P: PB \leqslant 2 \text{ units}\}$; (iii) $\{P: PB \geqslant 2 \text{ units}\}$.

8 $A(2, 1)$ and $B(4, 1)$ are fixed points. Draw diagrams to show the sets of points:

(i) $\{P: PA = PB\}$; (ii) $\{P: PA > PB\}$; (iii) $\{P: PA < PB\}$.

9 $\mathscr{E} = \{x: 0 \leqslant x \leqslant 90\}$. Write down the solution sets of each of the following:

(i) $\sin x° = 1$; (ii) $\sin x° \leqslant 0{\cdot}5$; (iii) $\cos x° > 1$; (iv) $0 \leqslant \sin x° \leqslant 0{\cdot}5$.

10 $P = \{(x, y): x^2 + y^2 = 1\}$; $Q = \{(x, y): x + y = 1\}$. Plot P and Q on a graph. List the members of $P \cap Q$.

11 $P = \{(x, y): x^2 + y^2 = 1\}$; $Q = \{(x, y): x + y = k\}$.

(*a*) If $n(P \cap Q) = 1$, what values can k take?
(*b*) If $n(P \cap Q) = 2$, what range of values can k take?
(*c*) If $n(P \cap Q) = 0$, what range of values can k take?

4 Linear programming

1 Introduction

We often need to consider problems involving large numbers of variables (unknowns) which have to obey many conditions. For example, the production manager of a car firm may have to decide on the number of each type of vehicle his firm should manufacture over a given period of time (the unknowns here are the numbers of each type of vehicle to be produced), and in doing this he will have to take into account not only public demand but also such factors as the number of craftsmen he needs, the time required for each manufacturing process on each type of vehicle, the production costs, and so on. Each condition of this kind will impose some restriction on the number of cars he can produce, and in the final event he will be concerned with minimizing production costs and maximizing profits. We will see in this chapter that the production manager can use a process called 'linear programming' to help him make his decision. In real life, linear programming usually involves the use of computers, because of the vast number of variables and conditions involved. However, problems can be solved graphically provided that no more than two variables are involved.

2 Processes of elimination

(*a*) In many of our everyday activities we consciously, or unconsciously, use a process of elimination to select items from a particular universal set. For example, suppose you are buying a record from a shop as a present for one of your friends.

Already in your mind you might have set various conditions on your eventual choice – these will depend largely upon the kind of music your friend enjoys, and your present financial state. You might therefore have decided that the record must be

(i) 'pop',
(ii) 'an L.P.',
(iii) 'vocal'.

By considering each condition in turn you can eliminate a certain section of the stock from your search, and eventually you will arrive at a solution set – which may have more than one member, or may even be empty! If it is empty then you will obviously have to select a new universal set (go to another shop), or relax some of your conditions.

Can you think of other everyday examples in which you use a process of elimination to select objects from sets? Try to list the conditions you impose in each case.

(*b*) List the members of the set $\{x\colon 2 \leqslant x < 20, x \text{ an integer}\}$.

Consider each of the following conditions in turn, and for each one strike out members of the set which do not satisfy the condition. Hence find the solution set of numbers which are:

(i) not multiples of 4;
(ii) factors of 30;
(iii) not primes;
(iv) even.

If you had considered (i)–(iv) in a different order, would your solution set still be the same?

Notice that the solution set is $\{x\colon x \text{ is not a multiple of } 4\} \cap \{x\colon x \text{ is a factor of } 30\} \cap \{x\colon x \text{ is not prime}\} \cap \{x\colon x \text{ is even}\}$.

(*c*) Suppose you are playing a game of dice which involves throwing a red die and a blue die simultaneously. To win on any particular throw you must satisfy the following conditions:

(i) the total score must be less than 8;
(ii) the score on the red die must not be greater than 4;
(iii) the score on the blue die must be greater than 2;
(iv) the score on the blue die must be not more than twice the score on the red die.

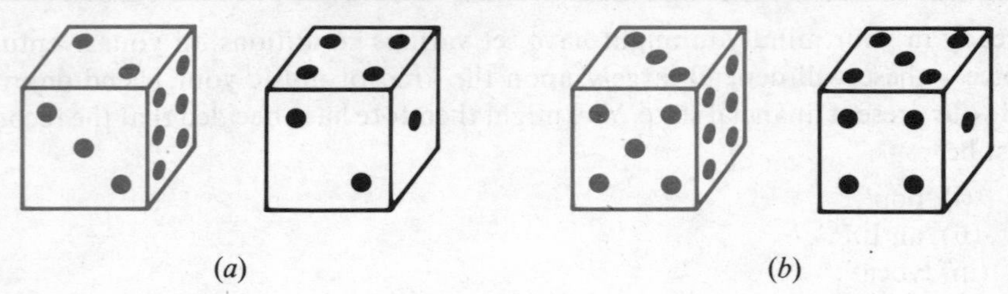

Fig. 1

Do Figures 1(*a*) and (*b*) represent winning combinations? List all the possible winning combinations. What is the probability you will win on any particular throw?

(*d*) Figure 2 shows how the set of winning combinations can be found graphically by eliminating combinations which do not satisfy each condition in turn (the 'ringed dots' are the unsatisfactory scores at each stage).

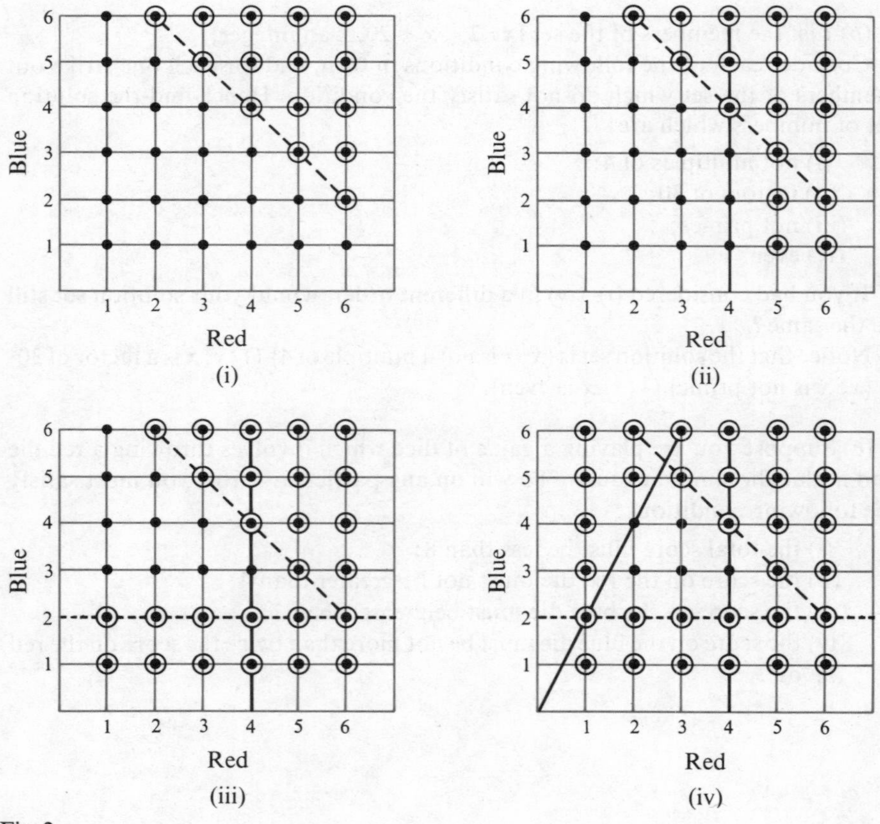

Fig. 2

We can see from (iv) that the winning combinations are represented by the ordered pairs (2, 3); (3, 3); (4, 3); (2, 4); (3, 4).

The ordered pairs (4,2) and (4,3) both appear to be on the 'boundary' of the solution set. Why is (4,3) a possible solution, but not (4,2)?

What does a broken line used as a boundary indicate? What does a continuous line indicate?

(*e*) On a diagram of the (x, y) plane *leave unshaded* the region which satisfies the conditions:

(i) $x+y<8$;
(ii) $x \leqslant 4$;
(iii) $y>2$;
(iv) $y \leqslant 2x$;

(that is, find the solution set

$$\{(x, y): x+y<8\} \cap \{(x, y): x \leqslant 4\} \cap \{(x, y): y>2\} \cap \{(x, y): y \leqslant 2x\}).$$

Compare your diagram with Figure 2(iv). What do you notice? Are the algebraic conditions equivalent to those stated in (*c*)?

In what ways are the two solution sets (i) the same (ii) different?

(*f*) We will now use the processes described above to help us solve a problem which, although artificial, helps to explain how 'linear programming' can be used to help us make decisions. Our main concern will be

(i) to express statements algebraically as relations;
(ii) to graph these relations (use a process of elimination to obtain the solution set).

(*g*) A car firm has contracted to deliver at least 60 cars per day to Dover over a long period of time, ready for export. The firm uses two types of carriers: (i) Type *A* which can carry 10 cars, and (ii) Type *B* which can carry 8 cars. There are 4 Type *A* carriers and 6 Type *B* carriers, but only 8 drivers available for the work. The carriers can only make one journey per day and each one carries a full load.

(i) How should the transport manager organize his carriers to meet the delivery demand?
(ii) What is the minimum number of drivers he needs?
(iii) What is the maximum number of cars he can deliver per day?

The two unknown quantities (variables) are:

(i) the number of Type A carriers to be used – call this x;
(ii) the number of Type B carriers to be used – call this y.

Because of the limitations on the number of carriers and drivers available we know that:

(i) $x \leqslant 4$;
(ii) $y \leqslant 6$;
(iii) $x + y \leqslant 8$.

Also, x Type A carriers can carry $10x$ cars, and y Type B carriers $8y$ cars. At least 60 cars have to be delivered, so that

(iv) $10x + 8y \geqslant 60$
(or, $5x + 4y \geqslant 30$).

The graphs of the inequalities (i) and (iii) are shown in Figure 3. Copy this figure and add graphs of the inequalities (ii) and (iv) shading the unrequired region.

(i) List the six number pairs which satisfy all four inequalities (remember that you can only have a whole number of carriers).

You should have found that the transport manager can use any of the following solutions to meet the delivery demand:

$$(2, 5); (2, 6); (3, 4); (3, 5); (4, 3); (4, 4).$$

(ii) If the first solution above is decided upon then 7 drivers will be needed. How many drivers are needed if the other solutions are used? Do you agree that a minimum of 7 drivers is needed?

(iii) If 2 Type A carriers and 5 Type B carriers are used, then $(2 \times 10) + (5 \times 8) =$ 60 cars can be delivered. Find the number of cars that can be delivered for the other five solutions that meet the delivery demand. What is the maximum number of cars that can be delivered?

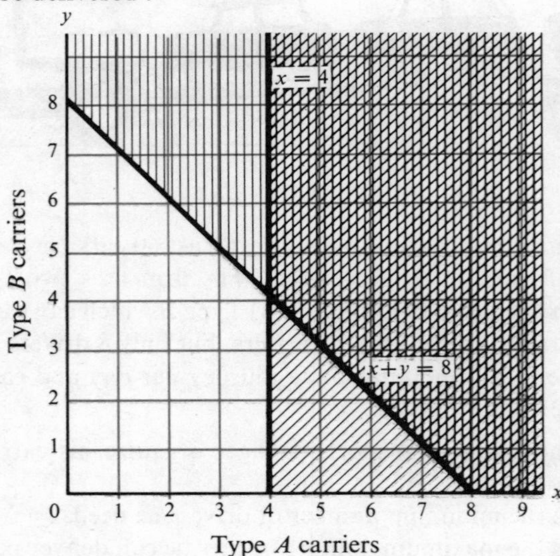

Fig. 3

Exercise A

1 On separate diagrams leave unshaded the regions representing the following inequalities:

(i) $x > 2$; (ii) $y \leqslant 6$; (iii) $x + y \geqslant 3$;
(iv) $2x + 3y \leqslant 12$; (v) $y + 2x \leqslant 50$; (vi) $xy \leqslant 144$;
(vii) $y \geqslant 2x$; (viii) $x \leqslant 2y$.

2 List the special kinds of polygons which belong to the set satisfying the following conditions:

(i) the polygons have four sides;
(ii) the polygons have half-turn symmetry;
(iii) the polygons have no lines of symmetry.

3 Leave unshaded the region of points whose coordinates satisfy all five of the following inequalities:

(i) $x \geqslant 0$; (ii) $y \geqslant 0$; (iii) $2x + y > 4$;
(iv) $x + 3y > 9$; (v) $x + y < 6$.

If x, y are integers mark each point of the solution set with a dot.

4 To make a 'perfect' cup of tea you must (i) make it in cups of capacity 50 cm^3; (ii) use at least ten times as much water as milk; (iii) use not more than 12 times as much water as milk; (iv) at least half fill the cup.

Express the four statements above as algebraic inequalities and then graph the information (use w, m to represent the number of cm^3 of water and milk respectively in a 'perfect' cup of tea).

Which of the following amounts of water and milk would make a 'perfect' cup of tea?

(i) 40 cm^3 of water and 5 cm^3 of milk;
(ii) 50 cm^3 of water and 5 cm^3 of milk;
(iii) 30 cm^3 of water and 11 cm^3 of milk;
(iv) 22 cm^3 of water and 2 cm^3 of milk.

5 A hotel caters for school parties of less than 30 people, and has the following rules: All parties must

(i) comprise more than 20 people;
(ii) include at least three adults;
(iii) include not more than six adults.

Take p to represent the number of pupils in the group and a to represent the number of adults. Write the four statements above in algebraic form and graph the information. If a school party includes 15 pupils, how many adults will there be in the party? What is the maximum number of pupils there can be in a party which includes only 4 adults?

6 A post office has to transport 900 parcels using lorries, which can take 150 at a time, and vans which can take 60.

(*a*) If l lorries and v vans are used, write down an inequality which must be satisfied.

(*b*) The costs of each journey are £5 by lorry and £4 by van and the total cost must be less than £44. Write down another inequality which must be satisfied by l and v.

(*c*) Represent these inequalities on a graph and dot in the members of the solution set.

(*d*) What is:

(i) the largest number of vehicles which could be used;
(ii) the arrangement which keeps the cost to a minimum;
(iii) the most costly arrangement?

7 In an airlift it is required to transport 600 people and 45 tons of baggage. Two kinds of aircraft are available: the Albatross which can carry 50 passengers and 6 tons of baggage, and the Buzzard which can carry 80 passengers and 3 tons of baggage.

(*a*) If a Albatrosses and b Buzzards are used, explain why

$$5a + 8b \geqslant 60 \qquad \text{and} \qquad 2a + b \geqslant 15.$$

(*b*) Only 8 Albatrosses and 7 Buzzards are available. Represent, on a graph, the possible arrangements of aircraft which can supply the necessary transport. Dot in the members of the solution set.

(*c*) What is the smallest number of aircraft that can be used?

3 Maximizing and minimizing

(*a*) We found in Section 2(*g*) that the maximum number of cars that could be transported to Dover per day could be found by considering each member of the solution set in turn. When the solution set contains a large number of elements, however, this method becomes impractical. We will now look again at the question 'what is the greatest number of cars that can be delivered' and answer it by a method which can be applied to solution sets with any number of elements.

(*b*) The number of cars that can be transported by x Type A carriers and y Type B carriers is given by the expression

$$10x + 8y.$$

The ordered pairs of whole numbers (x, y) which are solutions to the equation

$$10x + 8y = 60$$

tell us how many of each type of carrier could be used to transport 60 cars. If you look at the graph in Figure 4, you will see that the only two solutions are given by the ordered pairs $(6, 0)$ and $(2, 5)$. However, $(6, 0)$ is not in the solution set and so $(2, 5)$ is the only possible solution.

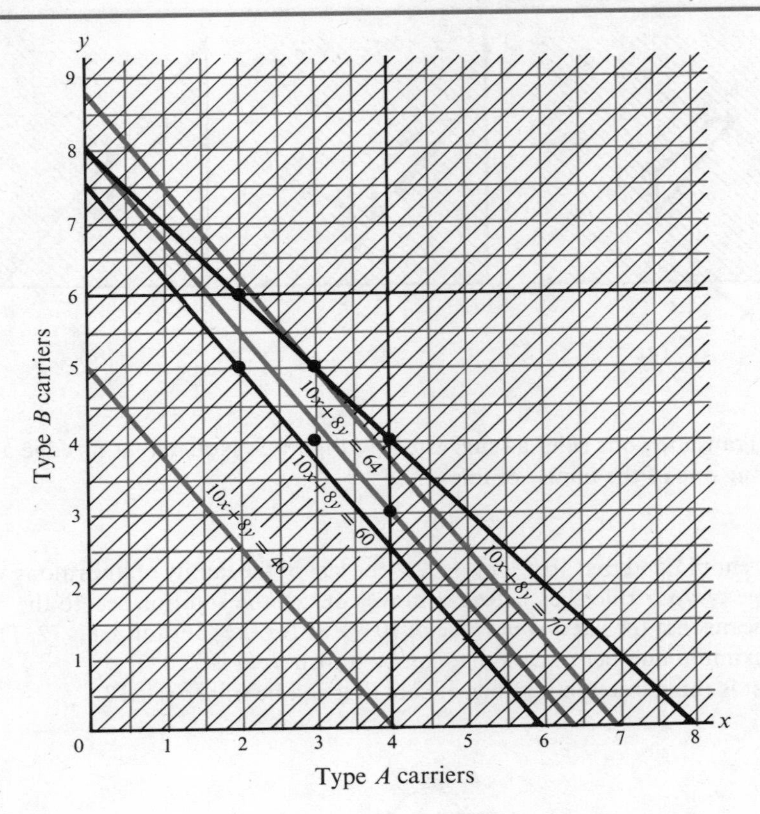

Fig. 4

The red lines in Figure 4 have equations

$$10x + 8y = 40; \qquad 10x + 8y = 64; \qquad 10x + 8y = 70.$$

How many of each type of carrier must be used to transport exactly

(i) 64 cars; (ii) 70 cars?

Can 40 cars be carried?

(*c*) Notice that the family of lines $10x + 8y = c$, for different values of c are parallel.

Place the edge of your ruler along the line $10x + 8y = 80$. Can 80 cars be delivered per day?

(*d*) What is the value of c for the lines which pass through

(i) $(3, 4)$; (ii) $(2, 6)$?

How many cars can be transported by (i) 7; (ii) 8 carriers?

(*e*) Place your ruler along the line of the family which passes through the point $(4, 4)$. What is the equation of this line?

What does the value of c in this equation tell you?

Do you agree that the maximum number of cars which can be transported is given by this value of c?

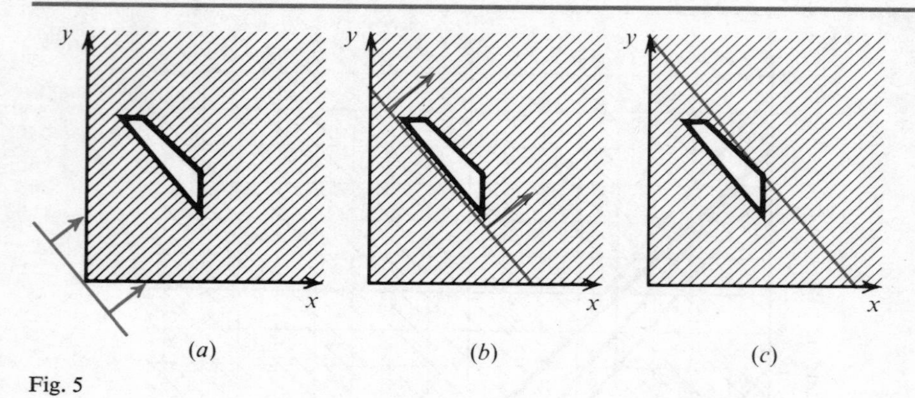

Fig. 5

(f) Translate your ruler as suggested by Figure 5. Notice that its edge always lies along one of the family of lines

$$10x + 8y = c.$$

You should find that $10x + 8y = 72$ is the last of the family of lines along which the edge of your ruler lies before it moves out of the solution set 'to the right'. This means that the *maximum* value of $10x + 8y$ for the solution set is 72. That is, the maximum number of cars that can be transported is 72.

What is the minimum number of cars that can be transported?

Example 1

A toy-firm manufactures two kinds of toy soldiers on a machine which can work for 10 hours per day. The 'Guard' takes 8 seconds to make and 8 g of metal is used to make it. The 'Cavalryman' takes 6 seconds to make and 16 g of metal is used to make it. Altogether 64 kg of metal is available per day.

If the profit on the 'Guard' is 5p and on the 'Cavalryman' is 6p, how many of each should be made to maximize the profits? What is the maximum profit that can be made per day?

The variables are

(i) the number of 'Guards' to be made – call this x;
(ii) the number of 'Cavalrymen' to be made – call this y.

The machine can work for $60 \times 60 \times 10$ or 36 000 s per day, so that

(i) $8x + 6y \leqslant 36\,000$.

Since 64 kg of metal is available,

(ii) $8x + 16y \leqslant 64\,000$.

If we now graph these relationships we obtain Figure 6.

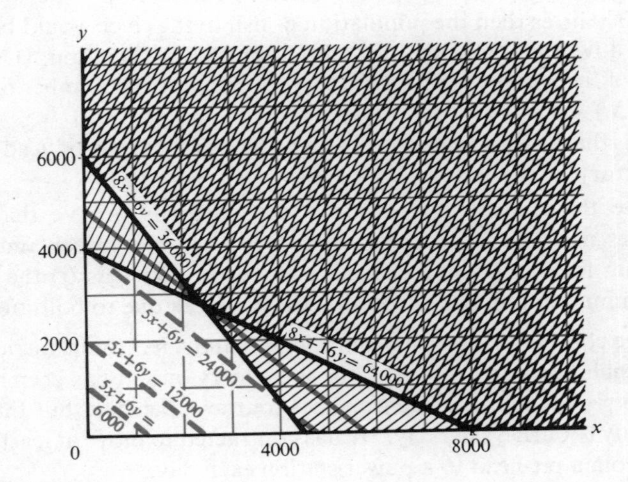

Fig. 6

Since the profit on the 'Guard' is 5p and that on the 'Cavalryman' is 6p, then the profit on making x 'Guards' and y 'Cavalryman' is given by the expression

$$(5x + 6y)\text{p}.$$

The integral points on the broken red lines in Figure 6 will tell us how many of each toy should be made to make a profit of (i) 6000p; (ii) 1200p; (iii) 24 000p. These lines are called 'lines of equal profit'.

The points on the continuous red line will tell us the number of each kind of toy that should be made to give us a maximum profit (why?). The only point of the solution set which lies on this line is (2400, 2800).

Hence, to make a maximum profit, 2400 'Guards' and 2800 'Cavalrymen' should be manufactured.

The maximum profit is then

$$(5 \times 2400)\text{p} + (6 \times 2800)\text{p} = (12\,000 + 16\,800)\text{p}$$
$$= £288.$$

Exercise B

1 Leave unshaded the set which satisfies all the following five inequalities:

(i) $x \geqslant 0$; (ii) $y \geqslant 0$; (iii) $x + y \geqslant 5$;
(iv) $x + y \leqslant 7$; (v) $x \geqslant 3y$.

Find (*a*) the maximum and (*b*) the minimum value of each of the following expressions for this set:

(i) x; (ii) y; (iii) $x + y$; (iv) $x - y$;
(v) $x + 2y$; (vi) $2x + y$; (vii) $3x + 2y$.

2 Two detergent factories pour their waste products into a river. Factory *A* always produces at least twice as much waste as Factory *B* and together they always produce at least 9000 litres of waste per week. Scientists estimate that if the total amount of waste entering the river per week was greater than 15 000 litres then the population of fish in the river would be in danger.

Write down three inequalities and graph the information. (Use x for the number of litres of waste from Factory *A* and y for the number of litres from Factory *B*.)

What is the maximum amount of waste that should be produced in any week by (i) factory *A*; (ii) factory *B* if the fish are to survive?

Despite the suggested restrictions the scientists believe that for every 1000 litres of waste from factory *A* two fish die every week and every 1000 litres from factory *B* three fish die every week. What is (i) the maximum; (ii) the minimum number of fish that die per week due to pollution?

3 A haulage contractor has 7 six-tonne lorries and 4 ten-tonne lorries. He has 9 drivers, each of whom stays with the same lorry once it has been allocated to him. The six-tonne lorries can make 8 journeys per day, but the ten-tonne lorries only 6 journeys per day. He has contracted to move at least 360 tonnes of coal from a pit-head to a power station each day.

If x denotes the number of six-tonne lorries in use, and y the number of ten-tonne lorries in use, explain the meaning of the expressions

(i) $48x + 60y \geqslant 360$
(ii) $x + y \leqslant 9$.

Write down two more inequalities involving x and y.

(i) What possible combinations of lorries can the contractor use to meet the delivery demand?
(ii) Which combination of lorries uses the smallest number of drivers?
(iii) Which combination of lorries carries the maximum tonnage?

4 For a camp of 70 children two types of tent are available on hire. The Patrol tent sleeps 7 and costs £5 a week; the Hike tent sleeps 2 and costs £1 a week. The total number of tents must not exceed 19.

Write down two inequalities connecting the number of Patrol tents (p) and the number of Hike tents (h); and an expression for the cost of hiring these numbers of tents for a week. Using a scale of 1 cm to 2 units, find the most economical cost of hire and the number of each kind of tent required.

5 Ten men are available to unload 7 lorries, but not more than two men can work together on a lorry. If x denotes the number of lorries being unloaded by one man at any time, and y the number of lorries being unloaded by two men at any time, write down the two inequalities satisfied by x and y (other than $x \geqslant 0$, $y \geqslant 0$) and graph the information.

Each lorry carries a tonne of goods. Experience shows that two men together can unload three times as fast as one man by himself. If one man can unload 2 tonnes per hour, show that the rate of unloading is $2x + 6y$ tonnes per hour. Use your graph to find x and y so that the first tonne of goods is unloaded as quickly as possible.

6 A bicycle manufacturer makes two models, a sports cycle and a racing cycle. The sports model takes 8 man-hours to make and the racing model 12 man-hours. There are 20 men available for the work and they work a 35-hour week, (so that the total number of man-hours available per week is 700). The sports model costs £5 in materials and the racing model £6, and the manufacturer has £400 worth of material for use per week. The firm has a contract to supply at least 30 of each type of bicycle per week. How many of each type of bicycle should be made to obtain the maximum profit if the profit on the sports model is £6 and that on the racing model £9?

7 A farmer wishes to stock his farm with cows and sheep. Cows cost £20 each and sheep £15 each. The farmer has accommodation for not more than 40 animals and he has £700 with which to buy the animals. When he resells the animals for slaughter he expects to make an overall profit of £12 per cow and £8 per sheep. How many cows and how many sheep should he purchase to maximize his profits? What is the maximum profit?

8 Two types of screw, A and B, are made on an automatic machine which takes 5 seconds to make a screw A and 4 seconds to make a screw B. Screw A requires 8 g of metal and screw B 16 g of metal. The profit from making screw A is 0·2p and from screw B 0·2p. If the machine can work for a maximum of 7 hours per day and 96 kg of metal is available, how many of each type of screw should be made to make the manufacturer's daily profit as large as possible? What is the maximum profit per day?

4 Non-linear conditions

So far we have only dealt with conditions which are linear (i.e. conditions which, when graphed, have straight lines as their boundaries).

However, the process we have described is easily extended to non-linear conditions as the following example shows:

Example 2

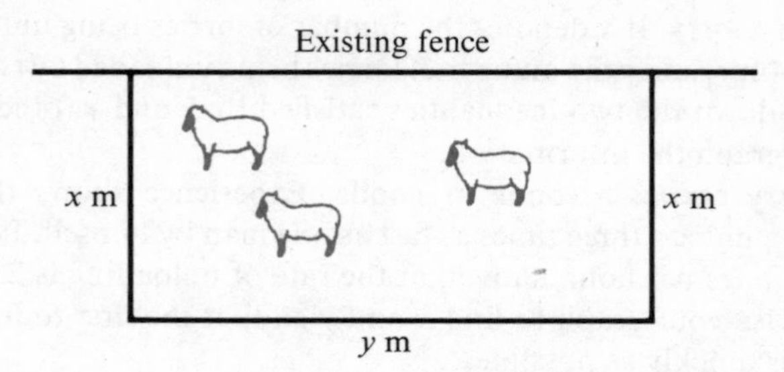

Fig. 7

A farmer wishes to make a rectangular enclosure of area 144 m². The enclosure is to be built against an existing straight fence (see Figure 7), and the farmer has 50 m of wire netting. What is the minimum amount of wire netting he can use to make the enclosure? What is the maximum length the existing fence needs to be? What is the minimum length it needs to be?

Suppose that the length of sides of the enclosure are x and y metres (see Figure 7). Then the area to be enclosed is xy m², so that

(i) $xy = 144$ (note that this is an *equation*).

The amount of wire netting to be used must not exceed 50 m, so that

(ii) $2x + y \leqslant 50$.

If we now graph these two relations, we obtain Figure 8.

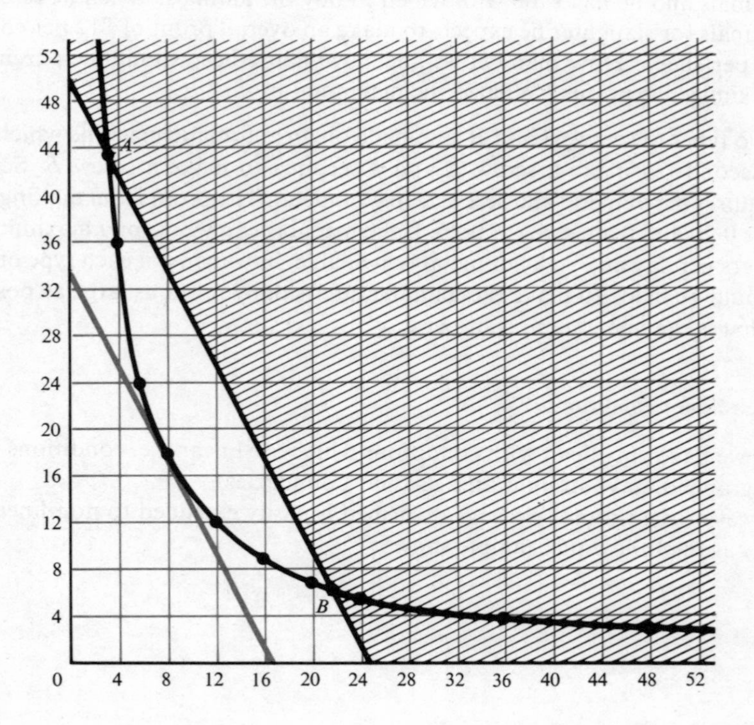

Fig. 8

The solution set for these two conditions is the set of points on the hyperbola $xy = 144$, for $3 \leqslant x \leqslant 22$.

The amount of wire netting required is given by the expression

$$(2x + y)\ \text{m}.$$

This is a minimum when $2x + y = 34$ (see the red line in Figure 8), and in this case $x = 9$, and $y = 16$. Therefore the minimum amount of netting is 34 m.

The maximum length the existing fence needs to be is given by the y-coordinate of point A – that is, 44 m.

The minimum length the existing fence needs to be is given by the y-coordinate of point B – that is, 6 m.

Exercise C

1 (*a*) Calculate the values of y when x equals 1, 2, 3, 4, 5, 6 for each of the following equations:

(i) $xy = 64$; (ii) $x^2 = y$; (iii) $y = x^2 + 4$.

(*b*) On separate axes draw the graphs of (i), (ii) and (iii) above. Shade in the regions $xy \geqslant 64$, $x^2 \leqslant y$, $y \geqslant x^2 + 4$.

2 Draw the graphs of $xy = 16$ and $x + y = 10$ on the same axes. Leave unshaded the region, R, where $R = \{(x, y) \colon x + y \leqslant 10\} \cap \{(x, y) \colon xy \geqslant 16\}$.

What is the maximum value of (i) x; (ii) y for the region R? What is the minimum value of $x + y$ for the region R?

3 A farmer wants to enclose a rectangular area for sheep. They require at least 480 m^2. For one side he will use a straight fence and for the other three sides he can use 'hurdles', each 2 m long, of which he has 40. Using x for the number of hurdles he uses for each of the equal sides, and y for the number of hurdles he uses for the remaining side, write down two algebraic inequalities satisfying the above information. Graph the information and hence find the smallest number of hurdles he needs.

4 A biscuit-container manufacturer wishes to make cylindrical containers. The curved surfaces of the containers are to be made from rectangular sheets of metal of area 900 cm^2. Taking x cm to be the circumference of the base of a container, and y cm the height of the container, write the second sentence as an algebraic equation, and represent it on a graph.

The manufacturer requires that the circumference of the container be less than 60 cm and the height less than 30 cm. By graphing two more relations find the range of values within which (*a*) the circumference and (*b*) the height of the containers must lie. What is the circumference and height of the container when the rectangular sheet has maximum perimeter?

5 The distance d km that can be travelled by a motor boat at a steady speed of v km/h without refuelling is given by the formula

$$d \leqslant 12v - v^2.$$

Draw up a table of values and plot $d = 12v - v^2$ for the domain $0 \leqslant v \leqslant 12$.

Taking suitable scales, shade the area representing distances impossible at various speeds. What is special about 12 km/h? How do you account for this?

A trip round the bay is to last 3 hours. Express the relation between d and v for such a trip and plot it on your graph. What is the greatest distance that can be covered in this time and what will be the speed?

Find also the greatest distance that can be covered at any speed. How long will a trip of this distance take?

6 A stationery firm is going to market an envelope which will take paper measuring 30 cm by 20 cm folded twice, and once at right-angles, and paper 24 cm by 14 cm folded into four (along the dotted lines in Figure 9).

If this envelope is to measure x cm long, y cm wide, write down inequalities for x and y. The designer considers that the length should not be more than $1\frac{1}{2}$ times the width. Express this by an inequality and also write the perimeter of an envelope in terms of x and y. Find a solution which satisfies all the conditions and makes the perimeter as small as you think possible. Is such a solution acceptable, do you think?

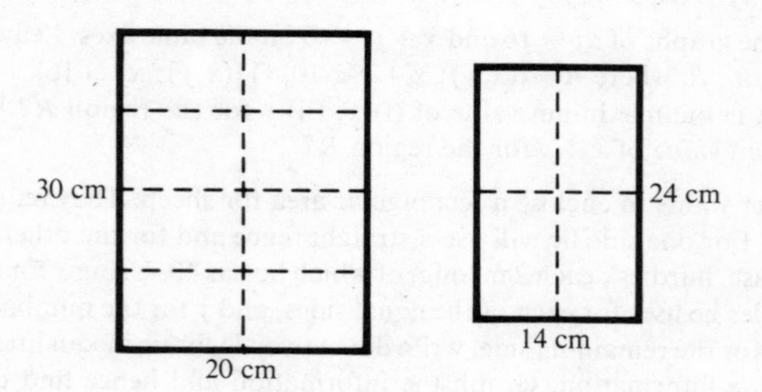

Fig. 9

Incidence matrices

(*a*) Write down the one-stage route matrix for the network in Figure 1. Call it **M**.

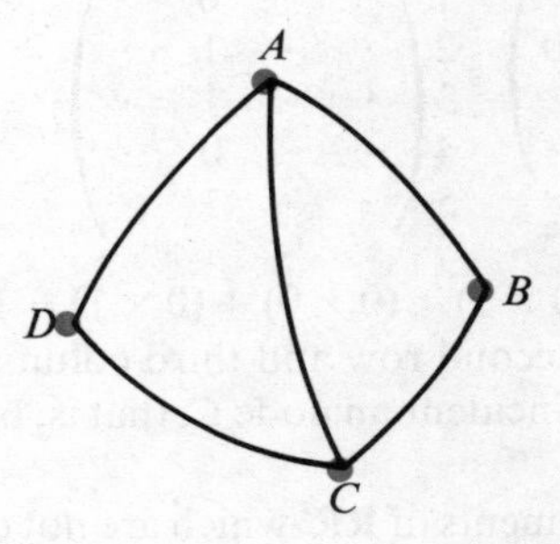

Fig. 1

(*b*) There is another way of compiling matrices to describe a network. In Figure 1, we labelled the nodes. Suppose we now label the arcs as well:

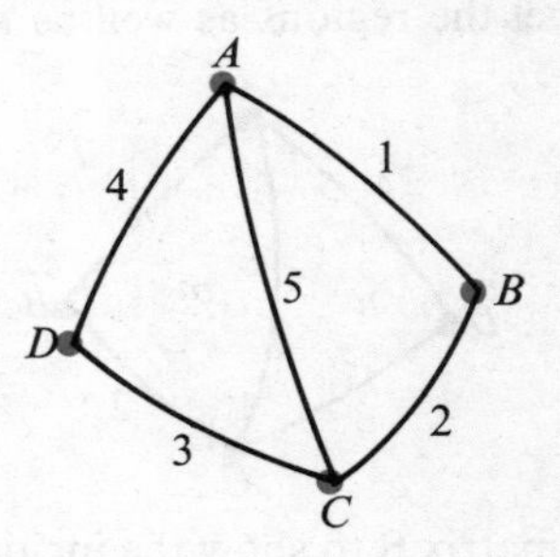

Fig. 2

Node A lies on arcs 1, 4 and 5 but not on arcs 2 and 3. We say that node A is *incident* on arcs 1, 4 and 5.

Copy and complete the matrix **R** to show which nodes are incident on which arcs.

$$\mathbf{R} = \begin{array}{c} \\ A \\ B \\ C \\ D \end{array} \begin{array}{c} \begin{array}{ccccc} 1 & 2 & 3 & 4 & 5 \end{array} \\ \left(\begin{array}{ccccc} 1 & 0 & 0 & 1 & 1 \\ & & & & \\ & & & & \\ & & & & \end{array} \right) \end{array} .$$

Now write down the 5 by 4 matrix $\mathbf{R}'$ which shows the incidence of arcs on nodes.

How are the matrices **R** and $\mathbf{R}'$ related?

Work out $\mathbf{RR}'$.

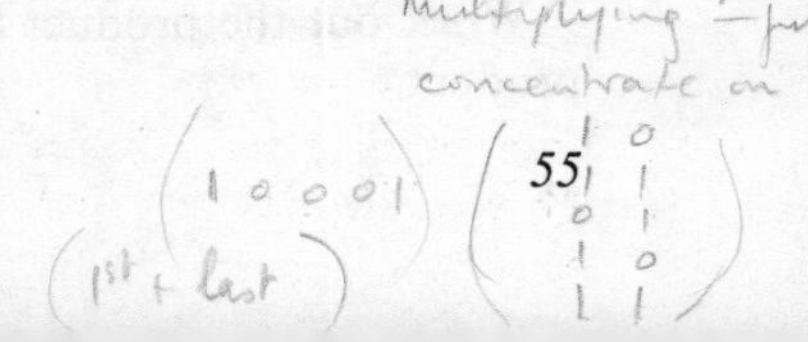

R could be described as 'a nodes by arcs' matrix and **R**′ as 'an arcs by nodes' matrix. How would you describe **RR**′? How would you describe the route matrix **M**?

Compare **RR**′ with **M**. What features do they have in common?

The result of combining the second row of **R** with the third column of **R**′ is shown below.

$$\begin{array}{c|ccccc|} & 1 & 2 & 3 & 4 & 5 \\ \hline A & \cdot & \cdot & \cdot & \cdot & \cdot \\ B & 1 & 1 & 0 & 0 & 0 \\ C & \cdot & \cdot & \cdot & \cdot & \cdot \\ D & \cdot & \cdot & \cdot & \cdot & \cdot \end{array} \quad \begin{array}{c|cccc|} & A & B & C & D \\ \hline 1 & \cdot & \cdot & 0 & \cdot \\ 2 & \cdot & \cdot & 1 & \cdot \\ 3 & \cdot & \cdot & 1 & \cdot \\ 4 & \cdot & \cdot & 0 & \cdot \\ 5 & \cdot & \cdot & 1 & \cdot \end{array} = \begin{array}{c|cccc|} & A & B & C & D \\ \hline A & \cdot & \cdot & \cdot & \cdot \\ B & \cdot & \cdot & 1 & \cdot \\ C & \cdot & \cdot & \cdot & \cdot \\ D & \cdot & \cdot & \cdot & \cdot \end{array}.$$

$$(1 \times 0) + (1 \times 1) + (0 \times 1) + (0 \times 0) + (0 \times 1) = 1.$$

We obtain a '1' in the second row and third column of **RR**′ because node B is incident on arc 2 which is incident on node C, that is, because there is a route from B to C.

Check that the other elements of **RR**′ which are not on the leading diagonal also count the number of routes between pairs of nodes.

The elements on the leading diagonal of **RR**′ are different from the corresponding elements of **M**. What do these elements count? Try to explain why this happens.

(*c*) Suppose we now label the regions as well as the nodes and the arcs. See Figure 3.

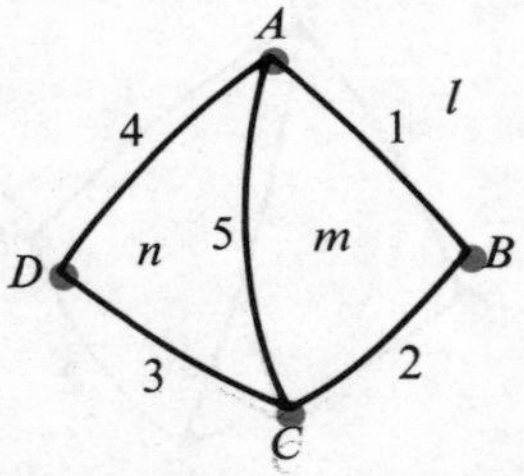

Fig. 3

Copy and complete the matrix **S** to show the incidence of arcs on regions and the matrix **T** to show the incidence of nodes on regions.

$$\mathbf{S} = \begin{array}{c|ccc|} & l & m & n \\ \hline 1 & 1 & 1 & 0 \\ 2 & 1 & & \\ 3 & 1 & & \\ 4 & 1 & & \\ 5 & 0 & & \end{array}; \qquad \mathbf{T} = \begin{array}{c|ccc|} & l & m & n \\ \hline A & & & \\ B & 1 & 1 & 0 \\ C & & & \\ D & & & \end{array}.$$

Work out **RS**.

Compare **RS** with **T**. What do you notice? Try to explain why this happens.

(*d*) Why is it impossible to form the product $\mathbf{S}^2$?

Write down the transpose of **S** and call it **S**′. What does **S**′ show?

Is it possible to form the products **SS**′ and **S**′**S**?

Work out the product **S**′**S** and explain the meaning of the elements in the product.

Exercise A

1 Draw the networks described by the following incidence matrices:

(*a*) nodes $\begin{matrix} & \text{arcs} \\ & \begin{matrix} 1 & 2 \end{matrix} \\ \begin{matrix} A \\ B \end{matrix} & \begin{pmatrix} 1 & 1 \\ 1 & 1 \end{pmatrix} \end{matrix}$; (*b*) nodes $\begin{matrix} & \text{arcs} \\ & \begin{matrix} 1 & 2 & 3 & 4 & 5 \end{matrix} \\ \begin{matrix} D \\ E \\ F \end{matrix} & \begin{pmatrix} 1 & 0 & 0 & 1 & 1 \\ 1 & 1 & 1 & 1 & 0 \\ 0 & 1 & 1 & 0 & 1 \end{pmatrix} \end{matrix}$;

(*c*) arcs $\begin{matrix} & \text{regions} \\ & \begin{matrix} a & b \end{matrix} \\ \begin{matrix} 1 \\ 2 \\ 3 \end{matrix} & \begin{pmatrix} 1 & 1 \\ 1 & 1 \\ 1 & 1 \end{pmatrix} \end{matrix}$; (*d*) nodes $\begin{matrix} & \text{regions} \\ & \begin{matrix} p & q & r & s \end{matrix} \\ \begin{matrix} X \\ Y \\ Z \end{matrix} & \begin{pmatrix} 1 & 1 & 1 & 0 \\ 1 & 1 & 1 & 1 \\ 0 & 1 & 0 & 1 \end{pmatrix} \end{matrix}$.

2 Find the matrices **R**, **S** and **T** for the network in Figure 4 and check that **RS** = 2**T**.

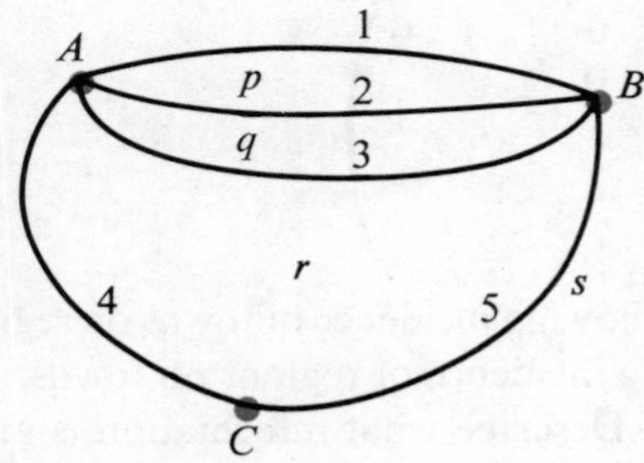

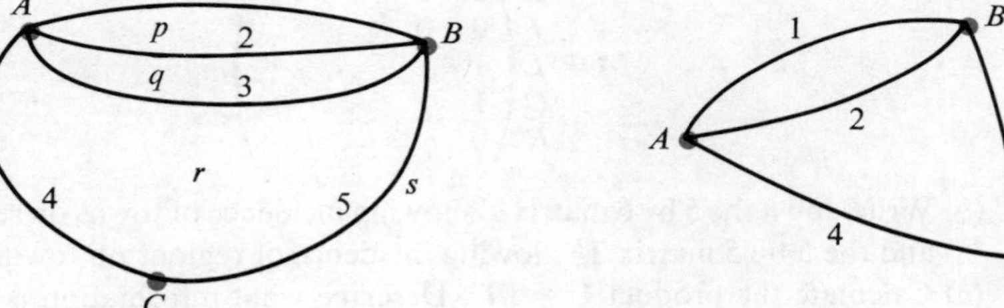

Fig. 4 Fig. 5

3 (*a*) Find the product **R**′**R** for the network in Figure 2. Is it the same as **RR**′?
(*b*) What is the meaning of the elements on the leading diagonal?

4 (*a*) Find the product **R**′**R** for the network in Figure 5.
(*b*) Explain the meaning of the elements which are not on the leading diagonal.

5 (*a*) For the network in Figure 6, write down
  (i) the matrix **P** giving the number of routes between any two nodes;
  (ii) the 3 by 6 incidence matrix **X** showing which arcs end at which nodes;
  (iii) the corresponding incidence matrix **Y** showing which nodes are at the end of which arcs.
(*b*) How are the matrices **X** and **Y** related?
(*c*) Calculate (i) **XY**; (ii) **XY** − **P**.
(*d*) State what information is given by the matrix **XY** − **P**.

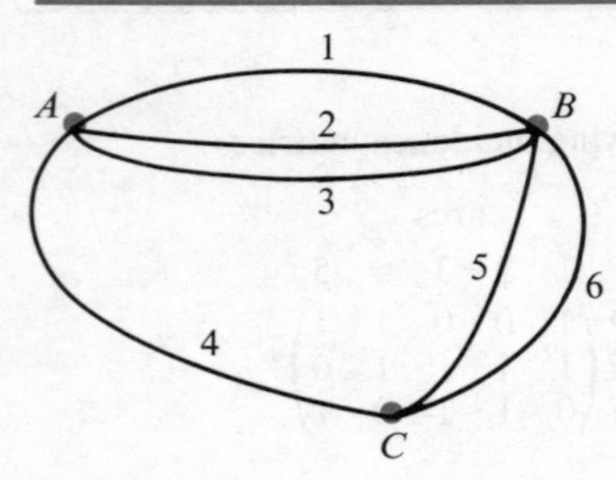

Fig. 6

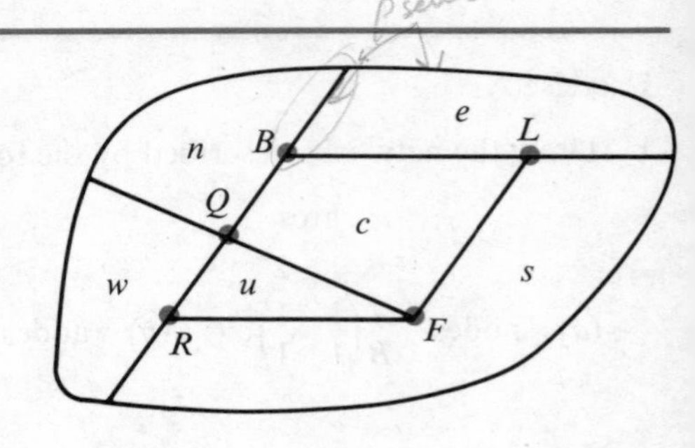

Fig. 7

6 For administrative purposes one of the English counties is divided into six regions (n, e, s, w, u, c) by trunk roads joining the five towns B, F, L, Q, R. See Figure 7.

(*a*) Copy and complete the route matrix **M** for the direct routes between the five towns:

$$\mathbf{M} = \begin{array}{c} \\ B \\ F \\ L \\ Q \\ R \end{array} \begin{array}{c} \begin{array}{ccccc} B & F & L & Q & R \end{array} \\ \left(\begin{array}{ccccc} 0 & 0 & 1 & 1 & 0 \\ 0 & 0 & & & \\ 1 & & & & \\ 1 & & & & \\ 0 & & & & \end{array} \right) \end{array}.$$

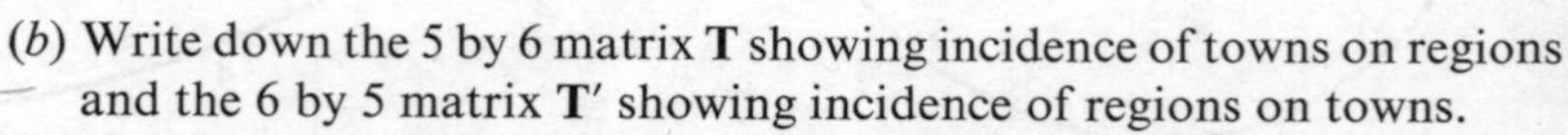

(*b*) Write down the 5 by 6 matrix **T** showing incidence of towns on regions and the 6 by 5 matrix **T′** showing incidence of regions on towns.

(*c*) Calculate the product $\mathbf{U} = \mathbf{TT'}$. Describe what information is given by the entry 2 of **U** relating B to Q.

(*d*) Explain why the entry relating Q to F in **U** is twice the corresponding entry in **M**.

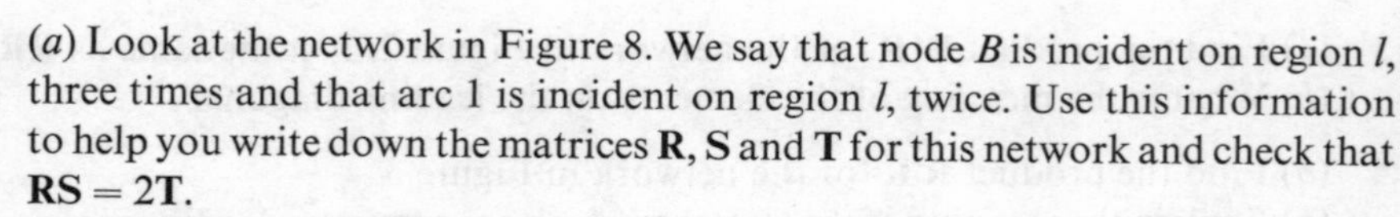

7 (*a*) Look at the network in Figure 8. We say that node B is incident on region l, three times and that arc 1 is incident on region l, twice. Use this information to help you write down the matrices **R**, **S** and **T** for this network and check that $\mathbf{RS} = 2\mathbf{T}$.

(*b*) Now find out whether $\mathbf{RS} = 2\mathbf{T}$ for the network in Figure 9. Explain how you can overcome any difficulties that arise.

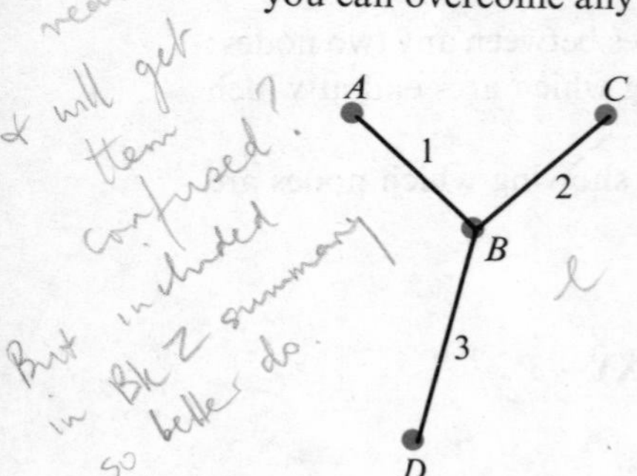

Fig. 8

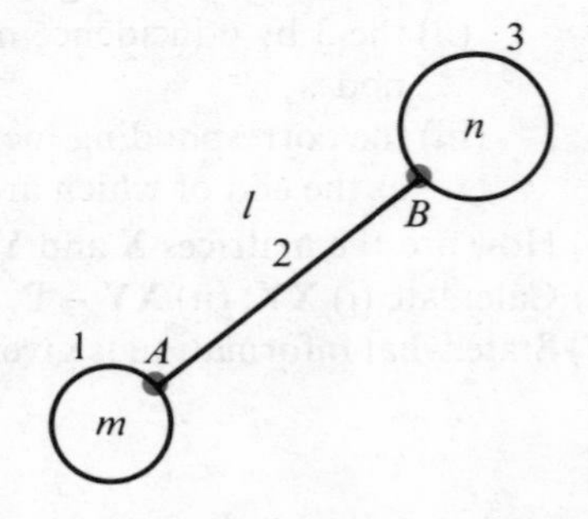

Fig. 9

Revision exercises

Slide rule session

Give the answers to the following as accurately as you can.

1 $2{\cdot}17 \times 19{\cdot}4$.

2 $37{\cdot}5 \div 6{\cdot}2$.

3 $\sqrt{174}$.

4 $(1{\cdot}62 \times 54) + (17{\cdot}7 \times 0{\cdot}8)$.

5 $(14{\cdot}8)^2 \div 7{\cdot}4$.

6 $\pi \times (1{\cdot}5)^2$.

7 $\dfrac{79}{16{\cdot}8 \times 4{\cdot}2}$.

8 $\sqrt{(36{\cdot}5 \times 2{\cdot}9)}$.

9 $(14{\cdot}2 \times 0{\cdot}9)^2$.

10 $45{\cdot}4 \times 3{\cdot}7 \times 0{\cdot}095$.

11 $\dfrac{8}{7{\cdot}4}$.

12 $\frac{1}{8} \div 43$.

13 $0{\cdot}7 \times \dfrac{1}{9{\cdot}2}$.

14 $(41 \div 23) \times \frac{1}{5} \div 5$.

15 $\sqrt{28{\cdot}4} \times \dfrac{1}{4{\cdot}6} \times (19{\cdot}4)^3$.

Computation 1

Estimation of answers.

Never perform a computation without considering whether your answer is reasonable. The following examples will give some practice at making reasonable estimates. Say which answers you consider are reasonable estimates, and in the case of the others, write what you think would be better.

1 $12{\cdot}3 \times 2{\cdot}9 = 36$ approx.

2 $0{\cdot}105 \times 0{\cdot}1 = 0{\cdot}1$ approx.

3 $9{\cdot}6 \times 26{\cdot}2 = 250$ approx.

4 $1023 \times 19 = 2000$ approx.

5 $(20{\cdot}2)^2 = 400$ approx.

6 $\sqrt{170} = 13$ approx.

7 $\sqrt[3]{0{\cdot}08} = 0{\cdot}2$ approx.

8 $\dfrac{16 \times 1{\cdot}1}{3{\cdot}4} = 5$ approx.

9 $\sqrt{(3{\cdot}1^2 + 6{\cdot}9^2)} = 10$ approx.

10 $62{\cdot}9 \times 0{\cdot}9 \times 0{\cdot}49 = 3$ approx.

Exercise A

1 If $ax > bx \Leftrightarrow a < b$, what can you say about x?

2 Express 0·0056 in standard index form.

3 What can you say about the sets A and B if $A \cup B = B$?

4 Find the value of t if $\frac{3}{4}(2t + 7) = 0$.

5 Find $f(^{-}2)$ if $f(x) = x^2 + x$.

6 State the area of a circle whose diameter is 6 cm, leaving π in your answer.

7 What is the probability of obtaining a single head when two coins are tossed?

8 Add the fractions $\frac{2}{5}$ and $\frac{3}{4}$.

Exercise B

1 Write down 63_{10} as a number in the scale of 8.

2 Write down the median of the numbers, 34, 36, 12, 23, 29, 87, 56.

3 In triangle ABC, angle $A = 45°$, angle $B = 90°$ and $BC = 6$ cm. Find AB.

4 List the set of prime numbers between 90 and 100.

5 What single transformation is equivalent to two successive reflections in two parallel mirror-lines?

6 Give the probability of throwing a total larger than 10 when 2 dice are thrown.

7 Simplify $\frac{1}{2} - \frac{2}{3} + \frac{5}{6}$.

8 Give the inverse of the function $x \rightarrow 3x + 1$.

Exercise C

1 State the point of intersection of the lines $y = x$ and $x + y = 4$.

2 Give the image of the point (3,4) when it is reflected in the line $y = x$.

3 (*a*) What is the probability that a card, drawn at random from a pack of 52 well-shuffled cards, is a black 2?
(*b*) Two boys pick their favourite colour from red, blue, yellow. Write down the probability that:

(i) both choose the same colour,
(ii) both choose yellow,
(iii) one chooses yellow, but not the other.

4 Find the coordinates of the vertices of the unit square after it has been enlarged by scale factor 3, centre the origin, followed by a half turn about the origin.

5 (*a*) Draw the lines $y = 2x - 1$, $x + y = 2$ and $y = {}^{-}1$ on the same graph taking values of both x and y from $^{-}2$ to 4.
(*b*) Give all the points with integral coordinates which satisfy all the inequalities $y \leqslant 2x - 1$, $x + y < 2$, $y > {}^{-}1$.

6 A certain reflection maps $(0, 1)$ onto $(4, 1)$ and $(2, {}^{-}2)$ onto itself. Give the equation of the mirror line.

7 Find the gradients of $2x - 3y = 6$, $4x - 6y = 9$, $2x - 3y = {}^{-}15$. What do you notice? Write down the equation of another member of the set of such lines.

8 Find the result if, within the set of integers, you:
(*a*) think of a number;
(*b*) add the next largest number;
(*c*) add 9;
(*d*) divide by 2;
(*e*) subtract the original number.

Exercise D

1 Find the points with integral coordinates that satisfy the orderings

$${}^{-}2 \leqslant x + y < 2; \qquad 0 \geqslant 3x - y; \qquad y - 3x \leqslant 2.$$

2 Draw a network for which this incidence matrix relates nodes and arcs:

$$\text{nodes} \quad \begin{matrix} & \text{arcs} \\ & \begin{matrix} 1 & 2 & 3 & 4 & 5 \end{matrix} \\ \begin{matrix} A \\ B \\ C \\ D \end{matrix} & \begin{pmatrix} 1 & 0 & 0 & 1 & 1 \\ 1 & 1 & 0 & 0 & 0 \\ 0 & 1 & 1 & 0 & 1 \\ 0 & 0 & 1 & 1 & 0 \end{pmatrix} \end{matrix}.$$

3 Find the area of the triangle whose vertices are $({}^{-}1, 2)$, $(1, 2)$, $(0, 4)$.

4 Simplify $\{(x, y) : y = 0\} \cap \{(x, y) : y = 1\}$.

5 An African boy has 15p in his pocket. He wants to spend his money on pineapples (60 cents each) and oranges (30 cents each). A pineapple takes $7\frac{1}{2}$ minutes to eat, an orange $2\frac{1}{2}$ minutes; the boy has 30 minutes to spare in which to consume his fruit (5p = 100 cents).

He wants to eat at least as many pineapples as oranges, and he also wants the total number of fruit eaten to be as large as possible.

If he eats x pineapples and y oranges, write down the orderings which x and y must satisfy. Find the solution satisfying all the conditions.

6 A box contains several hundred black pins and white pins – three times as many black pins as white pins. Two pins in succession are taken out in the dark. What is the probability that they are both black?

7 What angle does the plane $y = x$ make with the plane $x = 0$?

8 Give the equations of the planes which are determined by the following sets of points:

  (*a*) $(0,0,0)$, $(0,1,2)$, $(0,7,9)$;
  (*b*) $(1,2,3)$, $(3,2,1)$, $(5,2,4)$;
  (*c*) $(5,4,3)$, $(4,4,3)$, $(5,5,3)$.

Exercise E

1 Find the distance between points with coordinates $(5,6)$ and $(^-3,0)$. Write down the coordinates of a point whose distance from $(5,6)$ is twice this.

2 Find the distance between the points $(9,^-1,7)$ and $(10,1,9)$.

3 If the line $y = 3x + c$ passes through $(1,2)$, find c.

4 What positive values of x satisfy $4 < 12/x < 6$?

5 Find the gradients of the following lines:

  (*a*) $3y = {}^-2x + 4$; (*b*) $2x - y = 5$;
  (*c*) $x + y = 1$; (*d*) $y - 3x - 6 = 0$.

6 A triangle has vertices at $(2,1)$, $(5,2)$, $(3,3)$. Show that it is right angled and calculate its area.

7 Describe with the aid of a rough sketch the set of points $\{(x,y): x^2 + y^2 = 25\}$.

8 Find the Cartesian coordinates of points with the following polar coordinates:

$$(1,30°), (10,30°), (10,150°), (5,210°).$$

5 Tangents

1 Sine and cosine

(*a*) In *Book G* we defined the sine and cosine of an angle θ as the y and x coordinates respectively of a point P lying on the circumference of a unit circle, centre O, where OP made an angle $\theta°$ with the positive direction of the x axis.

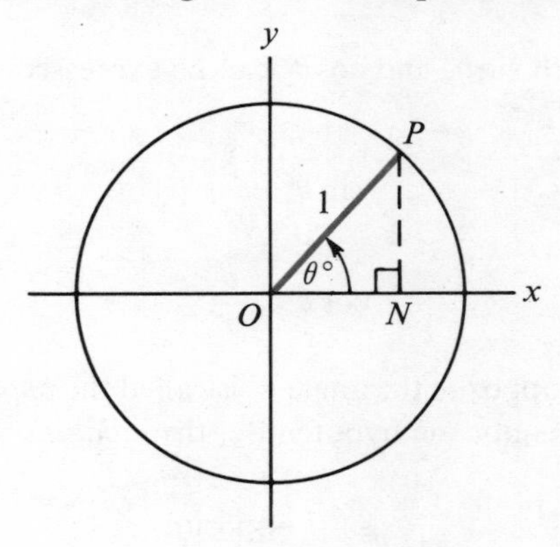

Fig. 1

Thus in Figure 1,

$$\cos\theta° = ON$$

and

$$\sin\theta° = PN.$$

The values of ON and PN for $0 \leqslant \theta \leqslant 90$ are tabulated for you in your cosine and sine tables, and you will see, for example, that when $\theta = 46{\cdot}5$, $\cos\theta° = 0{\cdot}688$ and $\sin\theta° = 0{\cdot}725$.

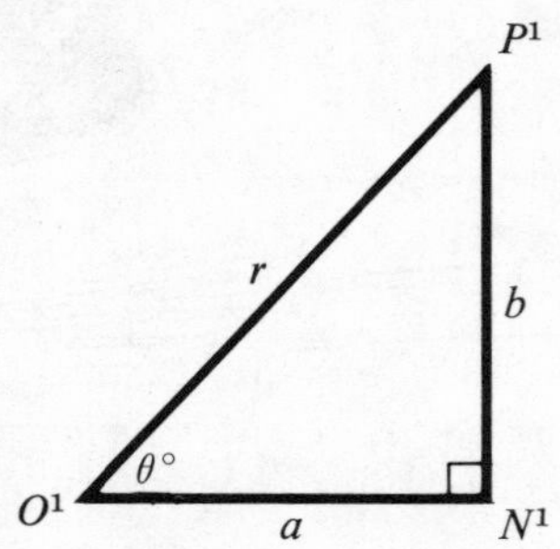

Fig. 2

(*b*) Figure 2 shows a triangle similar to triangle *OPN*. Figure 3 shows that this triangle has been obtained by enlarging triangle *OPN* with scale factor *r*.

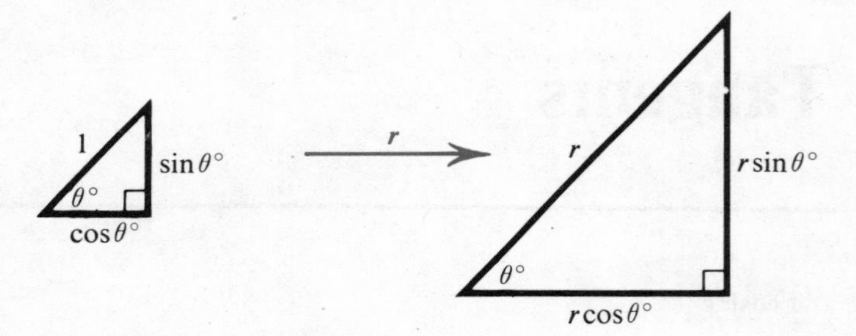

Fig. 3

By direct comparison we can see that

(i) $b = r\sin\theta°$,

(ii) $a = r\cos\theta°$.

This means that both $\sin\theta°$ and $\cos\theta°$ can be expressed as a ratio of two sides of a right-angled triangle:

$$\sin\theta° = \frac{b}{r},$$

$$\cos\theta° = \frac{a}{r}.$$

Sometimes the side opposite the angle $\theta°$ is called the *opposite* side, and the side adjacent to $\theta°$, which is not the hypotenuse, the *adjacent* side. We can therefore write:

$$\sin\theta° = \frac{\text{opposite}}{\text{hypotenuse}},$$

$$\cos\theta° = \frac{\text{adjacent}}{\text{hypotenuse}}.$$

(See Figure 4.)

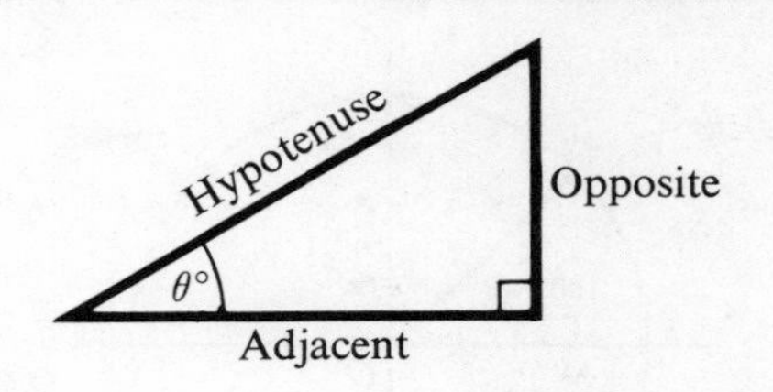

Fig. 4

(c)

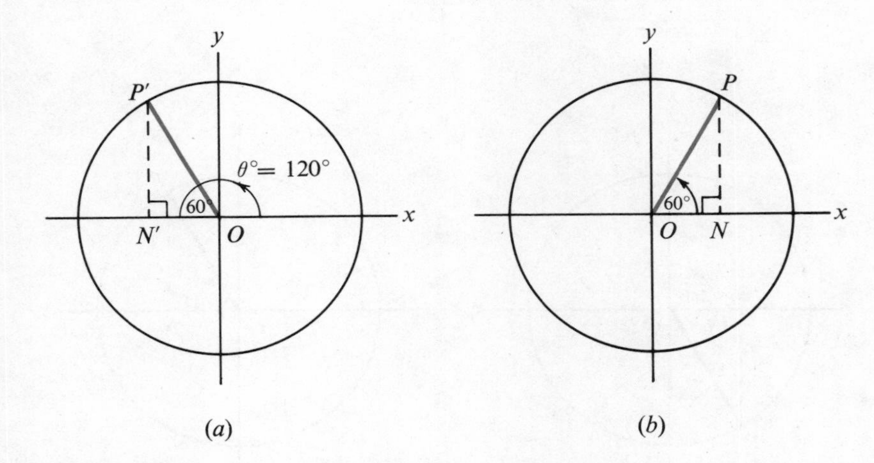

Fig. 5

Look at Figure 5(*a*) in which $\theta = 120$ and at Figure 5(*b*) in which $\theta = 60$. The y coordinates of P and P' in each figure are the same because the triangles OPN, $OP'N'$ are congruent (that is, they have the same shape and size).

Hence, $\sin 120° = P'N' = \sin 60° = 0{\cdot}867$.

However although the lengths ON and ON' are equal, the x coordinate of P' in Figure 5(*a*) is negative. Hence we have

$$\cos 120° = {}^{-}\cos 60° = {}^{-}0{\cdot}5.$$

Calculate:

(i) $\sin 150°$; (ii) $\cos 150°$;
(iii) $\sin 135°$; (iv) $\cos 135°$.

Do you agree that for $90 \leqslant \theta \leqslant 180$,

$$\sin \theta° = \sin (180° - \theta°)$$

and

$$\cos \theta° = {}^{-}\cos (180° - \theta°)?$$

(See Figure 6.)

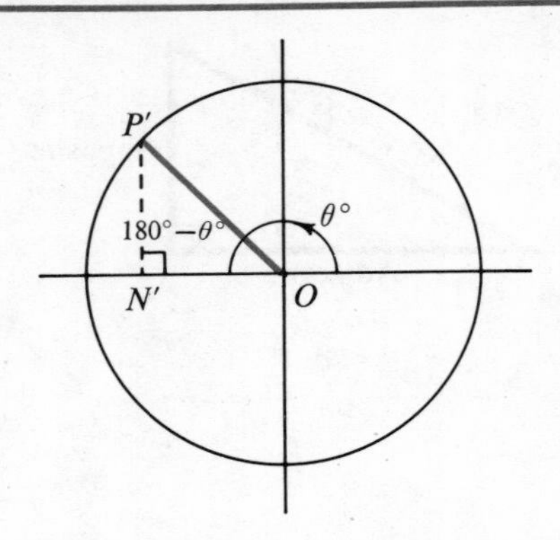

Fig. 6

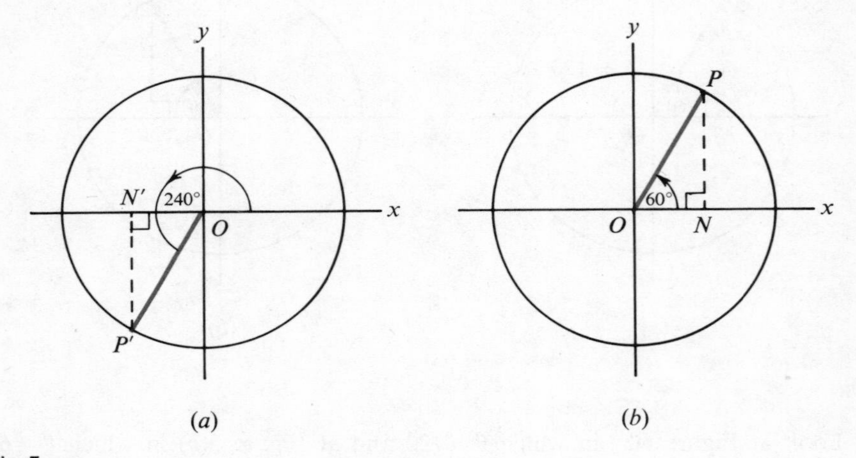

Fig. 7

(*d*) Look at Figure 7(*a*) in which $\theta = 240$, and at Figure 7(*b*) in which $\theta = 60$. In this case the triangles *OPN* and *OP′N′* are again congruent but *both* coordinates of *P′* are negative.

Hence

$$\sin 240° = {}^{-}\sin 60° = {}^{-}0{\cdot}867$$

and

$$\cos 240° = {}^{-}\cos 60° = {}^{-}0{\cdot}5.$$

Do you agree that if $180 \leqslant \theta \leqslant 270$, then

$$\sin \theta° = {}^{-}\sin (\theta° - 180°)$$

and

$$\cos \theta° = {}^{-}\cos (\theta° - 180°)?$$

(*e*)

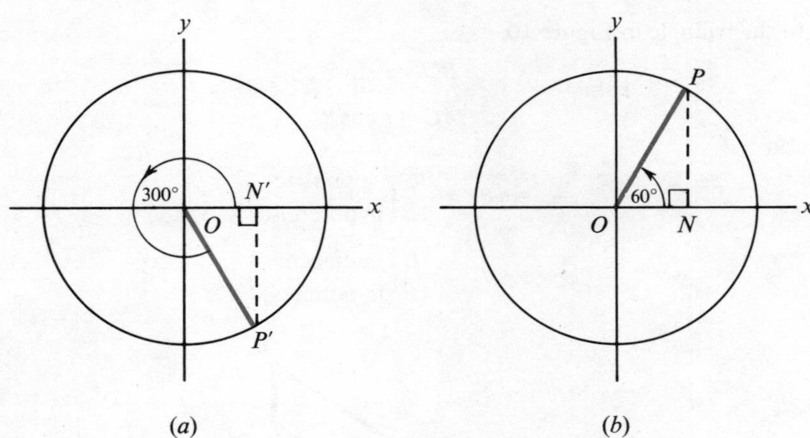

Fig. 8

Are the triangles $OP'N'$ and OPN in Figure 8 congruent? Use Figure 8 to help you find (i) sin 300°, and (ii) cos 300°.

Check that if $270 \leqslant \theta \leqslant 360$ then

$$\sin \theta° = {}^{-}\sin (360° - \theta°)$$

and

$$\cos \theta° = \cos (360° - \theta°).$$

(*f*) Figure 9 summarizes what we have found about the sign of sines and cosines of angles between 0° and 360°.

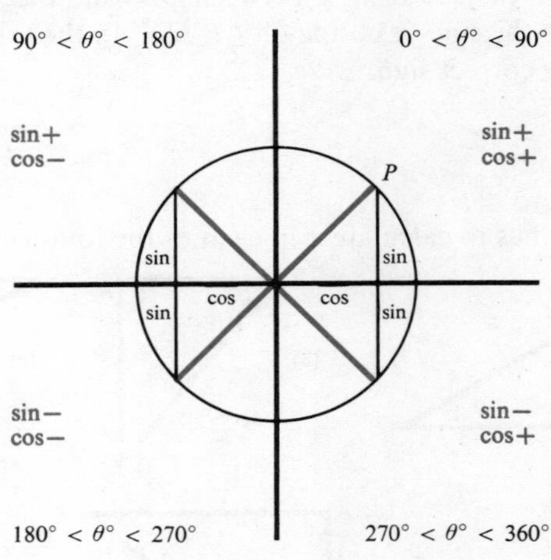

Fig. 9

What are the values of:

(i) sin 0°; (ii) cos 0°; (iii) sin 90°; (iv) cos 90°; (v) sin 180°; (vi) cos 180°; (vii) sin 270°; (viii) cos 270°; (ix) sin 360°; (x) cos 360°?

Tangents

Summary

1 In the triangle in Figure 10,

$$BC = r\sin\theta°,$$
$$AC = r\cos\theta°$$

and

$$\sin\theta° = \frac{BC}{AB}\left(\frac{\text{opposite}}{\text{hypotenuse}}\right),$$

$$\cos\theta° = \frac{AC}{AB}\left(\frac{\text{adjacent}}{\text{hypotenuse}}\right).$$

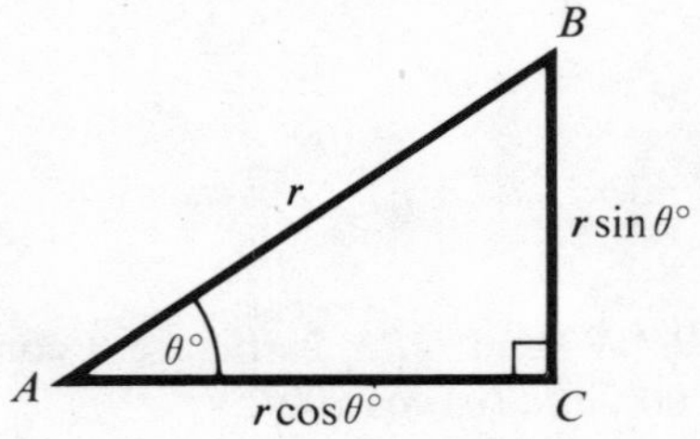

Fig. 10

2 The signs of the sine and cosine of angles between 0° and 360° are represented by Figure 9 above.

3 To calculate the sine or cosine of an angle greater than 90° we can

(i) draw a diagram like Figure 8;
(ii) calculate the acute angle between OP' and the x axis (i.e. $\angle PON$);
(iii) look up the sine or cosine of $\angle P'ON'$ in the tables;
(iv) give the correct sign.

Exercise A

1 Use sines or cosines to calculate x in each of the following triangles:

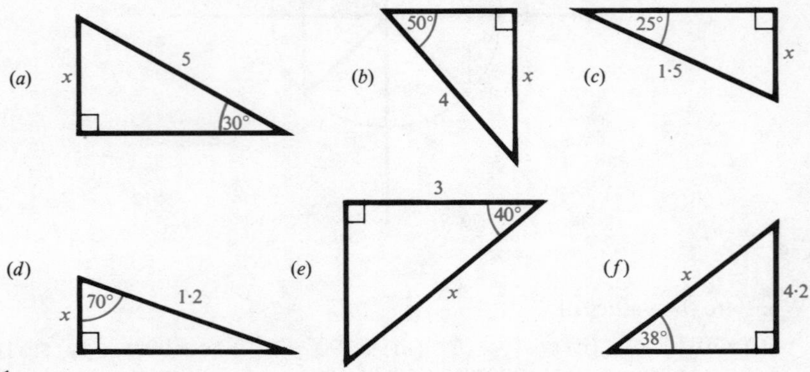

Fig. 11

2 Use your tables to find the value of:

(i) $\sin 30°$; (ii) $\cos 30°$; (iii) $\sin 57{\cdot}3°$;
(iv) $\sin 95°$; (v) $\cos 100°$; (vi) $\sin 170°$;
(vii) $\cos 160°$; (viii) $\sin 280°$; (ix) $\cos 280°$;
(x) $\sin 300°$; (xi) $\cos 330°$.

3 Find two angles for which (i) $\sin\theta° = 0{\cdot}5$; (ii) $\cos\theta° = 0{\cdot}5$; (iii) $\sin\theta° = {}^{-}0{\cdot}5$; (iv) $\cos\theta° = {}^{-}0{\cdot}5$. Are your answers the *only* possibilities?

4 Draw a circle of radius 10 cm, and draw the triangle OPN (see Figure 12). Measure PN and ON. Hence write down an approximate value for (i) $\sin 20°$; (ii) $\cos 20°$. Use your tables to check your answers. Use the same method to calculate $\sin\theta°$ and $\cos\theta°$ for $\theta = 0, 30, 60, 90, \ldots, 360$.

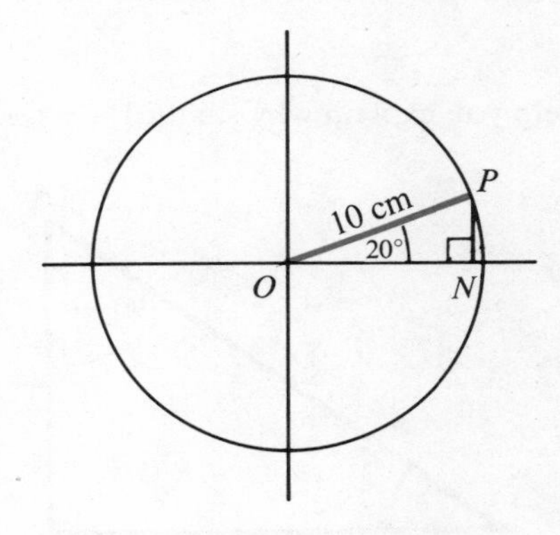

Fig. 12

Use your results to draw the graphs of $f\colon x \to \sin x°$ and $g\colon x \to \cos x°$ for $0 \leqslant x \leqslant 360$.

5 Figure 13 shows an equilateral triangle of side 2 units. Use Pythagoras' rule to show that $XA = \sqrt{3}$ units. Hence write down the value of $\sin 60°$ leaving square roots in your answer. Write down also the values of (i) $\sin 30°$; (ii) $\cos 60°$; (iii) $\cos 30°$.

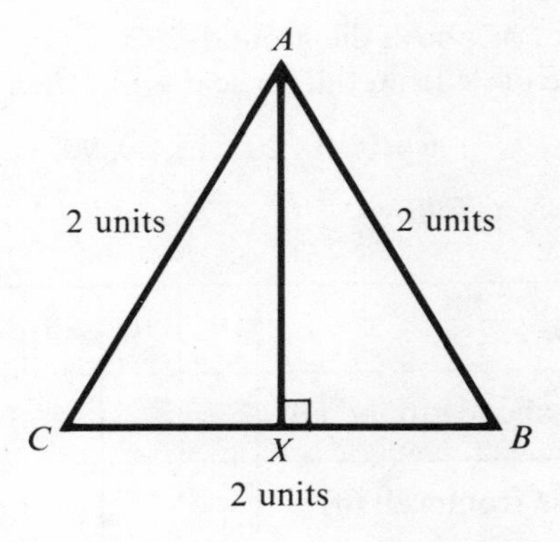

Fig. 13

6 Figure 14 shows an isosceles triangle in which $AB = BC = 1$ unit. Use Pythagoras' rule to calculate AC, giving your answer in the form $AC = \sqrt{x}$ units. Hence write down the value of (i) $\sin 45°$; (ii) $\cos 45°$ leaving square roots in your answer.

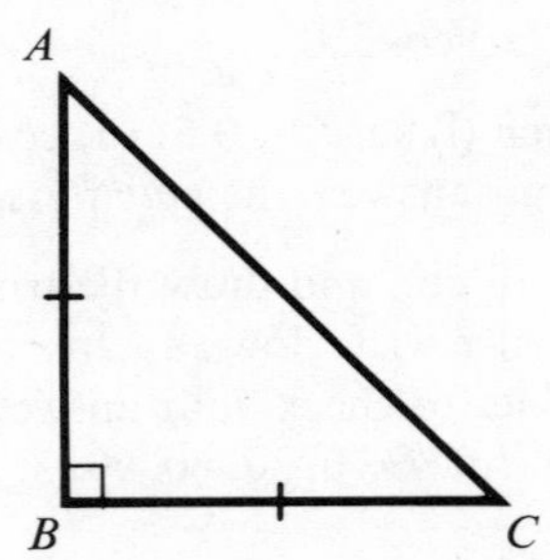

Fig. 14

7 Use Figure 15 to help you explain why $\sin (90° - \theta°) = \cos \theta°$.

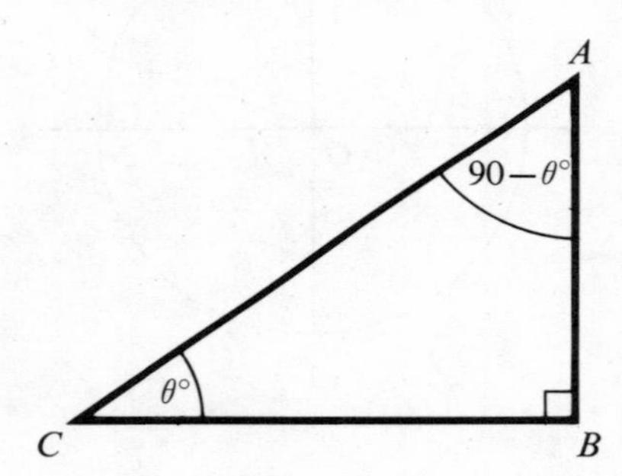

Fig. 15

If you only had sine tables, how would you find the value of $\cos 20°$?

8 If $\sin 2\theta° = 0{\cdot}417$, calculate θ.

9 Figure 16 shows a ladder, AB, of length 6 m leaning against a vertical wall. M is the mid-point of the ladder. The base of the ladder begins to slip away from the wall. Calculate:

(i) the height of M above the ground

and (ii) the distance of M from the vertical wall, when

$$\theta = 0, 10, 20, \ldots, 80, 90.$$

Tabulate your results as follows:

θ (in degrees)	0	10	20	30	. . .	90
Height of M above ground (p)						
Distance of M from wall (q)						

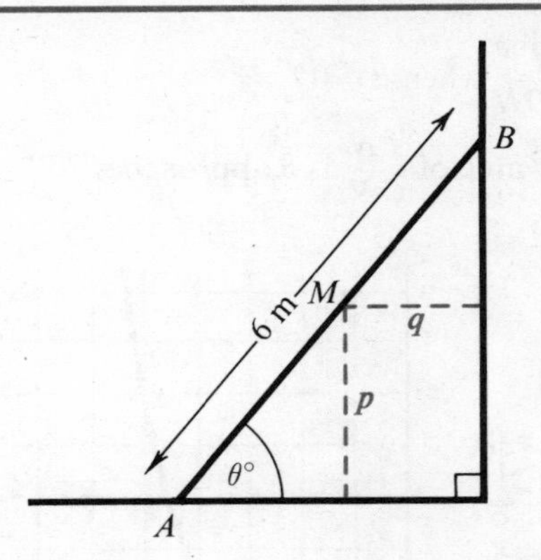

Fig. 16

Draw a graph of p against q using the same scale on both axes. What is the locus of M as the ladder slides down the wall?

2 Tangents of an angle

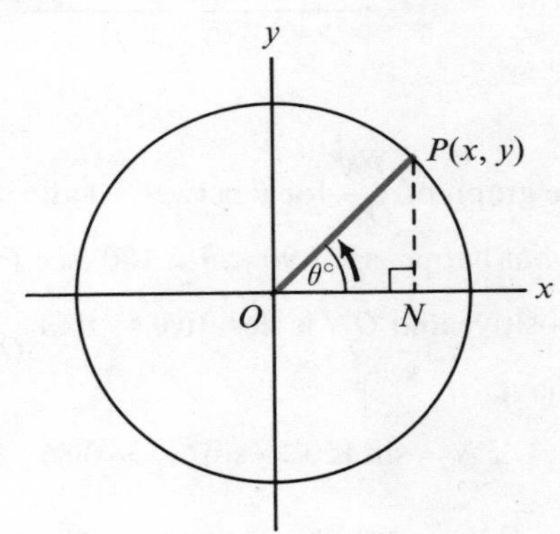

Fig. 17

(*a*) As the line segment OP in Figure 17 rotates in a positive sense from its initial position for which $\theta = 0$, the x and y coordinates of P vary. Thus the ratio $\frac{PN}{ON}$ varies as θ increases from 0 to 90.

At any time $\frac{PN}{ON}$ will have a particular value. For example, when $\theta = 30$,

$$PN = \sin\theta° = 0{\cdot}5$$

and

$$ON = \cos\theta° = 0{\cdot}867,$$

so that

$$\frac{PN}{ON} = 0{\cdot}577.$$

Use your sine and cosine tables to find the value of $\frac{PN}{ON}$ when $\theta = 10, 20, 30, 40, \ldots, 80$.

What is the value of $\frac{PN}{ON}$ when $\theta = 0$?

What happens to the value of $\frac{PN}{ON}$ as θ approaches 90?

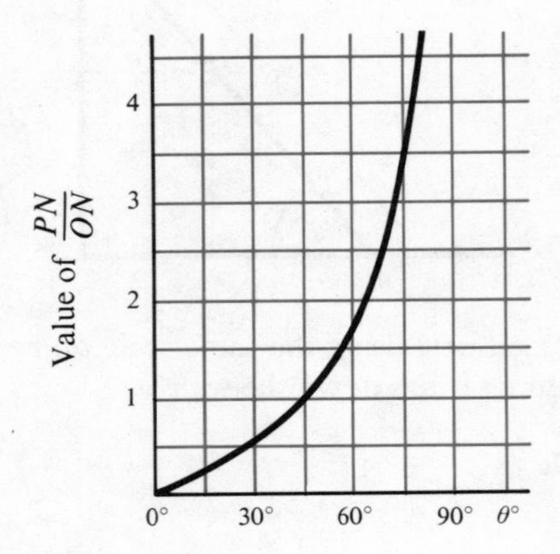

Fig. 18

Figure 18 shows the graph of $\frac{PN}{ON}$ for θ between 0 and 90.

(*b*) Now consider what happens for $90 < \theta < 180$ (see Figure 19).

In this case PN is positive and ON is negative so that $\frac{PN}{ON}$ is negative.

When $\theta = 120$, we have

$$PN = \sin 120° = \sin 60° = 0{\cdot}867$$

and

$$ON = \cos 120° = {}^{-}\cos 60° = {}^{-}0{\cdot}5.$$

Hence

$$\frac{PN}{ON} = -\frac{0{\cdot}867}{0{\cdot}5} = {}^{-}1{\cdot}734.$$

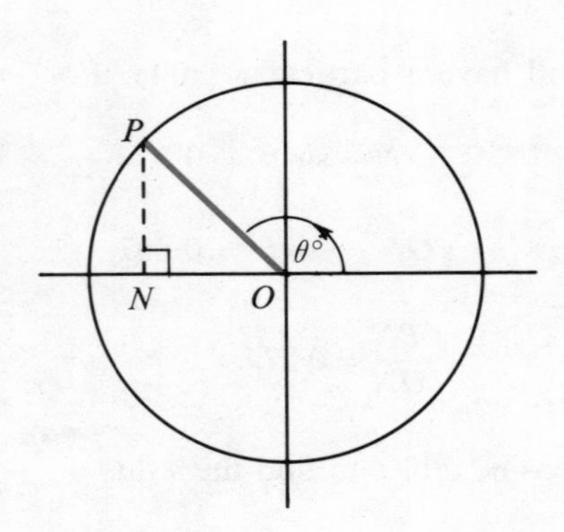

Fig. 19

Work out $\dfrac{PN}{ON}$ when θ is 100, 110, ..., 170 and draw a graph similar to Figure18.

What is the value of $\dfrac{PN}{ON}$ when $\theta = 180$?

(*c*) Work out the values of $\dfrac{PN}{ON}$ for $\theta = 190, 200, \ldots, 260$. Graph your results.

What happens to the value of $\dfrac{PN}{ON}$ as θ approaches 270?

Now work out the values of $\dfrac{PN}{ON}$ for $\theta = 290, 300, \ldots, 350$ and graph your results.

What is the value of $\dfrac{PN}{ON}$ when $\theta = 360$?

(*d*) If you graph the values of $\dfrac{PN}{ON}$ for $0 \leqslant \theta \leqslant 360$ (i.e. if you draw Figure 18 and the three graphs for $90 < \theta \leqslant 180$, $180 < \theta \leqslant 270$ and $270 < \theta \leqslant 360$ on the same axis) you will obtain Figure 20.

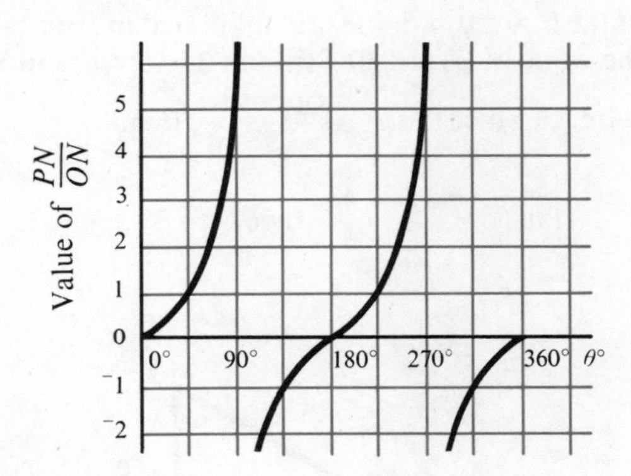

Fig. 20

The ratio $\dfrac{PN}{ON} = \dfrac{\sin\theta°}{\cos\theta°}$ is called the *tangent of* $\theta°$ and is written '$\tan\theta°$':

$$\tan\theta° = \frac{\sin\theta°}{\cos\theta°}.$$

(*e*) Figure 21 shows a triangle *OPN* and an enlargement of *OPN* scale factor *r*.

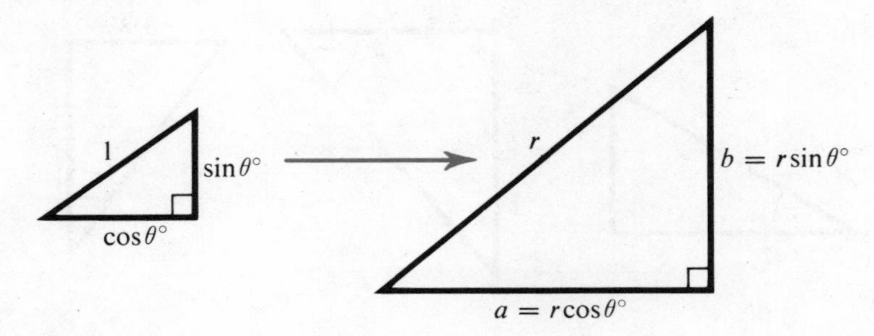

Fig. 21

Do you agree that $\tan\theta^\circ = \dfrac{r\sin\theta^\circ}{r\cos\theta^\circ} = \dfrac{b}{a}$?

$\left[\text{Using the terminology of Section 1}(b)\text{, } \tan\theta^\circ = \dfrac{\text{opposite}}{\text{adjacent}}.\right]$

Since $\tan\theta^\circ = \dfrac{b}{a}$, then, multiplying both sides of a, we have

$$b = a\tan\theta^\circ$$

(see Figure 22).

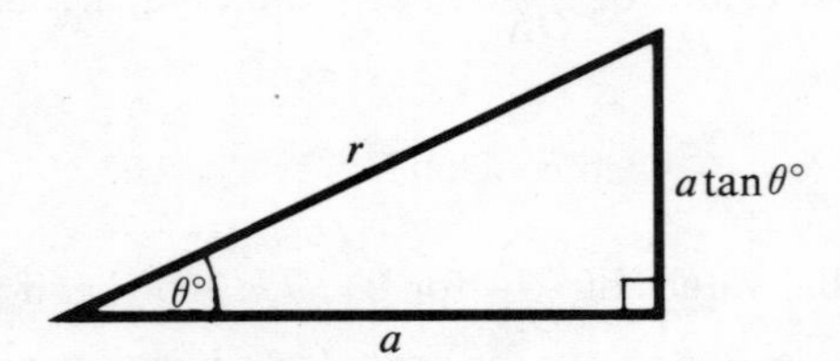

Fig. 22

(f) The values of $\tan\theta^\circ$ for $0 \leqslant \theta < 90$ are tabulated in your tangent tables. Use the tables to find the value of (i) tan 30°; (ii) tan 27·4°; (iii) tan 89°; (iv) tan 120°.

(g) Consider Figure 23. Since $\tan\theta^\circ = \dfrac{\text{opposite}}{\text{adjacent}}$, then

$$\tan\theta^\circ = \frac{AB}{BC} = \frac{4}{6} = 0{\cdot}667 \text{ (to 3 S.F.)}$$

$$\theta^\circ = 33{\cdot}7^\circ.$$

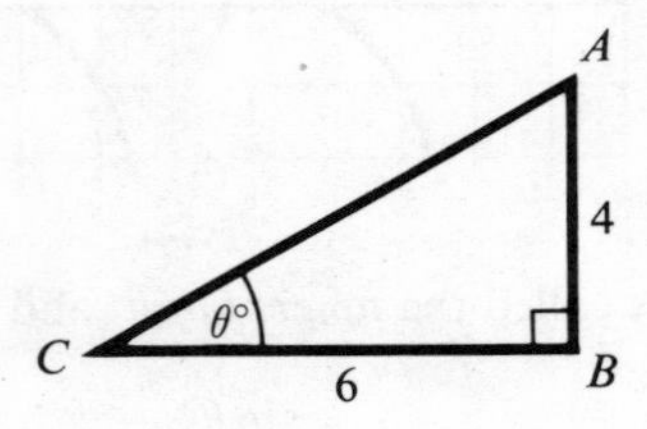

Fig. 23

Use this method to calculate θ in the three triangles in Figure 24.

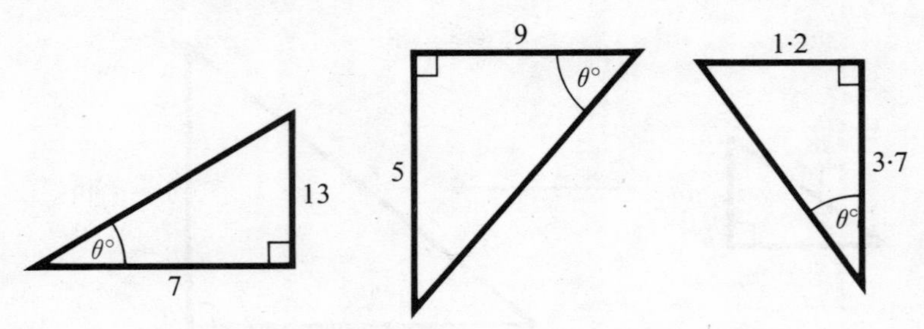

Fig. 24

(h) Use the equation $b = a \tan\theta°$ (see Figure 22) to calculate the length of AB in the triangles in Figure 25.

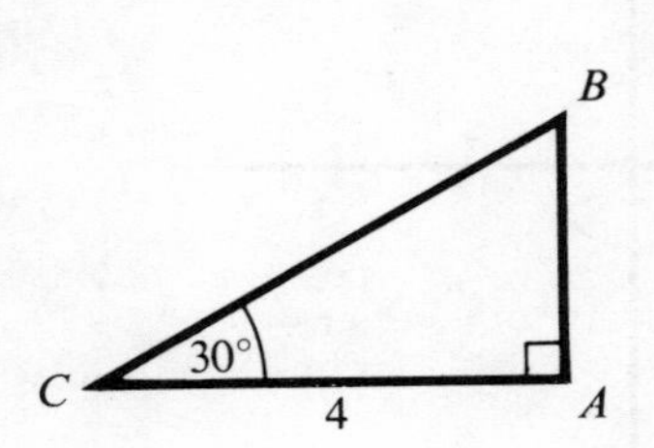

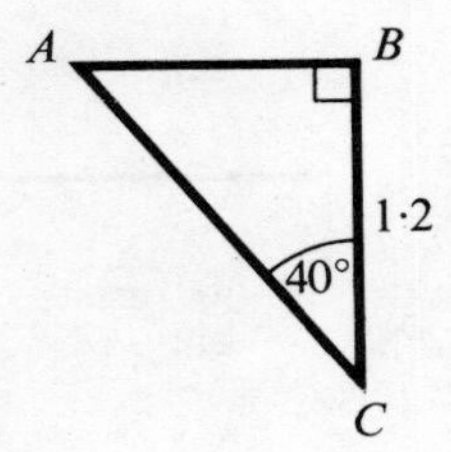

Fig. 25

Summary

1 $\tan\theta° = \dfrac{\sin\theta°}{\cos\theta°}$.

2 In triangle PQR, $\tan\theta° = \dfrac{b}{a}$, and $b = a\tan\theta°$.

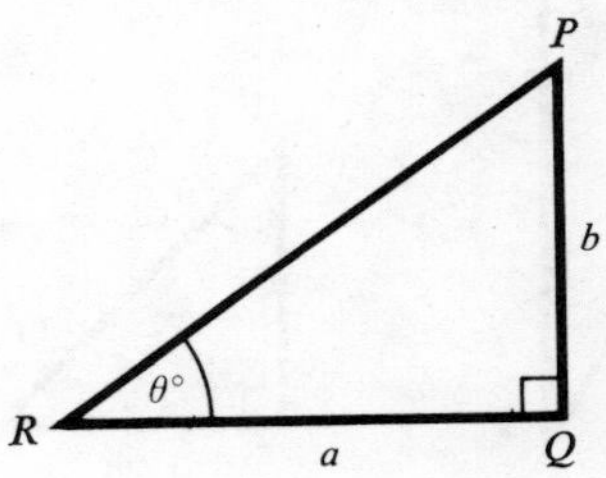

Fig. 26

3 The signs of $\tan\theta°$, for $0 \leqslant \theta \leqslant 360$, are given in Figure 27.

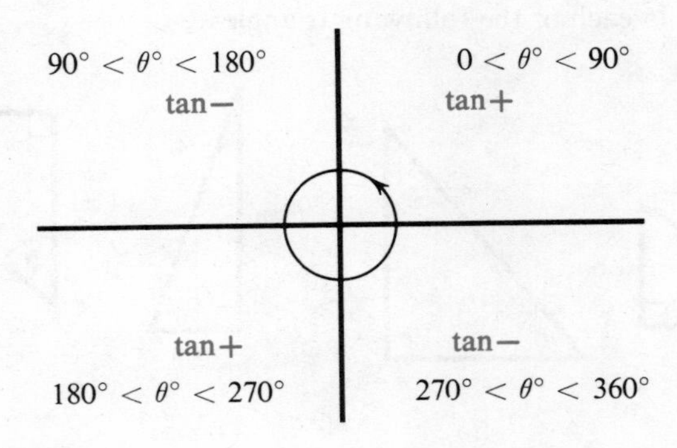

Fig. 27

Hence for $0 \leqslant \theta \leqslant 360$ we have:

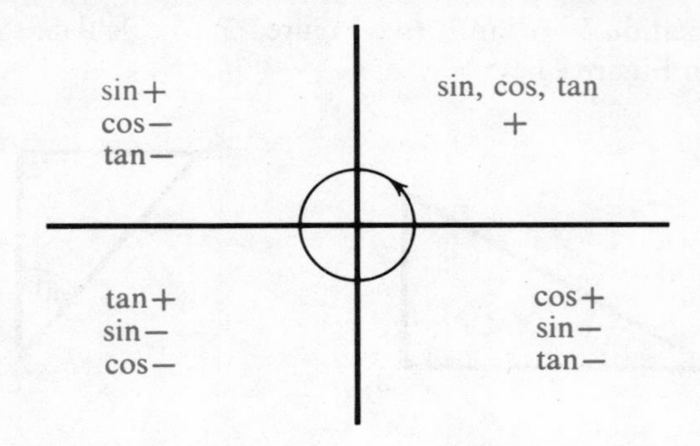

Fig. 28

Exercise B

1 Calculate θ in each of the triangles shown in Figure 29.

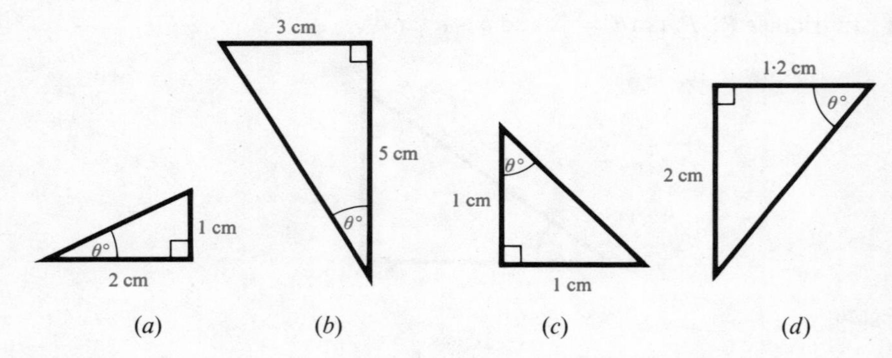

Fig. 29

2 Calculate x in each of the following triangles:

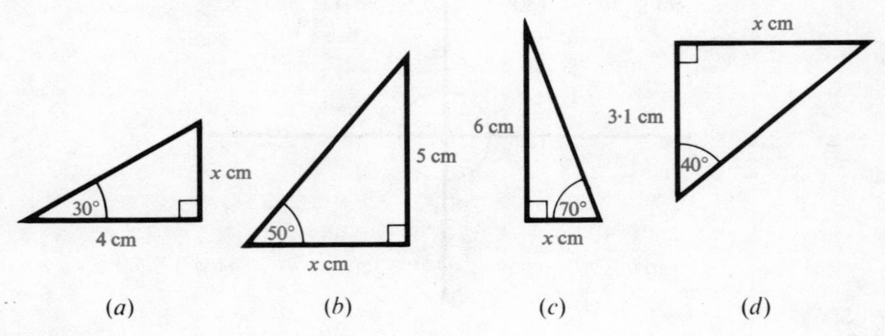

Fig. 30

3 Find two values of θ for which (i) $\tan\theta° = 0{\cdot}5$; (ii) $\tan\theta° = 1$; (iii) $\tan\theta° = 2$; (iv) $\tan\theta° = {}^{-}1$; (v) $\tan\theta° = {}^{-}0{\cdot}216$.

4 Find the value of (i) tan 110°; (ii) tan 200°; (iii) tan 170°; (iv) tan 230°.

5 Use Figures 13 and 14 (Exercise A) to write down the values of (i) tan 60°; (ii) tan 30°; (iii) tan 45°. (Leave $\sqrt{3}$ in your answers to (i) and (ii).)

6 Calculate (i) tan 30° × tan 60°; (ii) tan 20° × tan 70°; (iii) tan 55° × tan 35°. Use Figure 31 to help you write down the value of $\tan\theta° \times \tan(90° - \theta°)$. If $\tan\alpha° = x$, what is $\tan(90° - \alpha°)$ in terms of x?

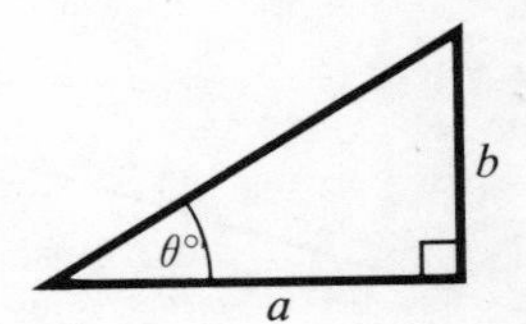

Fig. 31

7 If $\tan\theta° = \frac{1}{2}$, what is (i) $\sin\theta°$; (ii) $\cos\theta°$?

8 You are given that $\tan\theta° = \frac{3}{4}$. *Without using tables*, calculate (i) $\sin\theta°$; (ii) $\cos\theta°$. (Draw a right-angled triangle.)

9 If $\tan\theta° = \dfrac{a}{b}$, find (i) $\sin\theta°$; (ii) $\cos\theta°$ in terms of a and b.

10 Express BD in Figure 32 in terms of r and θ. Hence express DC in terms of r, θ and α. If $\theta = 40$, $\alpha = 35$ and $r = 3$ cm calculate (i) DC and (ii) AC.

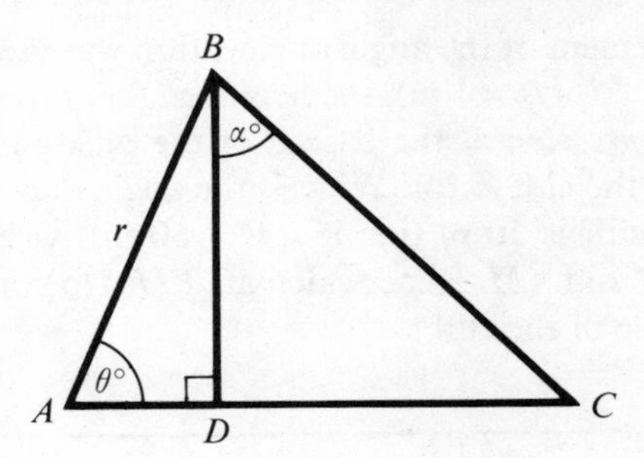

Fig. 32

11 If $0 \leqslant x \leqslant 90$, find a value of x for which

(i) $\sin x° = \tan x°$;
(ii) $\cos x° = \tan x°$;
(iii) $\sin x° = \cos x°$.

If $\sin x° = \cos x°$, what is the value of $\tan x°$?

12 (i) Solve the equation $\tan x° = 1$ for $0 \leqslant x \leqslant 360$. (The solution set contains two members.)
(ii) $f: x \to \tan x°$; $g: x \to 2x$. Find fg and gf in the form $fg: x \to \ldots$; $gf: x \to \ldots$.
(iii) Solve the equations (*a*) $\tan 2x° = 1$; (*b*) $2\tan x° = 1$, for $0 \leqslant x \leqslant 360$.

3 Using sin, cos and tan

(*a*) Tangents, sines and cosines can be used to solve problems which can be reduced to the study of right-angled triangles. Some examples are given below.

(*b*) A surveyor wishes to find the height of a church tower (see Figure 33). He stands at a distance 100 m from the base of the tower and sets up his measuring instruments. He finds that the angle between the horizontal DB and the line DC is 17°. This angle ($\angle CDB$) is known as *the angle of elevation* of the tower from D.

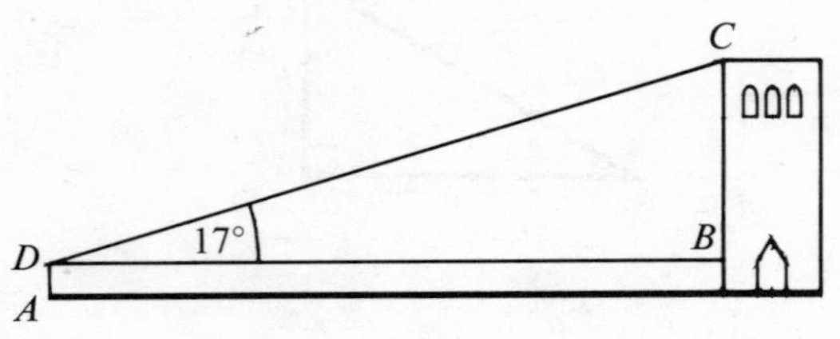

Fig. 33

To find the height of the tower we can use tangents. Since

$$CB = DB \tan 17°$$

then

$$CB = 100 \tan 17° \text{ m}$$
$$= 100 \times 0{\cdot}306 \text{ m}$$
$$= 30{\cdot}6 \text{ m}.$$

Hence, if the measurement of the angle of elevation was taken from a height 1 m above the ground (i.e. if $AD = 1$ m), the height of the tower is 31·6 m.

(*c*) *The angle of depression* of the base P of the oil rig in Figure 34 from the point A at the top of the cliff is 24°. What is the angle of elevation of A from P? If the height of the cliff is 50 m (i.e. if $AM = 50$ m), write down an equation involving tan 24°, PM and AM. Hence calculate PM. Do you agree that the oil rig is 112 m from the base of the cliff?

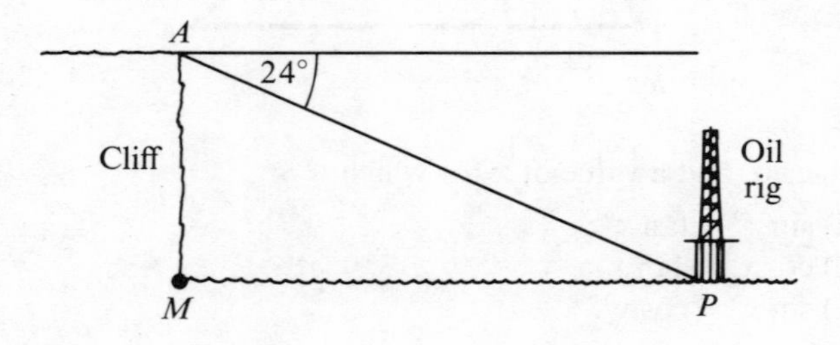

Fig. 34

(*d*) An aircraft flies from base for 600 km on a *bearing* 060° (bearings are measured in a clockwise direction from due North). It then flies 500 km on a bearing 090° (that is, due East). Suppose we wish to find the direction in which the aircraft must fly and the distance it must travel to return directly to base. The route back to base is represented by CA in Figure 35.

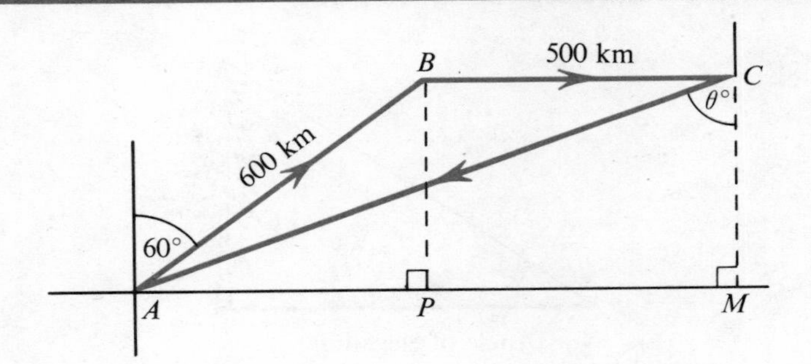

Fig. 35

To find the bearing we calculate CM and AM and then use tangents to calculate $\angle ACM$.

From triangle ABP, we have

$$\begin{aligned} BP &= 600 \sin 30^\circ \text{ km} \\ &= 600 \times 0{\cdot}5 \text{ km} \\ &= 300 \text{ km.} \end{aligned}$$

Hence

$$CM = BP = 300 \text{ km.}$$

To calculate AM:

$$\begin{aligned} AM = AP + PM &= AP + 500 \text{ km} \\ &= (600 \cos 30^\circ + 500) \text{ km} \\ &= (600 \times 0{\cdot}867 + 500) \text{ km} \\ &= 1020 \text{ km (to 3 S.F.).} \end{aligned}$$

Hence

$$\begin{aligned} \tan \theta^\circ &= \frac{AM}{CM} = \frac{1020}{300} \\ &= 3{\cdot}4, \end{aligned}$$

so that $\theta = 73{\cdot}6$.

The aircraft must therefore fly on a bearing of $73{\cdot}6^\circ + 180^\circ = 253{\cdot}6^\circ$.

To calculate CA, the distance the aircraft must fly, we can use

$$\cos 73{\cdot}6^\circ = \frac{CM}{CA} = \frac{300}{CA}$$

so that

$$\begin{aligned} CA &= \frac{300}{\cos 73{\cdot}6^\circ} \text{ km} \\ &= 1060 \text{ km (to 3 S.F.).} \end{aligned}$$

Summary

1 The angle of elevation of a point A from a point P (A is 'higher' than P) is the angle between the horizontal through P and the line through A and P. (Figure 36.)

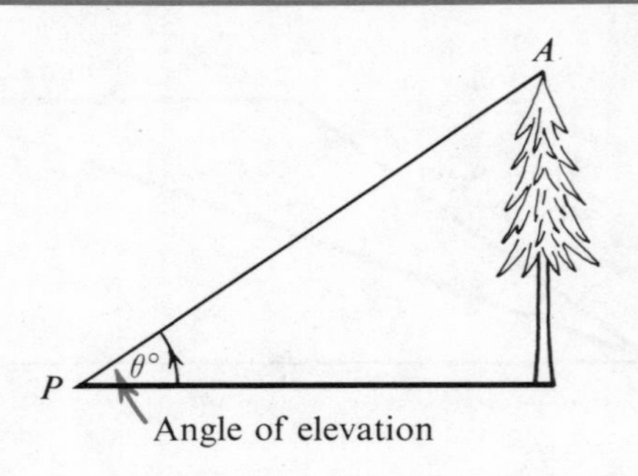

Fig. 36

2 The angle of depression (or the angle of declination) of a point A from a point P (A is 'lower' than P) is the angle between the horizontal through P and the line through A and P. (Figure 37.)

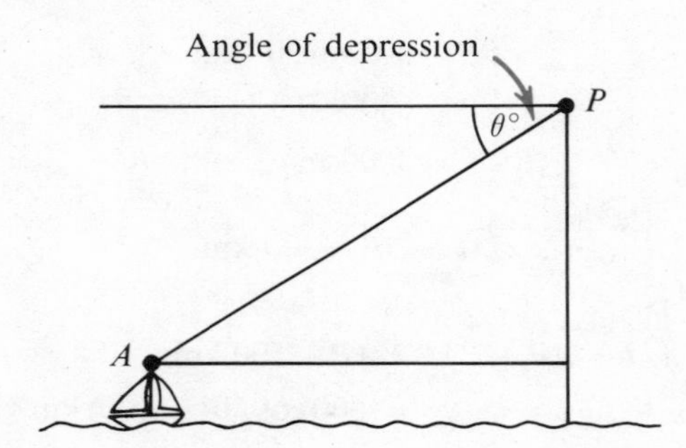

Fig. 37

Exercise C

1 The angle of elevation of the top of the Post Office tower, measured from a point 300 m from its base, is 32·2°. Calculate its height.

2 The angle of depression of a point at ground level 800 m from the Eiffel Tower, measured from the top of the tower, is 20·5°. Find the height of the tower.

3 A certain radar station can only detect aircraft with an angle of elevation, measured from the station, of more than 6°.

What is the nearest that an aircraft, flying at an altitude of 1500 m, can approach without being detected?

4 The angle of depression of the boat in Figure 38 from the point A at the top of the cliff 50 m above sea level is 17°. Calculate the distance of the boat from the base of the cliff. What is the angle of depression of the boat from A when it is a distance 100 m from the base of the cliff?

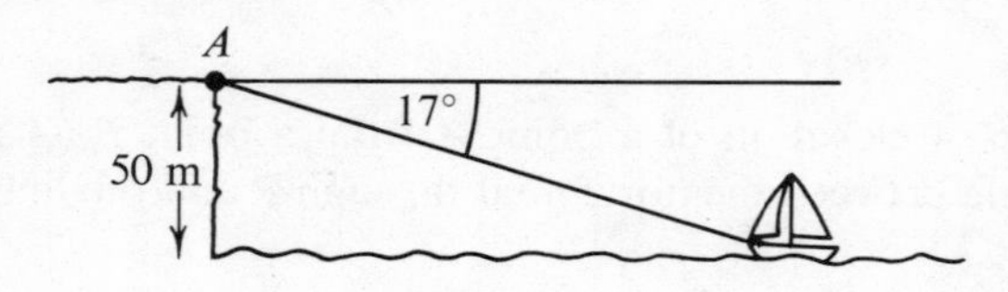

Fig. 38

5 The two towers in Figure 39 are 150 m apart. The angle of elevation of the top of Jodan tower from the top of Gable tower is 10°. If Gable tower is 30 m high, what is the height of Jodan tower?

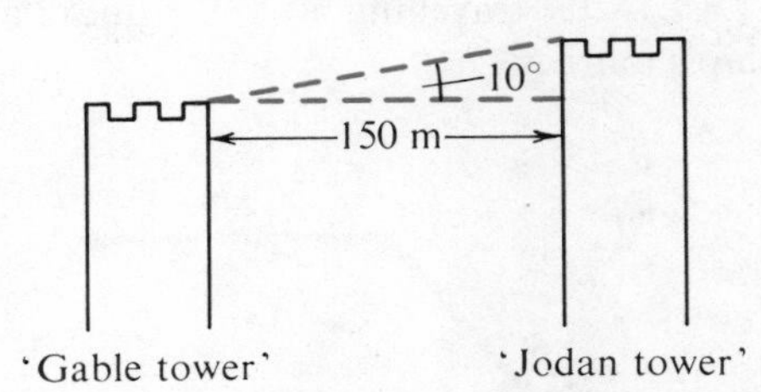

Fig. 39

6 A tree of height 20 m casts a shadow of length 35 m at a certain time of day. What is the angle of elevation of the sun at this time?
What length of shadow would a tree of height 25 m cast at the same time of day?

7 A man wishes to calculate the height of a tree on the opposite bank of a river. He stands at point A on the river bank (see Figure 40) and measures the angle

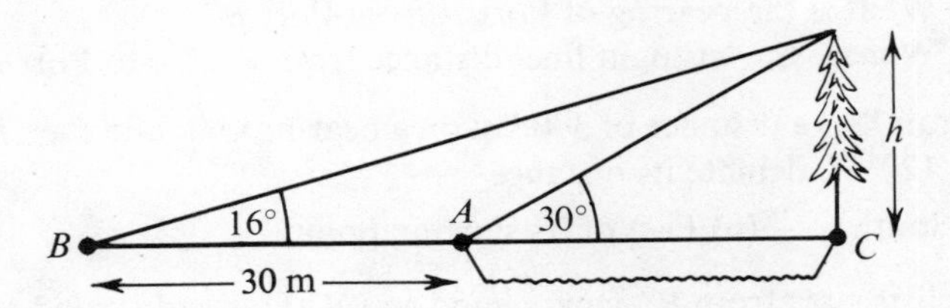

Fig. 40

of elevation of the top of the tree as 30°. He then walks 30 m from point A to point B, B lying on the straight line through A and C and measures the angle of elevation of the top of the tree as 16°. If the height of the tree is h, then $\tan 30° = \dfrac{h}{AC}$.

Write down a similar equation involving h, AC and 16°.

Solve the simultaneous equations and so find h. What is the width of the river?

8 What are the lengths of sides of an equilateral triangle of height 3 cm?

9 In Figure 41, $AB = 4$ cm, $\angle CAB = 30°$, $\angle CDB = 30°$, and $\angle DBE = 40°$. Calculate (i) CB; (ii) BD; (iii) DE.

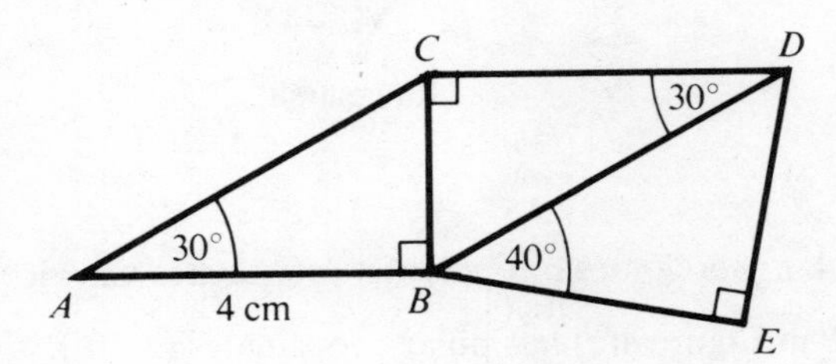

Fig. 41

10 A ship sets out from Port A on a bearing of 050° and travels in a straight line (assuming the earth to be flat!). What is the bearing of the Port from the ship once it has left Port? When the ship has travelled 50 km it changes direction and heads due East. After travelling 30 km it docks at Port B. Figure 42 represents the ship's journey.

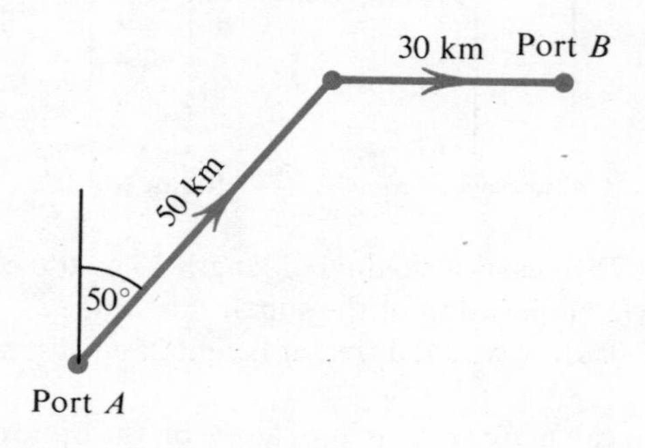

Fig. 42

(i) What is the bearing of Port A from Port B?
(ii) What is the 'straight line' distance from Port A to Port B?

11 An aircraft flies a distance of 300 km on a bearing 060° and then 500 km on a bearing 120°. Calculate its distance

(i) South; (ii) East of its starting point.

12 Two cyclists start from Rawmarsh and travel at a steady speed of 15 km per hour along straight roads which intersect each other at an angle of 30° (see Figure 43). Calculate the distance apart of the cyclists after (i) 2 hours; (ii) 3 hours. If road X runs due North, what is the bearing of the cyclist on road Y from the cyclist on road X at these two times?

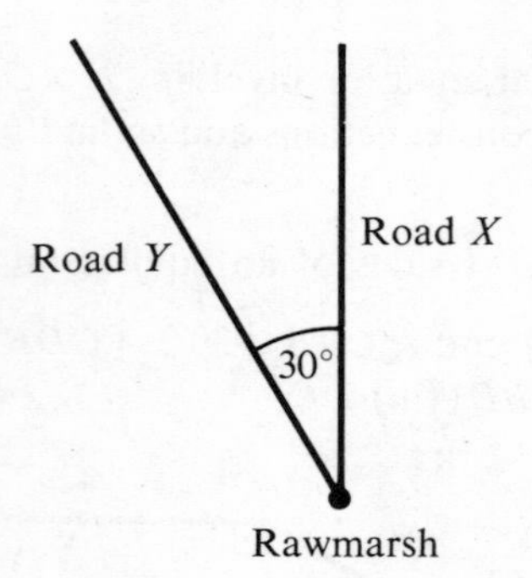

Fig. 43

4 Converting polar coordinates to Cartesian coordinates and vice versa

(a) The point P in Figure 4(a) has polar coordinates $(2, 30°)$.

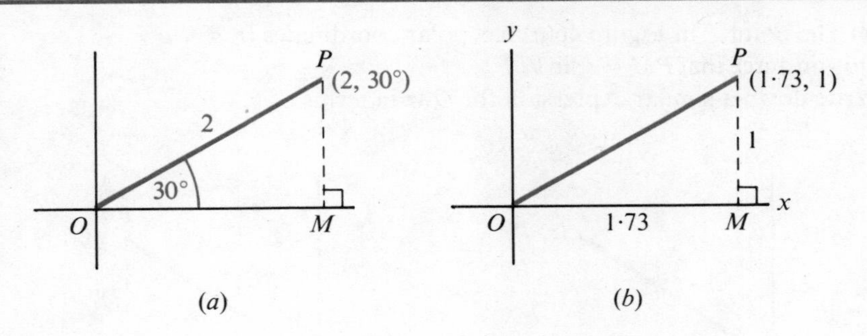

Fig. 44

In triangle PMO,

$$PM = 2 \sin 30° = 2 \times \tfrac{1}{2} = 1,$$

and

$$OM = 2 \cos 30° = 2 \times 0{\cdot}867 = 1{\cdot}73 \text{ (to 3 S.F.)}.$$

The *Cartesian coordinates* of P are therefore $(1{\cdot}73, 1)$ [see Figure 44(b)].

If the polar coordinates of a point are (i) $(2, 60°)$; (ii) $(1, 45°)$, what are the Cartesian coordinates?

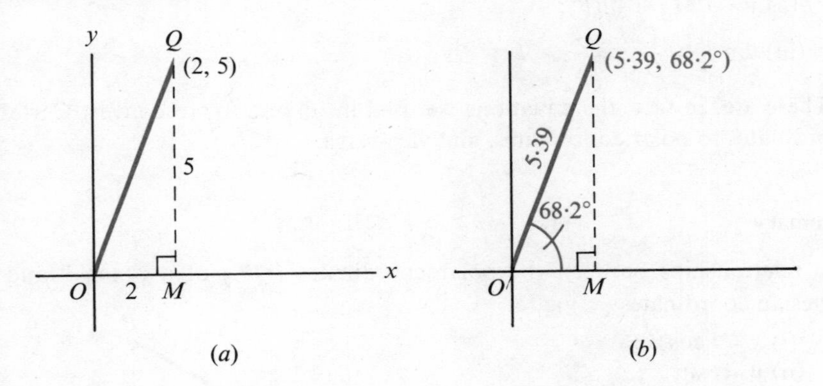

Fig. 45

(b) Figure 45(a) shows a point Q with Cartesian coordinates $(2, 5)$.

$$\tan \angle QOM = \tfrac{5}{2} = 2{\cdot}5$$

and so

$$\angle QOM = 68{\cdot}2°.$$

Also

$$QO^2 = 5^2 + 2^2 = 29.$$

Hence

$$QO = \sqrt{29} = 5{\cdot}39.$$

The polar coordinates of Q are therefore $(5{\cdot}39, 68{\cdot}2°)$ [see Figure 45(b)].

If the Cartesian coordinates of a point are (i) $(4, 4)$; (ii) $(3, 0)$; (iii) $(^-2, 1)$ what are the polar coordinates?

(c) The point P in Figure 46(a) has polar coordinates $(r, \theta°)$.
Do you agree that $PM = r \sin \theta°$?
Write down a similar expression for OM in terms of θ.

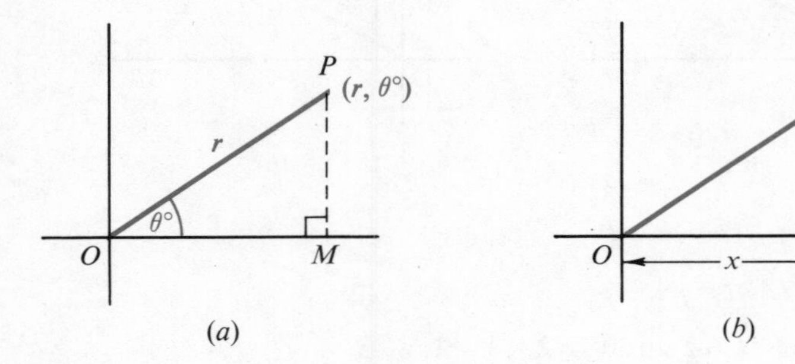

Fig. 46

Suppose P has Cartesian coordinates (x, y). Compare Figure 46(a) with Figure 46(b). We can see that:

(i) $x = OM = \mathrm{r} \cos \theta°$;
(ii) $y = PM = \mathrm{r} \sin \theta°$;
(iii) $\tan \theta° = \dfrac{PM}{OM} = \dfrac{y}{x}$.

These are, in fact, the equations we used in (c) and (d) to convert Cartesian coordinates to polar coordinates, and vice-versa.

Summary

The relationships between the polar coordinates $(r, \theta°)$ of a point P and its Cartesian coordinates (x, y) are:

(i) $x = r \cos \theta°$;
(ii) $y = r \sin \theta°$;
(iii) $\tan \theta° = \dfrac{y}{x}$.

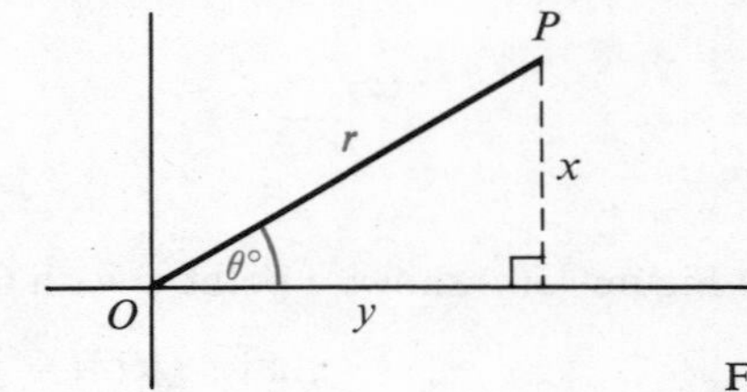

Fig. 47

Exercise D

1 Convert the following Cartesian coordinates to polar coordinates: (i) (2,0); (ii) (0,2); (iii) (2,2); (iv) $(\sqrt{3}, 1)$; (v) $(1, \sqrt{3})$; (vi) (4,2); (vii) (2,4); (viii) $({}^{-}2, 4)$; (ix) $({}^{-}2, {}^{-}4)$; (x) $(4, {}^{-}2)$; (xi) (3,7); (xii) $({}^{-}2, 9)$; (xiii) $(4, {}^{-}7)$.

2 Convert the following polar coordinates to Cartesian coordinates: (i) $(2, 90°)$; (ii) $(1, 0°)$; (iii) $(3, {}^{-}270°)$; (iv) $(2, 30°)$; (v) $(1, 45°)$; (vi) $(1, 135°)$; (vii) $(5, 60°)$; (viii) $(5, {}^{-}300°)$; (ix) $(3, 170°)$.

3 (*a*) The line l in Figure 48 has Cartesian equation $x = y$. Use the equations $x = r\cos\theta°$, $y = r\sin\theta°$, to show that its polar equation is $\sin\theta° = \cos\theta°$. Hence show that its polar equation can be written as $\tan\theta° = 1$. Would it be true to say that an alternative way of writing its equation is $\theta = 45$? Explain your answer.
(*b*) What is the polar equation of the line with Cartesian equation $y = {}^{-}x$?

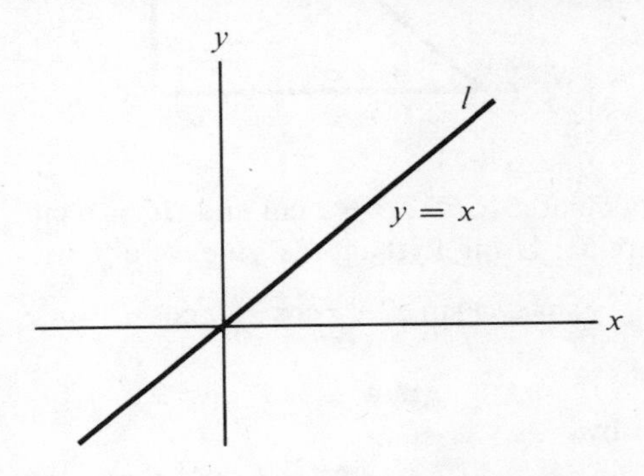

Fig. 48

4 Draw the straight lines with polar equation $\tan\theta° = \frac{1}{2}$. If the points with Cartesian coordinates $(x_1, 2)$, $(2, y_1)$, $(x_2, 4)$, $(4, y_2)$ lie on this line find x_1, y_1, x_2 and y_2. What is the Cartesian equation of the line?

5 (*a*) Graph the following: (i) $\tan\theta° = 1$; (ii) $\tan\theta° = 3$; (iii) $x = y$; (iv) $y = 3x$; (v) $\tan\theta° = {}^{-}2$.
(*b*) What is the angle $\theta°$ between the line $y = 3x$ and the positive x axis? (See Figure 49.) Write down the equation of the line in terms of $\tan\theta°$.

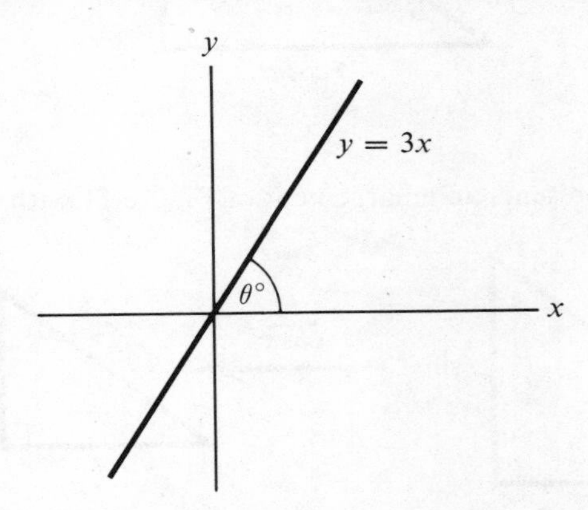

Fig. 49

6 Answer question 5(*b*) for the lines (i) $y = 5x$; (ii) $3y = 2x$.

4 Concerning Pythagoras

(*a*) For the triangle in Figure 50, Pythagoras' rule states that

$$AB^2 + BC^2 = AC^2.$$

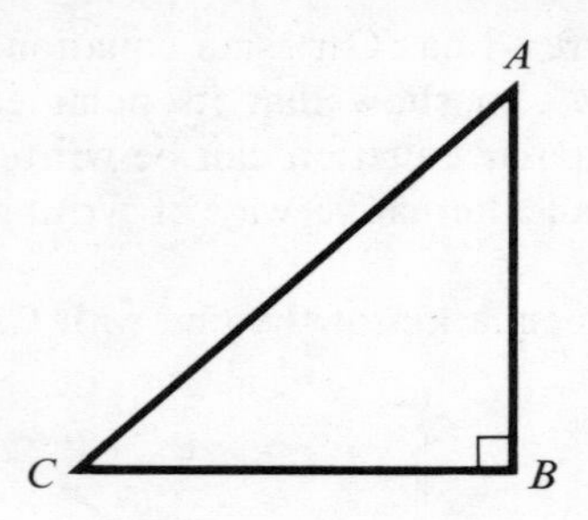

Fig. 50

Use the rule to calculate AC if $AB = 3$ cm and $BC = 6$ cm.

(*b*) Look at Figure 51. Using Pythagoras' rule we obtain

$$r \sin \alpha^\circ . r \sin \alpha^\circ + r \cos \alpha^\circ . r \cos \alpha^\circ = r . r,$$

that is,

$$r^2 \sin^2 \alpha^\circ + r^2 \cos^2 \alpha^\circ = r^2.$$

Dividing both sides by r^2,

$$\sin^2 \alpha^\circ + \cos^2 \alpha^\circ = 1.$$

[Notice that $(\sin \alpha^\circ)^2$ is written $\sin^2 \alpha^\circ$; $\sin \alpha^{\circ 2}$ would mean $\sin(\alpha^\circ)^2$.]

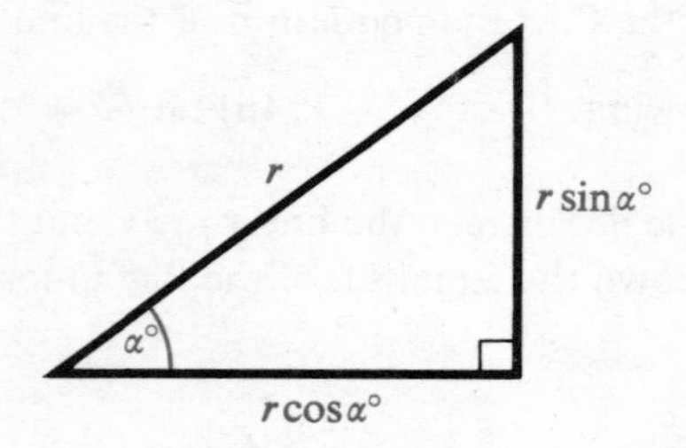

Fig. 51

(*c*) Figure 52 represents an enlargement of Figure 51 with scale factor $\dfrac{1}{\cos \alpha^\circ}$.

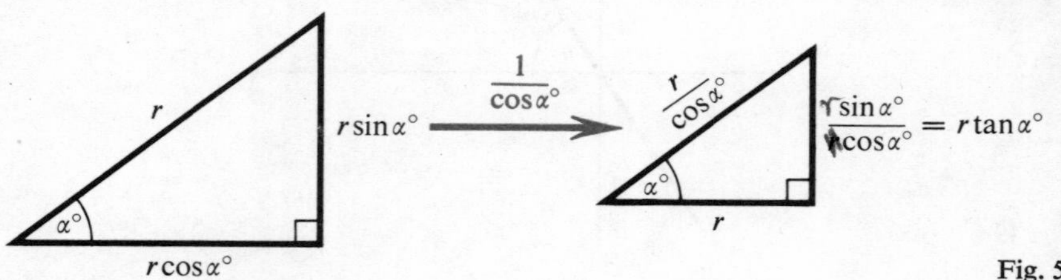

Fig. 52

Using Pythagoras' rule we have:

$$r^2 \tan^2 \alpha^\circ + r^2 = \frac{r^2}{\cos^2 \alpha^\circ}$$

$$\tan^2 \alpha^\circ + 1 = \frac{1}{\cos^2 \alpha^\circ}.$$

Divide both sides of $\sin^2\alpha° + \cos^2\alpha° = 1$ by $\cos^2\alpha°$. Did you obtain $\tan^2\alpha° + 1 = \dfrac{1}{\cos^2\alpha°}$?

By enlarging Figure 51 with scale factor $\dfrac{1}{\sin\alpha°}$ obtain a similar expression involving $\sin\alpha°$ and $\tan\alpha°$.

Divide both sides of $\sin^2\alpha° + \cos^2\alpha° = 1$ by $\sin^2\alpha°$. What do you notice?

Summary

Pythagoras' rule can be expressed in the forms:

(i) $\sin^2\alpha° + \cos^2\alpha° = 1$;

(ii) $1 + \tan^2\alpha° = \dfrac{1}{\cos^2\alpha°}$;

(iii) $1 + \dfrac{1}{\tan^2\alpha°} = \dfrac{1}{\sin^2\alpha°}$.

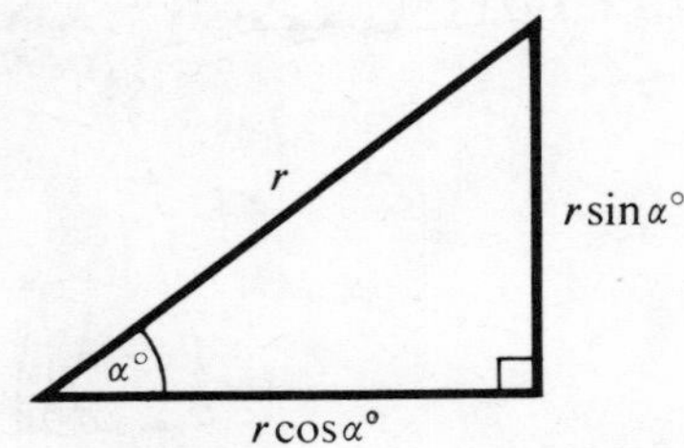

Fig. 53

Exercise E

1 From your tables find (i) $\sin 30°$; (ii) $\cos 30°$; (iii) $\sin 45°$; (iv) $\cos 45°$; (v) $\sin 70°$; (vi) $\cos 70°$. Calculate: (i) $\sin^2 30°$; (ii) $\cos^2 30°$; (iii) $\sin^2 45°$; (iv) $\cos^2 45°$; (v) $\sin^2 70°$; (vi) $\cos^2 70°$. Use your answer to calculate the value of $\sin^2\theta° + \cos^2\theta°$ for $\theta = 30$, 45 and 70. Is $\sin^2\theta° + \cos^2\theta° = 1$ in each case? Why wouldn't you expect to obtain the result $\sin^2\theta° + \cos^2\theta° = 1$ from your tables?

2 If $\sin\theta° = \frac{1}{2}$, use the expression $\sin^2\theta° + \cos^2\theta° = 1$ to calculate $\cos\theta°$ (you should obtain two answers).

3 Use equation (ii) in the summary to calculate the values of $\tan\theta°$ when $\cos\theta° = \frac{1}{2}$.

4 The equation of a circle of radius 2 in Cartesian form is $x^2 + y^2 = 4$ (see Figure 54). Use the equations $x = r\sin\theta°$, $y = r\cos\theta°$ to show that its equation can also be written $r = 2$.

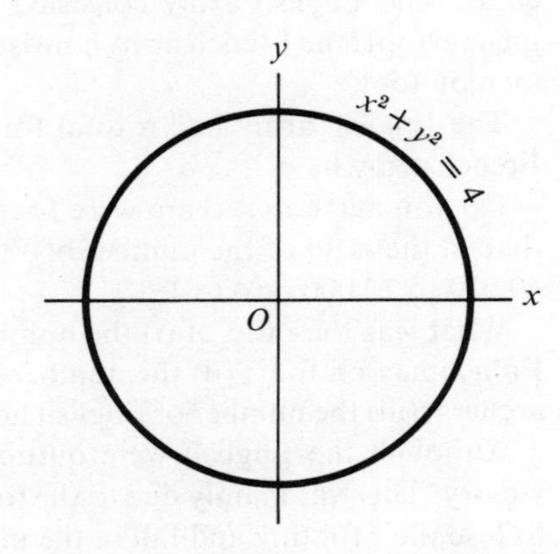

Fig. 54

'It appears that our troops are outnumbered by a good ratio.'

6 Ratio, proportion and application of slide rule

1 Ratio

(*a*) In 1415, Henry V led the English army to victory over the French at Agincourt. The English army consisted of 1000 horsemen, 6000 archers and 3000 men on foot; the French army consisted of 7500 horsemen, 500 archers and 32 000 men on foot.

The English army had a total force of 10 000 men. How many men did the French army have?

Do you agree that there were four times as many Frenchmen as Englishmen, that is, the ratio of the number of Frenchmen to the number of Englishmen was 40 000 to 10 000 or 4 to 1?

What was the ratio of (i) the number of Englishmen on foot to the number of Frenchmen on foot; (ii) the number of French archers to the number of English archers; (iii) the number of English horsemen to the number of French horsemen?

Although the English were outnumbered by 4 to 1, they won a convincing victory. This was mainly due to the torrential rain which caused the French horses to lose their footing and fall in the mud. The English archers, who outnumbered the French archers by 12 to 1, were then able to exercise their advantage to the full.

(*b*) If we wish to find the ratio of $\frac{1}{2}$ metre to 10 metres we can:

(i) work in metres so that the ratio is $\frac{1}{2}$ to 10, which equals 1 to 20;

(ii) work in centimetres so that the ratio is 50 to 1000, which also equals 1 to 20.

What is the ratio when working in millimetres?

Notice that ratio compares quantities measured in the *same units* and that it has *no units* itself.

Instead of using the word 'to' in stating a ratio we can use the symbol ':' and so write 1 to 20 as 1 : 20.

Since the ratio is 1 : 20, the first length is $\frac{1}{20}$ of the second length. This means that with every ratio there is an associated fraction. For example, the fraction associated with the ratio 4 : 14 is $\frac{4}{14}$ or $\frac{2}{7}$.

What fractions are associated with (i) 2 : 3, (ii) 3 : 2, (iii) $\frac{1}{2}$: 5, (iv) 5 to $\frac{1}{2}$?

(*c*) Figure 1 shows a rectangle *ABCD* and its image under an enlargement centre *O* and scale factor 3.

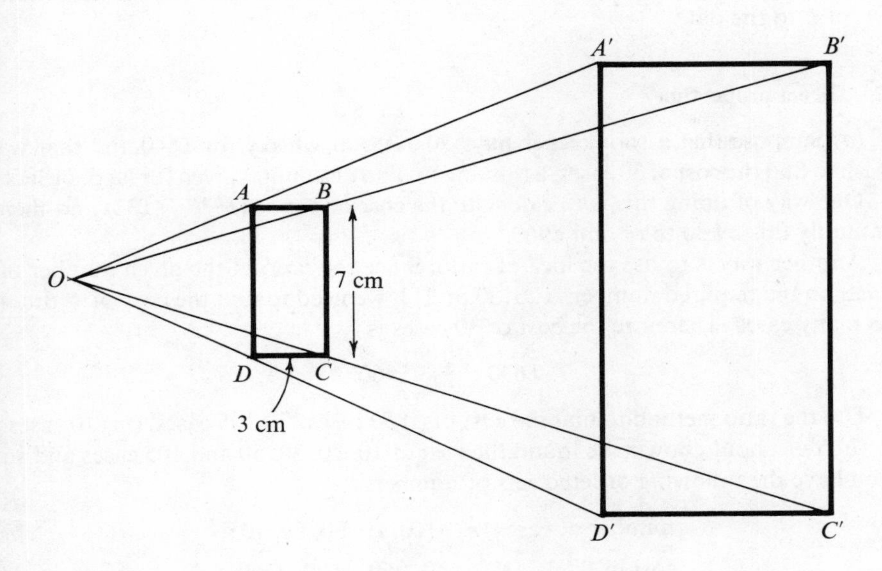

Fig. 1

Write down the associated fractions for the following ratios:

(i) $A'O : AO$;

(ii) $OD : OD'$;

(iii) $BC : B'C'$;

(iv) perimeter of $A'B'C'D'$ to the perimeter of $ABCD$;

(v) area of $ABCD$ to the area of $A'B'C'D'$.

Exercise A

1 Calculate the ratio and the associated fraction of the following quantities:

(*a*) 15 cm³ to 900 cm³; (*b*) 17 to 51;
(*c*) 250 g to 2 kg; (*d*) 56 p to £4·48;
(*e*) 15 cm to 200 m.

2 In a sale, prices are reduced in the ratio 3:5. Calculate the sale prices of the articles whose ordinary prices are:

(*a*) £2; (*b*) £1·25; (*c*) 35p.

3 In what ratio must 36 be increased to become 45?

4 In what ratio must 320 g be decreased to become 200 g?

5 Two distances are in the ratio 12:7. If the longer distance is 42 km, what is the shorter distance?

6 If $d = 1\frac{1}{3}L$ and the value of d is to be doubled, state the ratio of the new value of L to the old.

2 Direct proportion

(*a*) Suppose that a shopkeeper buys 20 cases of whisky for £640 and that we wish to find the cost of 30 cases, assuming that no discount is given for large orders.

One way of doing this is to calculate the cost of 1 case ($£\frac{640}{20} = £32$) and then multiply this by 30 to obtain £960.

Another way is to use the idea of ratio. Since the ratio of the given number of cases to the required number is 20:30 or 2:3, we need to find the cost of $\frac{3}{2}$ times as many cases. Therefore the cost of 30 cases is

$$£640 \times \tfrac{3}{2} = £960.$$

Use the ratio method to find the cost of (i) 50 cases, (ii) 105 cases, (iii) 10 cases.

(*b*) You should now have found the cost of 10, 20, 30, 50 and 105 cases and so you have the following ordered sets of numbers:

number of cases (n) {10, 20, 30, 50, 105}

cost in £ (c) {320, 640, 960, 1600, 3360}.

If the number of cases is multiplied by some number, the cost in £ is multiplied by the *same* number. For example, trebling the number of cases trebles the cost. When two quantities vary in this way they are said to be directly *proportional* to each other. Thus n is directly proportional to c and c is directly proportional to n.

(*c*) If y is directly proportional to x, fill in the blanks in the following table:

x	7	14	21	35	42
y	9			45	

Exercise B

1 If a set of 30 mathematics text books cost £24 how much would a set of (i) 20; (ii) 35; (iii) 12; (iv) 40 cost?

2 Assuming that y is directly proportional to x, fill in the blanks in the following table:

x	8	16	24	48	64
y	14		42		

3 Repeat Question 2 for the following table:

x	6	15	24	36	84
y	15		60		

4 Repeat Question 2 for the following table:

x	3		12		
y	5	15	20	30	45

5 The extension in the length of a spring is directly proportional to the mass applied to the spring. A spring 26 cm long stretches to a length of 29 cm when it supports a mass of 5 kg. What will be its length when it supports (i) 9 kg; (ii) 750 g?

6 1 cm³ of copper has a mass of 8·8 g and 1 cm³ of aluminium a mass of 2·7 g. What should the mass of a copper saucepan be if an aluminium saucepan of the same shape and size has a mass of 0·75 kg?

3 Scale factors

(*a*) Look at the following table:

x	4	2	3	7	5	6
y	10	5	$7\frac{1}{2}$	$17\frac{1}{2}$	$12\frac{1}{2}$	15

Do you think that x and y are directly proportional to each other?

Let us now consider the ratio of corresponding pairs of values.

The ratio for the first pair is 4:10 or 2:5. What is the ratio for each of the other pairs?

When a ratio is written in the form $1:n$, n can be thought of as a scale factor. If we write the ratio for the first pair, 4:10, in the form $1:n$ then we obtain the ratio $1:2\frac{1}{2}$ and so $2\frac{1}{2}$ is the scale factor.

What can you say about the scale factors for all the corresponding pairs of values?

When two quantities are directly proportional *every* value of one quantity is connected to the corresponding value of the other quantity by the *same scale factor*:

x	4	2	3	7	5	6
$\downarrow$	$\downarrow \times 2\frac{1}{2}$	$\downarrow \times 2\frac{1}{2}$	$\downarrow \times 2\frac{1}{2}$	$\downarrow \times 2\frac{1}{2}$	$\downarrow \times 2\frac{1}{2}$	$\downarrow \times 2\frac{1}{2}$
y	10	5	$7\frac{1}{2}$	$17\frac{1}{2}$	$12\frac{1}{2}$	15

In the example above, there is a constant scale factor of $2\frac{1}{2}$ and we can write

$$y = 2\tfrac{1}{2}x.$$

(*b*) Find the scale factors for the corresponding pairs of values in Table 1 and hence say whether or not *d* and *t* are directly proportional to each other.

TABLE 1

d	2	8	24	48	60
t	1	4	12	24	30

Express *t* in terms of *d*.

(*c*) Find the scale factors for the corresponding pairs of values in Table 2.

TABLE 2

u	8	2	6	12	4
v	10	2·5	7·5	16	5

Are *u* and *v* directly proportional to each other?

Can you express *v* in terms of *u*?

(*d*) In general, if two quantities, *p* and *q*, are directly proportional to each other then they can be linked by an equation of the form

$$p = kq,$$

where *k* is the *constant* scale factor.

Exercise C

1 State in which of the following examples the first quantity is directly proportional to the second:

(*a*) the mass of a rope, the length of the rope;
(*b*) the value of a 'silver coin', its mass;
(*c*) the height of a child, the mass of the child;
(*d*) the area of a wall, the cost of painting it.

2 In the following equations which are the quantities that are directly proportional to each other and what is the constant scale factor linking them?

(*a*) $P = 10Q$; (*b*) $C = 2\pi r$; (*c*) $V = 9{\cdot}8t$.

3 In the example of the whisky cases (the table is given below) find the scale factor and so form an equation linking c and n.

Number of cases (n)	10	20	30	50	105
Cost in £ (c)	320	640	960	1600	3360

4 Assuming that y is directly proportional to x, fill in the blanks in the following table:

x	4	8	24	28	32	44
y			42			

Form an equation between x and y.

4 Using the slide rule

(a) Look at Table 3. Investigate whether a and b are directly proportional to each other.

TABLE 3

a	1·2	2	2·5	4	4·5
b	2·64	4·4	5·5	8·8	9·9

It is unnecessary to calculate each scale factor individually; we can decide whether or not a and b are directly proportional by using a slide rule.

Place 2·64 on the C scale against 1·2 on the D scale of your slide rule. Keeping the rule fixed in this position, move the cursor so that the cursor line is over the 2 on the D scale as shown in Figure 2.

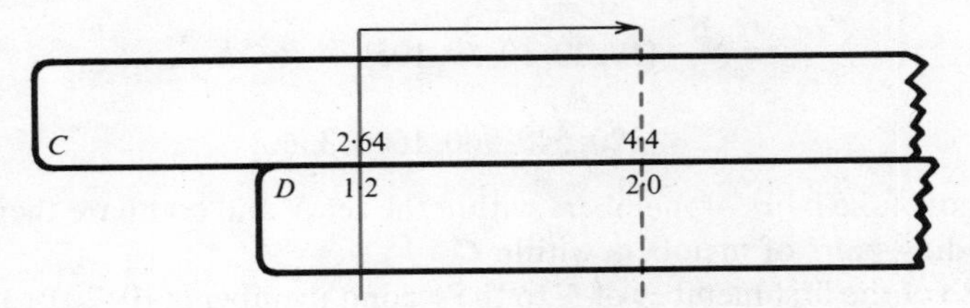

Fig. 2

You should find that 4·4 is the corresponding number on the C scale.

Move the cursor line over 2·5, 4 and 4·5 on the D scale in turn. What do you notice?

The constant scale factor can be obtained by reading the number on the C scale opposite the 1 on the D scale.

(b) Now consider the values of c and d given in Table 4.

TABLE 4

c	5	6·5	7·8	9·0	13·0	20·0
d	1·92	2·5	3·0	3·46	5·0	7·7

Use your slide rule to check that c and d are directly proportional to each other. What difficulty arises when you attempt to read the numbers on the C scale corresponding to 13·0 and 20·0 on the D scale?

What is the constant scale factor?

Exercise D

1 Use your slide rule to find whether the following pairs of quantities are directly proportional. If they are, state the scale factor.

(*a*)

p	4·5	6·3	8·1	19·5	24·3
q	3	4·2	5·4	13	16·2

(*b*)

x	1·8	2·7	7·5	9·55	14·6
y	1·32	1·98	5·5	7	10·7

(*c*)

s	1·45	4·2	5·1	6·1	9·0
t	2·38	2·56	8·5	9·4	14·8

(*d*)

u	1.55	4·0	6·2	6·7
v	5·0	12·9	20	28

5 Multipliers

(*a*) When discussing the example in Section 2 about the cost of whisky cases, we formed two ordered sets: the set of the numbers of cases, N, and the set of costs, C. So

$$N = \{10, 20, 30, 50, 105\}$$

and

$$C = \{320, 640, 960, 1600, 3360\}.$$

Let us now take pairs of members within the set N and compare them with the corresponding pairs of members within C.

The ratio of the first member of N to the second member is 10:20 or 1:2 and we say that the *multiplier* connecting the first and second members of N is 2.

Check that the multiplier connecting the second and third members is $\frac{3}{2}$, that connecting the third and fourth members is $\frac{5}{3}$ and that connecting the fourth and fifth members is $\frac{21}{10}$.

The red numbers over the arrows in Figure 3 are the multipliers for N:

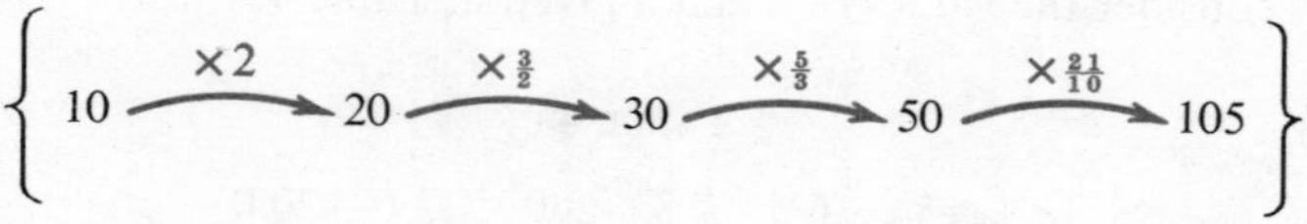

$$\left\{ 10 \xrightarrow{\times 2} 20 \xrightarrow{\times \frac{3}{2}} 30 \xrightarrow{\times \frac{5}{3}} 50 \xrightarrow{\times \frac{21}{10}} 105 \right\}$$

Fig. 3

These multipliers form a new set

$$M_N = \{2, \tfrac{3}{2}, \tfrac{5}{3}, \tfrac{21}{10}\}.$$

Find the multipliers for C, that is, find the set M_C.

Do you agree that $M_N = M_C$?

(*b*) We know that the numbers of cases of whisky and the costs in £ are directly proportional to each other. We therefore say that N and C are *proportional sets.*

We have seen that $M_N = M_C$ and this suggests that two sets are proportional when the sets of multipliers are equal.

(*c*) Find the multipliers for the ordered sets

$$A = \{4, 2, 3, 7, 5, 6\}$$

and

$$B = \{10, 5, 7\tfrac{1}{2}, 17\tfrac{1}{2}, 12\tfrac{1}{2}, 15\}$$

which we used in Section 3(*a*).

Is $M_A = M_B$?

Does your result support the suggestion in (*b*)?

(*d*) Check that the ordered sets

$$C = \{3, 5, 8, 12, 15\}$$

and

$$D = \{9, 15, 24, 38, 44\}$$

are not proportional sets and that $M_C \neq M_D$.

Example 1

Find the missing members of the following pairs of ordered sets given that the sets are proportional.

$$P = \{8, 4, 12, 9\}, \qquad Q = \{10, *, *, *\}.$$

Method 1 Scale Factors Since the sets are ordered we know that 8 corresponds to 10. The scale factor of the two sets is therefore 8:10 or $1:1\tfrac{1}{4}$.

The sets can be arranged like this:

$$\begin{array}{cccccccccc} \{ & 8 & & 4 & & 12 & & 9 & & \} \\ & \downarrow & \times 1\tfrac{1}{4} & \downarrow & \times 1\tfrac{1}{4} & \downarrow & \times 1\tfrac{1}{4} & \downarrow & \times 1\tfrac{1}{4} & \\ \{ & 10 & & & & & & & & \} \end{array}$$

The next member of the set Q is $1\tfrac{1}{4} \times 4 = 5$. What are the other members? Complete the set $Q = \{10, 5, \quad , \quad \}$.

Method 2 Multipliers In set P, the first member is 8 and the second member is 4 so the multiplier connecting these members is $\frac{1}{2}$. The member in set Q corresponding to 4 in set P must be $10 \times \frac{1}{2}$ which equals 5.

$$8 \xrightarrow{\times\frac{1}{2}} 4 \xrightarrow{\times 3} 12 \xrightarrow{\times\frac{3}{4}} 9$$

$$10 \xrightarrow{\times\frac{1}{2}} 5 \xrightarrow{\times 3} \quad \xrightarrow{\times\frac{3}{4}}$$

Work out the two remaining members of Q.

Do you get the same result as by method 1?

(*e*) The method of multipliers is very useful when comparing sets which are *not* measured in the same units. Consider the following example.

Four copper rods were measured and weighed. The results formed two ordered sets: L, the set of lengths measured in centimetres and W, the set of masses measured in grammes.

$L = \{1{\cdot}5, 2, 5, 6\}$ and $W = \{13{\cdot}5, 18, 45, 54\}$.

The set of multipliers for set L is:

$$M_L = \{\tfrac{4}{3}, \tfrac{5}{2}, \tfrac{6}{5}\}.$$

Form the set M_W.

Are the sets L and W directly proportional?

Find a relation between the lengths and masses of the copper rods.

Exercise E

In all the following examples, the sets are ordered.

1 Set $P = \{6, {}^{-}5\}$. Write down the set Q where the scale factor of the mapping $P \to Q$ is:

(i) 7; (ii) $\frac{3}{4}$; (iii) $^{-}8$.

2 $K = \{3, 6, 12\}$ and $L = \{5, x, y\}$ are proportional sets. Find x and y. Did you use multipliers in this question rather than the scale factor? If so, why? What is the scale factor?

3 Given the set $\{4, 12, 60, 30, 3000\}$ complete the following sets which are proportional to it:

(*a*) $\{5, *, *, *, *\}$, (*b*) $\{*, *, *, 5, *\}$, (*c*) $\{*, 5, *, *, *\}$.

Do not use your slide rule. Discuss in each case, whether it is better to use the scale factor or multipliers.

4 Use your slide rule to check whether the following table gives values consistent with proportional sets.

x	20	25	45	50
y	36	45	81	90

If the answer is yes, then state the scale factor.

5 Use your slide rule to discover which of the following sets are proportional. If they are proportional, state the scale factor.

(*a*) $\{3{\cdot}5, 4{\cdot}6, 6{\cdot}1, 7{\cdot}4\}$ and $\{5{\cdot}25, 6{\cdot}9, 9{\cdot}15, 11{\cdot}1\}$.
(*b*) $\{12{\cdot}7, 17{\cdot}4, 30, 63\}$ and $\{29{\cdot}2, 40, 69, 147\}$.
(*c*) $\{1{\cdot}6, 3{\cdot}0, 4{\cdot}6, 9{\cdot}5\}$ and $\{1{\cdot}12, 2{\cdot}1, 3{\cdot}22, 6{\cdot}65\}$.
(*d*) $\{2, 4{\cdot}8, 6{\cdot}5, 9{\cdot}0\}$ and $\{1{\cdot}5, 3{\cdot}6, 4{\cdot}95, 6{\cdot}75\}$.

6 The sets $\{a,b,c\}$ and $\{3a,q,r\}$ are proportional. Write down the scale factor and find expressions for q and r. If $q=6$, $r=8$, find b and c. Can you find a?

7 In Figures 4(*a*) and 4(*b*), arrows indicate parallel lines. Calculate x and y.

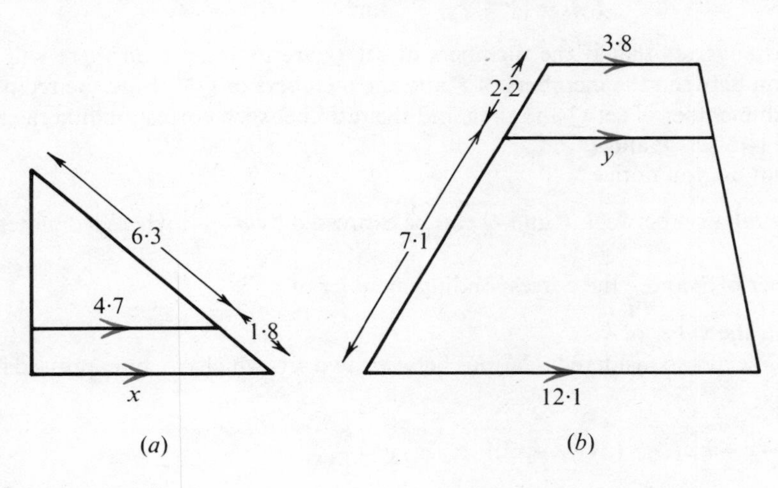

Fig. 4

6 Proportions

(*a*) Consider the two ordered sets:

$$X=\{5,8,9,11\} \quad \text{and} \quad Y=\{75,192,243,363\}.$$

Notice that $M_X=\{\frac{8}{5},\frac{9}{8},\frac{11}{9}\}$.
Find M_Y. Is $M_X=M_Y$?
Are these sets directly proportional?
You should have found that X and Y are not directly proportional.

(*b*) By squaring the elements of set X we form a new set: $\{25, 64, 81, 121\}$.

The notation we shall use is that if in a set P every element is raised to the kth power, the new set formed is denoted by P^k. Hence in this example $X^2=\{25, 64, 81, 121\}$.

List the members of M_{X^2} and compare M_{X^2} with M_Y. What do you notice? Can you suggest how the members of Y are related to the members of X^2?

(*c*) Since Y is directly proportional to X^2 there is a relation of the form $y=kx^2$ where x^2 denotes any member of X^2, and y denotes the corresponding member of Y. Find k by comparing corresponding members of X^2 and Y.

(*d*) P and Q are two ordered sets:

$$P = \{8, 6, 12, 2\}; \qquad Q = \{3, 4, 2, 12\}.$$

The sets of multipliers associated with these sets are

$$M_P = \{\tfrac{6}{8}, \tfrac{12}{6}, \tfrac{2}{12}\}; \qquad M_Q = \{\tfrac{4}{3}, \tfrac{2}{4}, \tfrac{12}{3}\}.$$

Does $M_P = M_Q$?

Can you see any relation between M_P and M_Q?

(*e*) You should have noticed that if each element of M_Q is inverted (turned upside down) we form the set:

$$M_{Q^{-1}} = \{\tfrac{3}{4}, \tfrac{4}{2}, \tfrac{2}{12}\} \qquad \text{and} \qquad M_{Q^{-1}} = M_P.$$

This suggests that if the members of set Q are inverted then there will be a relation between the members of P and the members of Q^{-1}. Find the reciprocal of each member of set Q and then find the ratio between corresponding members of the two sets P and Q^{-1}.

What do you notice?

The relation between P and Q can be expressed by $p = \frac{k}{q}$, where p denotes any member of P and $\frac{1}{q}$ the corresponding member of Q^{-1}.

Find the value of k.

(*f*) We have considered relations between two sets which can be expressed in the forms:

$$\text{(i) } y = kx; \qquad \text{(ii) } y = kx^2; \qquad \text{(iii) } y = \frac{k}{x}.$$

The main relations between two sets of values that you are likely to meet are:

(i) Direct function	y proportional to x	which can	$y = kx$
(ii) Square function	y proportional to x^2	be	$y = kx^2$
(iii) Reciprocal function	y proportional to $\frac{1}{x}$	expressed in the	$y = \frac{k}{x}$
(iv) Cube function	y proportional to x^3	form	$y = kx^3$

Summary

Ratio compares like quantities. For example, 5 minutes and one hour are in the ratio of $5:60 = 1:12$. Ratio has no units.

Proportion. The ordered sets $A = \{2, 3, 5, 8\}$ and $B = \{6, 9, 15, 24\}$ are proportional. The scale factor is the common ratio of corresponding members of the sets. It is used to map the members of A onto the members of B. In this example $A \xrightarrow{\times 3} B$; the scale factor is 3. A *multiplier* is used to map the first member of A onto the second member of A that is $\frac{3}{2} \times 2 = 3$. Since the sets are proportional the same multiplier maps the first member of B onto the second member of B that is $\frac{3}{2} \times 6 = 9$.

A different multiplier maps the second members onto the third, etc.

MULTIPLIERS

$$\text{Set } A\left\{2 \xrightarrow{\times\frac{3}{2}} 3 \xrightarrow{\times\frac{5}{3}} 5 \xrightarrow{\times\frac{8}{5}} 8\right\}$$

SCALE FACTOR $\downarrow\times 3 \quad \downarrow\times 3 \quad \downarrow\times 3 \quad \downarrow\times 3$

$$\text{Set } B\left\{6 \xrightarrow{\times\frac{3}{2}} 9 \xrightarrow{\times\frac{5}{3}} 15 \xrightarrow{\times\frac{8}{5}} 24\right\}$$

Slide rule exercise

1 How are the numbers on Scale D related to numbers on Scale A? Work out:

(*a*) $(4{\cdot}8)^2$; (*b*) $\sqrt{8{\cdot}4}$; (*c*) $\sqrt{19{\cdot}5}$;
(*d*) $(450)^2$; (*e*) $\sqrt{0{\cdot}071}$; (*f*) $(0{\cdot}034)^2$.

2 How are the numbers on Scale D related to the numbers on Scale K? Work out:

(*a*) $(1{\cdot}26)^3$; (*b*) $(58{\cdot}5)^3$; (*c*) $\sqrt[3]{61}$;
(*d*) $\sqrt[3]{350}$; (*e*) $(0{\cdot}36)^3$; (*f*) $\sqrt[3]{1{\cdot}7}$.

3 How are the numbers on the middle scale related to the numbers on Scale D? Work out:

(*a*) $1/4$; (*b*) $1/1{\cdot}5$; (*c*) $1/92$;
(*d*) $1/270$; (*e*) $1/1/8{\cdot}5$; (*f*) $1/0{\cdot}42$.

4 Work out:

(*a*) $3{\cdot}6/\sqrt{5{\cdot}7}$; (*b*) $(4{\cdot}5)^2/(2{\cdot}8)^3$; (*c*) $4{\cdot}6^3 \times \sqrt{5{\cdot}3}$;
(*d*) $\dfrac{35 \times 0{\cdot}08 \times 4{\cdot}8}{25}$; (*e*) $\sqrt{(15{\cdot}8)^3}$; (*f*) $\sqrt{45} \times 0{\cdot}045$.

Exercise F

1 If y is proportional to x^2, complete the following table:

x	8	12	15		24
y		252		700	1008

2 The radius of the 'outer' circle of a shooting target is 1·5 times the radius of the 'bull'. If the area of the bull is 12 cm², find the area of the inner ring (shaded in Figure 5).

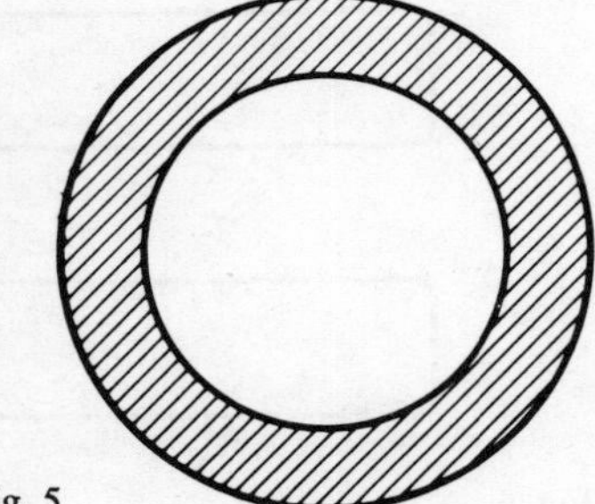

Fig. 5

3 Complete the following table of values on the assumption that:

(*a*) y is directly proportional to x;

(*b*) y is proportional to $\frac{1}{x}$.

x	6	24	54	150
y		100		

4 If y is proportional to x^3, complete the following table:

x	2	4	8	12	16
y			128		1024

5 Complete the following table of values on the assumption that:

(*a*) y is proportional to $\frac{1}{x}$;

(*b*) y is proportional to x^2;

(*c*) y is proportional to x^3.

x	0·7	2·1	2·8	3·5	4·9
y		10			

7 Testing relationships between sets of values

(*a*) Here are two ordered sets:

$$X = \{16, 54, 128, 250, 432\},$$
$$Y = \{2, 3, 4, 5, 6\}.$$

Can you spot a relation between the members of the sets X and Y?

The following sections show how to SEARCH for other relations using your slide rule.

(*b*) Consider set X. The multiplier between the first two members can be found by setting 54 on the D scale against 16 on the C scale. By moving the cursor to the appropriate end of the slide rule the multiplying factor 3·38 is obtained on scale D (see Figure 6).

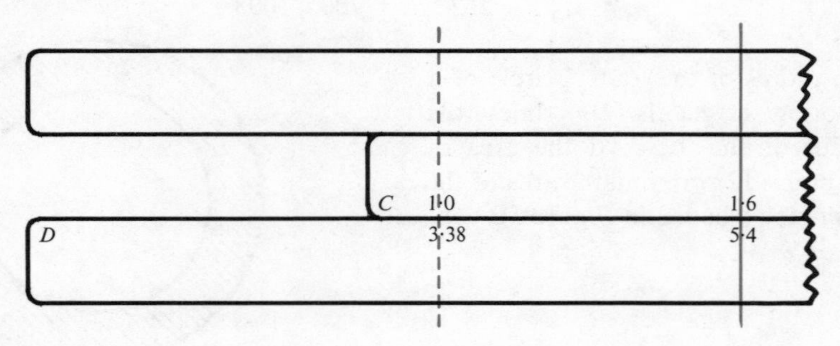

Fig. 6

If we now keep the *cursor fixed* and close the rule so that the 1 on the C scale is directly above the 1 on the D scale, the value under the cursor line on scale A gives the square of the first multiplying factor (see Figure 7).

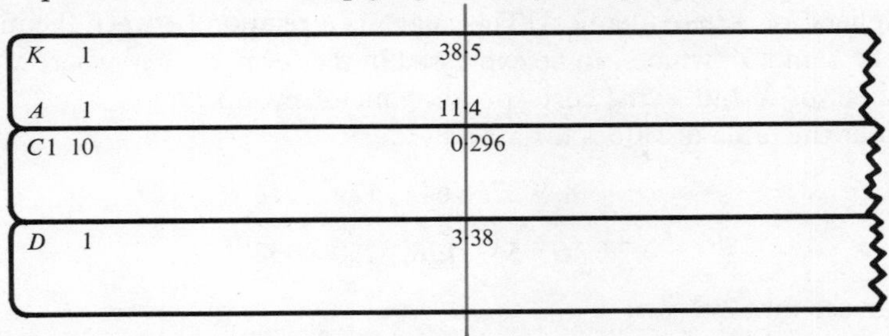

Fig. 7

What do the values 38·5 and 0·296 under the cursor lines on scales K (cube) and $C1$ (reciprocal) represent?

(*c*)

Members of X	Multipliers for X	Squares of multipliers for X	Cubes of multipliers for X	Reciprocals of multipliers for X
16				
	3·38	11·4	38·5	0·296
54				
128				
250				
432				

Fig. 8

Use the method of Section (*b*) to complete the table of multipliers in Figure 8 by considering in turn the pairs (54, 128), (128, 250) and (250, 432).

Repeat the process for the set of values for Y.

Members of Y	Multipliers for Y	Squares of multipliers for Y	Cubes of multipliers for Y	Reciprocals of multipliers for Y
2				
	1·5	2·25	3·38	0·667
3				
4				
5				
6				

Fig. 9

Look at the two tables. Are any two of the ten columns of values identical?

If two of the columns are identical then this suggests a relation between the two sets. In this example the columns headed 'Multipliers for X' and 'Cubes of multipliers for Y' are identical. This suggests a relation between the members of the sets X and Y^3 which can be expressed in the form $x = ky^3$ where x denotes any member of X and y^3 the corresponding member of Y^3.

Consider the table of values for x and y^3:

y^3	8	27	64	125	216
x	16	54	128	250	432

We can see that $2y^3 = x$.

8 The symbol '$\propto$'

(*a*) We use the symbol '$\propto$' to represent the words 'is proportional to'. A statement like: 'velocity, v, is proportional to the time, t' is written $v \propto t$. In the same way 'the stopping distance, d, of a car is proportional to the square of its speed, s' is written $d \propto s^2$, and 'the pressure, p, is inversely proportional to the volume, v' is written $p \propto \frac{1}{v}$.

Express as statements:

(i) $m \propto \frac{1}{n^2}$;

(ii) $q^2 \propto p^3$.

Write in symbols (using $\propto$):

(iii) 'The volume, v, is proportional to the cube of the length, l.'

(iv) 'A quantity x is inversely proportional to the square of a quantity y.'

(*b*) We have seen in previous sections that when two quantities are proportional to each other the relationship can be expressed in terms of a formula. For example:

$y \propto x$ can be written $y = kx$;

$y \propto x^2$ can be written $y = kx^2$;

$y \propto \frac{1}{x}$ can be written $y = \frac{k}{x}$.

Sometimes k is referred to as 'the constant of proportionality'.

Express each of the following as a formula:

(i) $p^2 \propto \frac{1}{q}$; (ii) $d^2 \propto b^3$; (iii) $m \propto \frac{1}{n^2}$.

(*c*) The heating power in watts of an electrical fire is proportional to the square of the current flowing through it.

Letting the power be p watts and the current be i amps, we can write

$$p \propto i^2.$$

This can be replaced by $p = ki^2$ where k is some constant.

If the power is 1000 watts when the current is 5 amps then substituting these values we obtain:

$$1000 = k \times 5 \times 5; \qquad k = 40.$$

The formula connecting p and i is therefore

$$p = 40i^2.$$

If the power is 2000 watts when the current is 4 amps and $p = ki^2$ what is the value of k?

Exercise G

1 Express the following statements by (i) using the $\propto$ notation; (ii) using a constant of proportionality k.

(*a*) The increase in length, d, of a rod is proportional to the increase in temperature, t.

(*b*) The circumference, c, of a circle is proportional to its radius, r.

(*c*) The mechanical energy of motion, e, of a motor-car is proportional to the square of its velocity, v.

(*d*) The volume, v, of a sphere is proportional to the cube of its radius.

(*e*) The distance to the horizon is proportional to the square root of the observer's height above the surface of the sea.

2 If $y \propto x$ and $y = 12$ when $x = 2$ find y when $x = 5$.

3 $x \propto \frac{1}{y^2}$ and $x = 3$ when $y = 4$. Find x when $y = 8$.

4 If $y \propto x^3$ complete this table of values

x	2	6	8
y	0.1		12·8

If $y = ax^3$ what is the value of a?

5 Some corresponding values of x and y are shown in the table:

x	1	5	10	20
y	5	125	500	2000

Which one or more of the following could be true?

(*a*) $y \propto x^2$;

(*b*) $y = 5x$;

(*c*) $y = 5x^2$;

(*d*) $y = 30x - 25$.

6 Some corresponding values of F and W are shown in the table:

F	100	200	300	500
W	30	15	10	6

Which one or more of the following is a possible relation between F and W?

(*a*) W is directly proportional to F;
(*b*) W is inversely proportional to F;
(*c*) $W = \frac{3}{10}F$;
(*d*) $FW = 3000$.

7 Use your slide rule to find the relations between the variables x and y in the following tables.

(*a*)

x	2	3	4	5
y	10	22·5	40	62·5

(*b*)

x	18	36	54	72
y	2	16	54	128

(*c*)

x	1·6	2·6	3·6	4·6
y	2·4	3·9	5·4	6·9

(*d*)

x	33·3	20·9	8·62	1·98
y	7·0	5·55	3·56	1·71

(*e*)

x	5	10	16	20	24
y	80	40	25	20	16·67

8 Given that $y \propto \dfrac{1}{x^2}$ complete the following table:

x	20	17	14		8
y		240		600	

9 Find the function for which the following are corresponding pairs:

(20, 35), (24, 42), (48, 84), (50, 87·5).

10 The speed of sound, s, in a gas is found to be proportional to the square root of the pressure, p, and inversely proportional to the square root of the density, d. If the speed is 33 000 cm per second for air when the pressure is 1030 g per cm^2 and the density 0·0013 g per cm^3, find the formula connecting s, p and d. Find the speed of sound in hydrogen (which is 0·07 times as dense as air) at the same pressure and temperature.

7 Looking for an inverse

We have seen how some transformations can be described algebraically using matrices. The mapping

$$\mathbf{T}:\begin{pmatrix}x\\y\end{pmatrix}\rightarrow\begin{pmatrix}2 & 4\\3 & 5\end{pmatrix}\begin{pmatrix}x\\y\end{pmatrix}$$

is an example of such a transformation.

If **T** maps $P\ (^{-}3,4)$ onto P', find the coordinates of P'.

When we know the coordinates of an object point, we can find the image. In this chapter we shall consider the reverse problem, that is, whether, knowing the image, we can find the object point(s).

For example, if **T** maps Q onto $Q'\ (^{-}5,3)$, can you find the coordinates of Q?

1 Thinking geometrically

(*a*) Figure 1 shows a flag F and its image F' after a positive (anticlockwise) rotation of 90° about the origin.

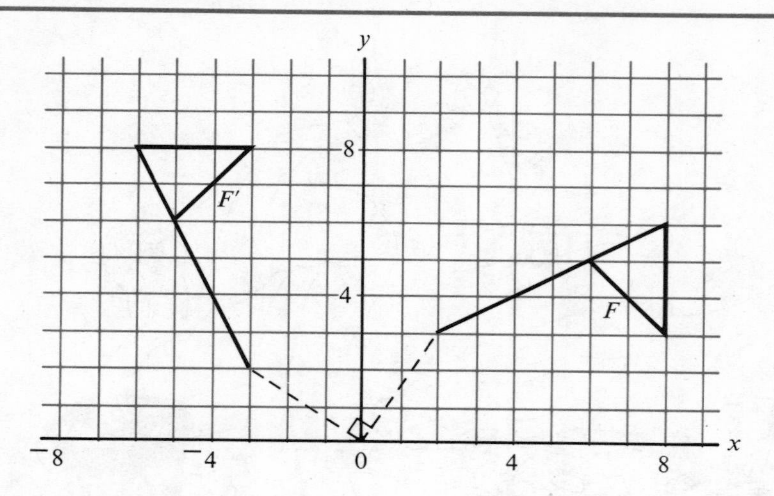

Fig. 1

By considering what happens to the base vectors $\begin{pmatrix}1\\0\end{pmatrix}$ and $\begin{pmatrix}0\\1\end{pmatrix}$, find the matrix **R** which represents this transformation.

Did you obtain the matrix $\begin{pmatrix}0 & ^-1\\1 & 0\end{pmatrix}$?

Describe the inverse transformation which maps F' onto F. By considering what happens to the base vectors, find the matrix **S** which represents this inverse transformation.

Work out **SR**. What can you say about the matrix **SR**? What transformation does it represent?

What can you say about the matrices **R** and **S**?

The matrix $\mathbf{I} = \begin{pmatrix}1 & 0\\0 & 1\end{pmatrix}$ is the identity for the set of 2 by 2 matrices under matrix multiplication.

Since

$$\mathbf{RS} = \mathbf{I} = \mathbf{SR},$$

the matrices **R** and **S** are multiplicative inverses, that is, **R** is the inverse of **S** and **S** is the inverse of **R**.

We often denote the multiplicative inverse of the matrix **R** by $\mathbf{R}^{-1}$. In this case $\mathbf{R}^{-1} = \mathbf{S} = \begin{pmatrix}0 & 1\\^-1 & 0\end{pmatrix}$.

(*b*) Describe the transformation **T** which is represented by the matrix $\mathbf{E} = \begin{pmatrix}2 & 0\\0 & 2\end{pmatrix}$.

Now describe $\mathbf{T}^{-1}$ and, by considering what happens to the base vectors, find the matrix $\mathbf{E}^{-1}$ which represents $\mathbf{T}^{-1}$.

Work out $\mathbf{E}\begin{pmatrix}^-1\\3\end{pmatrix}$. What is the image of the point $(^-1, 3)$ under **T**?

Work out $\mathbf{E}^{-1}\begin{pmatrix} ^-2 \\ 6 \end{pmatrix}$. What point is mapped by **T** onto the point $(^-2, 6)$?

What point is mapped by **T** onto the point $(^-5, 7)$?

(*c*) We shall now try to find the inverse of the mapping

$$\mathbf{M}: \begin{pmatrix} x \\ y \end{pmatrix} \to \begin{pmatrix} 0 & 3 \\ ^-1 & 0 \end{pmatrix}\begin{pmatrix} x \\ y \end{pmatrix}$$

by considering its geometrical description.

Since $\begin{pmatrix} 1 \\ 0 \end{pmatrix} \xrightarrow{\mathbf{M}} \begin{pmatrix} 0 \\ ^-1 \end{pmatrix}$ and $\begin{pmatrix} 0 \\ 1 \end{pmatrix} \xrightarrow{\mathbf{M}} \begin{pmatrix} 3 \\ 0 \end{pmatrix}$, the effect of **M** on the unit square is as shown in Figure 2.

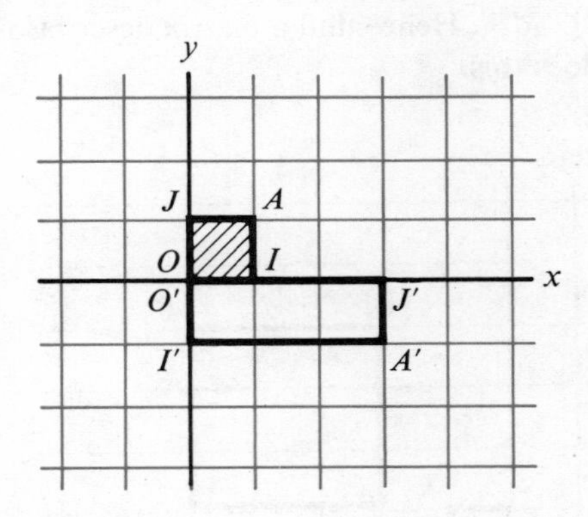

Fig. 2

We can see from this diagram that **M** is equivalent to a rotation about the origin followed by a one-way stretch with points on the line $x = 0$ invariant. What is (i) the angle of the rotation; (ii) the scale factor of the stretch?

The inverse transformation $\mathbf{M}^{-1}$ is equivalent to a one-way stretch with points on the line $x = 0$ invariant followed by a rotation about the origin. State the scale factor of this stretch and the angle of this rotation. By considering what happens to the base vectors, write down the matrices which represent these two transformations.

Do you agree that

$$\mathbf{M}^{-1}: \begin{pmatrix} x \\ y \end{pmatrix} \to \begin{pmatrix} 0 & ^-1 \\ 1 & 0 \end{pmatrix}\begin{pmatrix} \frac{1}{3} & 0 \\ 0 & 1 \end{pmatrix}\begin{pmatrix} x \\ y \end{pmatrix},$$

and therefore

$$\mathbf{M}^{-1}: \begin{pmatrix} x \\ y \end{pmatrix} \to \begin{pmatrix} 0 & ^-1 \\ \frac{1}{3} & 0 \end{pmatrix}\begin{pmatrix} x \\ y \end{pmatrix} ?$$

Check that

$$\begin{pmatrix} 0 & ^-1 \\ \frac{1}{3} & 0 \end{pmatrix}\begin{pmatrix} 0 & 3 \\ ^-1 & 0 \end{pmatrix} = \begin{pmatrix} 1 & 0 \\ 0 & 1 \end{pmatrix} = \begin{pmatrix} 0 & 3 \\ ^-1 & 0 \end{pmatrix}\begin{pmatrix} 0 & ^-1 \\ \frac{1}{3} & 0 \end{pmatrix}.$$

Find the coordinates of the point mapped by **M** onto $(^-3, ^-4)$.

(*d*) So far we have considered only transformations of the form

$$\begin{pmatrix} x \\ y \end{pmatrix} \rightarrow \begin{pmatrix} a & b \\ c & d \end{pmatrix}\begin{pmatrix} x \\ y \end{pmatrix},$$

that is, transformations which map the origin onto itself.

The following example shows you how to proceed when the origin is mapped onto some other point.

Example 1

Describe a sequence of transformations which maps the square $PQRS$ (see Figure 3) onto the square $P'Q'R'S'$. Hence find a matrix description of the single transformation **U** which does this.

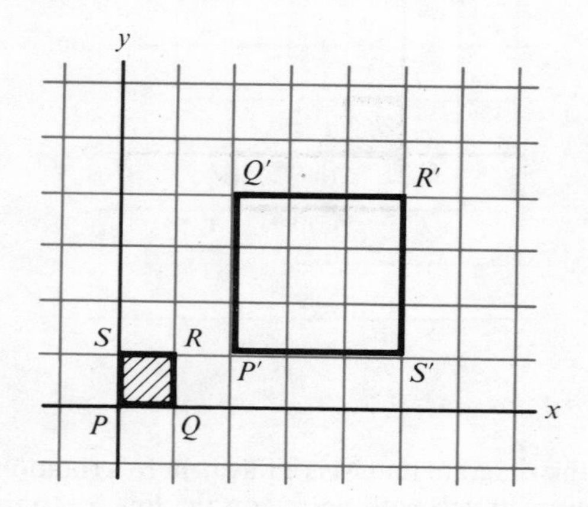

Fig. 3

A sequence of three transformations which maps $PQRS$ onto $P'Q'R'S'$ is the enlargement, **E**, centre the origin and scale factor 3 followed by the reflection, **M**, in the line $y = x$ followed by the translation, **T**, described by the vector $\begin{pmatrix} 2 \\ 1 \end{pmatrix}$. See Figure 4.

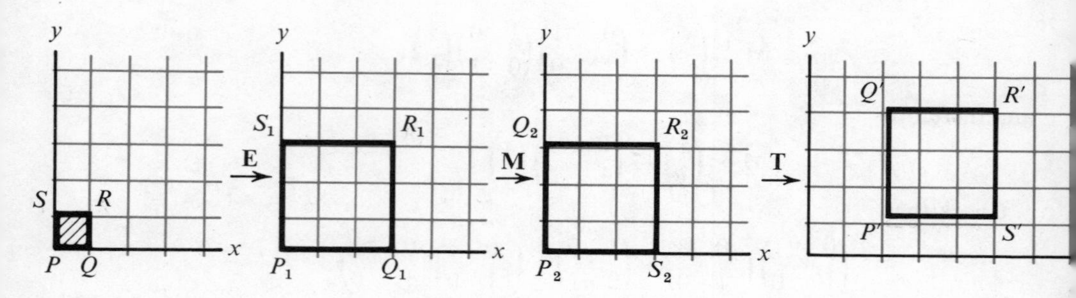

Fig. 4

These transformations are defined by the mappings:

$$\mathbf{E}:\begin{pmatrix}x\\y\end{pmatrix}\rightarrow\begin{pmatrix}3&0\\0&3\end{pmatrix}\begin{pmatrix}x\\y\end{pmatrix},$$

$$\mathbf{M}:\begin{pmatrix}x\\y\end{pmatrix}\rightarrow\begin{pmatrix}0&1\\1&0\end{pmatrix}\begin{pmatrix}x\\y\end{pmatrix},$$

$$\mathbf{T}:\begin{pmatrix}x\\y\end{pmatrix}\rightarrow\begin{pmatrix}x\\y\end{pmatrix}+\begin{pmatrix}2\\1\end{pmatrix}.$$

Therefore we can write:

$$\mathbf{E}:\begin{pmatrix}x\\y\end{pmatrix}\rightarrow\begin{pmatrix}3&0\\0&3\end{pmatrix}\begin{pmatrix}x\\y\end{pmatrix},$$

$$\mathbf{ME}:\begin{pmatrix}x\\y\end{pmatrix}\rightarrow\begin{pmatrix}0&1\\1&0\end{pmatrix}\begin{pmatrix}3&0\\0&3\end{pmatrix}\begin{pmatrix}x\\y\end{pmatrix},$$

$$\mathbf{TME}:\begin{pmatrix}x\\y\end{pmatrix}\rightarrow\begin{pmatrix}0&1\\1&0\end{pmatrix}\begin{pmatrix}3&0\\0&3\end{pmatrix}\begin{pmatrix}x\\y\end{pmatrix}+\begin{pmatrix}2\\1\end{pmatrix}.$$

Since $\mathbf{U}=\mathbf{TME}$,

$$\mathbf{U}:\begin{pmatrix}x\\y\end{pmatrix}\rightarrow\begin{pmatrix}0&3\\3&0\end{pmatrix}\begin{pmatrix}x\\y\end{pmatrix}+\begin{pmatrix}2\\1\end{pmatrix},$$

which can be written more simply as

$$\mathbf{U}:\begin{pmatrix}x\\y\end{pmatrix}\rightarrow\begin{pmatrix}3y+2\\3x+1\end{pmatrix}.$$

Describe geometrically the sequence of transformations which will map $P'Q'R'S'$ back onto $PQRS$ via $P_2Q_2R_2S_2$ and $P_1Q_1R_1S_1$.

Write down the mappings which define the transformations $\mathbf{E}^{-1}$, $\mathbf{M}^{-1}$ and $\mathbf{T}^{-1}$. Use the fact that $\mathbf{U}^{-1}=\mathbf{E}^{-1}\mathbf{M}^{-1}\mathbf{T}^{-1}$ to show that

$$\mathbf{U}^{-1}:\begin{pmatrix}x\\y\end{pmatrix}\rightarrow\begin{pmatrix}\frac{1}{3}&0\\0&\frac{1}{3}\end{pmatrix}\begin{pmatrix}0&1\\1&0\end{pmatrix}\left[\begin{pmatrix}x\\y\end{pmatrix}+\begin{pmatrix}^-2\\^-1\end{pmatrix}\right],$$

and express this in a simpler form.

Use your last answer to find the images of $P'(2,1)$, $Q'(2,4)$, $R'(5,4)$ and $S'(5,1)$ under $\mathbf{U}^{-1}$. Do you obtain $P(0,0)$, $Q(1,0)$, $R(1,1)$ and $S(0,1)$ respectively? If not, try again.

(*e*) Draw a diagram to show the image of the unit square under the mapping

$$\mathbf{M}:\begin{pmatrix}x\\y\end{pmatrix}\rightarrow\begin{pmatrix}1&0\\0&0\end{pmatrix}\begin{pmatrix}x\\y\end{pmatrix}.$$

Describe the transformation in your own words.

Find the images under $\mathbf{M}$ of (i) $(3,0)$, (ii) $(3,5)$, (iii) $(3,^-2)$.

If the domain of $\mathbf{M}$ is the set of all points in the plane, what is the range?

Is $\mathbf{M}$ a one–one mapping or a many–one mapping?

Does $\mathbf{M}$ have an inverse? Give a reason for your answer.

Looking for an inverse

If $\begin{pmatrix} x \\ y \end{pmatrix} \xrightarrow{\mathbf{M}} \begin{pmatrix} 3 \\ 0 \end{pmatrix}$, what can you say about x and y?

Exercise A

1 Describe, in words, the inverse of each of the following transformations:

(*a*) enlargement, centre the origin and scale factor $\frac{1}{2}$;
(*b*) positive rotation of 120° about the origin;
(*c*) reflection in the line $2x + 3y = 0$;
(*d*) one-way stretch with points on the line $y = 0$ invariant and scale factor 4.

2 Describe the effect on the unit square of each of the following mappings:

(i) $\mathbf{A}: \begin{pmatrix} x \\ y \end{pmatrix} \to \begin{pmatrix} 0 & 1 \\ 1 & 0 \end{pmatrix}\begin{pmatrix} x \\ y \end{pmatrix}$; (ii) $\mathbf{B}: \begin{pmatrix} x \\ y \end{pmatrix} \to \begin{pmatrix} {}^{-}2 & 0 \\ 0 & {}^{-}2 \end{pmatrix}\begin{pmatrix} x \\ y \end{pmatrix}$;

(iii) $\mathbf{C}: \begin{pmatrix} x \\ y \end{pmatrix} \to \begin{pmatrix} 2 & 0 \\ 0 & 1 \end{pmatrix}\begin{pmatrix} x \\ y \end{pmatrix}$; (iv) $\mathbf{D}: \begin{pmatrix} x \\ y \end{pmatrix} \to \begin{pmatrix} 1 & 3 \\ 0 & 1 \end{pmatrix}\begin{pmatrix} x \\ y \end{pmatrix}$.

Hence find $\mathbf{A}^{-1}$, $\mathbf{B}^{-1}$, $\mathbf{C}^{-1}$ and $\mathbf{D}^{-1}$.

3 The matrix which represents reflection in $x = 0$ is

$$\begin{pmatrix} {}^{-}1 & 0 \\ 0 & 1 \end{pmatrix}.$$

Explain (i) the geometrical meaning, (ii) the algebraic meaning of

$$\begin{pmatrix} {}^{-}1 & 0 \\ 0 & 1 \end{pmatrix}\begin{pmatrix} {}^{-}1 & 0 \\ 0 & 1 \end{pmatrix} = \begin{pmatrix} 1 & 0 \\ 0 & 1 \end{pmatrix}.$$

By considering other geometrical transformations which are self-inverse, find four more self-inverse matrices.

4 A transformation $\mathbf{T}$ is equivalent to enlargement centre the origin and scale factor 3 followed by reflection in $y = x$.

(*a*) What matrix represents $\mathbf{T}$?
(*b*) Describe $\mathbf{T}^{-1}$ and find the matrix which represents this transformation.
(*c*) $\mathbf{T}$ maps the point A onto A' $(5, {}^{-}2)$. What are the coordinates of A?

5 A translation $\mathbf{T}$ is defined by the mapping

$$\begin{pmatrix} x \\ y \end{pmatrix} \to \begin{pmatrix} x \\ y \end{pmatrix} + \begin{pmatrix} {}^{-}2 \\ 3 \end{pmatrix}.$$

(*a*) What mapping defines the inverse translation $\mathbf{T}^{-1}$?
(*b*) If $\mathbf{T}$ maps P, Q and R onto P' $(3, 5)$, Q' $({}^{-}6, 2)$ and R' $({}^{-}3, 0)$, what are the coordinates of P, Q and R?

6 Figure 5 shows the unit square $OIAJ$ and its image $O'I'A'J'$ under six different transformations. Break down each transformation into a sequence of simpler transformations and hence describe it in the form

$$\begin{pmatrix} x \\ y \end{pmatrix} \to \begin{pmatrix} a & b \\ c & d \end{pmatrix} \begin{pmatrix} x \\ y \end{pmatrix} + \begin{pmatrix} e \\ f \end{pmatrix}.$$

(*Hint*: make the translation the last transformation of the sequence.)

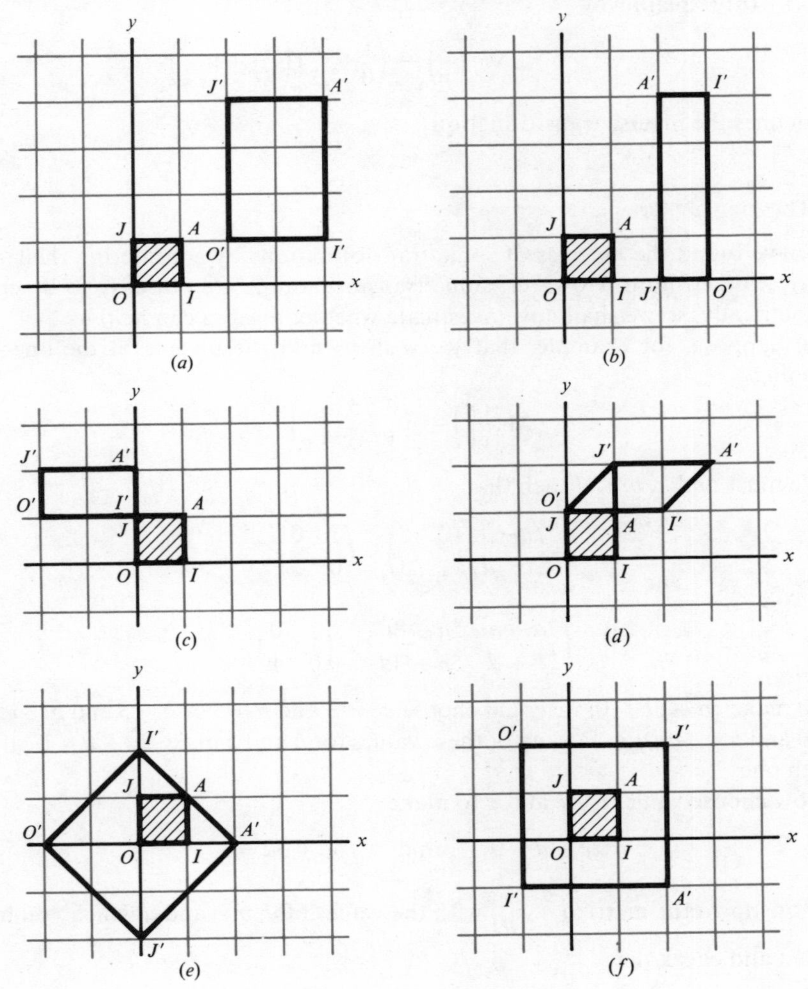

Fig. 5

7 Use the method of Section 1(d) to describe in the form

$$\begin{pmatrix} x \\ y \end{pmatrix} \to \begin{pmatrix} a & b \\ c & d \end{pmatrix} \begin{pmatrix} x \\ y \end{pmatrix} + \begin{pmatrix} e \\ f \end{pmatrix}$$

the inverse of the transformation shown in Figure 5(a).

8 A transformation $\mathbf{V}$ is defined by the mapping

$$\mathbf{V}: \begin{pmatrix} x \\ y \end{pmatrix} \to \begin{pmatrix} 1 & 1 \\ 0 & 1 \end{pmatrix}\begin{pmatrix} x \\ y \end{pmatrix} + \begin{pmatrix} 2 \\ -1 \end{pmatrix}.$$

(*a*) Find the image under $\mathbf{V}$ of the rectangle with vertices $P(^-1,0)$, $Q(^-1,1)$, $R(1,1)$, $S(1,0)$ and hence describe $\mathbf{V}$ as a combination of two simple transformations.

(*b*) Explain why

$$\mathbf{W}: \begin{pmatrix} x \\ y \end{pmatrix} \to \begin{pmatrix} 1 & ^-1 \\ 0 & 1 \end{pmatrix}\begin{pmatrix} x \\ y \end{pmatrix} + \begin{pmatrix} ^-3 \\ 1 \end{pmatrix}$$

defines the inverse transformation.

2 Thinking algebraically

We have found the inverses of some transformations by considering their geometrical description. However, some transformations are not easy to describe geometrically, so we shall now investigate whether algebra can help us.

(*a*) Suppose, for example, that we wish to find the inverse of the one–one mapping

$$\mathbf{M}: \begin{pmatrix} x \\ y \end{pmatrix} \to \begin{pmatrix} 2 & 5 \\ 1 & 3 \end{pmatrix}\begin{pmatrix} x \\ y \end{pmatrix}.$$

We must find a, b, c, d such that

$$\begin{pmatrix} a & b \\ c & d \end{pmatrix}\begin{pmatrix} 2 & 5 \\ 1 & 3 \end{pmatrix} = \begin{pmatrix} 1 & 0 \\ 0 & 1 \end{pmatrix},$$

that is

$$\begin{pmatrix} 2a+b & 5a+3b \\ 2c+d & 5c+3d \end{pmatrix} = \begin{pmatrix} 1 & 0 \\ 0 & 1 \end{pmatrix}.$$

To make $5a+3b=0$, we could choose $a = {^-3}$ and $b=5$ or $a=3$ and $b={^-5}$ or $a=6$ and $b={^-10}$, etc. Do any of these values for a and b make $2a+b=1$? If so, which ones?

Now choose values for c and d to make

$$2c+d=0 \qquad \text{and} \qquad 5c+3d=1.$$

Write down the matrix $\begin{pmatrix} a & b \\ c & d \end{pmatrix}$ with the values of a, b, c and d which you have chosen and check that

$$\begin{pmatrix} a & b \\ c & d \end{pmatrix}\begin{pmatrix} 2 & 5 \\ 1 & 3 \end{pmatrix} = \begin{pmatrix} 1 & 0 \\ 0 & 1 \end{pmatrix} = \begin{pmatrix} 2 & 5 \\ 1 & 3 \end{pmatrix}\begin{pmatrix} a & b \\ c & d \end{pmatrix}.$$

Write down the mapping which defines $\mathbf{M}^{-1}$.

(*b*) Compare the matrix you produced with the original matrix $\begin{pmatrix} 2 & 5 \\ 1 & 3 \end{pmatrix}$. What has happened to:

(i) the numbers in the leading diagonal;
(ii) the numbers in the other diagonal?

(*c*) Use the patterns which you noticed in (*b*) to write down the multiplicative inverses of:

(i) $\mathbf{P} = \begin{pmatrix} 7 & 5 \\ 4 & 3 \end{pmatrix}$; (ii) $\mathbf{Q} = \begin{pmatrix} 9 & 11 \\ 4 & 5 \end{pmatrix}$;

(iii) $\mathbf{R} = \begin{pmatrix} 3 & 2 \\ 7 & 5 \end{pmatrix}$; (iv) $\mathbf{S} = \begin{pmatrix} 2 & 5 \\ 3 & 8 \end{pmatrix}$.

Check, by multiplication, whether your answers are correct.

Now do the same with $\mathbf{T} = \begin{pmatrix} 4 & 5 \\ 1 & 3 \end{pmatrix}$. Have you found the inverse of **T**?

How can you adjust your matrix to give $\mathbf{T}^{-1}$?

(*d*) Figure 6 shows the effect on the unit square of the transformation represented by **T**.

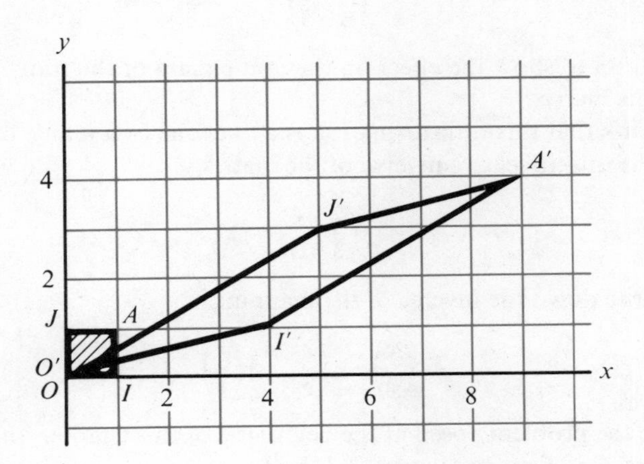

Fig. 6

What is the area factor of the transformation? You may remember that this factor can be found from the matrix by finding the difference between the products of the numbers in each diagonal:

$$(4 \times 3) - (1 \times 5) = 7.$$

We call 7 the *determinant* of **T** and write

$$|\mathbf{T}| = 7.$$

What are the determinants of **P**, **Q**, **R** and **S**?

(*e*) To find the multiplicative inverse of **T** we interchange the numbers in the leading diagonal and change the signs of the numbers in the other diagonal. This gives

$$\begin{pmatrix} 3 & ^-5 \\ ^-1 & 4 \end{pmatrix}.$$

We then divide each element by $|\mathbf{T}|$ and obtain

$$\mathbf{T}^{-1} = \begin{pmatrix} \frac{3}{7} & ^-\frac{5}{7} \\ ^-\frac{1}{7} & \frac{4}{7} \end{pmatrix}.$$

Find the multiplicative inverses of:

(i) $\begin{pmatrix} 4 & 2 \\ 5 & 3 \end{pmatrix}$; (ii) $\begin{pmatrix} 3 & 2 \\ 6 & 5 \end{pmatrix}$;

(iii) $\begin{pmatrix} 4 & 2 \\ 3 & 1 \end{pmatrix}$; (iv) $\begin{pmatrix} ^-1 & ^-2 \\ 1 & 3 \end{pmatrix}$.

Check, by multiplication, whether your answers are correct.

(*f*) What happens when you try to find the multiplicative inverse of

$$\begin{pmatrix} 4 & 2 \\ 6 & 3 \end{pmatrix}?$$

Draw a diagram to show the effect on the unit square of the transformation described by this matrix.

Do you think that this transformation is a one–one or a many–one mapping?

(*g*) Find the multiplicative inverse of the matrix

$$\begin{pmatrix} 2 & 4 \\ 3 & 5 \end{pmatrix}$$

and hence write down the inverse of the mapping

$$\mathbf{T}: \begin{pmatrix} x \\ y \end{pmatrix} \to \begin{pmatrix} 2 & 4 \\ 3 & 5 \end{pmatrix}\begin{pmatrix} x \\ y \end{pmatrix}.$$

Now solve the problem posed at the beginning of this chapter, that is, find the coordinates of Q if **T** maps Q onto Q' $(^-5, 3)$.

Exercise B

1 Find the multiplicative inverses of the following matrices. In each case check your result by multiplication.

(*a*) $\begin{pmatrix} 1 & 2 \\ 2 & 5 \end{pmatrix}$; (*b*) $\begin{pmatrix} 3 & 2 \\ 4 & 3 \end{pmatrix}$; (*c*) $\begin{pmatrix} 8 & 3 \\ 4 & 2 \end{pmatrix}$;

(*d*) $\begin{pmatrix} 7 & 3 \\ 5 & 2 \end{pmatrix}$; (*e*) $\begin{pmatrix} ^-2 & 2 \\ 2 & 3 \end{pmatrix}$; (*f*) $\begin{pmatrix} ^-4 & ^-3 \\ 5 & 3 \end{pmatrix}$;

(*g*) $\begin{pmatrix} 5 & 7 \\ ^-2 & 3 \end{pmatrix}$; (*h*) $\begin{pmatrix} ^-6 & 4 \\ 4 & ^-2 \end{pmatrix}$; (*i*) $\begin{pmatrix} ^-2 & ^-3 \\ ^-4 & ^-5 \end{pmatrix}$.

2 The transformation **W** defined by

$$\mathbf{W}: \begin{pmatrix} x \\ y \end{pmatrix} \to \begin{pmatrix} 3 & 4 \\ 1 & 2 \end{pmatrix} \begin{pmatrix} x \\ y \end{pmatrix}$$

maps the rectangle $ABCD$ onto the parallelogram $A'B'C'D'$ with vertices $A'(4,2)$, $B'(19,7)$, $C'(27,11)$, $D'(12,6)$.

(*a*) Find the multiplicative inverse of $\begin{pmatrix} 3 & 4 \\ 1 & 2 \end{pmatrix}$.

(*b*) Use the inverse mapping $\mathbf{W}^{-1}$ to find the coordinates of A, B, C, and D.

(*c*) Find the areas of $ABCD$ and $A'B'C'D'$ and explain the connection between these areas and (i) the determinant of $\begin{pmatrix} 3 & 4 \\ 1 & 2 \end{pmatrix}$; (ii) the determinant of the inverse of $\begin{pmatrix} 3 & 4 \\ 1 & 2 \end{pmatrix}$.

3 A matrix **M** is such that

$$|\mathbf{M}| = |\mathbf{M}^{-1}|.$$

What is the value of $|\mathbf{M}|$? Is there more than one possible value?

4 What point is mapped onto $(^-3, 7)$ by the transformation

$$\begin{pmatrix} x \\ y \end{pmatrix} \to \begin{pmatrix} 3 & 1 \\ 5 & 2 \end{pmatrix} \begin{pmatrix} x \\ y \end{pmatrix}?$$

5 A transformation **D** is defined by

$$\begin{pmatrix} x \\ y \end{pmatrix} \to \begin{pmatrix} 7 & 5 \\ 4 & 3 \end{pmatrix} \begin{pmatrix} x \\ y \end{pmatrix} + \begin{pmatrix} 2 \\ -5 \end{pmatrix}.$$

Use an algebraic method to find the inverse transformation and hence find the coordinates of the point mapped by **D** onto $(4, ^-4)$.

6 (*a*) Investigate some transformations represented by matrices whose determinant is 0 by finding their effect on the unit square or any other figures of your choice.

(*b*) Matrices whose determinant is zero have no multiplicative inverse. How does this link up with the geometrical properties which you found in (*a*)?

7 For what values of x does

$$\begin{pmatrix} x^2 & x \\ 1 & 1 \end{pmatrix}$$

have no inverse?

8 (*a*) Find the determinant of each of the following matrices:

(i) $\begin{pmatrix} 1 & 1 \\ 0 & 1 \end{pmatrix}$; (ii) $\begin{pmatrix} 0 & 1 \\ 1 & 0 \end{pmatrix}$; (iii) $\begin{pmatrix} 0 & 0 \\ 0 & 1 \end{pmatrix}$;

(iv) $\begin{pmatrix} 0 & 0 \\ 0 & 0 \end{pmatrix}$; (v) $\begin{pmatrix} 1 & 1 \\ 1 & 0 \end{pmatrix}$.

(*b*) Draw diagrams to show the effect on the unit square of the transformations represented by each of these matrices and describe the transformations as accurately as you can.
(*c*) Investigate the range of each of these transformations if the domain is the set of all points in the plane.
(*d*) In each case find all the points which are mapped onto (0, 4).

9 The numerical value of the determinant of a matrix gives the area factor of the associated transformation. Investigate whether the sign of a determinant has any geometrical significance.

Summary

For the 2 by 2 matrix

$$\mathbf{A} = \begin{pmatrix} a & b \\ c & d \end{pmatrix},$$

the number $ad - bc$ is called the determinant of $\mathbf{A}$ and is denoted by $|\mathbf{A}|$.

When $|\mathbf{A}| \neq 0$, $\mathbf{A}$ has a multiplicative inverse. This inverse is

$$\mathbf{A}^{-1} = \begin{pmatrix} \dfrac{d}{|\mathbf{A}|} & \dfrac{-b}{|\mathbf{A}|} \\ \dfrac{-c}{|\mathbf{A}|} & \dfrac{a}{|\mathbf{A}|} \end{pmatrix}.$$

When $|\mathbf{A}| = 0$, $\mathbf{A}$ has no multiplicative inverse.

If a one–one transformation is represented by $\mathbf{A}$, then the inverse transformation is represented by $\mathbf{A}^{-1}$.

If the transformations $\mathbf{S}$ and $\mathbf{T}$ have inverses $\mathbf{S}^{-1}$ and $\mathbf{T}^{-1}$ then $(\mathbf{TS})^{-1} = \mathbf{S}^{-1}\mathbf{T}^{-1}$.

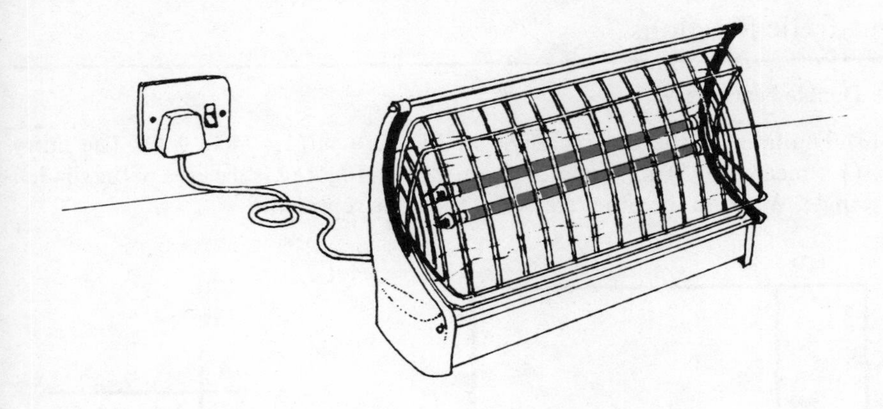

8 Quadratic functions

1 A reminder

(*a*) Since multiplication is distributive over addition for real numbers we know that:

(i) $a(b + c) = ab + ac$; for example $2(3 + \frac{1}{2}) = (2 \times 3) + (2 \times \frac{1}{2})$;
(ii) $(b + c)a = ba + ca$; for example $(\frac{1}{3} + 7)4 = (\frac{1}{3} \times 4) + (7 \times 4)$.

Explain why we can say that $p(q + r) = (q + r)p$. Multiply out, i.e. write without bracket (i) $2(3 + x)$; (ii) $(3x + 1)x$.

(*b*) Notice that

$$\begin{aligned} a(b - c) &= a(b + {}^{-}c) \\ &= a.b + a.{}^{-}c \\ &= ab + {}^{-}ac \\ &= ab - ac. \end{aligned}$$

Multiply out (i) $(b - c)a$; (ii) $3(2 - x)$; (iii) ${}^{-}3(2 - \frac{1}{2}x)$.

(*c*) We can use the distributive law in the 'opposite direction' to insert brackets (i.e. to *factorize* expressions). For example,

$$2ab + a^2 b = ab(2 + a).$$

Factorize (i) $a^2 + ab$; (ii) $3x - x^2$; (iii) $\frac{1}{2}x^2 - ax$.

2 Double brackets

(*a*) The area of the red rectangle in Figure 1 is $(4+\frac{1}{2})(1\frac{1}{2}+5)$. The entry $\frac{1}{2}.5$ ($\frac{1}{2}.5$ means $\frac{1}{2}\times 5$) in the combination table in Figure 2 is the area of the shaded rectangle. What do the other entries in the table represent?

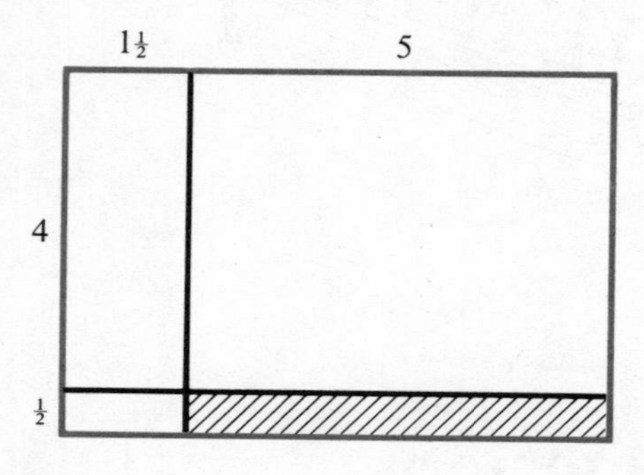

Fig. 1

•	$1\frac{1}{2}$	5
4	$4.1\frac{1}{2}$	4.5
$\frac{1}{2}$	$\frac{1}{2}.1\frac{1}{2}$	$\frac{1}{2}.5$

Fig. 2

Complete the statement $(4+\frac{1}{2})(1\frac{1}{2}+5) = 4.1\frac{1}{2} + \quad + \quad + \frac{1}{2}.5$ and check that your statement is correct by calculating both sides of the equation.

(*b*) Use Figures 3 and 4 to help you explain why

$$(a+b)(c+d) = ac+ad+bc+bd.$$

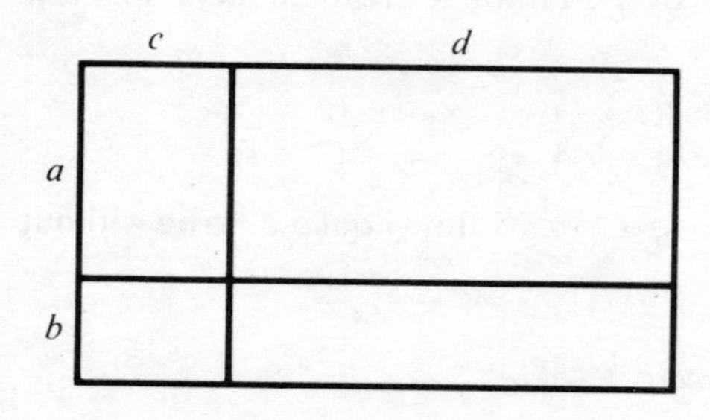

Fig. 3

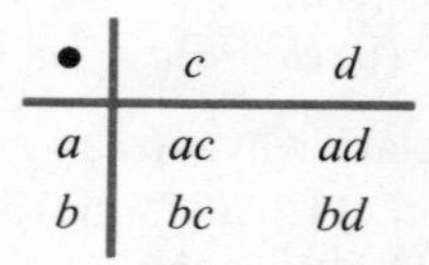

•	c	d
a	ac	ad
b	bc	bd

Fig. 4

Multiply out (i) $(x+2)(x+1)$; (ii) $(2+x)(3+x)$; (iii) $(x+1)(x+1)$; (iv) $(2x+1)(x+1)$.

(*c*) We can obtain the result in (*b*) by using the distributive law in two stages:

$$(a+b)(c+d) = a(c+d)+b(c+d)$$

[compare this with $\quad (a+b)z = az+bz$]

$$= ac+ad+bc+bd.$$

(*d*) Copy and complete

$$(2-x)(1+x) = 2(1+x)-x(1+x)$$

$$=$$

Explain how the combination table in Figure 5 helps you with this problem.

$\bullet$	1	x
2	2	$2x$
$-x$	$-x$	$-x^2$

Fig. 5

Multiply out (i) $(x-1)(x-2)$; (ii) $(x-1)(x+1)$; (iii) $(2x-1)(x+1)$.

(*e*) Your answers in (*b*) and (*d*) may begin to suggest to you that expressions of the form x^2+bx+c can be expressed as the product of two *factors* $(x+p)$ and $(x+q)$.

For example, $(x-1)$ and $(x-2)$ are the factors of x^2-3x+2.

Which of the following are the factors of $x^2+7x+12$:

(i) $(x+6)(x+2)$; (ii) $(x+12)(x+1)$; (iii) $(x+3)(x+4)$?

If $x^2+7x+12=(x+a)(x+b)$, what can you say about (i) $a+b$ and (ii) $a\times b$?

(*f*) Which of the following are the factors of x^2-x-12:

(i) $x+2)(x-6)$; (ii) $(x-12)(x+1)$; (iii) $(x-3)(x+4)$;
(iv) $(x-4)(x+3)$; (v) $(x-2)(x+6)$?

If $x^2-x-12=(x+a)(x+b)$, what can you say about (i) $a+b$ and (ii) $a\times b$?

Summary

1 $(a+b)(c+d)=(a+b)c+(a+b)d$
$\quad\quad\quad\quad\quad\;\;=ac+bc+ad+bd,$

and

$(a+b)(c+d)=a(c+d)+b(c+d)$
$\quad\quad\quad\quad\quad\;\;=ac+ad+bc+bd=ac+bc+ad+bd.$

2 If $x^2+bx+c=(x+p)(x+q)$, then $(x+p)$, $(x+q)$ are the factors of x^2+bx+c. Similarly if $ax^2+bx+c=(mx+p)(nx+q)$ then $(mx+p)$, $(nx+q)$ are the factors of ax^2+bx+c. For example, $2x^2+x-1=(2x-1)(x+1)$.

Exercise A

1 Copy Figure 3 and shade in the areas (i) $a(c+d)$; (ii) $b(c+d)$. Explain why $(a+b)(c+d)=a(c+d)+b(c+d)$ with reference to your diagram.

2 Copy Figure 3 and shade in the areas (i) $(a+b)c$; (ii) $(a+b)d$. Explain why $(a+b)(c+d)=(a+b)c+(a+b)d$ with reference to your diagram.

3 Copy Figure 6 twice. On your first copy, shade in the area $d(b-c)$. On the second copy, shade in the area $db - dc$. What do you conclude?

4 Explain carefully how you could use Figure 7 to show that

$$(x-2)(x-1) = x^2 - 3x + 2.$$

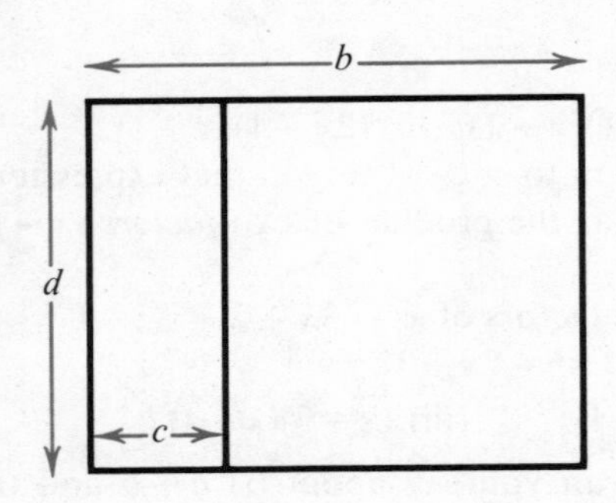

Fig. 6

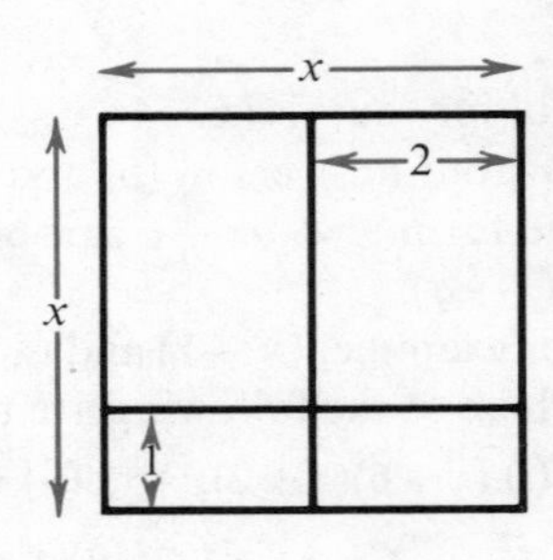

Fig. 7

5 Multiply out:

(i) $(x+2)(x+4)$; (ii) $(x+1)(x+6)$;
(iii) $(x-1)(x-1)$; (iv) $(6-x)(x+1)$;
(v) $(2x+1)(x+2)$; (vi) $\left(\frac{x}{2}+1\right)(x-1)$;
(vii) $(1-2x)(1-x)$; (viii) $(x+1)^2$;
(ix) $(x+4)^2$; (x) $(2x-1)^2$.

6 You are given one factor of each of the following expressions. Find the other factor.

(i) $x^2+7x+12$; $(x+3)$; (ii) x^2+4x+4; $(x+2)$;
(iii) x^2-x-6; $(x-3)$; (iv) $x^2+6x-16$; $(x+8)$;
(v) x^2-3x+2; $(x-1)$; (vi) $2x^2+5x+2$; $(2x+1)$;
(vii) $2x^2+x-1$; $(x+1)$; (viii) $3x^2+5x-2$; $(x+2)$;
(ix) $20x^2-19x+3$; $(5x-1)$; (x) $4x^2-25$; $(2x-5)$;
(xi) x^2-x; x; (xii) $2x^2-3x$; $(2x-3)$.

7 If $P(x) = x^2+x+2$, find (i) $P(1)$; (ii) $P(^-1)$; (iii) $P(0)$; (iv) $P(10)$; (v) $P(^-10)$.

3 The quadratic function

(*a*) In previous books we have discussed functions of the form $f\colon x \to ax+b$ (i.e. linear functions). We learned that their graphs are straight lines with gradient a and intercept on the y axis, b.

In this chapter we will discuss functions of the form $f\colon x \to ax^2+bx+c$ $(a \neq 0)$, which are known as *quadratic functions*.

(*b*) Copy and complete this table of values for the function $f: x \rightarrow x^2 + 2x - 8$:

x	$^-5$	$^-4$	$^-3$	$^-2$	$^-1$	0	1	2	3
x^2	25	16		4	1		1		
$2x$	$^-10$	$^-8$		$^-4$	$^-2$		2		
$x^2 + 2x - 8$	7	0		$^-8$	$^-9$		$^-5$		

(*c*) The arrow diagram in Figure 8 represents *f*. Is *f* one to one or many to one?

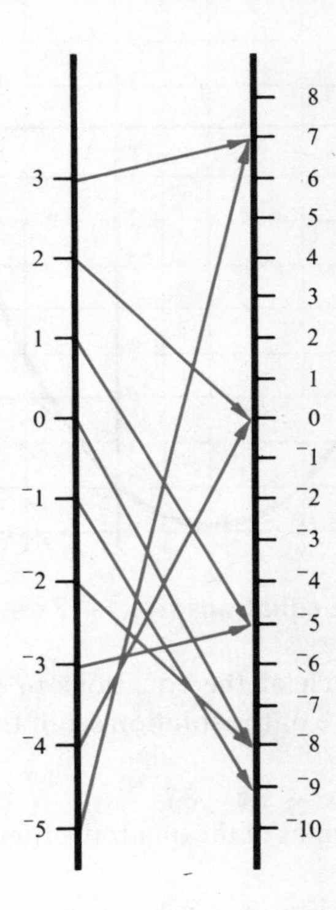

Fig. 8

Which member of the range of *f* is the image of one, and only one, member of the domain?

Notice that members of the domain of *f*, other than $^-1$, are mapped in pairs onto members of the range. For example, $^-4 \rightarrow 0$ and $2 \rightarrow 0$, i.e. if $f(x) = 0$, then $x = {}^-4$ or 2.

Notice also that $^-4$ and 2 are 'equidistant' from $^-1$ on the number line.

Are $^-5$ and 3 equidistant from $^-1$ on the number line? If x is (i) $^-5$; (ii) 3, what is $f(x)$?

What can you say about (i) $f(10)$ and $f(^-12)$; (ii) $f(^-1 + a)$ and $f(^-1 - a)$, where a is any real number?

What do your answers suggest to you about the graph of *f*?

(*d*) Figure 9 shows the graph of f for the domain $^-5 \leqslant x \leqslant 3$. The graph is symmetrical about the line $x = {}^-1$. The curve is called a parabola.

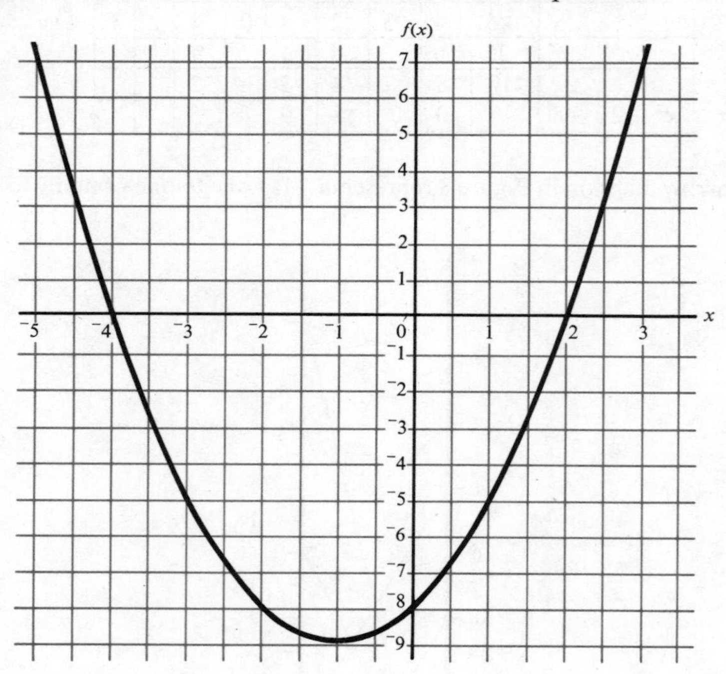

Fig. 9

Use the graph to solve the equations (i) $x^2 + 2x - 8 = 0$; (ii) $x^2 + 2x - 8 = 7$; (iii) $x^2 + 2x - 8 = \frac{1}{2}$.

(*e*) The solution set of each of the equations in (*d*) contains two members. How many members are there of the solution set of the equation $x^2 + 2x - 8 = k$ if k is (i) $^-9$; (ii) $^-4$; (iii) $^-12$?

What is the *least* value of $x^2 + 2x - 8$?

(*f*) Figure 10 shows the graphs of the quadratic functions (*a*) $f: x \rightarrow x^2 + 4x + 4$; (*b*) $g: x \rightarrow x^2 + 1$.

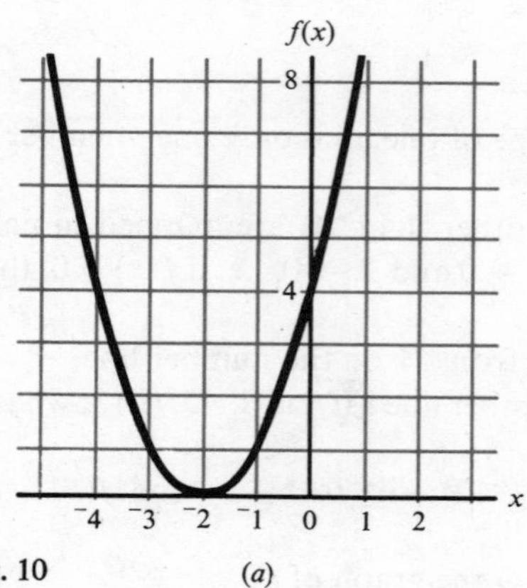

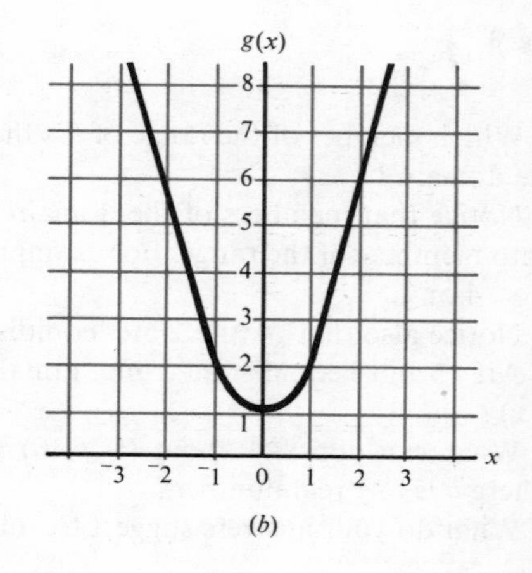

Fig. 10 (*a*) (*b*)

What is the equation of the line of symmetry of each graph?
How many members are there in the solution sets of the following equations:

(i) $x^2 + 4x + 4 = 0$; (ii) $x^2 + 1 = 0$?

Summary

1 A quadratic function f is of the form $f\colon x \to ax^2 + bx + c$. Its graph is a parabola, and it has a line of symmetry.

2 $f\colon x \to ax^2 + bx + c$ is a many to one function. Only one member of the range is the image of exactly one member of the domain. All other members of the domain are mapped in pairs onto their images.

3 The solution set of the equation $ax^2 + bx + c = 0$ may contain (i) 2 members; (ii) 1 member; (iii) 0 members. These cases are illustrated by graphs (*a*), (*b*) and (*c*) in Figure 11.

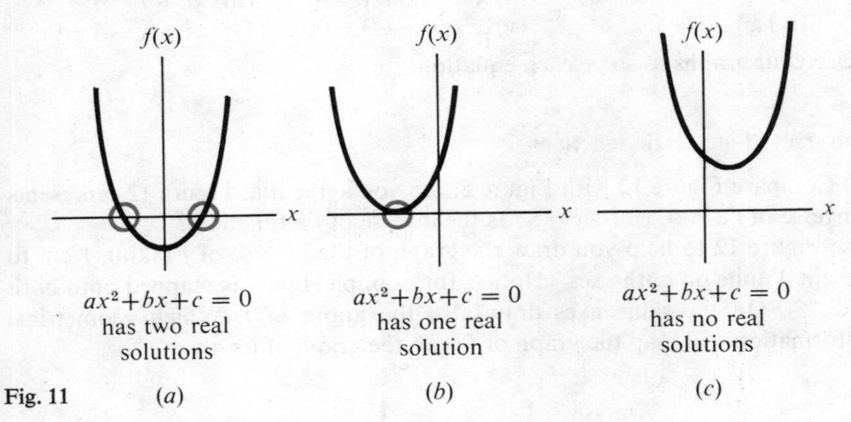

Fig. 11

Exercise B

1 Draw the graph of each of the following functions for the domain $^-4 \leqslant x \leqslant 4$. (Use the same scale for each graph.)

(i) $f\colon x \to x^2$; (ii) $g\colon x \to x^2 - 1$; (iii) $h\colon x \to (x-1)^2$.

What is the equation of the line of symmetry of each graph? Which geometrical transformation will map the graph of (i) f onto g; (ii) g onto h; (iii) f onto h?

2 On the same axes draw the graphs of (i) $f\colon x \to x^2 + 2x + 1$; (ii) $f\colon x \to 2x + 5$. Use your graphs to solve the equations:

(*a*) $x^2 + 2x + 1 = 2x + 5$; (*b*) $x^2 - 4 = 0$.

3 If $f\colon x \to x^2 - 1$, show that $f(a) = f(^-a)$. What does your answer suggest about the graph of f?

4 $f\colon x \to x^2 - 2x + 1$. Show that $f(1+a) = f(1-a)$. [*Hint*: $f(1+a) = (1+a)^2 - 2(1+a) + 1$]. Interpret your answer geometrically.

5 A stone is thrown over a cliff and the distance in metres it has fallen in t seconds is approximately

$$5t^2 + 20t.$$

Draw a graph showing the distance fallen as a function of time. Estimate the time when the stone has fallen 200 m.

6 (*a*) By drawing graphs find the least values of the expressions:

(i) $x^2 + 7x + 4$; (ii) $x^2 + 2x$.

(*b*) By drawing graphs find the greatest values of the expressions:

(i) $^-x^2$; (ii) $^-x^2 + 2x + 6$.

7 By drawing graphs find the number of elements in the solution sets of the equations:

(i) $x^2 - 2x - 3 = 0$; (ii) $x^2 + 2x + 1 = 0$; (iii) $x^2 + 7 = 0$;
(iv) $x^2 - 7 = 0$; (v) $x^2 + 4x + 3 = 0$; (vi) $x^2 + x + 4 = 0$.

Use your graphs to solve each equation.

4 Inverses of quadratic functions

(*a*) Compare Figure 12 with Figure 8. Do you agree that Figure 12 represents the inverse of $f: x \to x^2 + 2x - 8$? Is the inverse of f a function?

Use Figure 12 to help you draw the graph of the inverse of f taking 1 cm to represent 2 units on both axes. (Notice, for example, that 7 is mapped onto both 3 and $^-5$). On the same axes draw also the graph of f. Which geometrical transformation will map the graph of f onto the graph of its inverse?

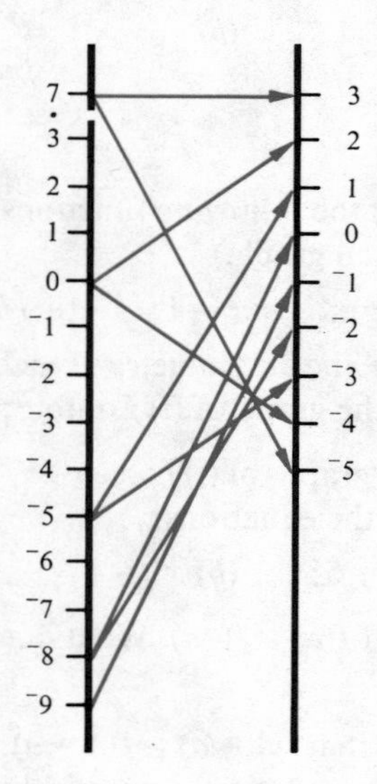

Fig. 12

(b) You will remember that we can use the flow diagram method to help us express the inverse of a function in the form $x \rightarrow \ldots$.

For example, consider the function $f\colon x \rightarrow x^2 - 4$:

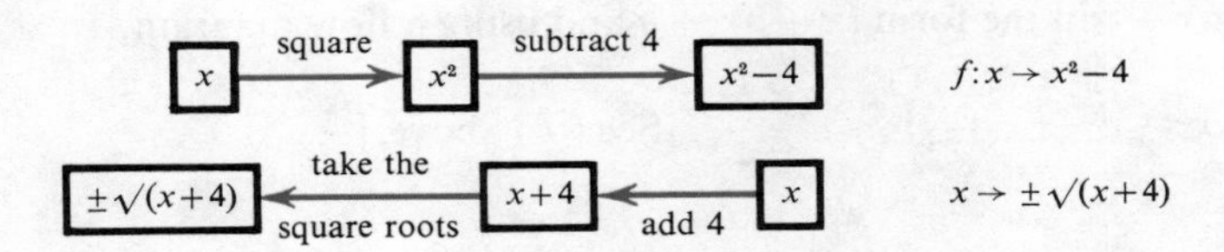

Fig. 13

The inverse relation is $x \rightarrow \pm\sqrt{(x+4)}$.

Use Figure 13 to help you solve the equations

(i) $x^2 - 4 = 0$;
(ii) $x^2 - 4 = 12$.

(c) Use the method above to try to express the inverse of $f\colon x \rightarrow x^2 + 2x - 8$ in the form $x \rightarrow \ldots$. If you think you have succeeded, use your mapping to find the images of (i) 0; (ii) $^-8$. Check your answers using Figure 12 and then read on.

(d) Unless you first expressed $x^2 + 2x - 8$ in a different form you will have found it impossible to construct f by a flow diagram.

Now check that $x^2 + 2x - 8 = (x+1)^2 - 9$, and study Figure 14.

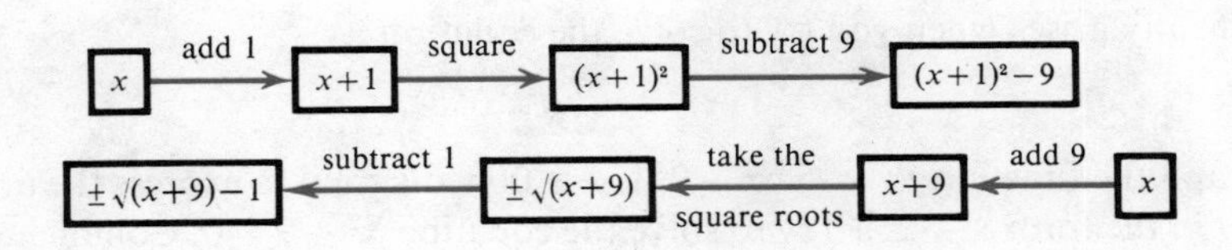

Fig. 14

The inverse of f is $x \rightarrow \pm\sqrt{(x+9)} - 1$. Use the mapping to find the images of (i) 0; (ii) $^-8$ and check your answers using Figure 12.

Use the flow diagram to solve the equations:

(i) $x^2 + 2x - 8 = 0$; (ii) $x^2 + 2x - 8 = 7$.

(e) Section (d) suggests that if we can express a quadratic function in the form $f\colon x \rightarrow (x+p)^2 + q$, then we can find its inverse. Notice that if

$$x^2 + 4x - 6 = (x+p)^2 + q, \text{ then } p = 2.$$

$$\text{So } x^2 + 4x - 6 = (x+2)^2 + q.$$

What is the value of q?

Express the inverse of $f\colon x \rightarrow x^2 + 4x - 6$ in the form $x \rightarrow \ldots$.

(f) If $x^2 + bx + c = (x+p)^2 + q$, what is p in terms of b? Find q in terms of b and c.

Summary

1 The inverse of a quadratic function is a one to many relation.

2 The inverse relation can be expressed in the form $x \to \ldots$ by first expressing $x^2 + bx + c$ in the form $(x+p)^2 + q$ and using a flow diagram.

3 $x^2 + bx + c = \left(x + \frac{b}{2}\right)^2 + c - \frac{b^2}{4}$. [See ($f$) above.]

Exercise C

1 Complete each of the following:

(i) $x^2 + 8x + 6 = (x+4)^2 + \cdots$; (ii) $x^2 - 6x - 3 = (x-3)^2 + \cdots$; (iii) $x^2 + x - 1 = (x + \frac{1}{2})^2 + \cdots$.

2 (a) Draw flow diagrams to form the functions:

(i) $f: x \to x^2 + 3x + 2$; (ii) $f: x \to x^2 - 3$; (iii) $f: x \to x^2 + 2x$; (iv) $f: x \to (x-3)^2$; (v) $f: x \to x^2 + 4x + 4$.

Hence find the inverse relation of each and write it in the form $x \to \ldots$.
(b) Solve the equations:

(i) $x^2 + 3x + 2 = 0$; (ii) $x^2 - 3 = 0$; (iii) $x^2 + 2x = 0$; (iv) $(x-3)^2 = 0$; (v) $x^2 + 4x + 4 = 0$.

3 Write the function $x \to x^2 + 4x + 6$ in the form $x \to (x+2)^2 + \cdots$. What difficulty arises when you try to solve the equation

$$x^2 + 4x + 6 = 0?$$

4 Graph the function $f: x \to x^2 + 9$. Use a flow diagram to express the inverse of f in the form $x \to \ldots$. Try to solve the equation $x^2 + 9 = 0$. Comment with reference to your graph.

5 Express $x^2 + bx + c$ in the form $(x+p)^2 + q$. Hence express the inverse of $f: x \to x^2 + bx + c$ in the form $x \to \ldots$. Solve the equation $x^2 + bx + c = 0$.

5 Factors and solutions of quadratic equations

(a) Calculate (i) 2×0; (ii) $\frac{1}{2} \times 0$; (iii) $0 \times 1{\cdot}7$; (iv) 0×2; (v) 0×0; (vi) 1×0; (vii) 0×1.

(b) (i) If a and b are numbers and $a \times b = 0$, what can you say about a and b?
(ii) If $(x+1)(x-2) = 0$, what can you say about $(x+1)$ and $(x-2)$?
(iii) If $(x+1) = 0$, what is x? If $(x-2) = 0$ what is x?
(iv) What values of x make the expression $(x+1)(x-2)$ zero?
(v) Multiply out $(x+1)(x-2)$.
(vi) What values of x make the expression $x^2 - x - 2$ zero?
(vii) Solve the equation $x^2 - x - 2 = 0$.
(viii) Where does the graph of $f: x \to x^2 - x - 2$ cut the x axis?

(*c*) You should have discovered in (*b*) that if we can express $x^2 + bx + c$ in the form $(x + p)(x + q)$ then we can solve the equation $x^2 + bx + c = 0$ and the solutions are $x = {}^-p$ and $x = {}^-q$. For example, consider the equation

$$x^2 - 3x - 4 = 0.$$

Check that

$$(x - 4)(x + 1) = x^2 - 3x - 4.$$

Hence

$$(x - 4)(x + 1) = 0.$$

This is true if (i) $(x - 4) = 0$ or if (ii) $(x + 1) = 0$, that is, if $x = 4$ or $^-1$.

Where does the graph of $f\colon x \to x^2 - 3x - 4$ cut the x axis?

Example 1

Solve the equation

$$2x^2 + 5x - 3 = 0.$$

$$2x^2 + 5x - 3 = (2x - 1)(x + 3),$$

so

$$(2x - 1)(x + 3) = 0$$

and hence

$$x = \tfrac{1}{2} \text{ or } {}^-3.$$

Example 2

Solve the equation

$$(x - 1)^2 = x + 1.$$

Multiplying out the brackets,

$$x^2 - 2x + 1 = x + 1.$$

Subtracting x from both sides,

$$x^2 - 3x + 1 = 1.$$

Subtracting 1 from each side

$$x^2 - 3x = 0.$$

Factorizing:

$$x(x - 3) = 0,$$

and hence

$$x = 0 \text{ or } 3.$$

(*d*) It is often difficult to spot the factors of a quadratic expression. In such cases, quadratic equations can be solved by writing expressions like $x^2 + bx + c$ in the form $(x + p)^2 + q$ and proceeding as below:

Consider the equation $x^2 - 4x + 1 = 0$.

Can you spot the factors of $x^2 - 4x + 1$?

Check that $x^2 - 4x + 1 = (x - 2)^2 - 3$.

Hence $(x - 2)^2 - 3 = 0$.

Adding 3 to each side, we have

$$(x-2)^2 = 3$$

and taking the square root of each side,

$$x - 2 = \pm\sqrt{3}.$$

Hence

$$x = 2 \pm \sqrt{3}$$

$$x = 2 + 1{\cdot}732 \text{ or } 2 - 1{\cdot}732$$

that is

$$x = 3{\cdot}732 \text{ or } 0{\cdot}268.$$

[Notice that the factors of $x^2 - 4x + 1$ are $(x - 3{\cdot}732)$ and $(x - 0{\cdot}268)$.]

Now solve the equation $2x^2 - 12x + 10 = 0$ by first dividing both sides by 2. Did you get $x = 5$ or 1?

Notice that if we were to write the factors of $2x^2 - 12x + 10$, we must re-introduce the 2 which was 'lost' from the expression when we divided by 2:

$$2x^2 - 12x + 10 = 2(x-1)(x-5).$$

Exercise D

1 Solve the following equations by first of all expressing the left-hand side in terms of its factors:

(i) $x^2 - 7x + 12 = 0$; (ii) $x^2 + 6x + 9 = 0$;
(iii) $x^2 - 4x - 12 = 0$; (iv) $x^2 - 4 = 0$;
(v) $x^2 - 4x = 0$; (vi) $2x^2 + x - 1 = 0$;
(vii) $2x^2 + 5x + 3 = 0$; (viii) $3x^2 + 4x + 1 = 0$;
(ix) $2x^2 - 7x + 6 = 0$; (x) $x^2 - 25 = 0$;
(xi) $2x^2 + 3x = 0$.

2 Solve the equations in Question 1 by first expressing the left-hand side in the form $(x+p)^2 + q$, and continuing as in section (*d*) above.

3 The solutions of a quadratic equation $x^2 + bx + c = 0$ are $x = 2$ and $x = 3$. (i) Write down the factors of the expression $x^2 + bx + c$; (ii) multiply the factors and rearrange the terms in the form $x^2 + bx + c = 0$.

Answer the same question if the solutions are (i) $x = 1$, $x = {}^-1$; (ii) $x = 3$, $x = {}^-1$; (iii) $x = \frac{1}{2}$, $x = 10$; (iv) $x = 0$, $x = 1$; (v) $x = 0$, $x = {}^-1$.

4 Solve the equations:

(i) $x^2 + x - 1 = x$;
(ii) $x^2 + 3x - 1 = {}^-1$;
(iii) $(x+1)^2 = 1$;
(iv) $x^2 + 7x - 4 = 7x$.

5 Where do the graphs of $f\colon x \to x^2 + 2x + 1$; $g\colon x \to 2x^2 + 2x$ intersect? Solve algebraically the equation $x^2 + 2x + 1 = 2x^2 + 2x$.

Some properties of the circle

1 Introduction

In this interlude we are going to look at some of the properties of a circle. Before we do this, however, let us remind ourselves of the angle properties of a triangle.

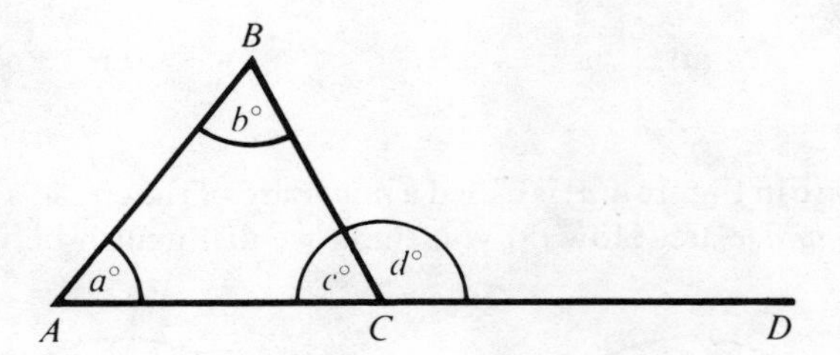

Fig. 1

In Figure 1 the interior angles of the triangle ABC are $a°$, $b°$ and $c°$. Figure 2 will help remind you that the angle sum of a triangle is 180°, i.e. $a° + b° + c° = 180°$

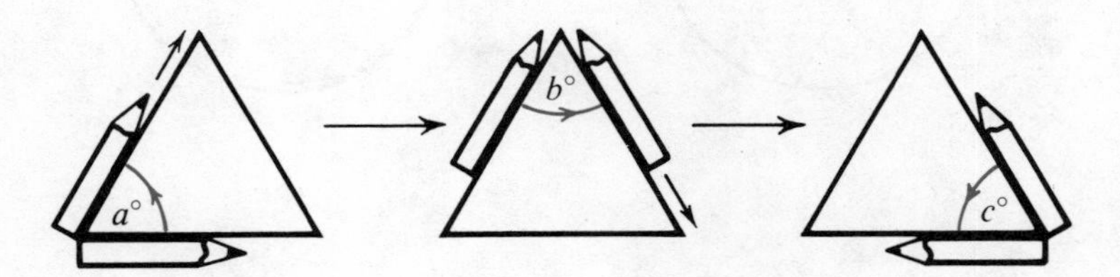

Fig. 2

(in turning through the three angles of the triangle the pencil has been given a half-turn).

However, $d° + c° = 180°$, because the two angles lie on a straight line.

Hence $d° = a° + b°$.

That is, the exterior angle ($d°$) of a triangle equals the sum of the two opposite interior angles ($a° + b°$). What does Figure 3 suggest to you about the angles marked $p°$, $q°$ and $r°$?

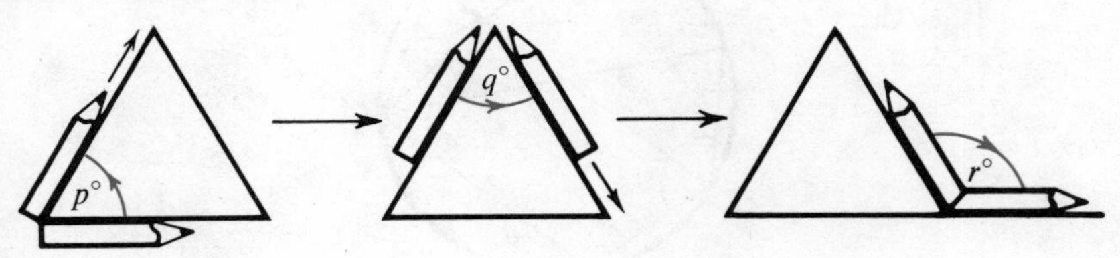

Fig. 3

2 Some descriptive terms about the circle

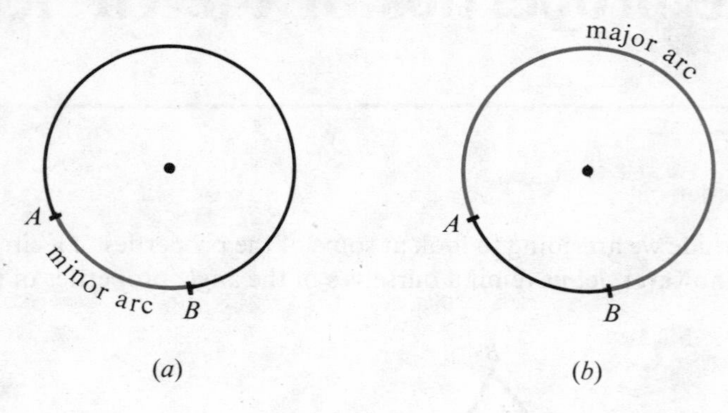

Fig. 4

(*a*) The red arc in Figure 4(*a*) is called a *minor* arc of the circle, and the red arc in Figure 4(*b*), a *major* arc. How do you think we distinguish between minor and major arcs?

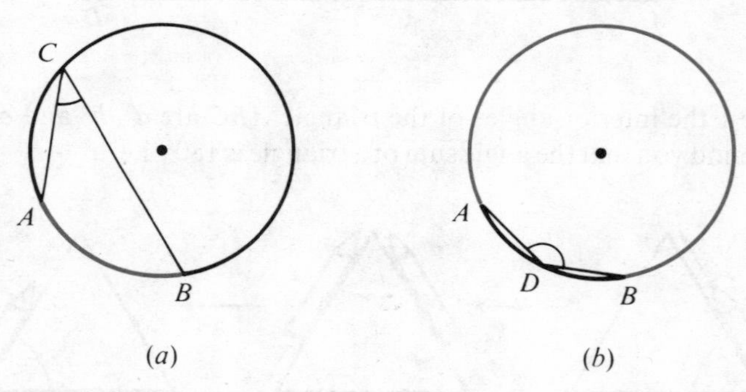

Fig. 5

(*b*) Figure 5(*a*) shows an angle ACB 'standing on' the (minor) arc AB. Notice that C lies on the circumference of the circle. We say that angle ACB is an angle 'subtended at the circumference of the circle by the (minor) arc AB'. How would you describe angle ADB in Figure 5(*b*)? Which arcs subtend the angles marked $a°$, $b°$, $c°$, $d°$, $e°$, $f°$ and $g°$ in Figure 6?

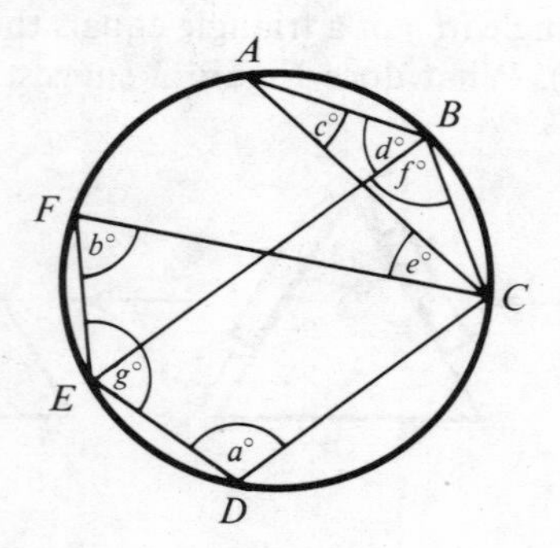

Fig. 6

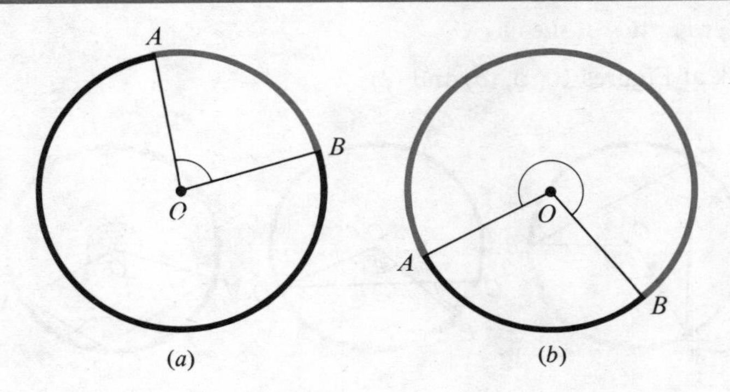

Fig. 7

(*c*) O is the centre of the circles in Figure 7. In Figure 7(*a*) angle AOB is the angle 'subtended at the centre by the (minor) arc AB'. Would you describe the reflex angle AOB in Figure 7(*b*) in exactly the same way?

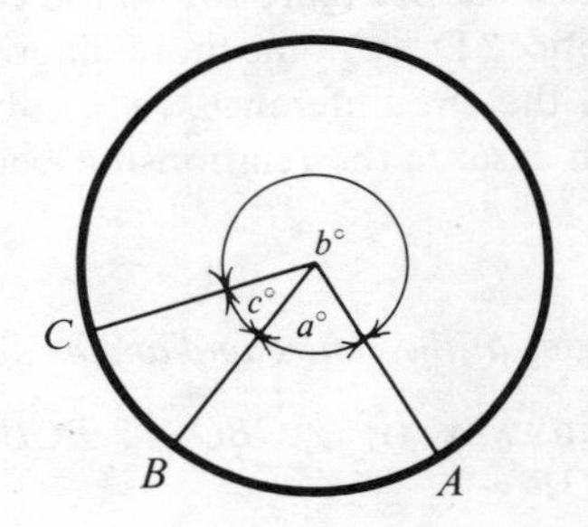

Fig. 8

Describe the angles marked $a°$, $b°$ and $c°$ in Figure 8.

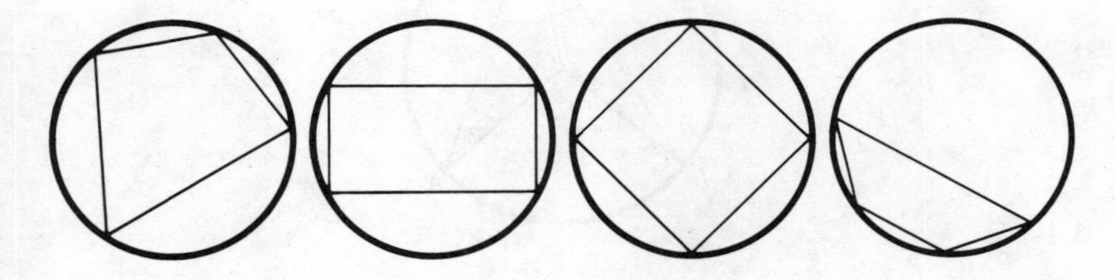

Fig. 9

(*d*) Quadrilaterals whose vertices all lie on the circumference of a circle are called *cyclic quadrilaterals*. Figure 9 shows some examples of cyclic quadrilaterals.

Are all rectangles cyclic quadrilaterals? Give reasons for your answer. Which of the following are always cyclic quadrilaterals: (i) squares; (ii) parallelograms; (iii) rhombuses; (iv) kites?

3 Some properties of the circle

(*a*) Look at Figures 10(*a*), (*b*) and (*c*).

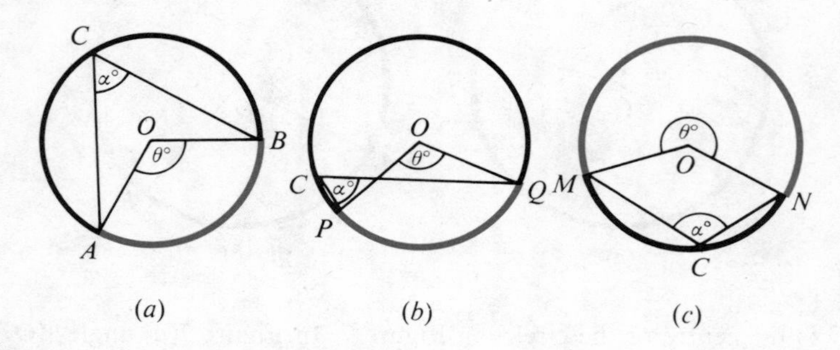

Fig. 10

Describe the angles $\theta°$ and $\alpha°$ in each case by referring to the red arcs.

Draw some diagrams like those in Figure 10 and use your protractor to measure $\theta°$ and $\alpha°$. What do you notice? Draw some more diagrams showing angles at the centre,($\theta°$), and angles on the circumference, ($\alpha°$), subtended by the same arc. Measure $\theta°$ and $\alpha°$ in each case. Is the relationship between θ and α always the same?

Property 1: concerning angles at the centre and angles at the circumference

Study Figure 11. Explain why (i) $\angle CBO = \angle BCO$; (ii) $\angle OCA = \angle CAO$. (Notice that $CO = OA = OB$.)

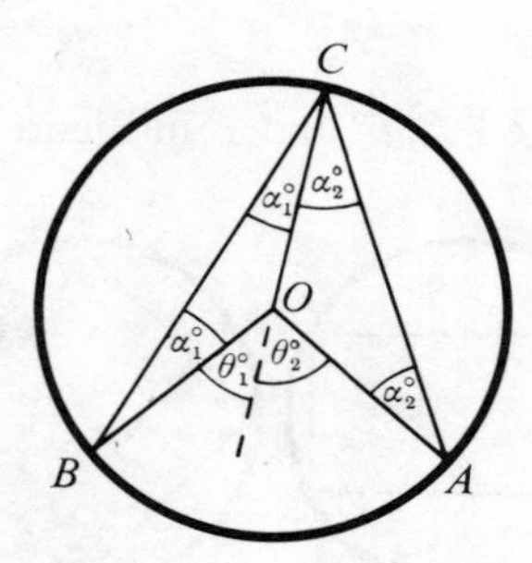

Fig. 11

Since the exterior angle of a triangle is equal to the sum of the two opposite interior angles, we have in $\triangle BOC$,

$$\theta_1° = \alpha_1° + \alpha_1° = 2\alpha_1°,$$

and in $\triangle COA$,

$$\theta_2° = \alpha_2° + \alpha_2° = 2\alpha_2°.$$

Hence

$$\theta_1° + \theta_2° = 2(\alpha_1° + \alpha_2°)$$

i.e.

$$\theta° = 2\alpha°, \text{ where } \angle BOA = \theta° \text{ and } \angle BCA = \alpha°.$$

It can be shown by similar arguments that $\theta^\circ = 2\alpha^\circ$ in the other cases represented by Figures 10(*b*) and (*c*). You might like to do this yourself. We can conclude that 'the angle subtended at the centre of a circle by an arc is twice the angle subtended at the circumference by the same arc'. We will call this Property 1.

There are several properties of the circle, and of cyclic quadrilaterals which can be derived using this result. Use the suggestions that follow to derive the properties for yourself.

Property 2: concerning angles at the circumference subtended by the same arc

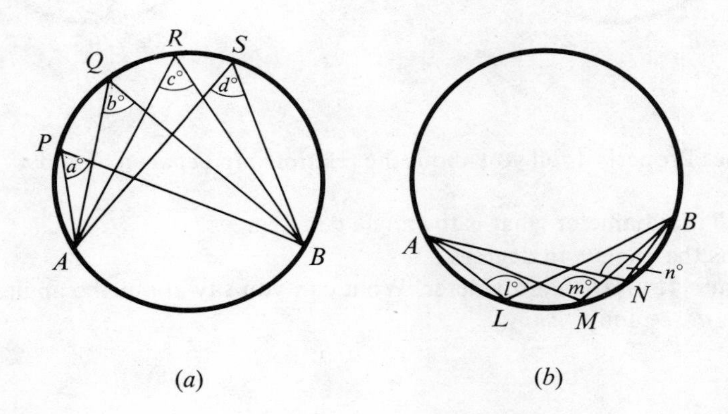

Fig. 12

Copy Figures 12(*a*) and 12(*b*) and measure the angles marked a°, b°, c°, d° and l°, m°, n° (the exact positions of A, B, P, Q, R, S, L, M, N are not important). What do you notice?

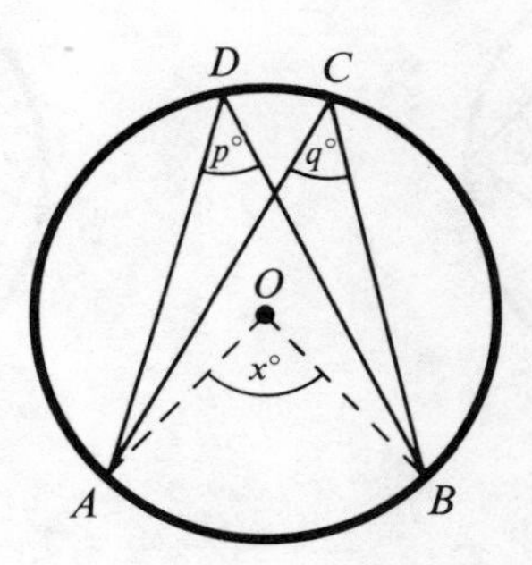

Fig. 13

In Figure 13, $x^\circ = 2p^\circ$ (by Property 1). Write a similar equation involving x° and q°. What can you say about p° and q°?

Does this confirm your findings about the angles marked a°, b°, c° and d° in Figure 12(*a*), and l°, m° and n° in Figure 12(*b*)?

What do your results tell you about angles at the circumference subtended by the same arc?

Property 3: concerning angles in a semi-circle

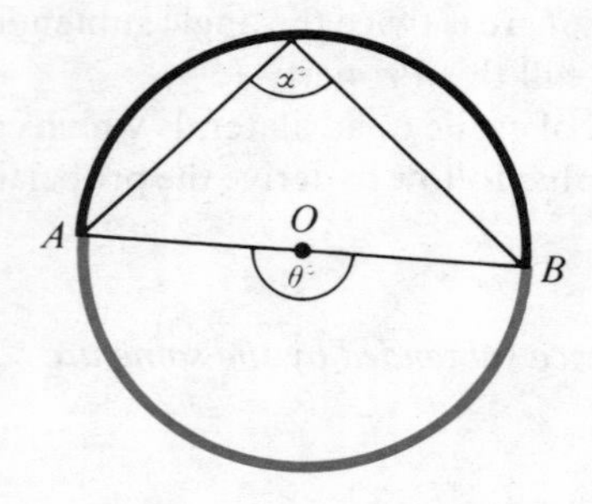

Fig. 14

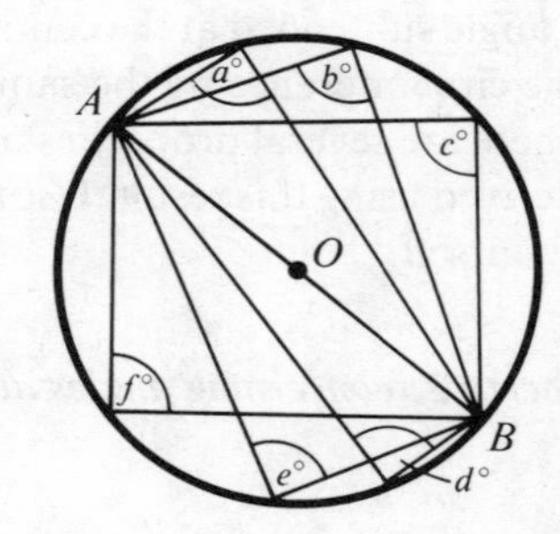

Fig. 15

What does Property 1 tell you about the relationship between $\theta°$ and $\alpha°$ in Figure 14?

If AOB is a diameter what is the angle θ in degrees?

What is the angle α in degrees?

In Figure 15 AOB is a diameter. What can you say about the angles marked $a°$, $b°$, $c°$, $d°$, $e°$ and $f°$?

Property 4: concerning opposite angles of a cyclic quadrilateral

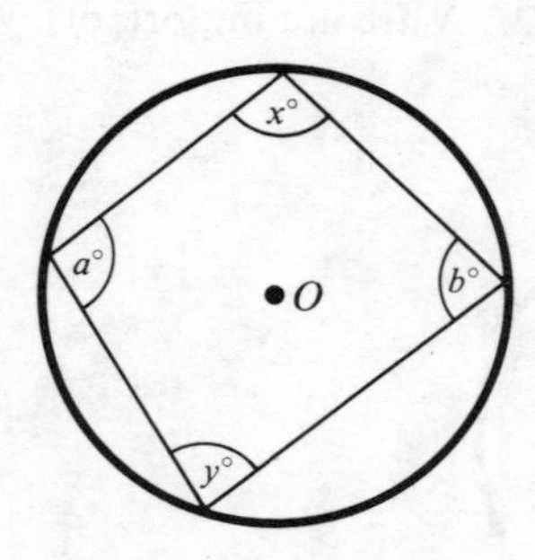

Fig. 16

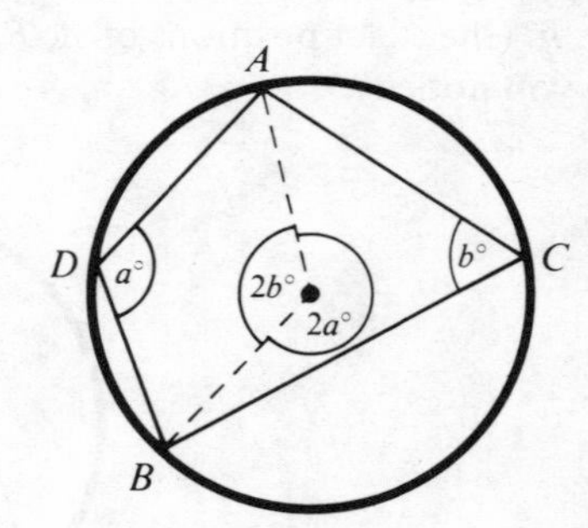

Fig. 17

Draw some diagrams of cyclic quadrilaterals. Measure the interior angles (the angles marked $a°$, $b°$, $x°$, $y°$ in Figure 16). What do your results suggest about the values of $x° + y°$ and $a° + b°$?

Explain why the labelling of the angles in Figure 17 as $a°$, $b°$, $2a°$ and $2b°$ is correct. What is the value of $2a° + 2b°$? Can you deduce the value of $a° + b°$?

What is the sum in degrees of the opposite angles of any cyclic quadrilateral?

Property 5: concerning the exterior angle of a cyclic quadrilateral

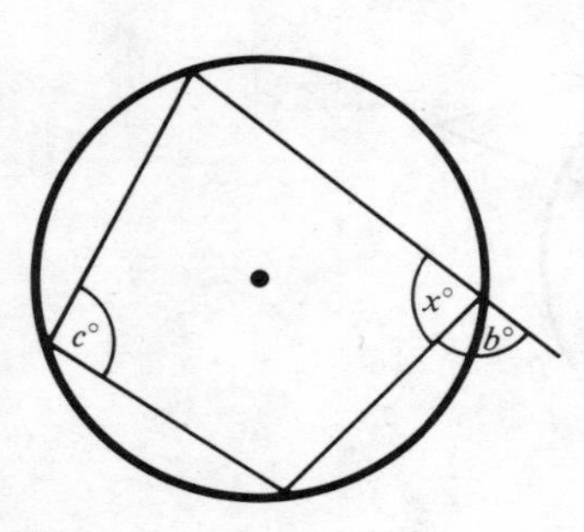

Fig. 18

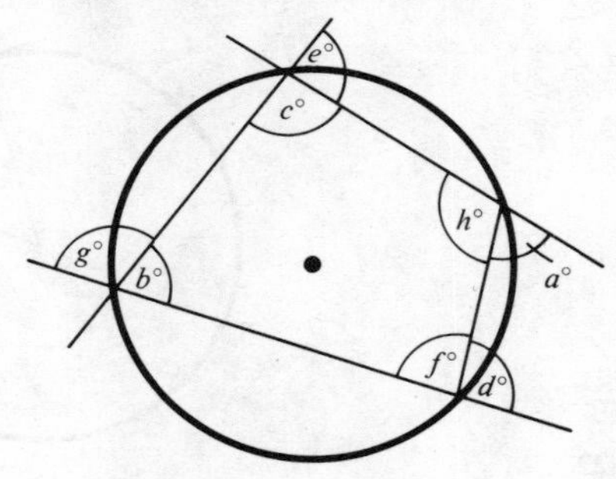

Fig. 19

In Figure 18,

$$x^\circ + c^\circ = 180^\circ \text{ (why?)}$$

and

$$x^\circ + b^\circ = 180^\circ \text{ (why?)}$$

What do these equations tell you about the exterior angle of a cyclic quadrilateral (b°) and the opposite interior angle (c°)?

Write some equations connecting pairs of angles marked in Figure 19. (For example $a^\circ = b^\circ$.)

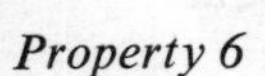

Property 6

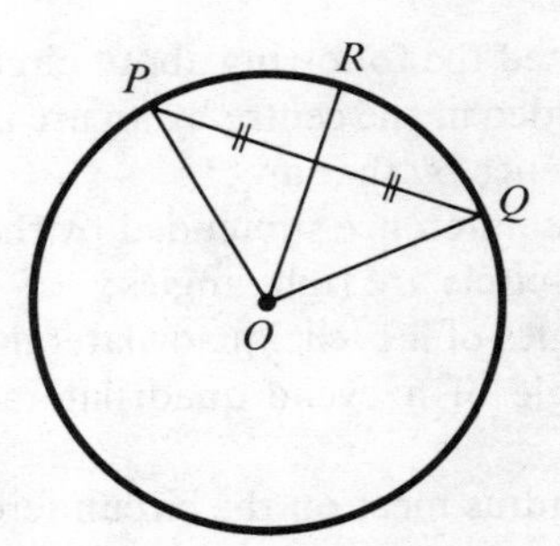

Fig. 20

Explain why triangle POQ in Figure 20 is isosceles.

By the symmetry of an isosceles triangle we know that PQ is perpendicular to OR. Now consider the sequence of diagrams in Figure 21. Angle BOA is decreasing at each stage until OA and OB are coincident. What does this sequence suggest about the angle in which a radius meets a tangent to a circle (i.e. about the angle marked θ°)?

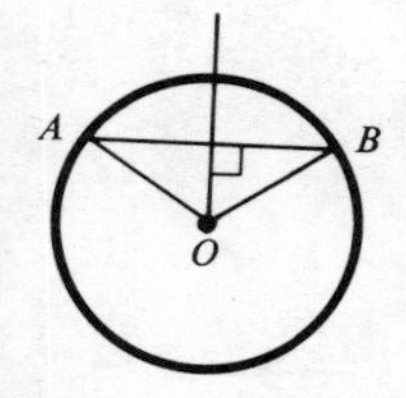

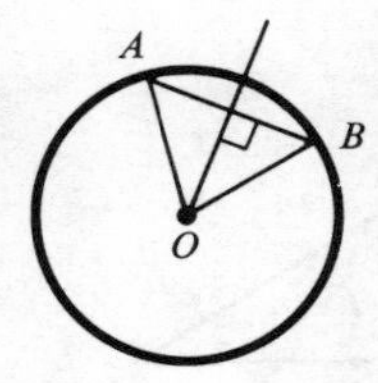

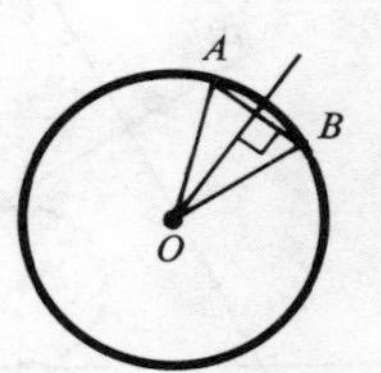

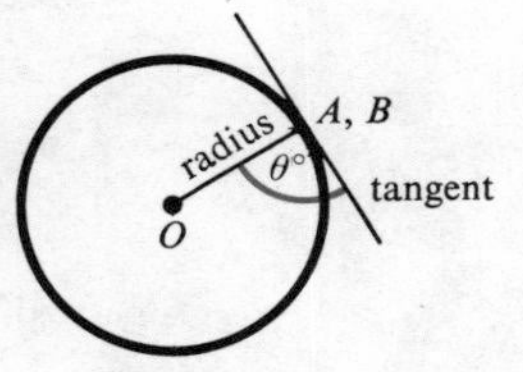

Fig. 21

Property 7

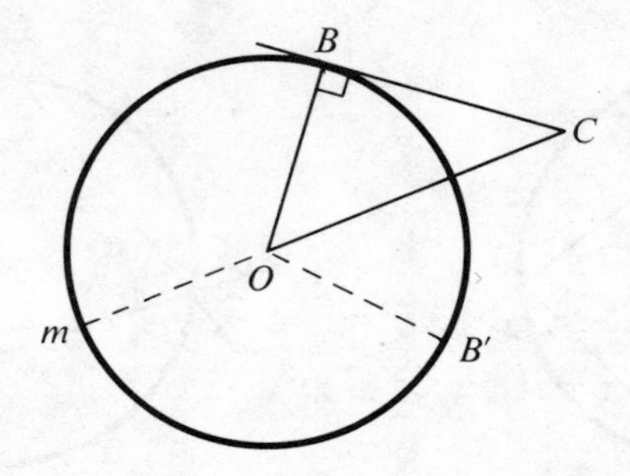

Fig. 22

Since the line m in Figure 22 is a line of symmetry of the circle, B', the image of B after a reflection in m, will be on the circumference of the circle. Where is C', the image of C, under this reflection? How many degrees is the angle $OB'C$? Is the line through $B'C'$ (i.e. through $B'C$) a tangent to the circle? What can you say about the lengths of the line segments CB and CB'? If you draw two tangents from a point to a circle what can you say about the tangents?

Summary

You should have discovered the following about circles:

(i) the angle subtended at the centre by an arc is twice the angle subtended at the circumference by that arc;

(ii) angles at the circumference subtended by the same arc are equal;

(iii) angles in a semi-circle are right angles;

(iv) the opposite angles of a cyclic quadrilateral sum to 180°;

(v) the exterior angle of a cyclic quadrilateral is equal to the opposite interior angle;

(vi) a tangent and radius meet on the circumference of the circle in a right angle;

(vii) if two tangents are drawn to a circle from a point P, to touch the circle at A and B, then $PA = PB$.

Investigation 1

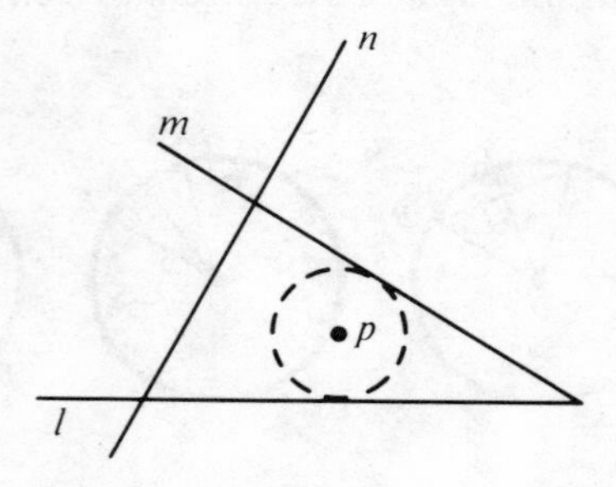

Fig. 23

Copy Figure 23 and draw several more circles which have the lines l and m as tangents. What is the locus of the centres of all the circles which have the lines l and m as tangents?

What is the locus of the centres of the circles which have (i) m and n; (ii) l and n as tangents?

Draw a triangle and construct its in-circle (that is, the circle which touches the three sides of the triangle).

Investigation 2

Find the radius in cm of the largest metal disc that can be cut from a triangular metal plate with sides 12 cm, 8 cm and 10 cm.

Investigation 3

Use any circular object (such as a 2p piece) to draw a circle. Use a pair of compasses and a ruler to find the centre of the circle.

Investigation 4

Figure 24 shows a plan of a circular art-gallery. P is a hidden camera which scans 60°. You have been asked to install cameras as an anti-theft device. How many cameras would you use and exactly where would you fix them (they must be fixed to the walls)?

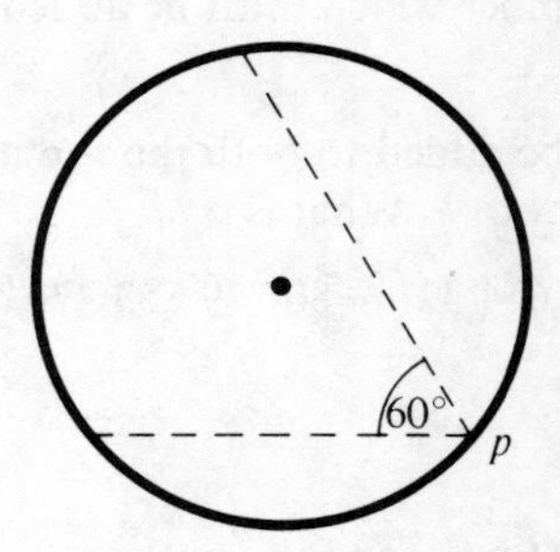

Fig. 24

How many cameras would you need if each scanned (*a*) 40°; (*b*) 90°?

Investigation 5

A circle has radius a cm. What is the maximum area of any quadrilateral inscribed in this circle?

Revision exercises

Computation 2

1 It takes me between 12 and 17 minutes to drive from Abingdon to Oxford station. I must allow at least 5 minutes for parking and buying my ticket. The train I want to catch departs at 16.33 hours. What is the latest time (given in the ordinary way, not on the '24 hour' system) that I can safely leave Abingdon?

2 Calculate in your head 999×23.

3 Calculate in your head $(64 \times 29) - (32 \times 58)$.

4 Calculate in your head $28\,640\,082 \times 50$.

5 Calculate in your head $0{\cdot}125 \times 684$.

6 A mathematics teacher wins £382·75 in a game of roulette. He generously shares it out equally amongst 25 pupils, keeping nothing for himself. How much does each pupil receive?

7 What is the smallest number which must be added to 3370 to make it a square number?

8 What is the smallest number which must be added to 1700 to make it a cube number?

9 A certain number may be added to both the top and bottom of the fraction $\frac{101}{245}$ to make it equivalent to $\frac{1}{2}$. What is it?

10 Is it true that $123\,456 \times 654\,321 = 80\,779\,853\,376$?

Computation 3

1 Evaluate the following. (Look for easy methods.)

(*a*) $15 + 66 + 92 + 34 + 8 + 85$;
(*b*) $1 + 2 + 3 + \cdots + 97 + 98 + 99$;
(*c*) $1{\cdot}21 + 4{\cdot}89 + 6{\cdot}11 + 8{\cdot}79 + 7{\cdot}84 - 2{\cdot}3 - 7{\cdot}7$.

2 Check the following additions:

(*a*) (in binary);
$$\begin{array}{r} 1110 \\ +111 \\ \hline 10101 \\ \hline \end{array}$$

(*b*) (in base three);
$$\begin{array}{r} 1010 \\ 101 \\ 211 \\ \hline 1322 \\ \hline \end{array}$$

(*c*) $99 + 38 + 88 + 14 = 230$ (in base twelve).

3 A cricketer scores 320 runs in 15 innings, and takes 24 wickets for 371 runs. What are his batting and bowling averages?

4 Boulogne is 243 km from Paris? How long will it take to drive from Boulogne to Paris at an average speed of 35 km/h?

5 Is the record of 10·1 seconds for 100 m faster or slower than the record 9·2 seconds for 100 yards? (A 'yard' is an old unit of length, 1 yard = 0·914 m.)

6 A cube has a volume of 1·679 cm³. What is the length of each of its edges?

7 Is it true that $342 - 89 = 342 + 11 - 100$? Compute by easy methods:

(*a*) $534 - 89$; (*b*) $6345 - 89$; (*c*) $652 - 96$;
(*d*) $796 - 99$; (*e*) $1462 - 989$; (*f*) $7231 - 994$.

Exercise F

1 Give the inverse of the function $x \to 2x - 1$.

2 Simplify $\frac{1}{2} - \frac{1}{3} + \frac{1}{4}$.

3 Give the equation of the straight line passing through $(3,4)$ and $(6,7)$.

4 How many square centimetres are there in 1 m²?

5 Find t if $t(t+7) = 0$.

6 Find the number base of the correct subtraction $234 - 45 = 167$.

7 Find $12\frac{1}{2}\%$ of £2.

8 Give the value of $\sin 90° + \sin 30°$.

9 Solve $x^2 + 2x + 1 = 0$.

10 Give the value of $(^-2)^3 + 2^3$.

Exercise G

1 Find y if $y = \dfrac{1}{x}$ and $x = \frac{3}{4} \div \frac{1}{2}$.

2 Give the value of $(^-3)^2 + 3^2$.

3 Sphere A has three times the diameter of sphere B. How many times greater is its volume?

4 Find p if $\frac{1}{2}(2p - 1) = 7$.

5 A is the point $(1,2)$. Find the coordinates of point C if

$$\mathbf{AB} = \begin{pmatrix} ^-2 \\ 1 \end{pmatrix} \quad \text{and} \quad \mathbf{BC} = \begin{pmatrix} 3 \\ ^-4 \end{pmatrix}.$$

6 Perform the matrix multiplication

$$\begin{pmatrix} 2 & ^-1 \\ ^-1 & 0 \end{pmatrix}\begin{pmatrix} 1 & 2 \\ 3 & 4 \end{pmatrix}.$$

7 Calculate the gradient of the line joining (2, 5) and (7, 4).

8 What can you say about the sets A and B if $A \cap B = B$?

9 Find $f(^-3)$ if $f(x) = x^3 - x - 3$.

10 What is the probability of obtaining two heads when three coins are tossed?

Exercise H

1 (*a*) What geometrical transformation is associated with the matrix $\begin{pmatrix} 0 & 1 \\ 1 & 0 \end{pmatrix}$?

(*b*) What is the multiplicative inverse of $\begin{pmatrix} 0 & 1 \\ 1 & 0 \end{pmatrix}$?

2 If $\mathbf{R}$ denotes an anticlockwise rotation of 240° about the origin, describe (i)$\mathbf{R}^2$; (ii) $\mathbf{R}^3$; (iii) $\mathbf{R}^{-1}$.

3 What point is mapped onto (0, $^-5$) by the transformation

$$\begin{pmatrix} x \\ y \end{pmatrix} \to \begin{pmatrix} 2 & 4 \\ ^-1 & 3 \end{pmatrix}\begin{pmatrix} x \\ y \end{pmatrix}?$$

4 (*a*) Write down an equation involving x which has the two possible solutions $x = 0$ or $x = 2$.
(*b*) If $p + q = 7$ and $p^2 + q^2 = 25$, find the value of pq.
(*c*) If $r + s = 7$ and $r^2 - s^2 = 28$, find the value of $r - s$.

5 Ludlow is 35 kilometres due north of Hereford, and 32 kilometres due west of Kidderminster. Find the bearing of Kidderminster from Hereford.

6 The sizes of two acute angles in a right-angled triangle are in the ratio 7 to 11. How big are they?

7 The two cylinders in Figure 1 are similar.
(*a*) If the diameter of the circular base of the smaller cylinder is 9 cm what is the diameter of the base of the larger one?
(*b*) What is the ratio of their diameters?
(*c*) What is the ratio of (i) their surface areas; (ii) their volumes?

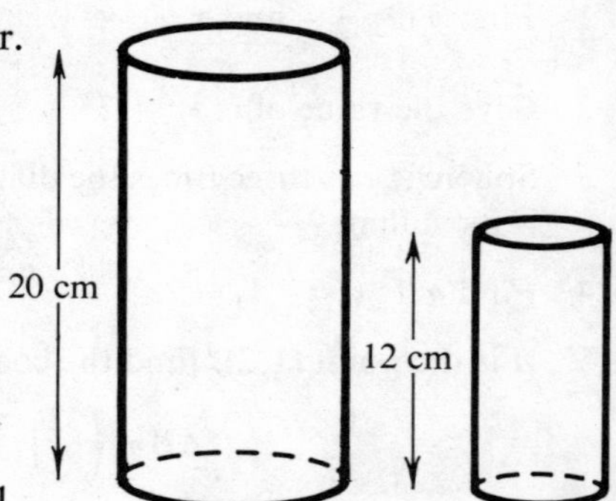

Fig. 1

8 The square $OABC$ is enlarged and then sheared to $OA'B'C'$ as shown in Figure 2. Find the matrix which represents this transformation.

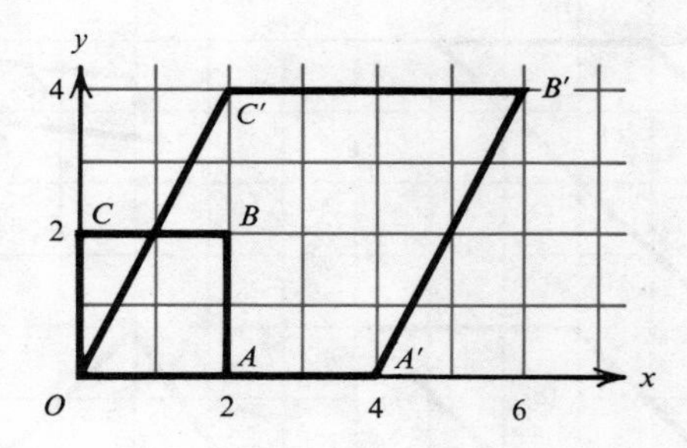

Fig. 2

Exercise I

1 Write down the inverse, $\mathbf{B}$, of the matrix

$$\mathbf{A} = \begin{pmatrix} 5 & 3 \\ 2 & 1 \end{pmatrix}.$$

Simplify $\mathbf{A}^2\mathbf{B}^2$.

2 (*a*) Find the determinants of the matrices:

(i) $\begin{pmatrix} ^-2 & 3 \\ 1 & 5 \end{pmatrix}$; (ii) $\begin{pmatrix} ^-2 & ^-3 \\ ^-4 & ^-6 \end{pmatrix}$.

(*b*) Write down, if possible, the multiplicative inverses of the above matrices.

3 Each of the following matrices represents a transformation. In each case illustrate the transformation by a sketch showing the image of the unit square and describe the transformation geometrically.

(*a*) $\begin{pmatrix} 3 & 0 \\ 0 & 1 \end{pmatrix}$; (*b*) $\begin{pmatrix} 0 & 0 \\ 0 & 1 \end{pmatrix}$; (*c*) $\begin{pmatrix} ^-1 & 0 \\ 0 & 1 \end{pmatrix}$;

(*d*) $\begin{pmatrix} 0 & ^-1 \\ ^-1 & 0 \end{pmatrix}$; (*e*) $\begin{pmatrix} 1 & 0 \\ 0 & 2 \end{pmatrix}$; (*f*) $\begin{pmatrix} 1 & 0 \\ 0 & 0 \end{pmatrix}$.

4 Calculate the areas of the shapes in Figure 3.

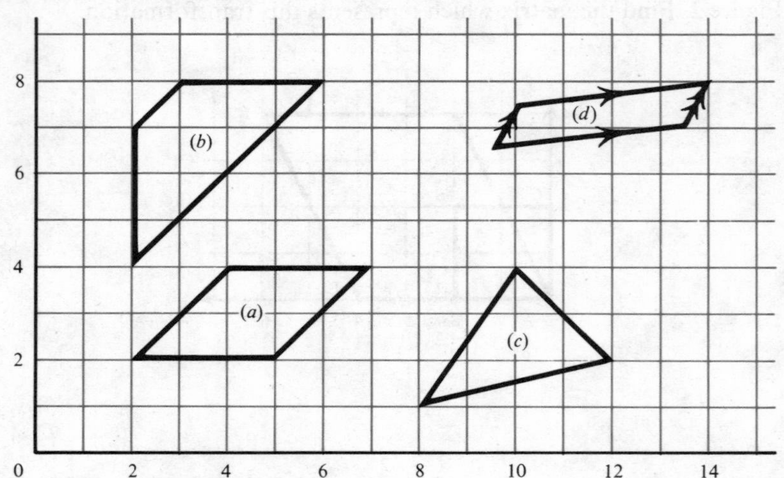

Fig. 3

5 Calculate the volume and surface area of a cube with an edge of 7 cm.

6 Draw a right-angled triangle ABC in which $\tan A = \frac{3}{4}$. Without using tables find $\sin A$ and $\cos A$.

7 A ship sets out from Port Barnacle and sails along a straight course. After a time its position is recorded as 60 km north and 80 km east of the port. How far has it then sailed, and what angle does its course make with a northerly direction?

8 A certain fly is sitting at a point A on the lower rim of an upright cylinder. It climbs on a path of constant slope round the outside of the cylinder, and reaches a point B on the upper rim vertically above A after one rotation. If the circumference of the cylinder is 36 cm and its height is 15 cm, find the length of the fly's journey from A to B.

Exercise J

1 Multiply:

(*a*) $(x+2)(x+7)$; (*b*) $(x-2)(x-7)$;
(*c*) $(x+2)(x-7)$; (*d*) $(x-2)(x+7)$;
(*e*) $(2x+1)(x-4)$; (*f*) $(2x+1)(2x-1)$;
(*g*) $\left(\frac{x}{2}+1\right)(x+1)$; (*h*) $\left(\frac{x}{2}-1\right)(2x+1)$.

2 Solve the following equations:

(*a*) $x^2+2x+2=1$; (*b*) $x^2+5x+8=2$;
(*c*) $x^2+x-6=0$; (*d*) $x^2-x={}^{-}6$;
(*e*) $x^2-9=0$; (*f*) $x^2+9=6x$;
(*g*) $x^2+4={}^{-}4x$; (*h*) $x^2=1$.

3 Solve the following equations:

(*a*) $x^2 - 9x = 0$; (*b*) $2x^2 - 5x + 3 = 0$;
(*c*) $3x^2 + 8x + 5 = 0$; (*d*) $4x^2 - 64 = 0$;
(*e*) $2x^2 + 5x = 0$; (*f*) $7x^2 + 6x - 1 = 0$;
(*g*) $9x^2 - 81 = 0$; (*h*) $2x^2 + 6x + 18 = 0$.

4 A transformation maps the unit square $OIAJ$ onto the rhombus with vertices at $O(0,0)$, $I'(2,4)$, $A'(^-2,2)$ and $J'(^-4,^-2)$. Write down the matrix which represents this transformation and hence find the area of $OI'A'J'$.

5 If $\mathbf{A} = \begin{pmatrix} 3 & ^-1 \\ 2 & 1 \end{pmatrix}$, calculate $\mathbf{A}^{-1}$.

6 A transformation **T** is defined by

$$\mathbf{T}: \begin{pmatrix} x \\ y \end{pmatrix} \to \begin{pmatrix} 2 & 1 \\ 3 & 2 \end{pmatrix} \begin{pmatrix} x \\ y \end{pmatrix} + \begin{pmatrix} ^-1 \\ 3 \end{pmatrix}.$$

Find the inverse transformation and hence find the coordinates of the point mapped by **T** onto (2, 5).

7 If

$$\mathbf{A} = \begin{pmatrix} 1 & 2 & 1 \\ 3 & 0 & ^-2 \\ 2 & 1 & 0 \end{pmatrix} \quad \text{and} \quad \mathbf{B} = \begin{pmatrix} ^-2 & ^-1 & 4 \\ 4 & 2 & ^-5 \\ ^-3 & ^-3 & 6 \end{pmatrix}$$

find **AB** and hence give the multiplicative inverses of **A** and **B**.

8 Show that the mapping

$$\begin{pmatrix} x \\ y \\ 1 \end{pmatrix} \to \begin{pmatrix} 3 & ^-2 & 4 \\ 1 & 0 & 6 \\ 0 & 0 & 1 \end{pmatrix} \begin{pmatrix} x \\ y \\ 1 \end{pmatrix}$$

has the same effect as the mapping

$$\begin{pmatrix} x \\ y \end{pmatrix} \to \begin{pmatrix} 3 & ^-2 \\ 1 & 0 \end{pmatrix} \begin{pmatrix} x \\ y \end{pmatrix} + \begin{pmatrix} 4 \\ 6 \end{pmatrix}.$$

9 Simultaneous equations and inequalities

1 Equations with two variables

Consider the equation

$$2x + y = 12.$$

If $x = 2$, what is the value of y?

One solution of the equation $2x + y = 12$ is $x = 2$, $y = 8$. Write down some other solutions of this equation. How many solutions are there?

Now write down some solutions of the equation

$$x - y = 3.$$

How many solutions does this equation have?

Can you find a solution of the equations

$$\begin{cases} 2x + y = 12 \\ x - y = 3 \end{cases}$$

taken together, that is, can you find values of x and y which satisfy both these equations simultaneously?

Is there more than one such solution?

2 Solving simultaneous equations and inequalities

(*a*) Using graphs

(*a*) We already know one method of solving simultaneous equations such as

$$\begin{cases} 2x + y = 12 \\ x - y = 3. \end{cases}$$

Figure 1 should remind you of this method.

The graphs of $2x + y = 12$ and $x - y = 3$ intersect in the point $(5, 2)$. Therefore $x = 5$, $y = 2$ satisfies both the equation $2x + y = 12$ and the equation $x - y = 3$.

Do the graphs intersect in any other point?

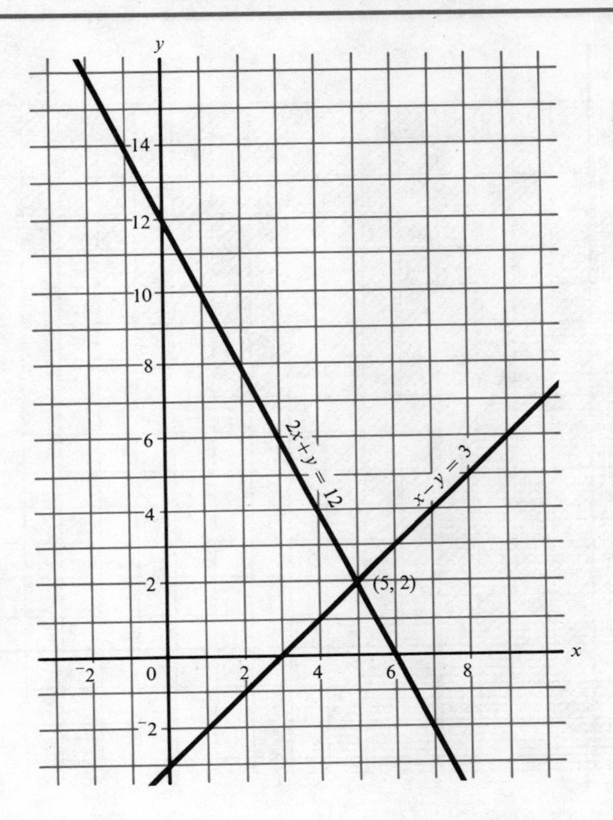

Fig. 1

(*b*) Figure 2 shows the graphs of the simultaneous inequalities

$$\begin{cases} 2x + y > 12 \\ x - y > 3. \end{cases}$$

The coordinates of any point in the unshaded region satisfy both inequalities. Is there a solution such that (i) $x = 10$, (ii) $x = {}^-2$, (iii) $x = 5$?
Is there a solution such that (i) $y = 10$, (ii) $y = {}^-4$, (iii) $y = 100$?

Figure 2 shows that *every* solution is such that $x > 5$. There is no similar restriction on the value of y; solutions can always be found for any desired value of y.

(*c*) Sketch the graphs of the simultaneous inequalities

$$\begin{cases} 2x + y < 12 \\ x - y > 3. \end{cases}$$

Is it possible to find a solution for (i) any value of x, (ii) any value of y?

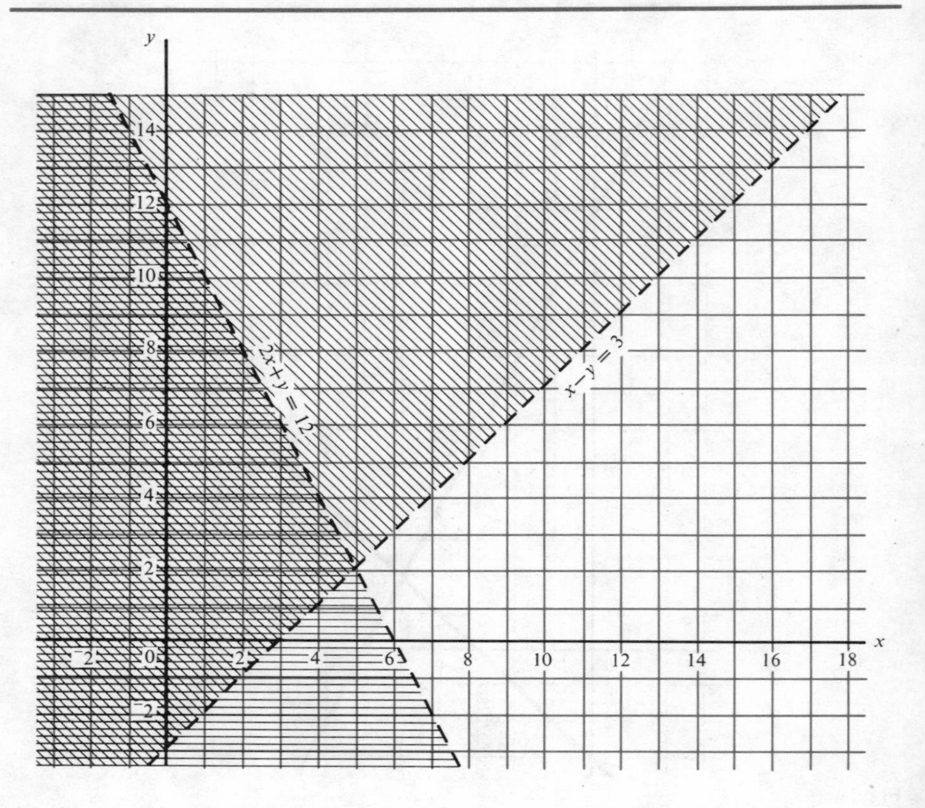

Fig. 2

Exercise A

1 Draw the graphs of $y = x + 5$ and $x + 2y = 1$ from $x = {}^{-}5$ to $x = 2$ on the same axes.

(*a*) Give the solution of the simultaneous equations

$$\begin{cases} y = x + 5 \\ x + 2y = 1. \end{cases}$$

(*b*) By shading out the unwanted regions, show the solution set of the simultaneous inequalities

$$\begin{cases} y > x + 5 \\ x + 2y < 1. \end{cases}$$

2 Use a graphical method to solve:

(*a*) $\begin{cases} y - 4x = 1 \\ 2x + y = 4; \end{cases}$ (*b*) $\begin{cases} 2x - y = 4 \\ x + 2y + 3 = 0. \end{cases}$

3 Sketch the graphs of the simultaneous inequalities:

(a) $\begin{cases} y - x > 0 \\ y + x > {}^{-}2; \end{cases}$ (b) $\begin{cases} y - x \leqslant 0 \\ {}^{-}2y - x \leqslant 3. \end{cases}$

4 Sketch the graphs of the inequalities $\begin{cases} y < x + 1 \\ y > 1 - 4x. \end{cases}$

Is it possible to find a solution for (i) any value of x, (ii) any value of y?

5 Draw the graphs of the inequalities $\begin{cases} x - y \geqslant 2 \\ {}^{-}2x + y \geqslant {}^{-}2. \end{cases}$

For what values of x is it possible to find a solution?
For what values of y is it possible to find a solution?

(*b*) Using substitution

(*a*) Look again at the simultaneous equations

$$\begin{cases} 2x + y = 12 & (1) \\ x - y = 3. & (2) \end{cases}$$

If we rearrange equation (2) so that x is the subject of the equation, we have

$$x = y + 3. \qquad (3)$$

Since x and $y + 3$ are equal, we can substitute $y + 3$ for x in equation (1) and obtain

$$\begin{aligned} 2(y + 3) + y &= 12 \\ 2y + 6 + y &= 12 \\ 3y + 6 &= 12 \\ 3y &= 6 \\ y &= 2. \end{aligned}$$

Substituting 2 for y in equation (3), we have

$$x = 5.$$

Therefore the solution is $x = 5$, $y = 2$.
(*b*) Solve the simultaneous equations

$$\begin{cases} y - 2x = 7 \\ x + 2y = {}^{-}1 \end{cases}$$

by making y the subject of the first equation and then substituting for y in the second.

Check that your answer satisfies the equations.

Would it be equally sensible to make y the subject of the second equation and then substitute for y in the first?

Exercise B

Use the method of Section 2(*b*) to solve the simultaneous equations in Questions 1–6.

1 $\begin{cases} x+2y=10 \\ x-y=4. \end{cases}$
2 $\begin{cases} x-y=2 \\ y={}^-x+6. \end{cases}$
3 $\begin{cases} 2x+3y=7 \\ y={}^-3x. \end{cases}$

4 $\begin{cases} 2x+y=1 \\ 2x+5y=9. \end{cases}$
5 $\begin{cases} x+y=10 \\ y+11x=35. \end{cases}$
6 $\begin{cases} 2x+3y=1 \\ 3x-y=1. \end{cases}$

7 For each of the following simultaneous equations, make b the subject of one of the equations and then substitute for b in the other to obtain an equation connecting a and c.

(*a*) $\begin{cases} a=2+b \\ c=3b; \end{cases}$
(*b*) $\begin{cases} 2a=1+b \\ 3c=2b-4; \end{cases}$
(*c*) $\begin{cases} a=b^2 \\ c=1-b. \end{cases}$

8 By eliminating d from the simultaneous equations

$$\begin{cases} 4=\dfrac{d}{t} \\ d=3s, \end{cases}$$

obtain an equation connecting t and s.

9 By eliminating k from the simultaneous equations

$$\begin{cases} D=5k \\ P=14k, \end{cases}$$

obtain an equation connecting D and P.

(*c*) More graphs

(*a*) Figure 3 shows the graphs of $y=x^2$ and $y=2x+3$ for values of x from ${}^-4$ to 4.

What are the coordinates of the points where the graphs intersect?

Does $x={}^-1$, $y=1$ satisfy both the equation $y=x^2$ and the equation $y=2x+3$?

One solution of the simultaneous equations

$$\begin{cases} y=x^2 & (1) \\ y=2x+3 & (2) \end{cases}$$

is $x={}^-1$, $y=1$. There is another solution. What is it?

If we substitute $2x+3$ for y in (1), we obtain

$$2x+3=x^2.$$

What are the two solutions of this equation in x? Check that your answer is correct.

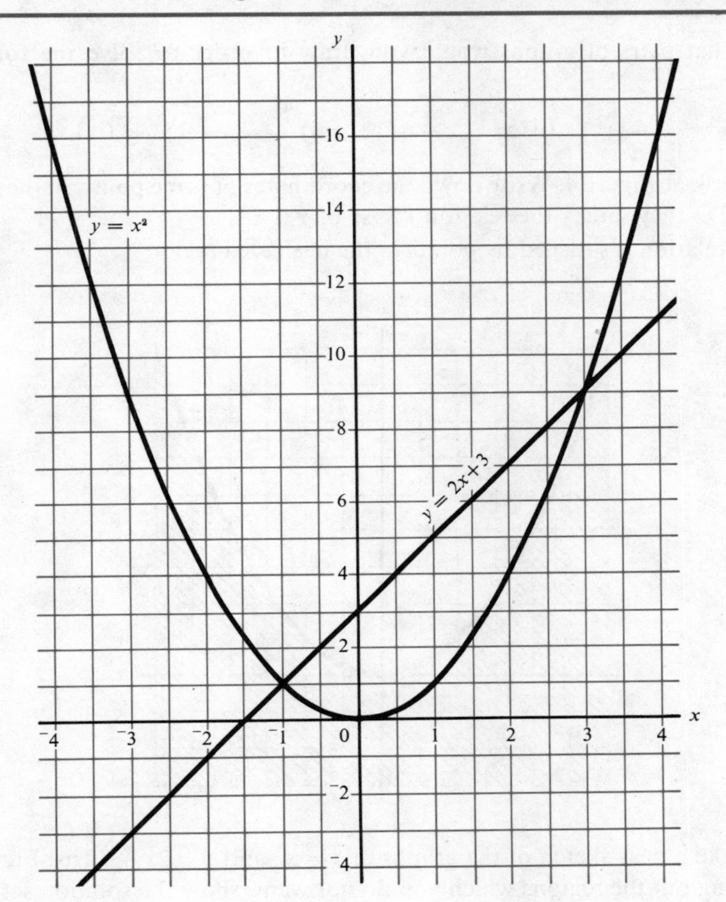

Fig. 3

(*b*) We can solve the equation

$$x^2 = 2 - 2x$$

by first drawing the graphs of the simultaneous equations

$$\begin{cases} y = x^2 \\ y = 2 - 2x. \end{cases}$$

Sketch these graphs to find the approximate position of the point(s) of intersection.

Now carefully draw a portion of these graphs which will include these points. (You should have found that it will be sufficient to draw the graphs for values of x from $^-4$ to 2.)

Use your graphs to find as accurately as you can the x-coordinates of the points of intersection.

What can you now say about the values of x which satisfy the equation

$$x^2 = 2 - 2x?$$

(*c*) What pairs of graphs would you draw in order to solve the following equations:

(i) $x^2 = x - \frac{1}{4}$; (ii) $x^3 = x + 6$; (iii) $\frac{1}{x} = 2x - 1$; (iv) $x^2 - x = 2$?

(*d*) Look at Figure 4. Write down the coordinates of some points in the shaded region. Do they satisfy the relation $y < x^2$ or $y > x^2$?

What relation is satisfied by points in the unshaded region?

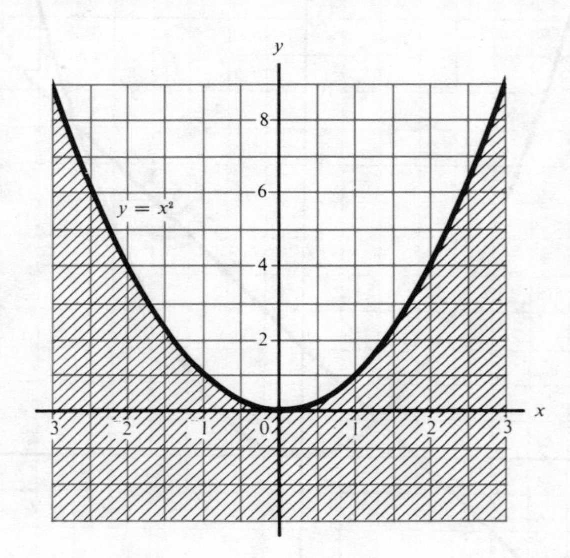

Fig. 4

(*e*) Make a neat sketch of the graphs of $y = x^2$ and $y = 2x + 3$ (see Figure 3). By shading out the regions which you do not want, show the solution set of the simultaneous inequalities

$$\begin{cases} y > x^2 \\ y < 2x + 3. \end{cases}$$

Exercise C

In Questions 1–4, remember to draw sketch graphs to find the approximate position of the point(s) of intersection.

1 Use a graphical method to solve the following simultaneous equations to 2 S.F.

(*a*) $\begin{cases} y = x^2 \\ y = x; \end{cases}$ (*b*) $\begin{cases} y = x^2 \\ y = \frac{3}{4} - x; \end{cases}$ (*c*) $\begin{cases} y = x^3 \\ y = 4x; \end{cases}$ (*d*) $\begin{cases} y = \frac{1}{2}x + 2 \\ y = x^2. \end{cases}$

2 By drawing suitable pairs of graphs, solve the following equations to 2 S.F.

(*a*) $x^2 = x - \frac{1}{4}$; (*b*) $x^2 - x = 3$;

(*c*) $x^3 = x + 6$; (*d*) $x^3 = 1 - x$.

3 Figure 5 shows the graph of $y = \frac{1}{x}$. Use a graphical method to solve the following equations to 2 S.F.

(*a*) $\frac{1}{x} = 3 - x$; (*b*) $\frac{1}{x} = 2x - 1$.

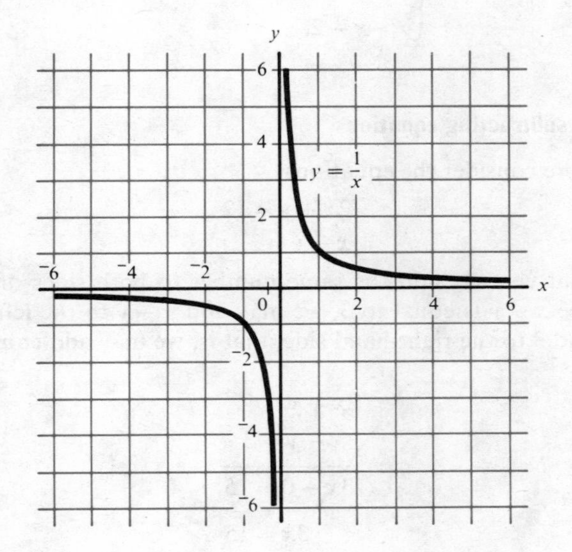

Fig. 5

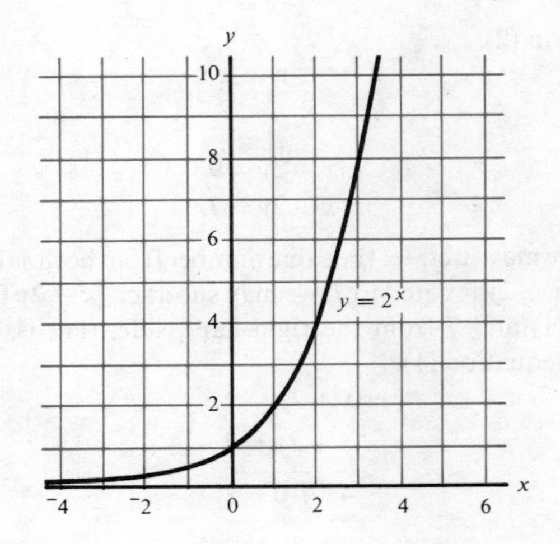

Fig. 6

4 Figure 6 shows the graph of $y = 2^x$. Use a graphical method to solve the following equations to 2 S.F.

(*a*) $2^x = \frac{1}{x}$; (*b*) $2^x = x^2$.

5 Sketch the graphs of the simultaneous inequalities:

(a) $\begin{cases} y < x^2 \\ y > x; \end{cases}$ (b) $\begin{cases} y \geqslant x^2 \\ y \leqslant 4 - 2x; \end{cases}$

(c) $\begin{cases} y < x^3 \\ y > 4x; \end{cases}$ (d) $\begin{cases} y < \dfrac{1}{x} \\ y > x - 2. \end{cases}$

(*d*) Adding and subtracting equations

(*a*) Once more consider the equations

$$\begin{cases} 2x + y = 12 & (1) \\ x - y = 3. & (2) \end{cases}$$

We know that we may add the same number to both sides of an equation. Therefore, since $x - y$ is equal to 3, we may add $x - y$ to the left-hand side of equation (1) and 3 to the right-hand side, that is, we may add equations (1) and (2).

$$2x + y = 12 \quad (1)$$
$$x - y = 3 \quad (2)$$

Adding,

$$3x + 0 = 15$$
$$3x = 15$$
$$x = 5.$$

Substituting for x in (2),

$$y = 2.$$

(*b*)

$$\begin{cases} 3x + 2y = 9 & (1) \\ {}^{-}x + 2y = 7. & (2) \end{cases}$$

We know that we may subtract the same number from both sides of an equation. Therefore, since ${}^{-}x + 2y$ is equal to 7, we may subtract ${}^{-}x + 2y$ from the left-hand side of equation (1) and 7 from the right-hand side, that is, we may subtract equation (2) from equation (1).

$$3x + 2y = 9 \quad (1)$$
$${}^{-}x + 2y = 7 \quad (2)$$

Subtracting,

$$4x + 0 = 2$$
$$4x = 2$$
$$x = \tfrac{1}{2}.$$

Substituting for x in (2),

$$2y = 7\tfrac{1}{2}$$
$$y = 3\tfrac{3}{4}.$$

Therefore the solution is $x = \frac{1}{2}$, $y = 3\frac{3}{4}$.

(*c*) In (*a*) the coefficients of y in the two equations are numerically equal but opposite in sign, so adding the equations eliminates y and gives a linear equation in x which we are able to solve.

In (*b*) the coefficients of y in the two equations are numerically equal and equal in sign. This time, subtracting the equations eliminates y and gives a linear equation in x.

The following example shows how we can always arrange for the coefficient of one of the variables in two simultaneous equations to be numerically equal.

$$\begin{cases} 2x + 5y = 3 & (1) \\ 3x - 2y = 1. & (2) \end{cases}$$

We can multiply both sides of equation (1) by 3 to obtain $6x + 15y = 9$ and both sides of equation (2) by 2 to obtain $6x - 4y = 2$.

$$6x + 15y = 9$$

$$6x - 4y = 2$$

Subtracting,

$$0 + 19y = 7$$

$$19y = 7$$

$$y = \tfrac{7}{19}.$$

Substituting for y in (2),

$$3x = 1 + \tfrac{14}{19}$$

$$3x = \tfrac{33}{19}$$

$$x = \tfrac{11}{19}.$$

Therefore the solution is $x = \tfrac{11}{19}$, $y = \tfrac{7}{19}$.

Now solve the equations

$$\begin{cases} 2x + 5y = 3 & (1) \\ 3x - 2y = 1 & (2) \end{cases}$$

by multiplying (1) by 2 and (2) by 5 and then adding.

Do you obtain the same answer as before?

(*d*) We can eliminate x from the equations

$$\begin{cases} 3x - y = 7 & (1) \\ {}^{-}x + 2y = 5 & (2) \end{cases}$$

by multiplying (2) by 3 and then adding.

How would you eliminate x from each of the following pairs of equations:

(i) $\begin{cases} x + y = 8 \\ x - y = 3; \end{cases}$ (ii) $\begin{cases} 6x - 3y = {}^{-}1 \\ 2x + 5y = 10; \end{cases}$ (iii) $\begin{cases} 2x + y = 8 \\ 5x + 2y = 3? \end{cases}$

How would you eliminate y from each of the above pairs of equations?

Exercise D

Use the method of Section 2(d) to solve the simultaneous equations in Questions 1–8.

1 $\begin{cases} x+2y=7 \\ 3x-2y={}^-3. \end{cases}$ 2 $\begin{cases} 4x-3y=1 \\ x-2y=4. \end{cases}$ 3 $\begin{cases} 4x+3y=9 \\ 2x+5y=15. \end{cases}$

4 $\begin{cases} 5x+3y=1 \\ 2x+3y={}^-5. \end{cases}$ 5 $\begin{cases} 3x-2y=4 \\ 2x+3y={}^-6. \end{cases}$ 6 $\begin{cases} 3x=2y+1 \\ 5x=3y+3. \end{cases}$

7 $\begin{cases} 5x-3y=1 \\ 3x=y+5. \end{cases}$ 8 $\begin{cases} ax+by=c \\ x-y=1. \end{cases}$

(e) Using matrices

The simultaneous equations

$$\begin{cases} 2x+y=12 & (1) \\ x-y=3 & (2) \end{cases}$$

can be written in the form

$$\begin{pmatrix} 2 & 1 \\ 1 & {}^-1 \end{pmatrix}\begin{pmatrix} x \\ y \end{pmatrix} = \begin{pmatrix} 12 \\ 3 \end{pmatrix}.$$

The problem of solving the equations is therefore equivalent to that of looking for the point (x, y) of the domain which is mapped onto the point $(12, 3)$ of the range by the one–one mapping

$$\begin{pmatrix} x \\ y \end{pmatrix} \to \begin{pmatrix} 2 & 1 \\ 1 & {}^-1 \end{pmatrix}\begin{pmatrix} x \\ y \end{pmatrix}.$$

Hence the inverse of this mapping applied to the point $(12, 3)$ will give the required solution.

Check that the inverse mapping is

$$\begin{pmatrix} x \\ y \end{pmatrix} \to \begin{pmatrix} \frac{1}{3} & \frac{1}{3} \\ \frac{1}{3} & {}^-\frac{2}{3} \end{pmatrix}\begin{pmatrix} x \\ y \end{pmatrix}.$$

(If you have difficulty, look back at Chapter 7, Section 2.)

So

$$\begin{pmatrix} x \\ y \end{pmatrix} = \begin{pmatrix} \frac{1}{3} & \frac{1}{3} \\ \frac{1}{3} & {}^-\frac{2}{3} \end{pmatrix}\begin{pmatrix} 12 \\ 3 \end{pmatrix} = \begin{pmatrix} 5 \\ 2 \end{pmatrix},$$

that is,

$$x=5, \qquad y=2.$$

The inverse mapping could also be written in the form

$$\begin{pmatrix} x \\ y \end{pmatrix} \to {}^-\tfrac{1}{3}\begin{pmatrix} {}^-1 & {}^-1 \\ {}^-1 & 2 \end{pmatrix}\begin{pmatrix} x \\ y \end{pmatrix}$$

and you will find that the arithmetic is often simpler if division by the value of the determinant is delayed, thus:

$$\begin{pmatrix} x \\ y \end{pmatrix} = {}^-\tfrac{1}{3}\begin{pmatrix} ^-1 & ^-1 \\ ^-1 & 2 \end{pmatrix}\begin{pmatrix} 12 \\ 3 \end{pmatrix}$$

$$= {}^-\tfrac{1}{3}\begin{pmatrix} ^-15 \\ ^-6 \end{pmatrix} = \begin{pmatrix} 5 \\ 2 \end{pmatrix}.$$

Exercise E

Use the method of Section 2(*e*) to solve the simultaneous equations in Questions 1–10.

1 $\begin{cases} 2x+y=1 \\ 5x+3y=2. \end{cases}$ 2 $\begin{cases} x+2y=10 \\ 2x+3y=13. \end{cases}$ 3 $\begin{cases} x+y=2 \\ 5x+7y=4. \end{cases}$ 4 $\begin{cases} 3x+2y=1 \\ 7x+3y={}^-1. \end{cases}$

5 $\begin{cases} x+4y={}^-17 \\ 3x-2y=19. \end{cases}$ 6 $\begin{cases} 5x-2y=4 \\ 6x-y=9. \end{cases}$ 7 $\begin{cases} 3x-5y=7 \\ 2x+4y=1. \end{cases}$ 8 $\begin{cases} 2x-4y=1 \\ x+3y=2. \end{cases}$

9 $\begin{cases} 3x=2y \\ 4x=5y+7. \end{cases}$ 10 $\begin{cases} 4x+3y=4 \\ 2x=5y+15. \end{cases}$

3 Solution sets

(*a*) Several times in this chapter we have solved the simultaneous equations

$$\begin{cases} 2x+y=12 \\ x-y=3. \end{cases}$$

The solution set, which consists of a single pair of elements, is $\{(5,2)\}$.

(*b*) The graphs of the simultaneous equations

$$\begin{cases} x+2y=3 \\ 2x+4y=10 \end{cases}$$

are shown in Figure 7. What do the graphs tell you about the number of pairs of elements which belong to the solution set?

What happens when you apply each of the methods of Sections 2(*b*), 2(*d*) and 2(*e*) to the equations

$$\begin{cases} x+2y=3 \\ 2x+4y=10? \end{cases}$$

Does each method lead you to the conclusion that the solution set is $\varnothing$?

(*c*) Investigate what happens when you apply each of the methods of Sections 2(*a*), 2(*b*), 2(*d*) and 2(*e*) to the simultaneous equations

$$\begin{cases} x+2y=3 \\ 2x+4y=6. \end{cases}$$

Do these equations have a solution?

What can you say about the number of pairs of elements in the solution set?

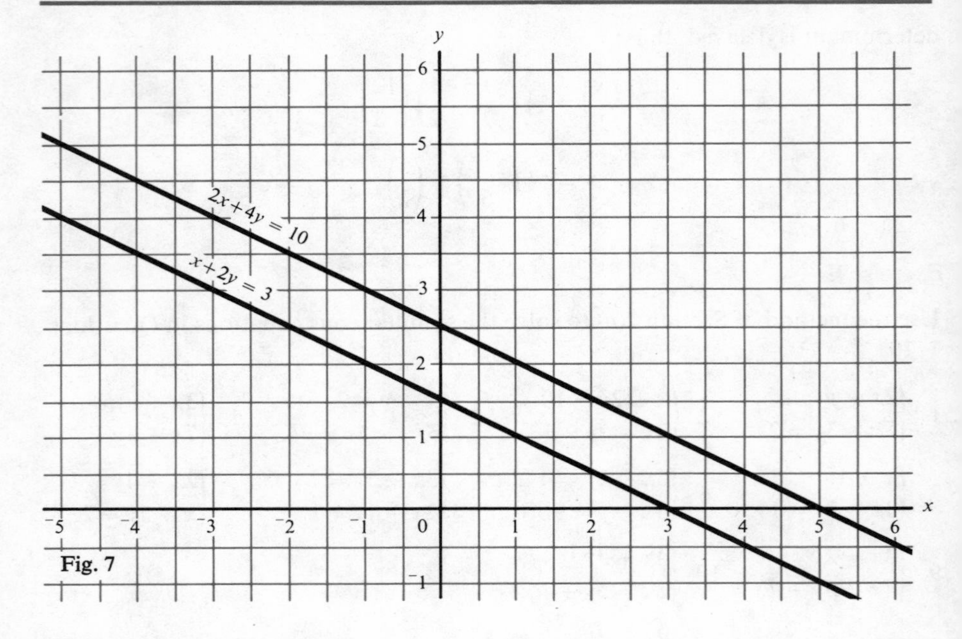

Fig. 7

Miscellaneous exercise F

1 Solve the simultaneous equations:

(a) $\begin{cases} x+y=14 \\ y-x=0; \end{cases}$ (b) $\begin{cases} 12x-10y=7 \\ 12x-6y=1. \end{cases}$

2 Sketch the graphs of the simultaneous inequalities:

(a) $\begin{cases} 2y \geqslant 1-x \\ y \leqslant \frac{1}{3}x+4; \end{cases}$ (b) $\begin{cases} y > \frac{1}{2}x \\ 3y > x-2. \end{cases}$

3 The values of x and y are connected by the equation $y = mx + c$. When $x = 3$, $y = 4$ and when $x = {}^-7$, $y = {}^-1$. Find m and c. What is the value of y when $x = {}^-1$?

4 The straight line $y = mx + c$ passes through the points $(4, 5)$ and $(1, 14)$. Find m and c.

5 The equation $x^2 + px + q = 0$ is satisfied by $x = {}^-1$ and $x = 2$. Find the values of p and q.

6 Discuss whether the following equations have a solution set with (i) no elements, (ii) one pair of elements, (iii) an infinite number of pairs of elements.

(a) $\begin{cases} x+y=1 \\ 2x+2y=2; \end{cases}$ (b) $\begin{cases} x+y=1 \\ 2x+2y=3; \end{cases}$ (c) $\begin{cases} x+y=1 \\ x-y=0; \end{cases}$ (d) $\begin{cases} 4x+6y=5 \\ 6x+9y=7\frac{1}{2}. \end{cases}$

7 Sketch the graphs of the simultaneous inequalities:

(a) $\begin{cases} x+2y>4 \\ x+2y<8; \end{cases}$ (b) $\begin{cases} x+2y>4 \\ x+2y>8; \end{cases}$ (c) $\begin{cases} x+2y<4 \\ x+2y<8; \end{cases}$ (d) $\begin{cases} x+2y<4 \\ x+2y>8. \end{cases}$

10 More mensuration

1 A reminder

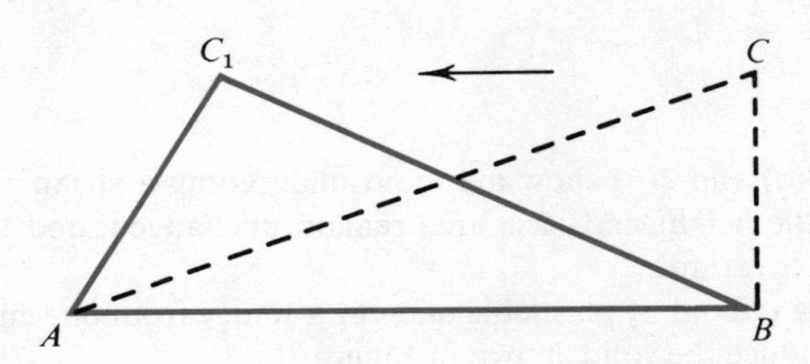

Fig. 1

(*a*) Figure 1 illustrates a shear in 2 dimensions under which the triangle ABC has been transformed into the triangle ABC_1.

Which line segment is invariant under this transformation?

What can you say about the areas of the two triangles?

(*b*) In two dimensions, a shear is a transformation which has a line of invariant points, and which preserves area.

2 Shearing in three dimensions

(*a*) Figure 2(*a*) shows a pack of playing cards.
Figure 2(*b*) shows the same pack after it has been sheared parallel to AB.

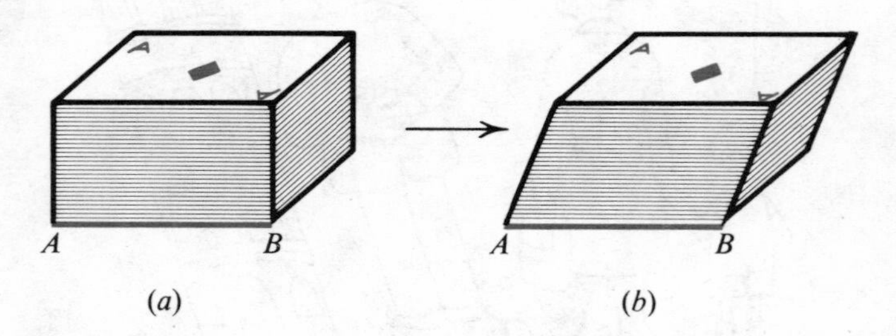

Fig. 2

Has the height of the pack changed?
Has the base area of the pack changed?
If the volume of the pack in Figure 2(*a*) is 50 cm³, what is the volume of the pack in Figure 2(*b*)?

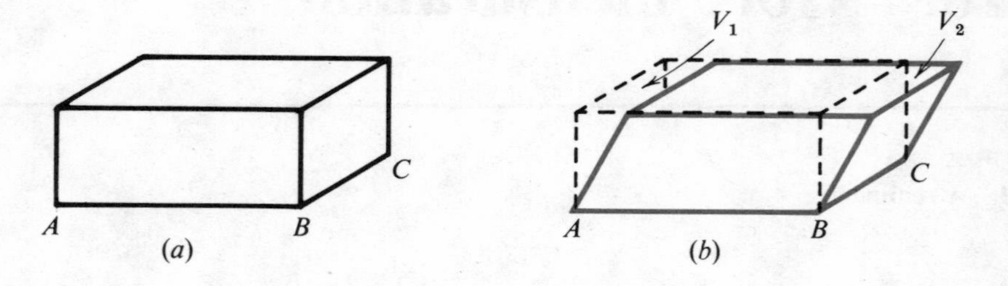

Fig. 3

(*b*) Figures 3(*a*) and 3(*b*) show a cuboid undergoing a shear parallel to AB. Notice that the height and base area remain unchanged, and that all points on the base are invariant.

If you make a cuboid of plasticine and cut a wedge from one end, you should be able to reproduce the solid shown in Figure 3(*b*).

What can you say about the wedge-shaped volumes V_1 and V_2 in Figure 3(*b*)?

Can you explain why the volume of the solid in Figure 3(*a*) is the same as that of the solid in Figure 3(*b*)?

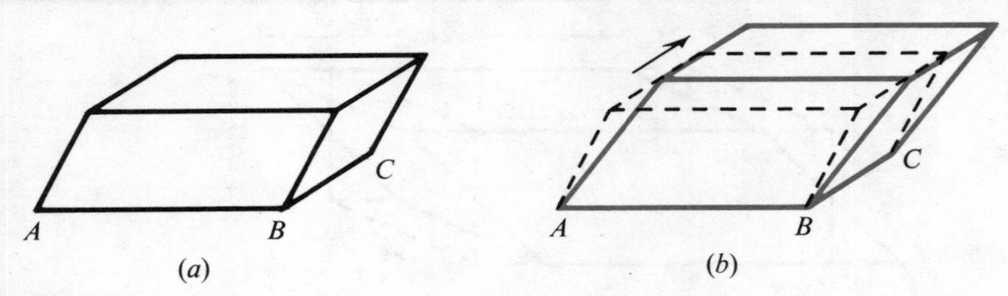

Fig. 4

(*c*) Figures 4(*a*) and 4(*b*) show the solid in Figure 3(*b*) undergoing a further shear parallel to *BC*.

You might like to try reproducing this with your plasticine.

Notice that all points of the base of the solid have remained invariant under both shears.

What other properties of the solid in Figure 3(*a*) have remained invariant?

If the volume of the cuboid in Figure 3(*a*) is 50 cm³, what is the volume of the solid in Figure 4(*b*)?

Summary

A shear in three dimensions is a transformation which (*a*) leaves points in one plane invariant and (*b*) preserves volume.

Exercise A

1 By considering a shear of a simple solid find the volumes of each of the following:

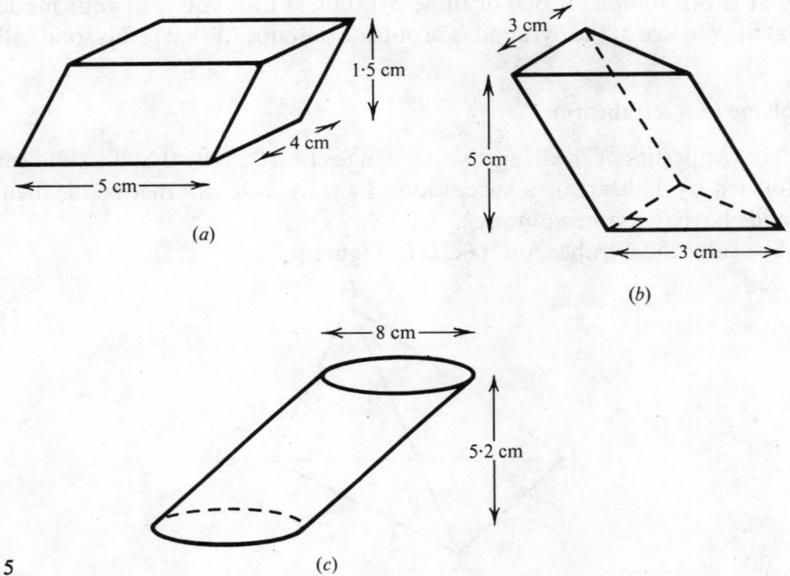

Fig. 5

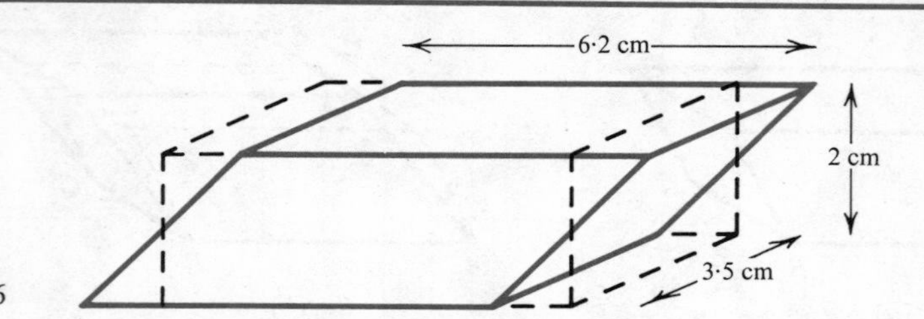

Fig. 6

2 Figure 6 shows a cuboid which has undergone a shear. Which plane is invariant under this transformation? What is the volume of the solid formed?

3 Do you think that surface area is invariant under shearing in three dimensions? Give an example to support your answer.

4 Figure 7 shows a cube with side length 6 cm, and illustrates how the cube can be divided into six congruent square-based pyramids. Figure 8 shows the net for such a pyramid.

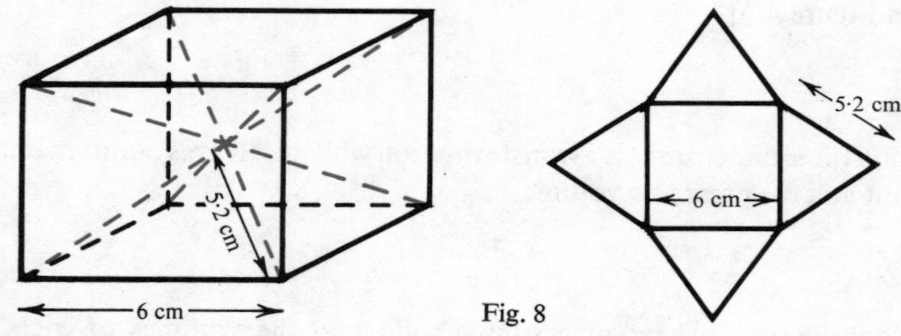

Fig. 7

Fig. 8

Make six pyramids and tape them so that they fold together to form the cube. What is the volume of one of these pyramids? Can you fold your model so that the vertices of the pyramids are outside the cube? What is this solid called?

3 Volume of a tetrahedron

(*a*) The properties of shearing discussed in Section 2 imply that if a solid can be transformed by a shear, or a succession of shears, into another solid, then the two solids have the same volume.

(*b*) Consider the tetrahedron *ABCD* in Figure 9.

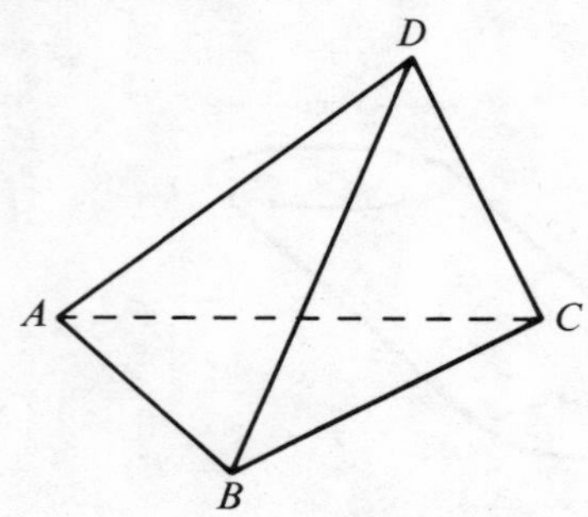

Fig. 9

We can now transform this as follows:

(i) Keeping the face ADC invariant (see Figure 10) shear so that B moves in the plane ABC to B_1, where $B_1AC = 90°$.

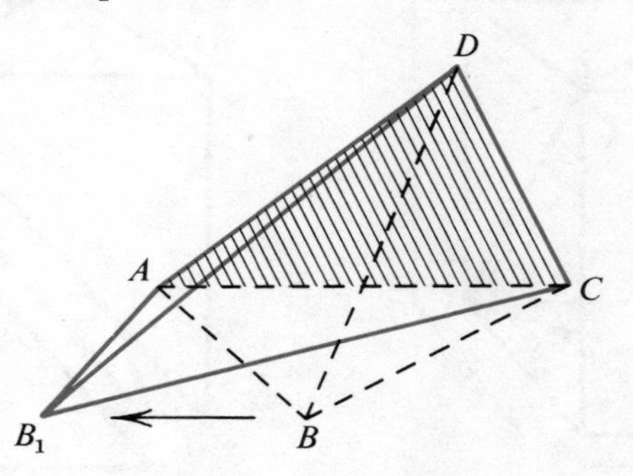

Fig. 10

Has the base area of the tetrahedron changed?

Has the height of the tetrahedron changed?

Do you agree that the volume of AB_1CD is the same as the volume of $ABCD$?

(ii) Keeping the face AB_1C invariant (see Figure 11) shear D to D_1, so that AD_1 is perpendicular to the plane AB_1C.

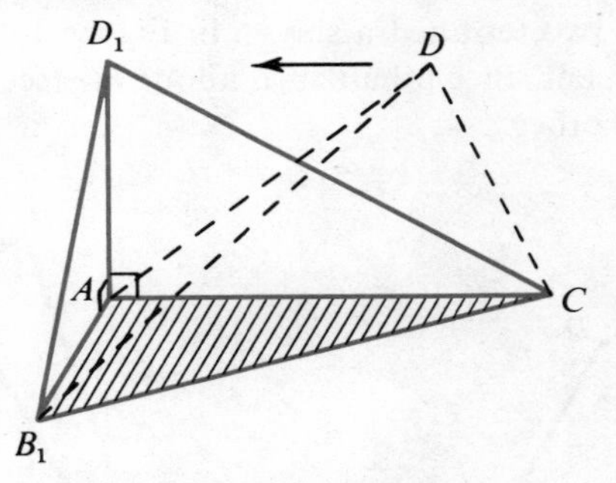

Fig. 11

Has the base area of the tetrahedron changed?

Has the height of the tetrahedron changed?

Do you agree that the volume of AB_1CD_1 is the same as the volume of $ABCD$?

(*c*) We have seen in Section (*b*) that we can transform any tetrahedron into a tetrahedron of the type shown in Figure 12(*b*), without changing its height, base area or volume.

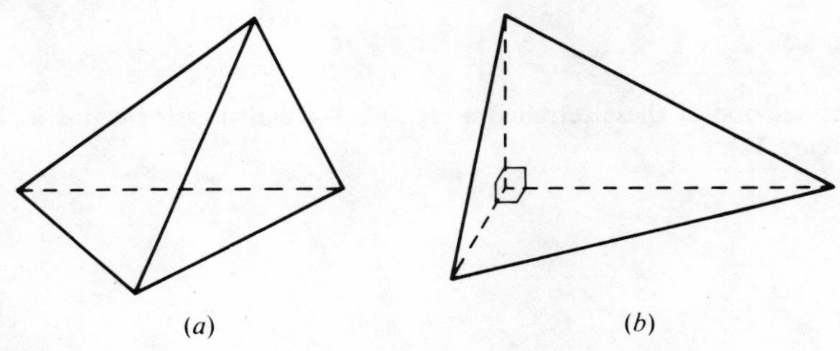

Fig. 12 (*a*) (*b*)

(d) Now consider the cuboid in Figure 13.

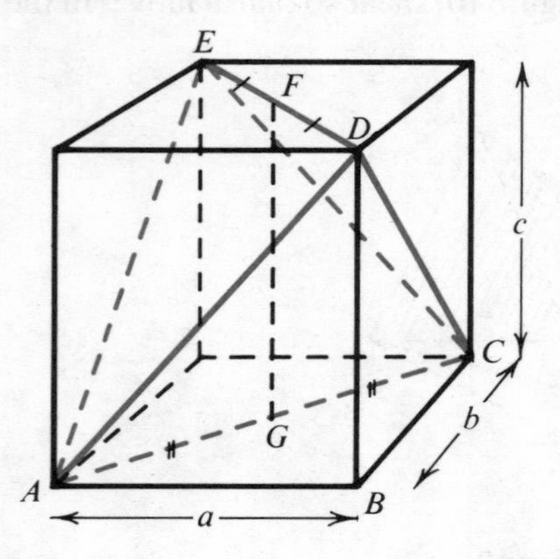

Fig. 13

Fig. 14

If we cut from this four tetrahedra (see Figure 14) each congruent to the tetrahedron $ABCD$, then we are left with the solid shown in Figure 15.

(e) If we divide the solid in Figure 15 along the plane ACF, where F is the midpoint, we obtain the two tetrahedra shown in Figure 16. We can see that these tetrahedra are congruent since a half-turn about FG (see Figure 13) maps each tetrahedron onto the other.

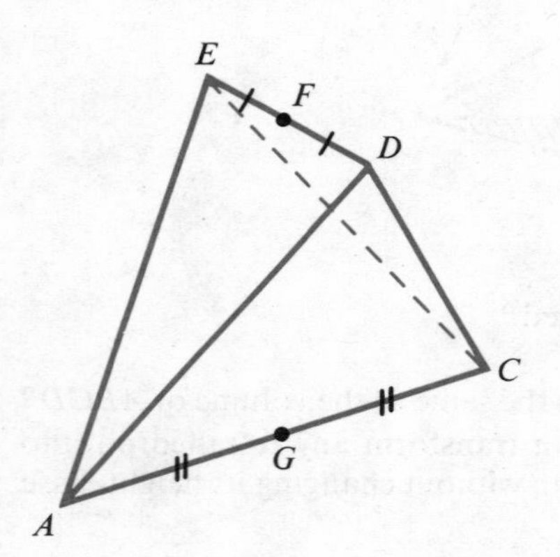

Fig. 15

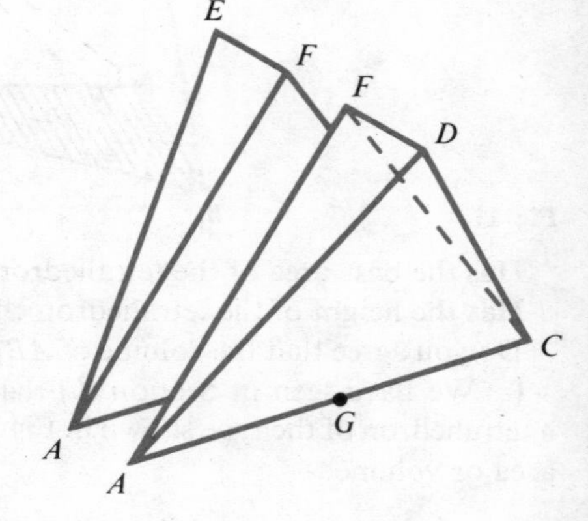

Fig. 16

Consider one of these tetrahedra $ACDF$. We can transform this as follows (see Figure 17).

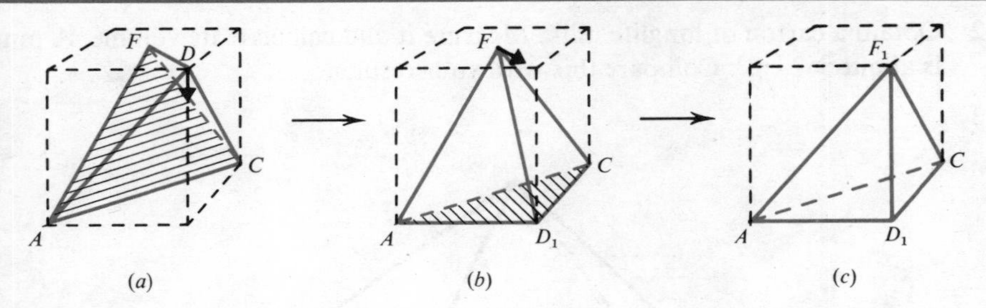

Fig. 17

(i) Keeping the face AFC invariant, shear D to D_1,
(ii) Keeping the face AD_1C invariant, shear F to F_1.

Do you agree that the volume of AD_1CF_1 equals the volume of $ADCF$?

What do you notice about AD_1CF_1 and the tetrahdron $ABCD$ in Figure 13?

Into how many tetrahedra of equal volume have we now dissected the cuboid?

(*f*) Do you agree that the volume of the cuboid in Figure 13 is abc? You should have found that the cuboid has been divided into six tetrahedra of equal volume. Hence the volume of each tetrahedron is one-sixth of the volume of the cuboid, that is, $\frac{1}{6}abc$.

Look again at the tetrahedron $ABCD$ in Figure 13.

Do you agree that its base area is $\frac{1}{2}ab$, and its height is c?

The volume of the tetrahedron is

$$\begin{aligned}&\tfrac{1}{6}abc\\&= \tfrac{1}{3} \times (\tfrac{1}{2}ab) \times c\\&= \underline{\tfrac{1}{3} \times \text{base area} \times \text{height.}}\end{aligned}$$

(*g*) In Section 3(*b*) and 3(*c*) we found that any tetrahedron can be transformed into a tetrahedron of the above type, having the same height, base area and volume.

Can you now write down the formula for the volume of *any* tetrahedron?

Exercise B

1 Copy and complete the following table for volumes of various tetrahedra:

Base area (cm^2)	Height (cm)	Volume (cm^3)
20	15	
27	6·2	
8·9	2·7	
32		65
	17	39·2
0·53	0·16	

2 Obtain a carton of longlife milk. Measure it and calculate its volume. A pint is about 567 cm³. Compare this with your results.

3

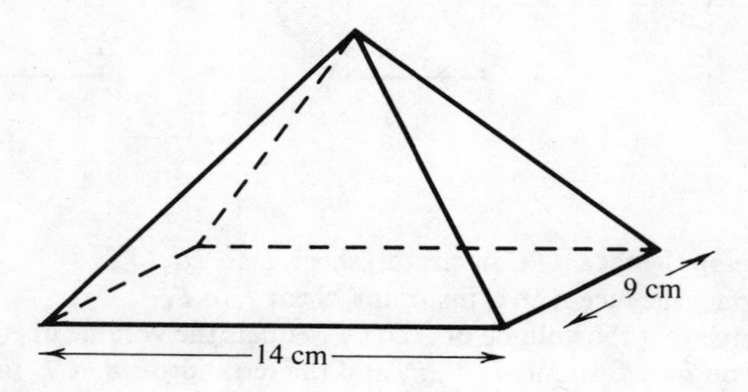

Fig. 18

Figure 18 shows a pyramid of height 6 cm having a rectangular base. The vertex is vertically above the centre of the rotational symmetry of the base (such a pyramid is called a right pyramid). By considering the pyramid divided into tetrahedra calculate its volume.

4 A tetrahedron has all edges of length 10 cm. Find its base area, height and volume.

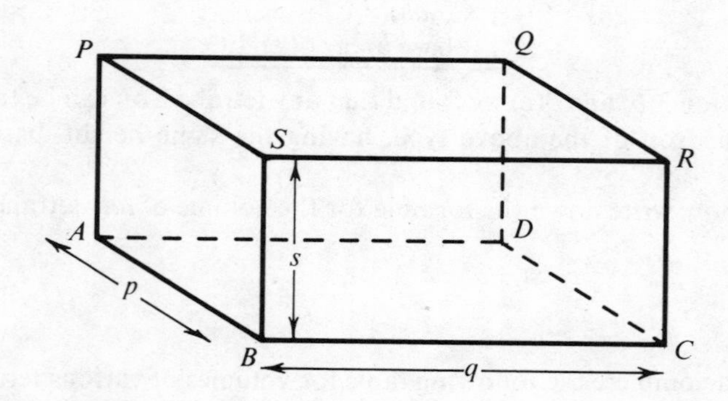

Fig. 19

5 $ABCDPQRS$ is a cuboid with $AB = p$ cm, $BC = q$ cm, $BS = s$ cm.
(i) Calculate the volume of tetrahedron $ABCS$.
(ii) Find as many tetrahedra as you can that have the same volume as $ABCS$.

6 *VABCD* in Figure 20 is a right pyramid. Calculate its volume.

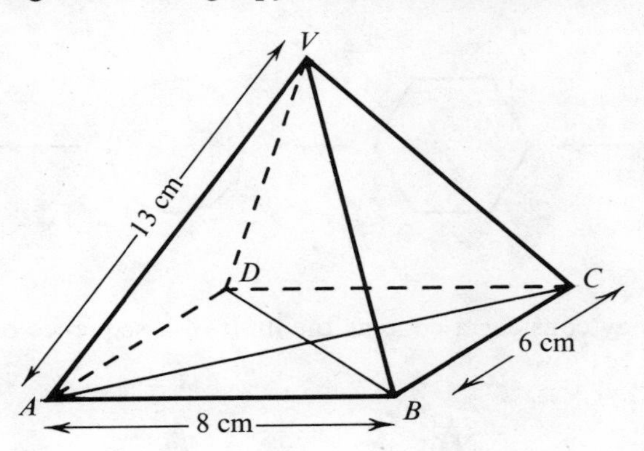

Fig. 20

4 Volume of a pyramid and cone

(*a*) Figure 21 shows a pentagonal-based pyramid, which has been divided into three tetrahedra *VPQR*, *VPRT* and *VRST*.

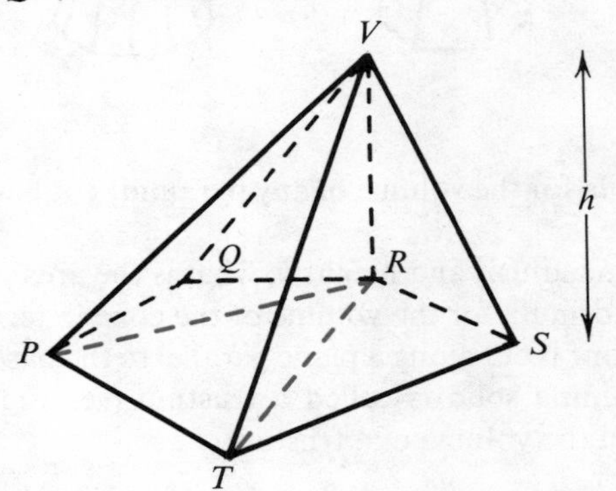

Fig. 21

Can any pyramid be divided into tetrahedra in this way?

If the base of the pyramid had six sides, how many tetrahedra would there be?

How many for seven sides?

How many for *n* sides?

(*b*) In Figure 21 we will call the areas of the triangles *PQR*, *PRT* and *RTS*, A_1, A_2, and A_3 respectively.

What is the base area of the pyramid in terms of A_1, A_2, A_3?

We can see that the volume of tetrahedron $VPQR = \frac{1}{3}A_1 h$,

the volume of tetrahedron $VPRT = \frac{1}{3}A_2 h$,

and the volume of tetrahedron $VRST = \frac{1}{3}A_3 h$.

Hence the volume of the pyramid $= \frac{1}{3}A_1 h + \frac{1}{3}A_2 h + \frac{1}{3}A_3 h$

$= \frac{1}{3}(A_1 + A_2 + A_3)h.$

But $A_1 + A_2 + A_3$ is the base area of the pyramid. So the volume of the pyramid

$= \frac{1}{3} \times$ base area $\times$ height.

Is this formula true for any pyramid? Why?

(*c*) We may consider a circle as the limit of a sequence of regular polygons (see Figure 22).

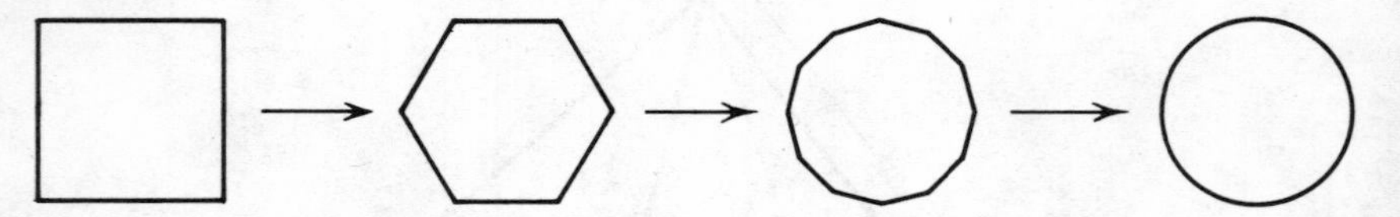

Fig. 22

Similarly, we may consider a cone as the limit of a sequence of pyramids (see Figure 23).

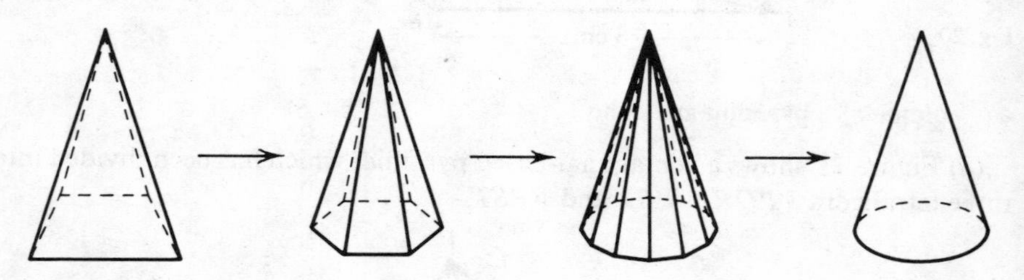

Fig. 23

Hence, the formula for the volume of any pyramid, $\frac{1}{3} \times$ base area $\times$ height, can be applied to a cone.

If a cone has base radius r and height h, what is the area of its base?

Write down the formula for the volume of the cone in terms of r and h.

(*d*) If a circular cone is cut along a plane parallel to the base, and the small cone removed, the remaining solid is called a frustum (see Figure 24). Example 1 explains how to find the volume of a frustum.

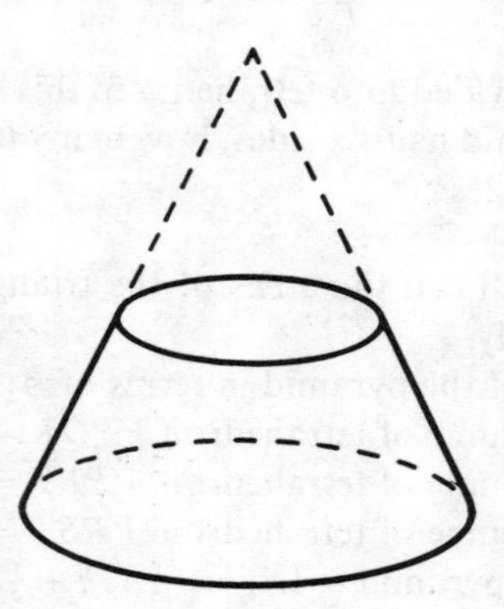

Fig. 24

Example 1

A frustum has base radius 16 cm, upper radius 6 cm and height 5 cm. Calculate its volume.

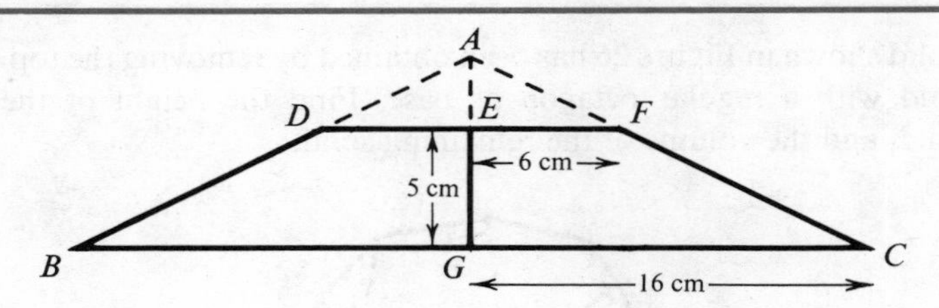

Fig. 25

Figure 25 shows a plane of symmetry of the frustum. What can you say about the triangles AEF and AGC?

The ratios $\frac{AE}{AG}$ and $\frac{EF}{GC}$ are equal. Can you see why?

If the height of the original cone is h,

then
$$\frac{AE}{AG} = \frac{EF}{GC},$$

so
$$\frac{h-5}{h} = \frac{6}{16}.$$

Solve this equation for h.
Did you find $h = 8$?
Since the volume of frustum (v) is the volume of the larger cone minus the volume of the small cone we have:

$$\begin{aligned} v &= \tfrac{1}{3}\pi \times 16^2 \times 8 - \tfrac{1}{3}\pi \times 6^2 \times 3 \quad \text{cm}^3 \\ &= \tfrac{1}{3}\pi(256 \times 8 - 36 \times 3) \quad \text{cm}^3 \\ &= \frac{1940\pi}{3} \quad \text{cm}^3 \\ &\approx 2030 \text{ cm}^3. \end{aligned}$$

Summary

1 The volume of any pyramid $= \frac{1}{3} \times$ base area $\times$ height.

2 The volume of a cone with base radius r and height h

$$\begin{aligned} &= \tfrac{1}{3} \times \text{base area} \times \text{height} \\ &= \tfrac{1}{3}\pi r^2 h. \end{aligned}$$

Exercise C

1 Find the volume of a pyramid of height 7 cm having a rectangular base 10·3 cm long and 6·8 cm wide.

2 Find the volume of a pyramid of height 9 cm, having as base a regular hexagon with side length 3 cm.

3 The solid shown in Figure 26 has been obtained by removing the top 4 cm of a pyramid with a regular octagon as base. Find the height of the original pyramid, and the volume of the remaining solid.

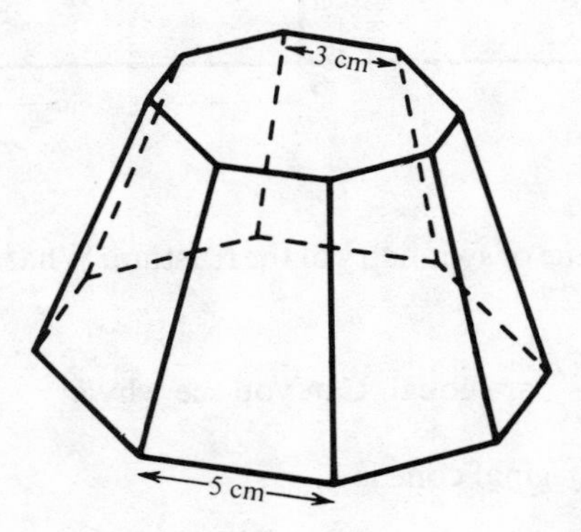

Fig. 26

4 Find the volume of a circular cone with base radius 4 cm and height 9 cm.

5 Find the height of a circular cone with base radius 1·5 cm and volume 8 cm³.

6 Find the volume of a frustum of a circular cone with base radius 12 cm, top radius 5 cm and height 7 cm.

7 Justify the use of the formula $\frac{1}{3} \times$ base area $\times$ height in finding the volume of a cone with an elliptical base.

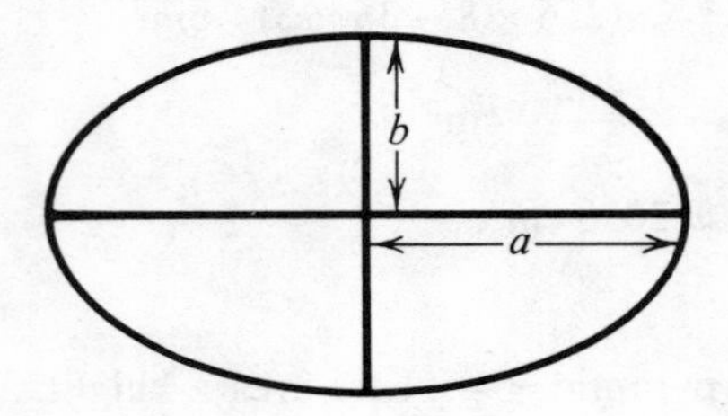

Fig. 27

Given that the area of the ellipse shown in Figure 27 is πab, find the volume of a cone of height 6 cm, having as base an ellipse with $a = 5$ cm, $b = 4$ cm.

5 Surface area of pyramid and cone

(*a*) The surface area of any pyramid can be found by calculating the area of each of its faces and adding the results. The process is explained in the following example.

Example 2

Find the total surface area of a right pyramid of height 10 cm having a square base with sides 6 cm (see Figure 28).

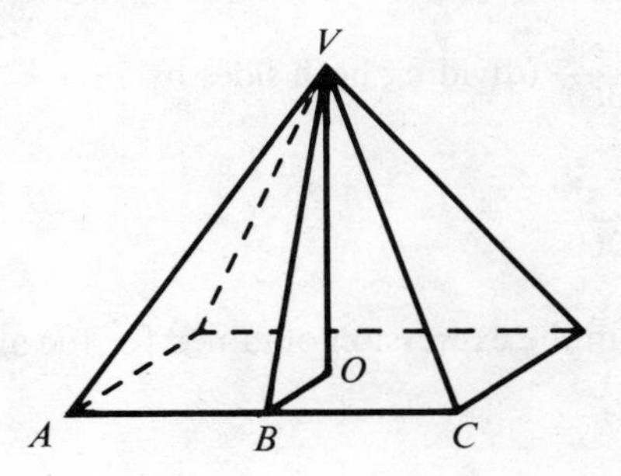

Fig. 28

The area of the base $= 6 \times 6\ \text{cm}^2 = 36\ \text{cm}^2$.

By Pythagoras' rule $VB = \sqrt{(100 + 9)}\ \text{cm} = 10{\cdot}4\ \text{cm}$.

Hence the area of triangle $AVC = (\frac{1}{2} \times 6 \times 10{\cdot}4)\ \text{cm}^2 = 31{\cdot}2\ \text{cm}^2$.

Hence total surface area $= (36 + 4 \times 31{\cdot}2)\ \text{cm}^2$

$\approx \underline{160{\cdot}8\ \text{cm}^2}$

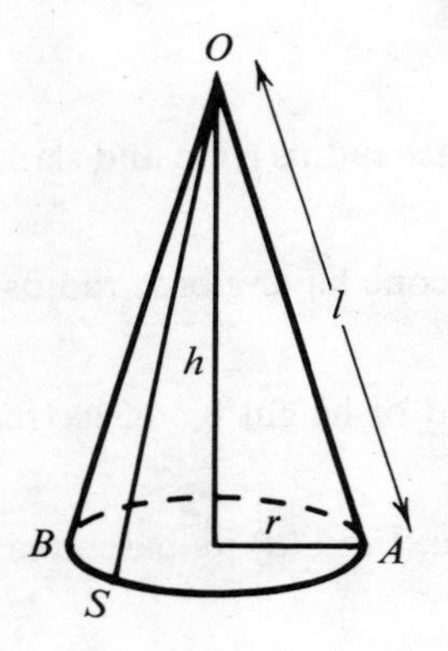

Fig. 29

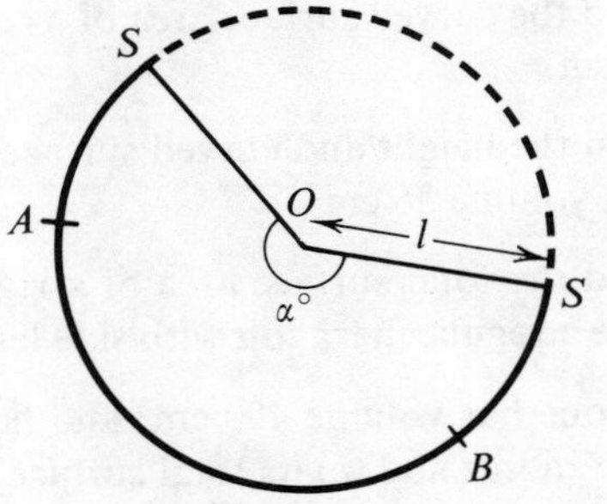

Fig. 30

(*b*) If we slit the surface of the right circular cone shown in Figure 29 along a slant line OS we obtain the sector of the circle of radius OS shown in Figure 30.

By calculating the area of the sector we can find the curved surface area of the cone.

$$\text{The area of sector} = \frac{\alpha^\circ}{360^\circ} \times \text{area of circle}$$

$$= \frac{\alpha^\circ}{360^\circ} \times \pi l^2.$$

Arc $SABS$ is the circumference of the base of the cone and equals $2\pi r$. So

$$2\pi r = \frac{\alpha^\circ}{360^\circ} \times 2\pi l$$

$$\frac{2\pi r}{2\pi l} = \frac{\alpha^\circ}{360^\circ} \text{ (dividing both sides by } 2\pi l\text{)}$$

and

$$\frac{r}{l} = \frac{\alpha^\circ}{360^\circ}.$$

Substituting for $\frac{\alpha^\circ}{360^\circ}$ in the expression obtained for the area of sector,

$$\text{area of sector} = \frac{r}{l} \times \pi l^2$$

$$= \pi r l.$$

That is, the curved surface area of a cone
= π × base radius × slant height.

What is the base area of the cone? Write down a formula for the *total* surface area of the cone.

Exercise D

1 Find the curved surface area of a cone with base radius 6 cm and slant height 11 cm.

2 Find the height and curved surface area of a cone having base radius 2·5 cm and volume 36 cm^3.

3 Find the total surface area of a right pyramid of height 8 cm, having as its base a regular hexagon with side length 4 cm.

4 A cone has volume 400 cm^3 and height 12 cm. Find (*a*) its base area, (*b*) its base radius and (*c*) its total surface area.

5 A cone has base radius r and height h. Find a formula for its total surface area in terms of r and h.

6 Find the area of material needed to make the lampshade shown in Figure 31.

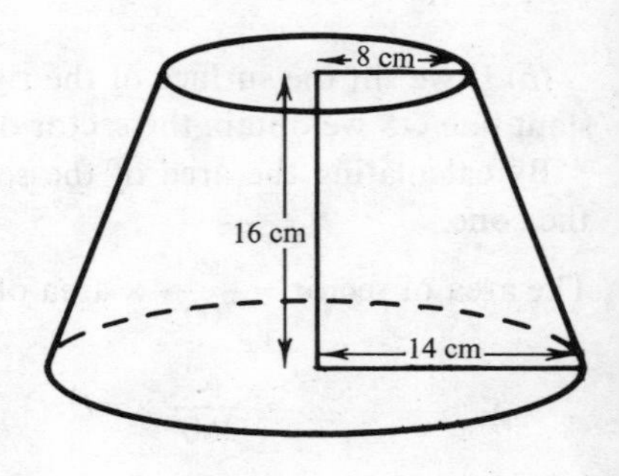

Fig. 31

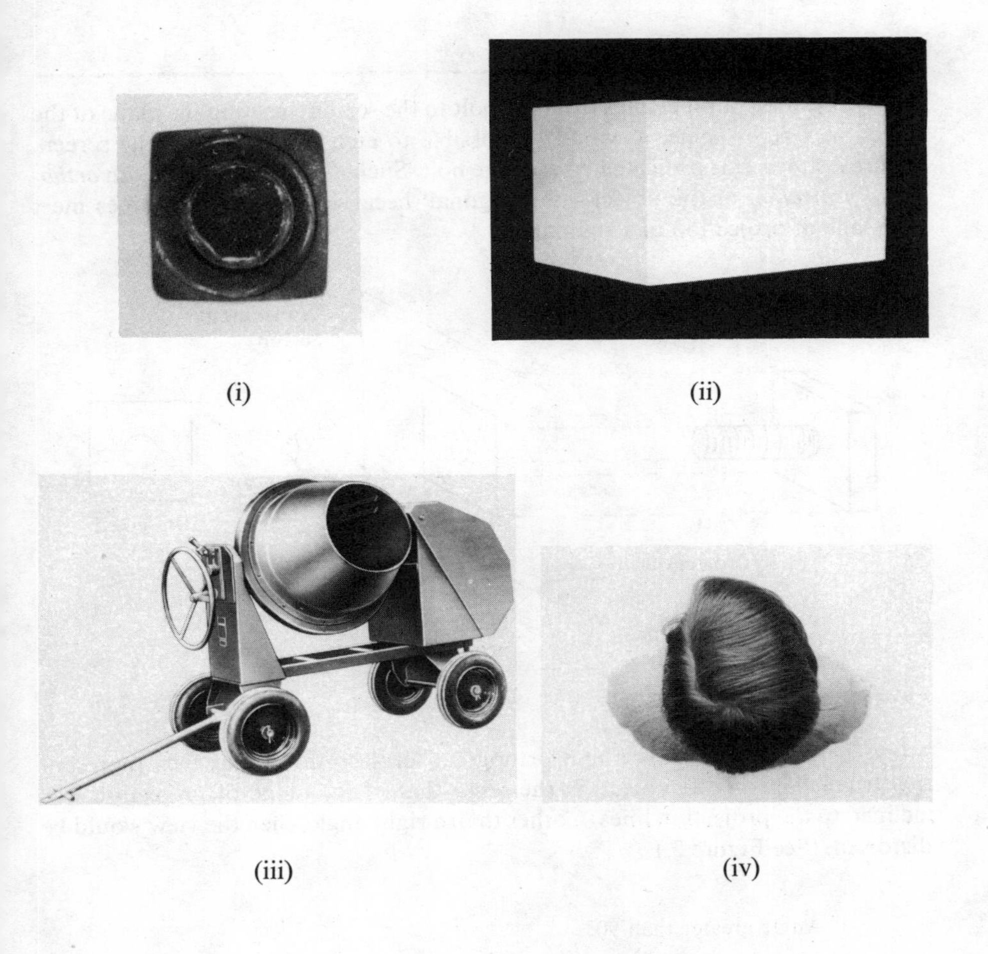

(i) (ii) (iii) (iv)

11 Plans and elevations

1 Orthogonal projections

(*a*) The photographs at the beginning of this chapter are of four everyday objects. Can you guess what they are?

(*b*) Photograph 1 is of a square-headed bolt with the screw shaft pointing directly towards the camera. Suppose you are viewing the bolt from this angle. Try to visualize a transparent screen placed between you and the bolt as suggested by Figure 1.

By projecting parallel lines from the bolt to the screen, meeting the plane of the screen in a right angle, it would be possible to etch the view onto the screen. Figure 2 shows this projected view of the bolt. Such a view is said to be an *orthogonal projection* of the object – 'orthogonal' because the projection lines meet the plane of projection in a right angle.

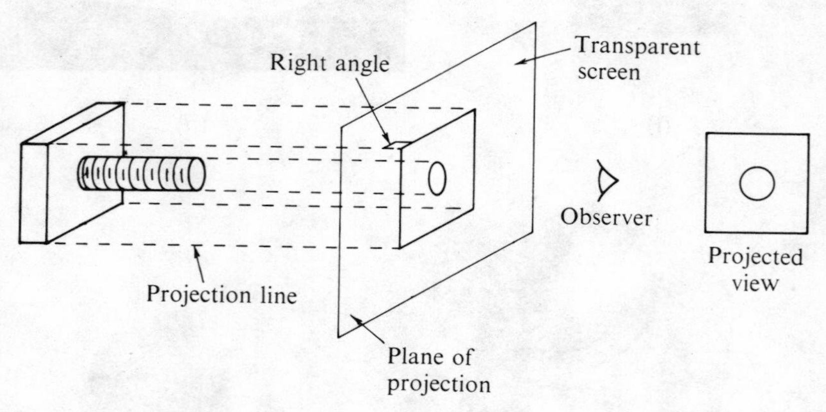

Fig. 1

Fig. 2

(*c*) The important thing about orthogonal projections is that they represent the 'true' shape of the object, in the sense that, if the plane of projection was inclined to the projection lines in other than a right angle, then the view would be distorted. (See Figure 3.)

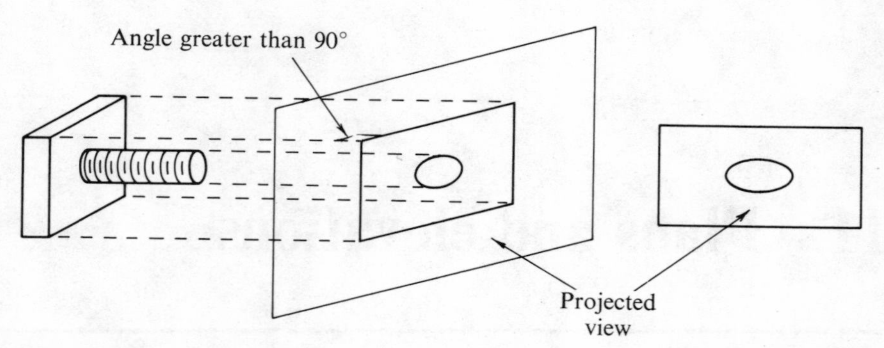

Fig. 3

Provided that the faces of the object we draw are parallel to the plane of projection we can use our drawing to measure otherwise unknown lengths and angles on these faces.

The length of side of the head of the bolt in Figure 2 is 2 cm. Measure the greatest distance across the head. Check your answer using Pythagoras' rule.

(*d*) Is the photograph of the bolt a true orthogonal projection? (Think about a bolt with a screw shaft 1 m long!)

Would you say that (i) maps of the world; (ii) contour maps are orthogonal projections? Study an atlas and see if you can decide how maps of the world have been produced from a globe.

(*e*) Figure 4 shows an orthogonal projection of one of the solids in Figure 5. Can you say which solid?

Fig. 4

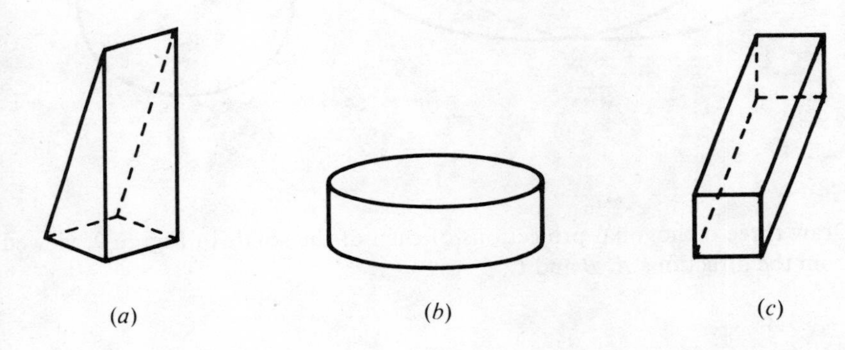

Fig. 5

(*f*) Figure 6 shows a second view of the same solid, viewed from a position orthogonal to the first view. Does this help you to decide which solid is being represented?

Fig. 6

(*g*) A third projected view of the solid is shown in Figure 7. The plane of projection is orthogonal to the planes of projection for the first two views. Only now can you be quite sure that it is (*b*) that we are drawing. Now draw three more solids which have two identical projections.

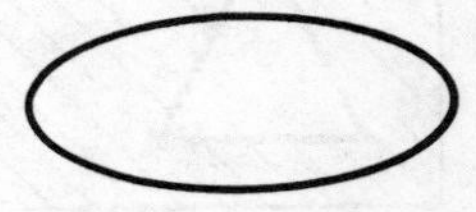

Fig. 7

(*h*) We can obviously draw as many orthogonal projections of a solid as we wish, but we usually find that three mutually orthogonal projections are sufficient. Two more views of the solid (*b*) are shown in Figure 8. Where do you think the planes of projection are situated?

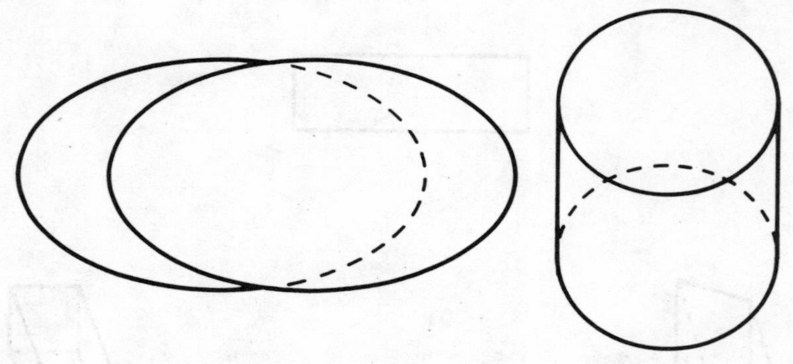

Fig. 8

Exercise A

1 Draw three orthogonal projections of each of the solids in Figure 9, viewed from the directions *A*, *B* and *C*.

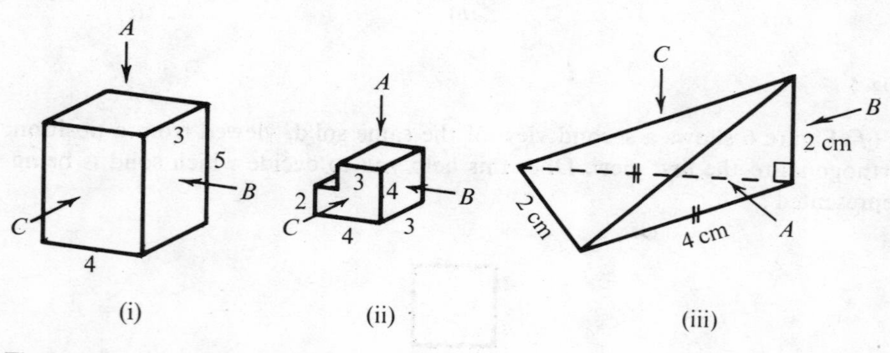

Fig. 9

2 The cone will slot exactly through the two holes in the piece of wood represented by Figure 10.

Fig. 10

Draw solids which will fit exactly through each hole for the following:

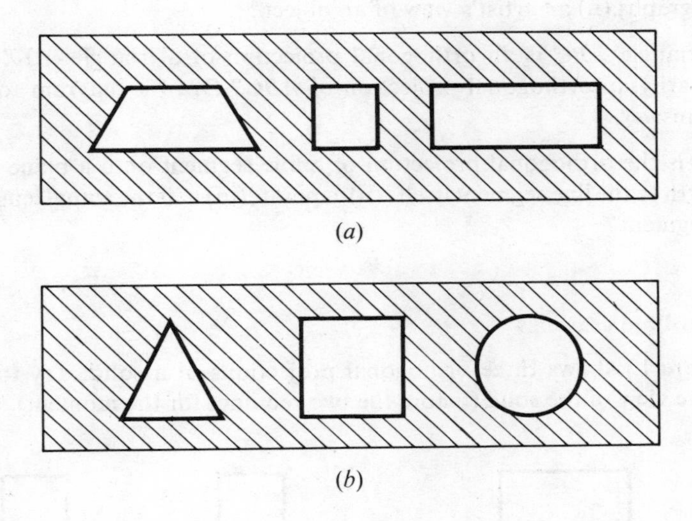

Fig. 11

3 Which of the drawings in Figure 12 do you think *could* be orthogonal projections of a cuboid? The broken lines represent hidden features.

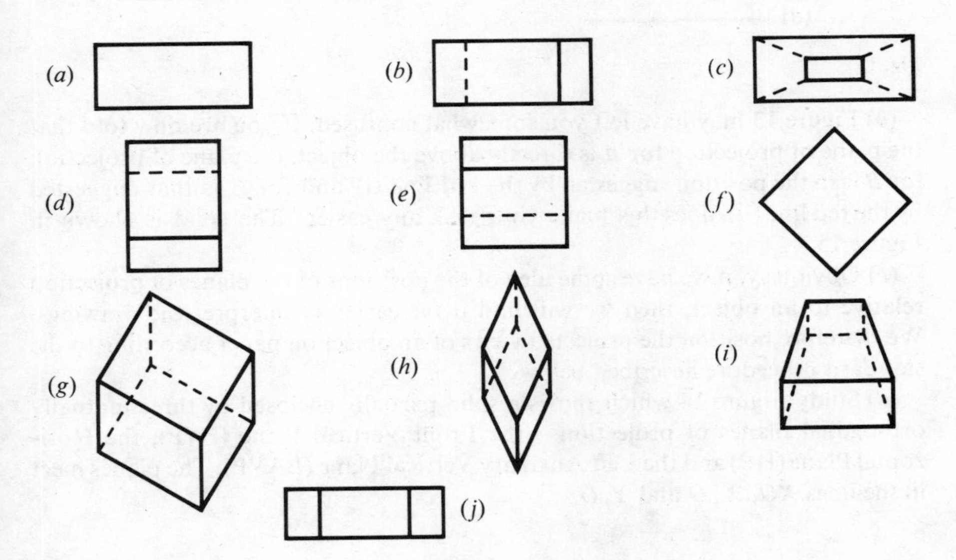

Fig. 12

4 Considering orthogonal projection as a transformation of 3D space into 2D space, which properties of figures are invariant?

5 Does the distance of the plane of projection from an object have any effect on the orthogonal projection of the object? Is the same principle true of (i) a photograph; (ii) an artist's view of an object?

6 If a triangle ABC is an orthogonal projection of a triangle XYZ is XYZ necessarily an orthogonal projection of ABC? Draw a diagram to explain your answer.

7 When is the orthogonal projection of a line segment onto a plane (i) equal in length to the line segment itself; (ii) a point; (iii) half the actual length of the line segment?

2 Standardized drawings

(*a*) Figure 13 shows three orthogonal projections of a solid. Try to draw a perspective view of the solid (ignore the two red lines for the moment).

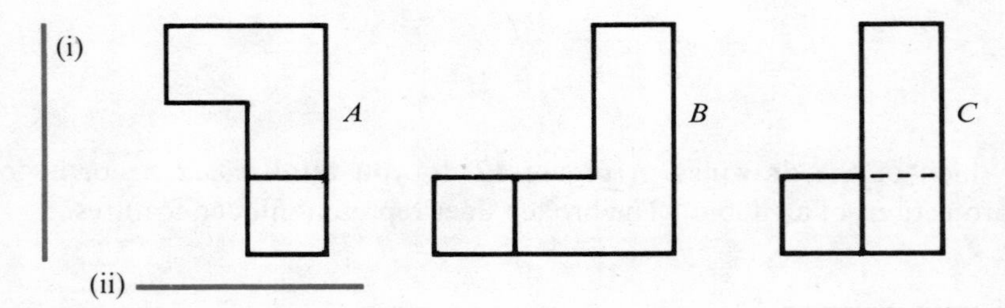

Fig. 13

(*b*) Figure 13 may have left you somewhat confused. If you are now told that the plane of projection for A is directly above the object, the plane of projection for B is in the position suggested by the red line (i) and for C is that suggested by the red line (ii), does this make your task any easier? The solid is shown in Figure 15.

(*c*) Obviously, if we have some idea of the positions of the planes of projection relative to an object, then we will find it far easier to interpret the drawings. We therefore position the projected views of an object on paper according to the standard procedure described below.

(*d*) Study Figure 14 which shows a solid partially enclosed by three mutually orthogonal planes of projection – the Front Vertical Plane (FVP), the Horizontal Plane (HP) and the Left Auxiliary Vertical Plane (LAVP). The planes meet in the lines XO, X_1O and Y_1O.

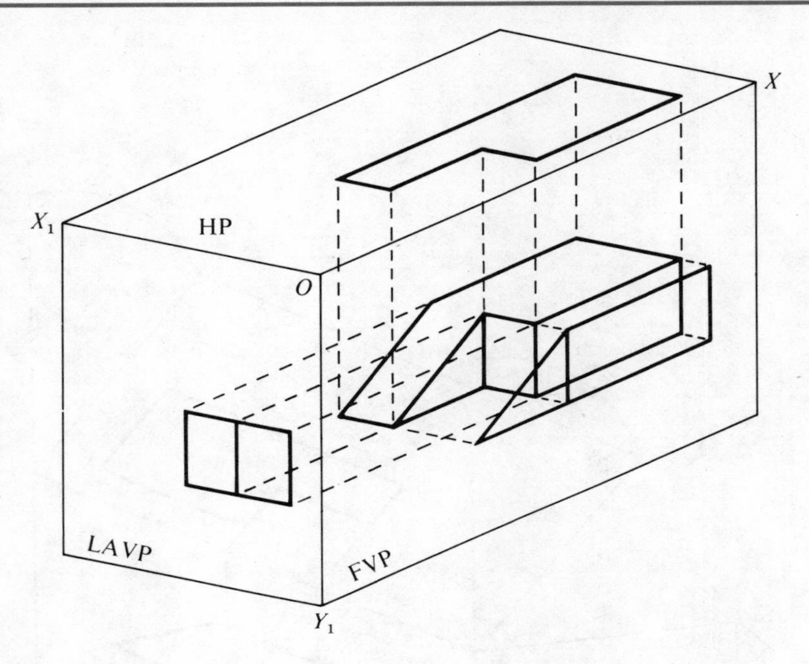

Fig. 14

If we project the object onto each of these planes and then 'open' the planes out so that they all lie in the same plane as FVP (see Figure 16) the projections will finally be positioned as in Figure 17.

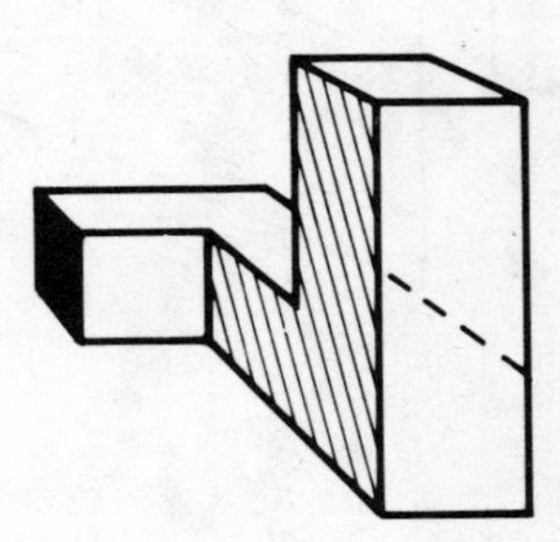

Fig. 15

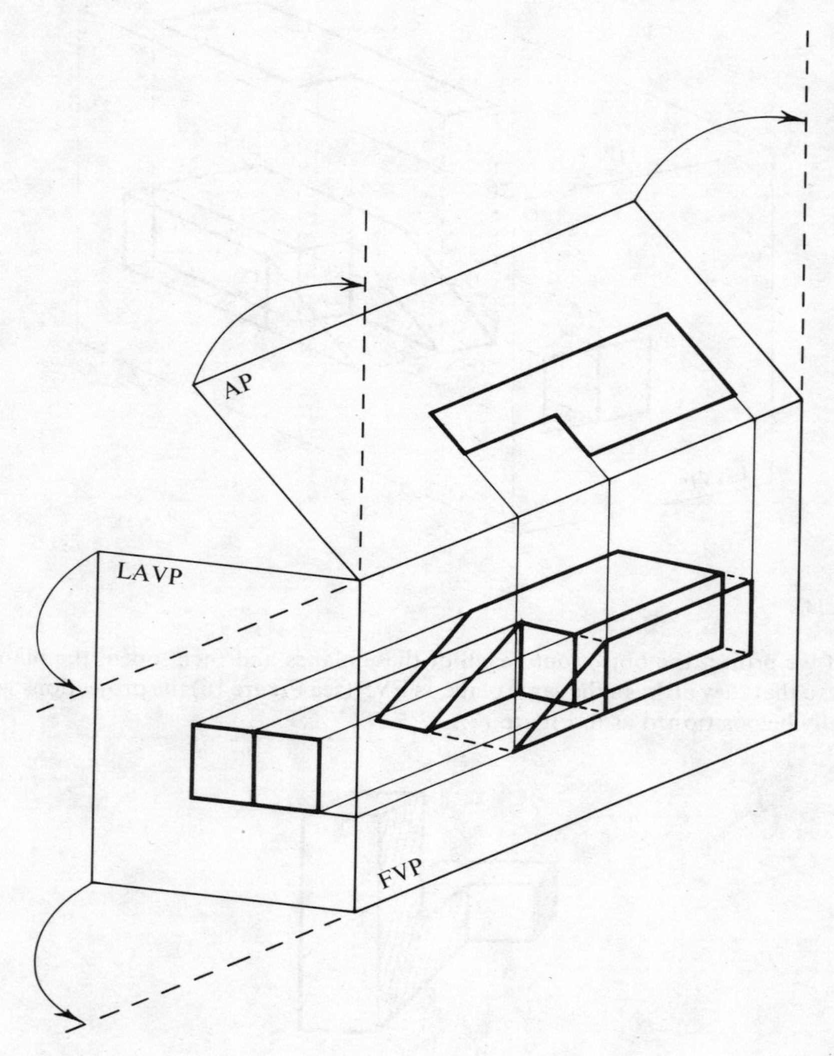

Fig. 16

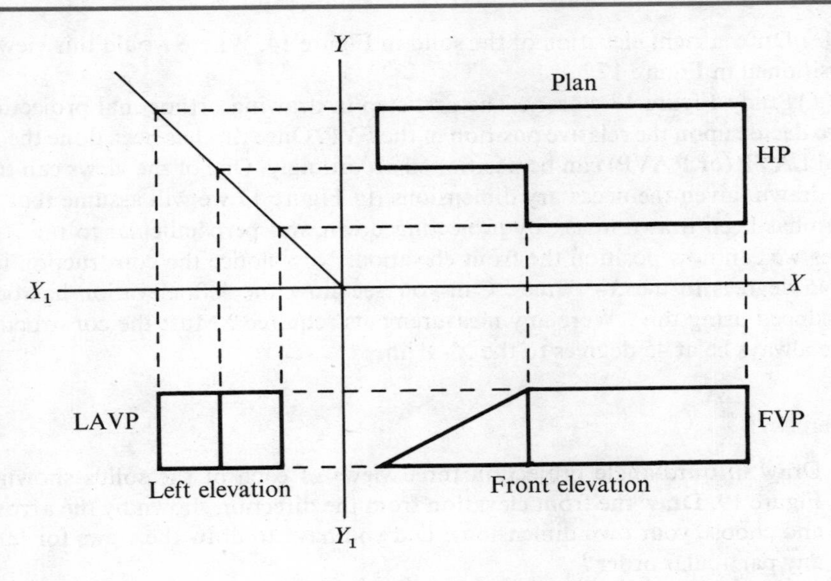

Fig. 17

Projections onto the vertical planes are called ELEVATIONS, and those onto the horizontal plane, PLANS. The particular system of standard planes of projection which leads to the plan being drawn directly above the *front elevation* and the auxiliary elevations (*left and right elevations*) to the left and right of the front elevation is called *third-angle projection*. The system was originally devised in America and is now almost universally approved. Figure 18 will suggest to you the reason for calling it third-angle projection.

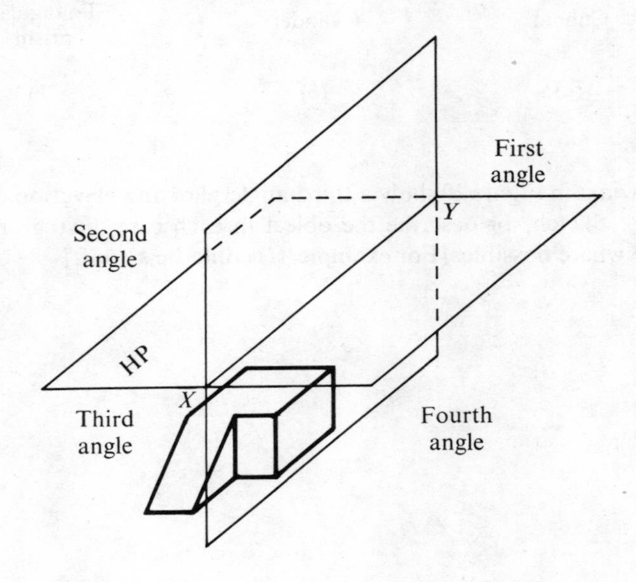

Fig. 18

(*e*) Draw a right elevation of the solid in Figure 14. Where would this view be positioned in Figure 17?

(*f*) Study Figure 17 closely. The first step in drawing orthogonal projections is to decide upon the relative position of the FVP. Once this has been done the HP and LAVP (or RAVP) can be positioned accordingly. One of the views can then be drawn, given the necessary dimensions (in Figure 17 we will assume that the plan has been drawn first). By projecting down and perpendicular to the *X–Y* lines we can now position the front elevation. Now notice the construction line at 45 degrees to the *X–Y* lines. Can you see how the left elevation has been produced using this? Were any measurements required? Must the construction line always be at 45 degrees to the *X–Y* lines?

Exercise B

1 Draw in third-angle projection three views of each of the solids shown in Figure 19. Draw the front elevation from the direction shown by the arrows, and choose your own dimensions. Did you have to draw the views for (*c*) in any particular order?

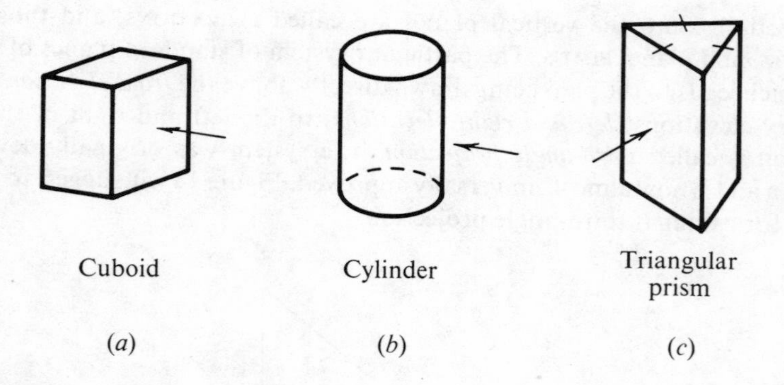

Fig. 19

2 The drawings in Figure 20 show a third-angle plan and elevation of a number of solids. Sketch, or describe the object in each case, giving an 'everyday' example where possible. [For example, (*i*) could be a roof.]

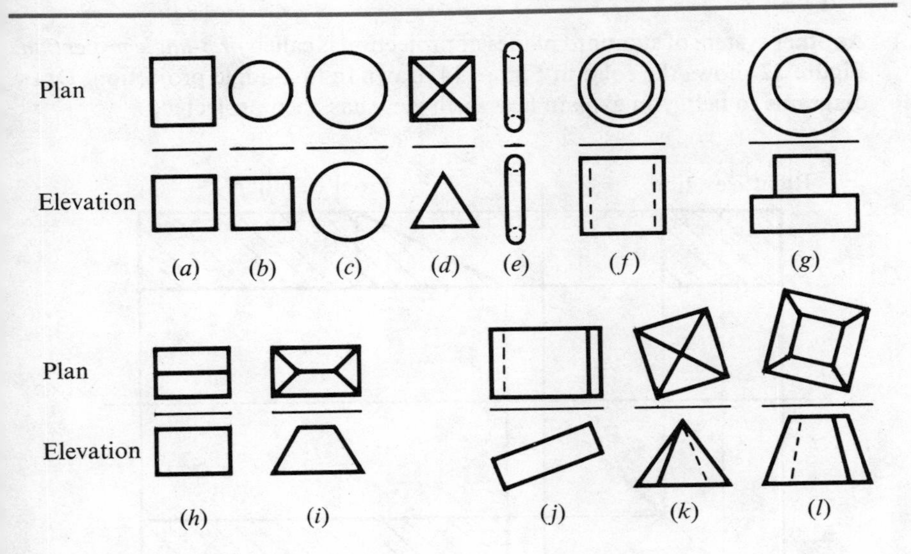

Fig. 20

3 **In Figure 21, one view out of three has been omitted. Copy what is given and sketch the missing view.**

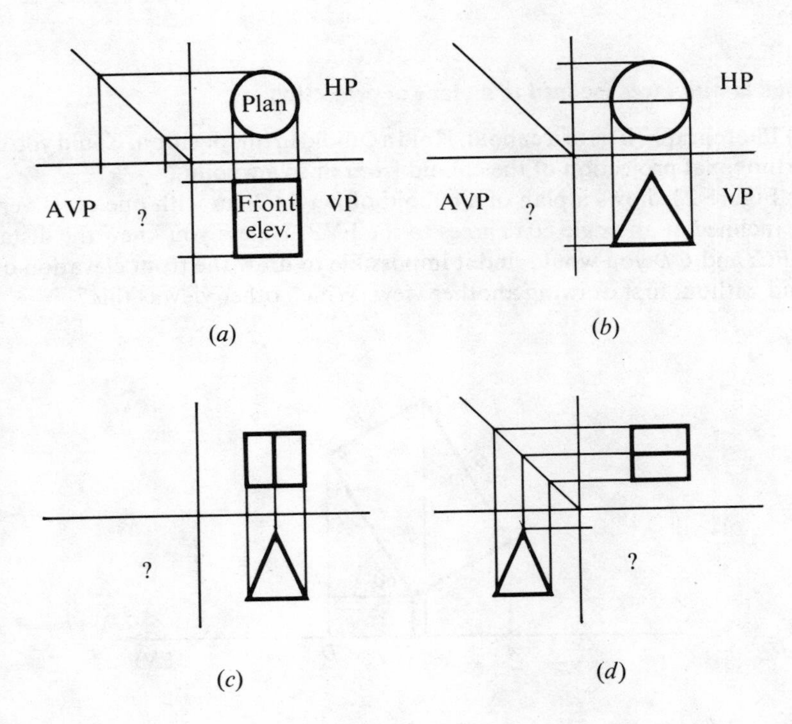

Fig. 21

4 Another system of standard planes of projection is called *first-angle projection.* Figure 22 shows the solid in Figure 14 drawn in first-angle projection. Draw diagrams to help you explain how each view has been projected.

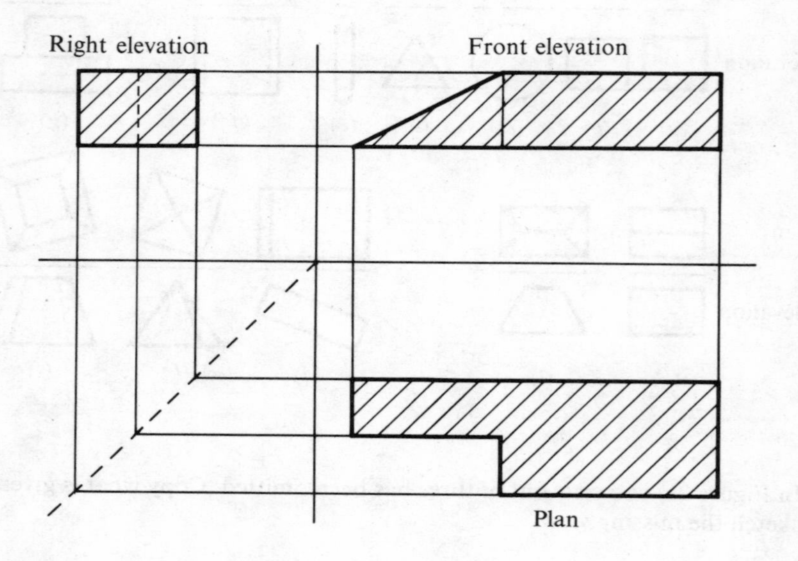

Fig. 22

3 Solids with faces inclined to a plane of projection

(*a*) Photograph (ii) is of a cuboid. Hold a cuboid in this position. Could you draw an orthogonal projection of the cuboid from this viewpoint?

(*b*) Figure 23 shows a plan of a cuboid of height 3 cm with one of its vertical faces inclined at an angle 60 degrees to the FVP. Unless you knew the distances *AB*, *BC*, and *CD* you would find it impossible to draw the front elevation of the cuboid without first drawing another view. Which other view is this?

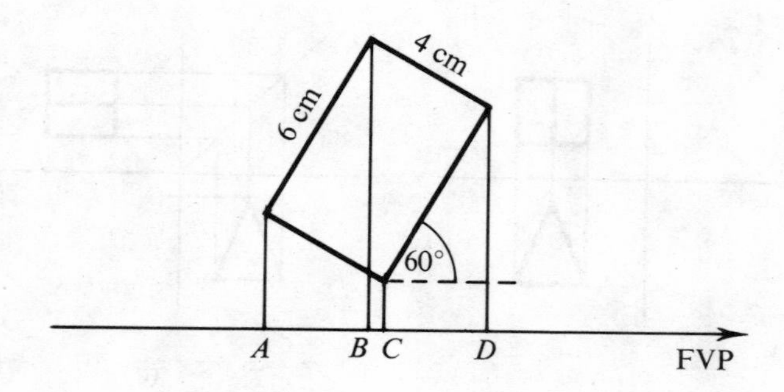

Fig. 23

(*c*) Figure 24 shows that by first drawing the *plan* of the cuboid the front elevation can be produced immediately. The plan is easily drawn because it projects a 'true' shape onto the horizontal plane.

Whenever we wish to draw a view of an inclined solid we must always first draw the view which projects such a 'true' shape.

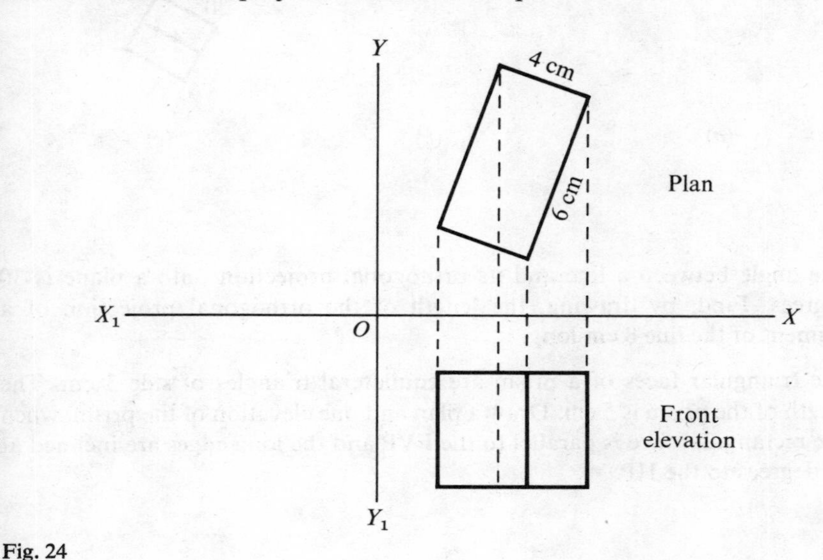

Fig. 24

Exercise C

1 Draw the left elevation of the cuboid in Figure 24.

2 The solid in Figure 25 is shown in either plan or elevation in Figures 26 (*a*)–(*c*). In each case copy what is drawn and project the missing view.

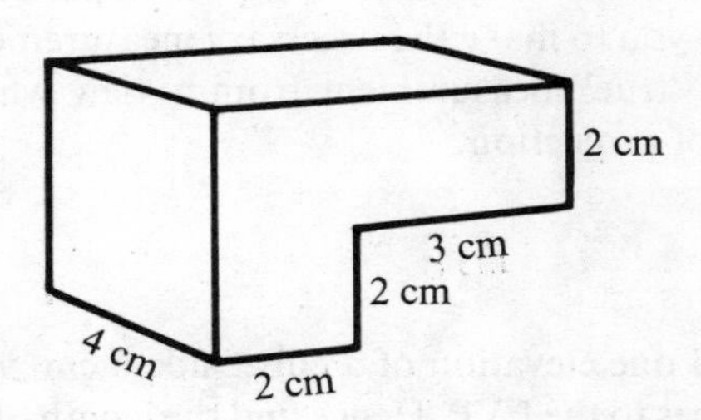

Fig. 25

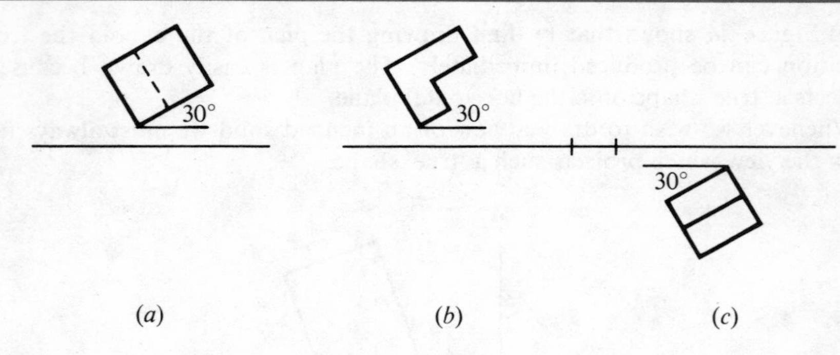

Fig. 26

3 The angle between a line and its orthogonal projection onto a plane is 30 degrees. Find, by drawing, the length of the orthogonal projection of a segment of the line 8 cm long.

4 The triangular faces of a prism are equilateral triangles of side 3 cm. The length of the prism is 5 cm. Draw a plan and one elevation of the prism, when one rectangular face is parallel to the FVP and the long edges are inclined at 30 degrees to the HP.

4 Solving 3D problems by drawing

In Exercise D you will be asked to find unknown lengths and angles of solids by drawing. Each of the problems can be solved using

(i) orthogonal projections, and
(ii) scale drawings.

Your main problem will be that of deciding which particular projections you should draw to enable you to make the necessary measurements. Remember that you can only make a 'true' measurement from a view which projects a 'true' shape onto the plane of projection.

Exercise D

1 Draw the plan and one elevation of a cube, side 8 cm, with its vertical faces inclined at 45 degrees to the FVP. Hence find the length of the diagonals of the cube.

2 $VABCD$ is the square-based pyramid shown in Figure 27. $AB = 6$ cm, $XA = 1$ cm, $CY = 2$ cm, $VB = 8$ cm. By drawing the triangles VAB and VDC and the square base find the length of:

(i) VX; (ii) VY; (iii) XY.

Hence find the angle YVX.

3 In Figure 27, find by drawing:

(i) the height of the pyramid;

(ii) the angle between VC and VA.

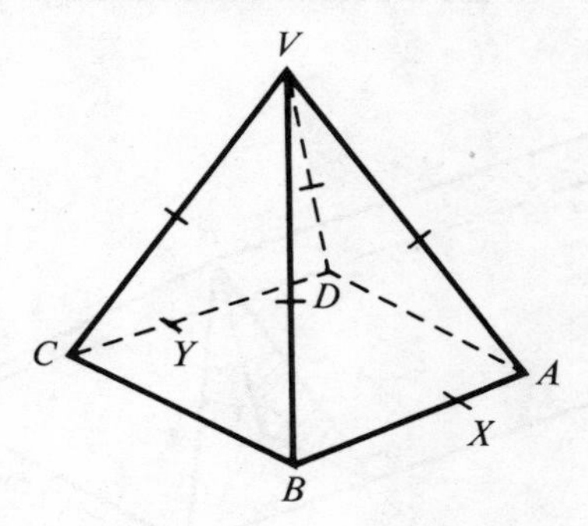

Fig. 27

4 Figure 28 shows a roof section of a house (the section has two planes of symmetry). Find by scale drawing:

(i) BC; (ii) angle BCD.

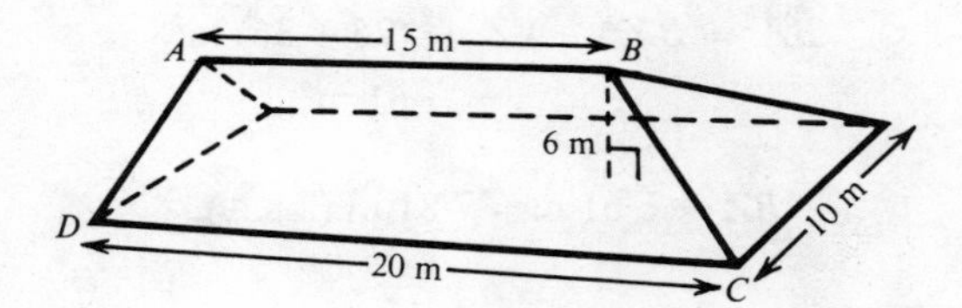

Fig. 28

5 Find, by drawing, the radius of the smallest sphere which will accommodate a right squared-based pyramid of height 6 cm, and length of base 4 cm.

6 Find, by drawing, the magnitude of the vector $\begin{pmatrix} 2 \\ 1 \\ 3 \end{pmatrix}$.

5 Another method of solving 3D problems

(*a*) If we can reduce the problem of finding unknown angles and lengths of solids to that of finding the angles and lengths of right-angled triangles then we can use

(i) Pythagoras' rule, and

(ii) trigonometrical ratios

to help us calculate the answers. In general, they will be more accurate than those obtained by drawing.

The following example shows how Question 4 in Exercise D can be solved using this method.

Example 1

Find (i) the length BC; (ii) the angle BCD, of the roof section shown in Figure 28.

(i) To calculate BC we can use the right-angled triangles BXZ and BZC shown in Figure 29.

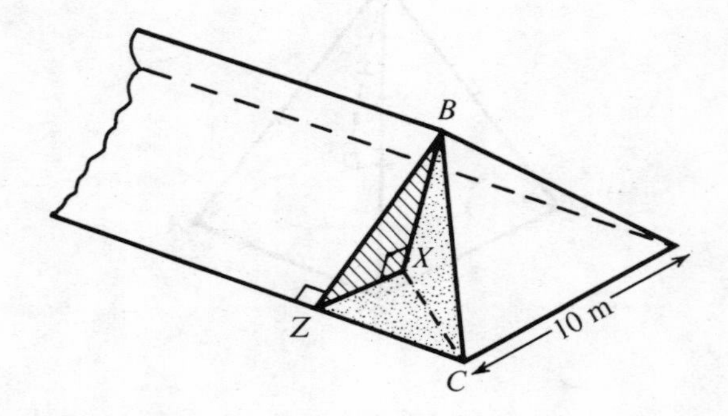

Fig. 29

By studying the symmetry of the roof section we can see that $XZ = 5$ m. Since $BX = 6$ m and BXZ is a right-angled triangle (see Figure 30) then

$$BZ^2 = BX^2 + XZ^2 = (25 + 36)\ \text{m}^2$$
$$= 61\ \text{m}^2.$$

Hence

$$BZ = \sqrt{61}\ \text{m} = 7{\cdot}81\ \text{m (3 s.f.)}.$$

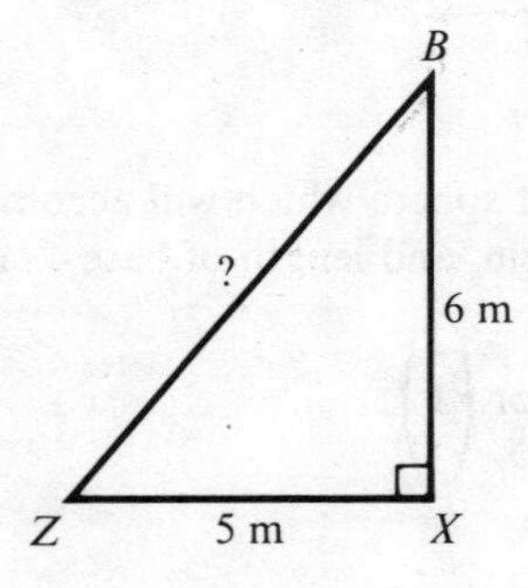

Fig. 30

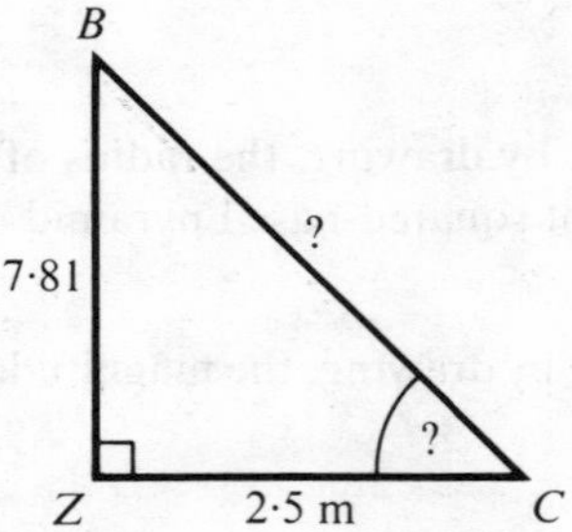

Fig. 31

Again using the symmetry of the roof section, $ZC = 2{\cdot}5$ m, and using triangle BZC (see Figure 31);

$$BC^2 = ZC^2 + BZ^2 = (2{\cdot}5^2 + 61)\ \text{m}^2$$
$$= 67{\cdot}25\ \text{m}^2.$$

(Why did we use $BZ^2 = 61$ and not $7{\cdot}81^2$?)

Hence

$$BC = \sqrt{67{\cdot}25}\ \text{m} = 8{\cdot}20\ \text{m (3 s.f.)}.$$

(ii) To calculate angle BCD we can use triangle BCZ (Figure 31).
We have

$$\tan \angle BCD = 7{\cdot}81/2{\cdot}5 = 3{\cdot}12 \text{ (3 s.f.)}$$
$$\text{angle } BCD = 72{\cdot}3^\circ.$$

Did you obtain approximately these answers by drawing?

(*b*) When you are solving 3D problems by calculation:

(i) sketch the solid and decide upon the right-angled triangles you should use to help you solve the problem;

(ii) if necessary draw each right-angled triangle separately and mark on the diagram the lengths of sides and angles you already know;

(iii) use Pythagoras' rule and your knowledge of trigonometrical ratios to calculate the answers.

Exercise E

1 For the cuboid in Figure 32 calculate:

(i) DB; (ii) AE; (iii) angle DAC; (iv) angle EAB.

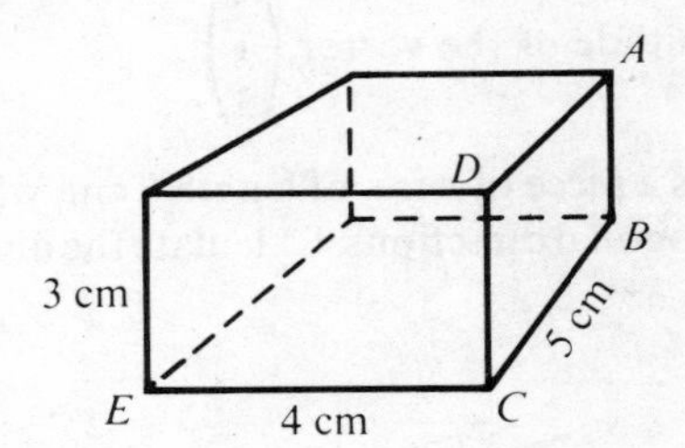

Fig. 32

2 Figure 33 represents a beam of light from a torch. Calculate the area of a circle of light cast upon a wall 60 m away.

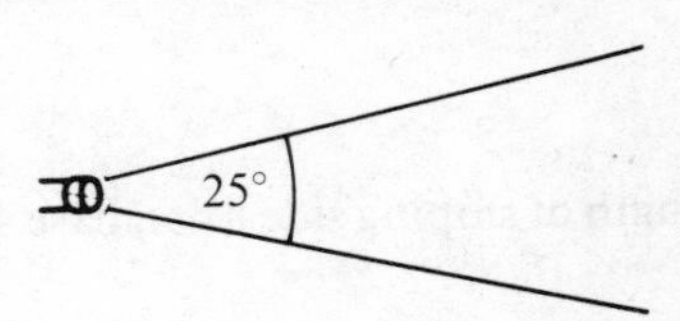

Fig. 33

3 A goldfish bowl in the form of a truncated sphere (Figure 34) has diameter 40 cm. The base and the opening at the top each have a radius of 10 cm. Calculate the height of the bowl.

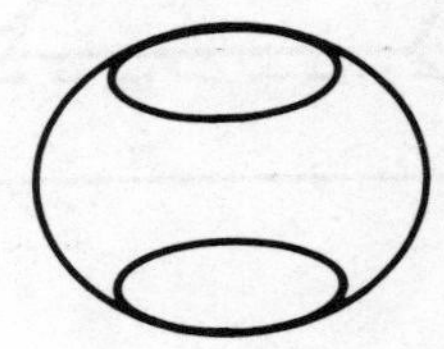

Fig. 34

4 Figure 35 shows a tetrahedron $ABCD$ in which $\angle ACB = \angle ACD = 90°$. CBD is an equilateral triangle of length of side 4 cm. $AC = 3$ cm. Calculate

(i) AB;
(ii) the length of the perpendicular from A to BD;
(iii) $\angle ABD$.

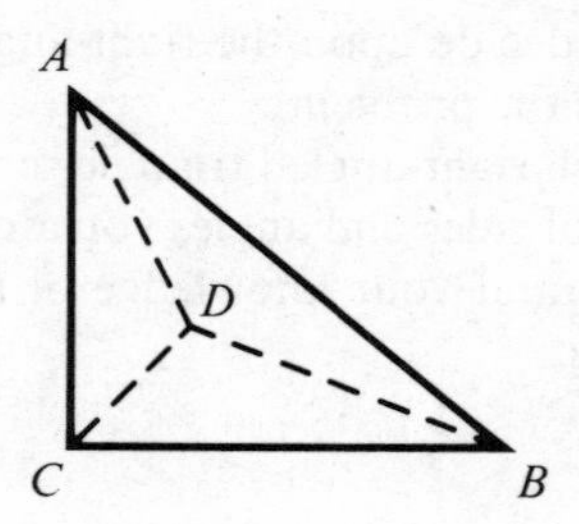

Fig. 35

5 Calculate the magnitude of the vector $\begin{pmatrix} 2 \\ 1 \\ 3 \end{pmatrix}$.

6 Figure 36 represents a piece of wire, of length 8 cm, which has been bent into three mutually orthogonal directions. Calculate the distance between the ends of the wire.

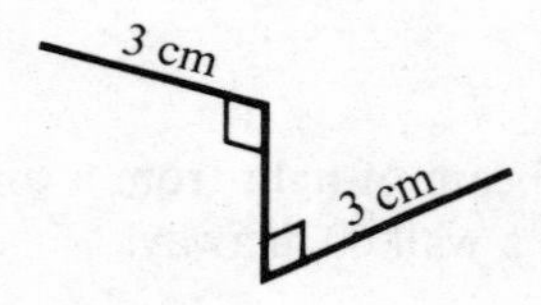

Fig. 36

7 Will a right pyramid, length of sloping side 8 cm, base 4 cm, fit inside a sphere of radius 5 cm?

8 The framework in Figure 37 has two planes of symmetry. $AC = 20$ m, $DF = AF = DE = 8$ m and $DB = 30$ m. Calculate (i) AB; (ii) $\angle AEC$.

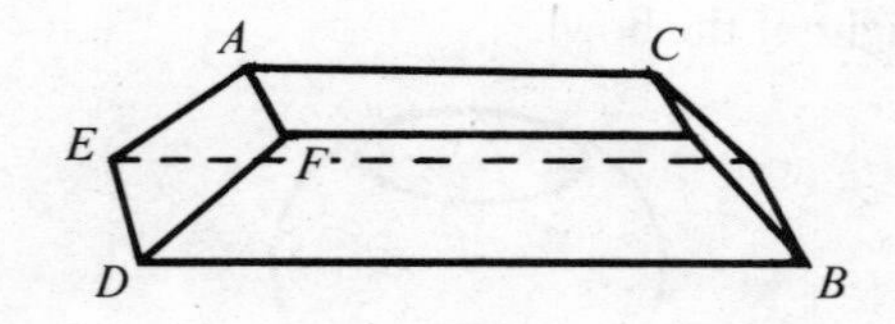

Fig. 37

9 A church spire is in the form of a hexagonal pyramid (Figure 38). The sides of the hexagon are 20 m, and the sloping faces of the tower are isosceles triangles with the two equal sides 80 m long. Calculate

(i) the height of the tower;
(ii) the angle at which each face of the tower slopes to the horizontal;
(iii) the area of lead needed to cover the exterior of the tower.

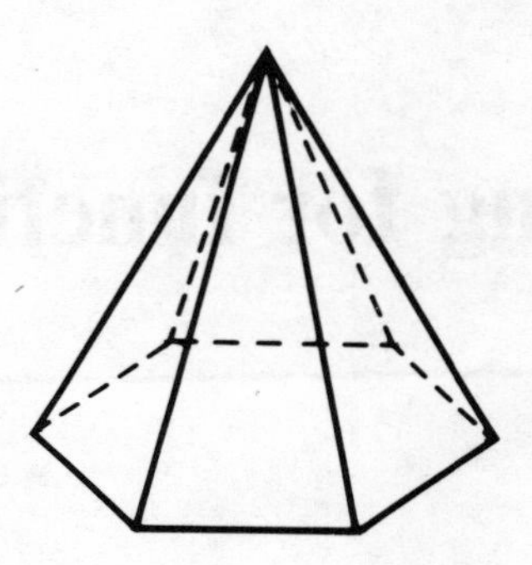

Fig. 38

10 Three vertical poles of height 60 m, 50 m and 40 m are situated in flat country at the vertices of an equilateral triangle of side 100 m. The tops of the poles are joined by straight wire cables. Calculate (i) the length of each cable; (ii) the angle between the plane in which the cables lie and the horizontal.

11 Two planes P_1 and P_2 meet in an angle of 30° (see Figure 39). $ABCD$ is a square drawn on plane P_1 with AD parallel to the line of intersection of the two planes. What kind of figure is the orthogonal projection of $ABCD$ onto P_2? Calculate the area of this figure if $AB = 4$ cm.

12 If the side AD of a square as in Question 11 was produced and met the line of intersection of the two planes in an angle of 45° what kind of quadrilateral would the orthogonal projection onto P_2 be? What would be the area of this quadrilateral?

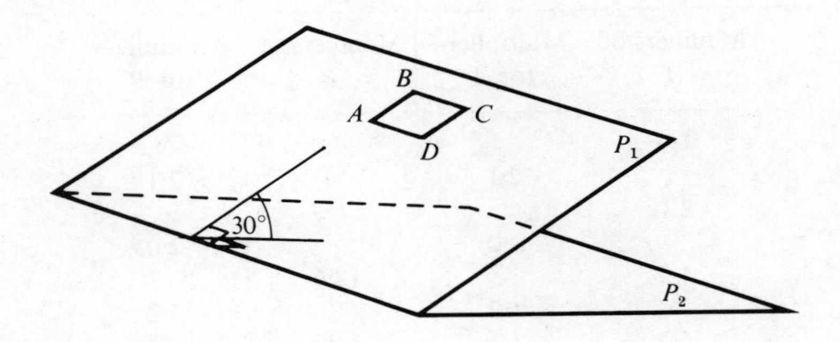

Fig. 39

$$P = \{1 \xrightarrow{\times 4} 4 \xrightarrow{\times 2\cdot 5} 10 \xrightarrow{\times 2\cdot 2} 22\}$$

$$\downarrow \times 3\cdot 5 \quad \downarrow \times 3\cdot 5 \quad \downarrow \times 3\cdot 5 \quad \downarrow \times 3\cdot 5$$

$$Q = \{3\cdot 5 \xrightarrow{\times 4} 14 \xrightarrow{\times 2\cdot 5} 35 \xrightarrow{\times 2\cdot 2} 77\}$$

12 Looking for functions

1 The linear rule

(*a*) The diagram above reminds us that proportional sets have the following properties:

(i) corresponding pairs of numbers are related by a constant scale factor;
(ii) sets of multipliers are identical.

In this case the scale factor is 3·5 and

$$M_P = M_Q = \{4, 2\cdot 5, 2\cdot 2\}.$$

(*b*) Let us look at the ordered sets

$$A = \{1, 2\cdot 1, 4\cdot 2, 5\},$$
$$B = \{3, 6\cdot 3, 12\cdot 6, 15\}.$$

Are *A* and *B* proportional sets? Does the table of multipliers confirm your answer?

Members of *A*	Multipliers for *A*	Members of *B*	Multipliers for *B*
1·0		3·0	
	2·1		2·1
2·1		6·3	
	2·0		2·0
4·2		12·6	
	1·2		1·2
5·0		15·0	

Corresponding members of A and B are related by the rule $x \to 3x$. This function is shown in the arrow diagram [Figure 1(a)] and graph [Figure 1(b)].

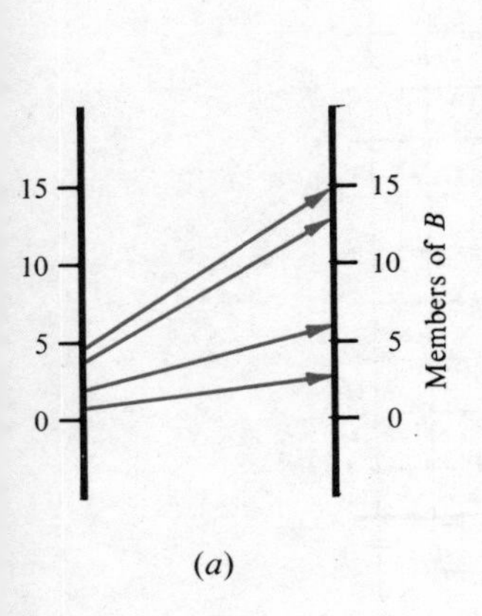

(a)

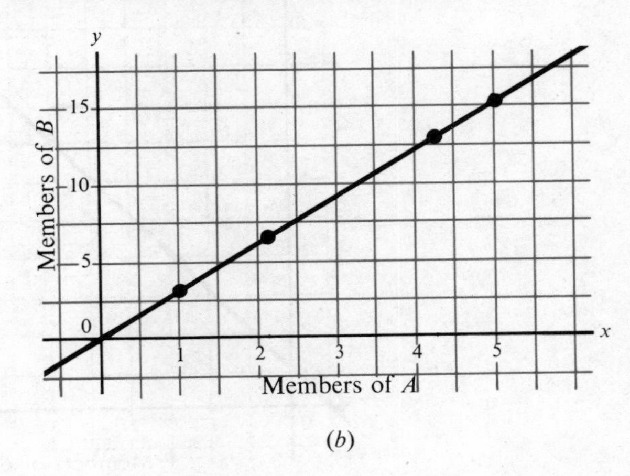

(b)

Fig. 1

The line through the four points in Figure 1(b) passes through the origin and has gradient 3. Do you agree that its equation is $y = 3x$? Notice that in this case sets A and B are proportional with scale factor 3.

(c) Let us now consider the ordered sets

$$C = \{1{\cdot}4, 3{\cdot}0, 3{\cdot}9, 4{\cdot}6\}$$
$$D = \{3{\cdot}8, 7{\cdot}0, 8{\cdot}8, 10{\cdot}2\}.$$

The table of multipliers for C and D shows that they are not proportional sets.

Members of C	Multipliers for C	Members of D	Multipliers for D
1·4		3·8	
	2·14		1·84
3·0		7·0	
	1·30		1·26
3·9		8·8	
	1·18		1·16
4·6		10·2	

But if we graph the members of C and D we obtain a straight line which does not pass through the origin (see Figure 2).

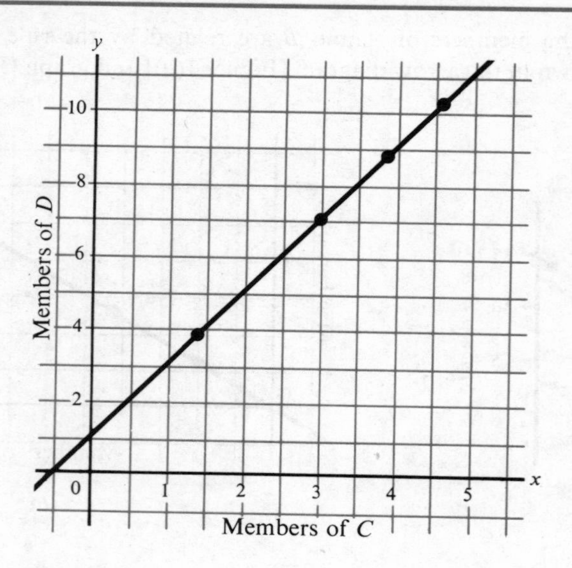

Fig. 2

What is the gradient of the line and where does it meet the y-axis? Do you agree that the equation of the line is

$$y = 2x + 1?$$

(*d*) Copy and complete the multipliers table for the sets E and F shown below.

Members of E	Multipliers for E	Members of F	Multipliers for F
0·8		1·6	
	1·75		
1·4		1·3	
2·0		1·0	
			0·4
3·2		0·4	

Are E and F proportional sets?
Corresponding members of E and F are graphed in Figure 3.

The graph is again a straight line which does not pass through (0,0). Check that its equation is

$$y = {}^{-}\tfrac{1}{2}x + 2.$$

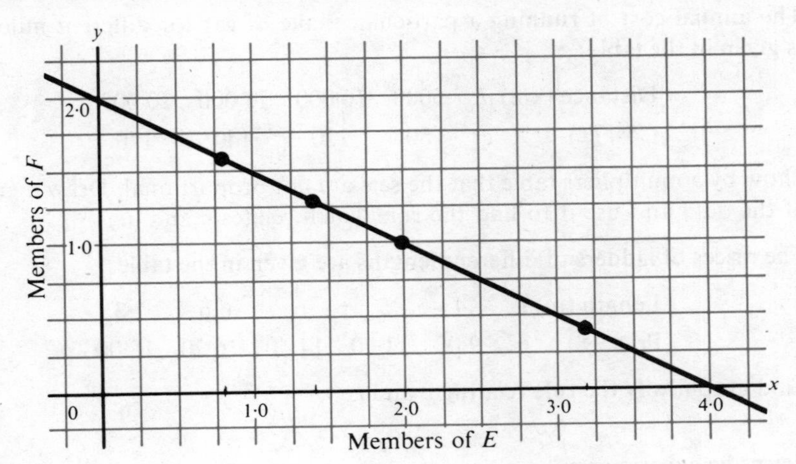

Fig. 3

(*e*) In sections (*b*), (*c*) and (*d*) we have found the following rules relating members of pairs of sets:

(i) $y = 3x$;
(ii) $y = 2x + 1$;
(iii) $y = {}^{-}\frac{1}{2}x + 2$.

These are all of the form $y = mx + c$ and we call them *linear* rules. Can you explain why this name is used? Rule (i) is a special case and describes the relation between proportional sets. When a rule is of the form $y = mx$ we sometimes write it as $y \propto x$.

Exercise A

1 R and S are proportional sets and $R = \{1, 2{\cdot}5, b\}$, $S = \{4, a, 15\}$.

(*a*) Write down the scale factor.
(*b*) Find the values of a and b.
(*c*) List the members of M_R and M_S.

2 This question refers to the sets C and D of Section 1(*c*).

(*a*) List the set D_1 whose members are each one less than the corresponding members of D.
(*b*) Complete the multipliers table for sets C and D_1.
(*c*) Are C and D_1 proportional sets?

3 The following table was produced in an experiment:

s:	1·2	1·4	1·6	1·8	2·0	2·2
f:	34	39·6	45·3	51	56·6	62·3

Make a table showing the multipliers for each set. What is the rule which relates f and s?

4 The annual cost of running a particular make of car for different mileages is given in the table:

Distance (km) d:	5000	10 000	15 000	20 000
Cost (£) c:	80	110	140	170

Show by a multipliers table that the sets are not proportional. Draw a graph of the data and use it to find the rule which relates c and d.

5 The prices of ladders of different lengths are given in the table:

Length (m) l:	4	5	6	6·6	7·3
Price (£) p:	9·00	11·70	14·40	16·20	18·00

Find graphically the rule relating p and l.

2 Scientific experiments

Many experiments which are performed in school science laboratories and elsewhere, involve taking readings of two quantities and comparing the sets of data in an attempt to find a simple relation between them. Here are some examples:

Length and time of swing of a pendulum.
Object and image distance of a lens.
Voltage and current in an electrical circuit.
Age and mass of rabbits.
Height and diameter of base of stem (of seedling).

Can you think of others?

In some cases the rule will be a linear one and confirmation of this will appear when the data is graphed. 'Real life' experiments however do not produce perfectly straight lines. Here is a table obtained by suspending different masses on a spring:

Mass (kg) x:	1	2	3	4	5
Length of spring (cm) y:	22·4	27·6	29·8	35·0	38·4

Graph the results and draw the straight line which you think is suggested by the five marked points. What is the gradient of your line and where does it meet the y axis? Write down the equation of your line. What length of spring do you think would result if a mass of 3·5 kg were suspended? Compare your answers with those obtained by others. Do you expect them to be the same?

Many experiments on the other hand will produce sets of results which are not even approximately related by a linear rule. Here is an example. The table shows the distance travelled by a freely falling object at 1-second intervals for the first 5 seconds.

Time of fall (seconds) t:	1	2	3	4	5
Distance travelled (metres) d:	5	20	45	80	125

The data is graphed in Figure 4 which shows that the rule is not a linear one. Can you suggest a name for the shape of the graph?

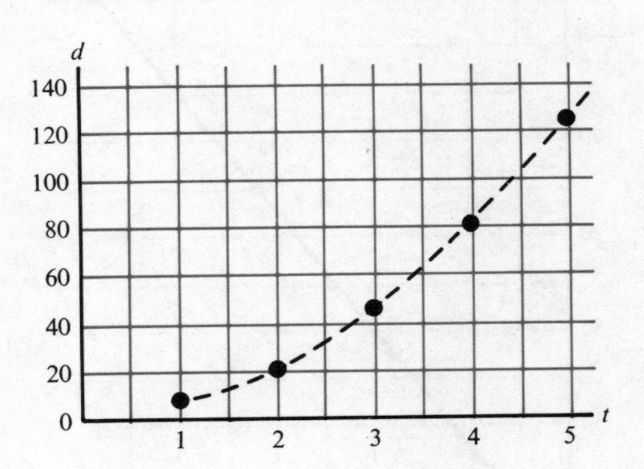

Fig. 4

We now look more closely at the two sets of data. In the multipliers table the squares of multipliers for t and the multipliers for d are identical showing that d is proportional to t^2.

t	Multipliers for t	Squares of multipliers for t	d	Multipliers for d
1			5	
	2·0	4·0		4·0
2			20	
	1·5	2·25		2·25
3			45	
	1·33	1·78		1·78
4			80	
	1·25	1·56		1·56
5			125	

Corresponding values of t^2 and d are:

t^2:	1	4	9	16	25
d:	5	20	45	80	125

What is the scale factor?

Do you agree that the rule is $d = 5t^2$ and that the curve in Figure 4 is a *parabola* whose equation is $d = 5t^2$?

Notice that when d is plotted against t^2 a straight line is obtained (see Figure 5). What is its gradient?

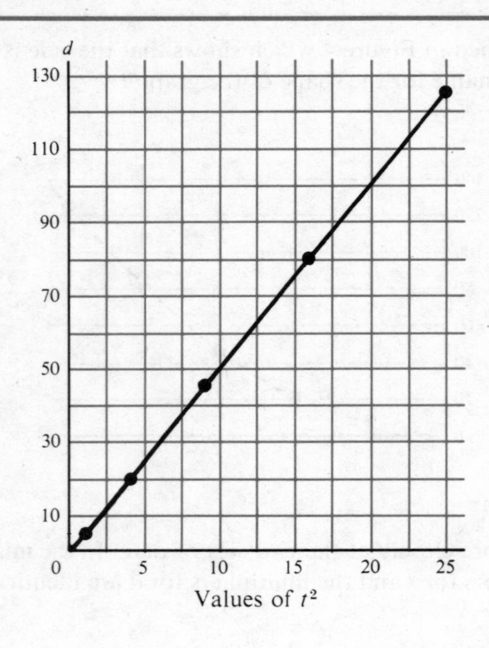

Fig. 5

Summary

1 If the function rule connecting two sets of numbers is of the form $y = ax^2$ it is called a square rule.

2 The graph of y against x^2 in this case is a straight line through the origin.

3 The value of a is measured by the gradient of this line.

4 The graph of $y = ax^2$ is a parabola.

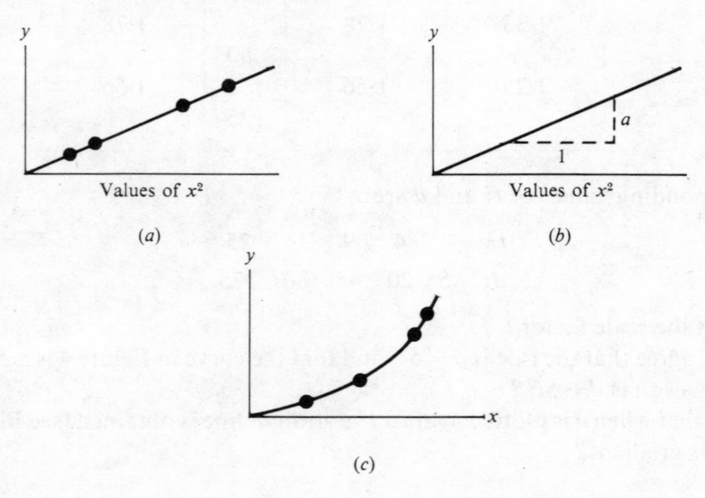

Fig. 6

Exercise B

1 Here are two corresponding sets of experimental data:

e:	4	7	10	13	16	19
c:	15	20	24	27·4	30·4	33

Show by a table of multipliers that the sets are approximately related by a square rule.

2 Figure 7 shows a curve which looks like part of a parabola. Take readings from the graph and complete the table:

x:	5	10	15	20	25	30
y:						

Make a table of the sets of multipliers for x and y and the squares of multipliers for x. Is the curve a parabola? What is its equation?

3 Figure 8 shows the graph of A against r^2 where A is the area of a circle and r is its radius (r is measured in cm and A in cm²). What should the gradient of the line be? Check your answer by measurement. Sketch the graph of A against r.

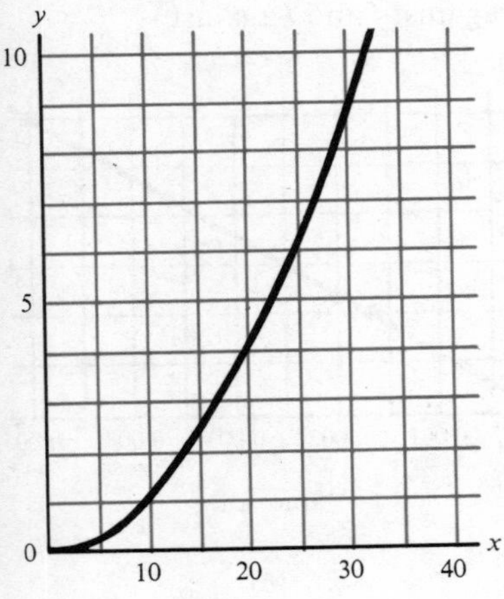

Fig. 7

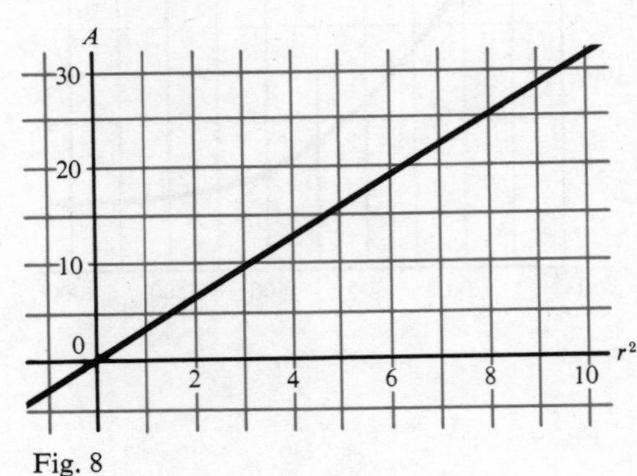

Fig. 8

3 The reciprocal rule

The dial of a transistor set is marked in metres and kilohertz. Here are some corresponding numbers:

Wave-length l:	300	500	600	1500
Frequency f:	1000	600	500	200

The multipliers table has two identical columns showing that l is proportional to $\frac{1}{f}$.

l	Multipliers for l	f	Multipliers for f	Reciprocals of multipliers for f
300		1000		
	1·67		0·60	1·67
500		600		
	1·20		0·83	1·20
600		500		
	2·50		0·4	2·50
1500		200		

Here is the table of values of l and $\frac{1}{f}$:

l: 300 500 600 1500

$\frac{1}{f}$: 0·001 0·0017 0·002 0·005

Figure 9 shows the graphs obtained by plotting l against f and l against $\frac{1}{f}$.

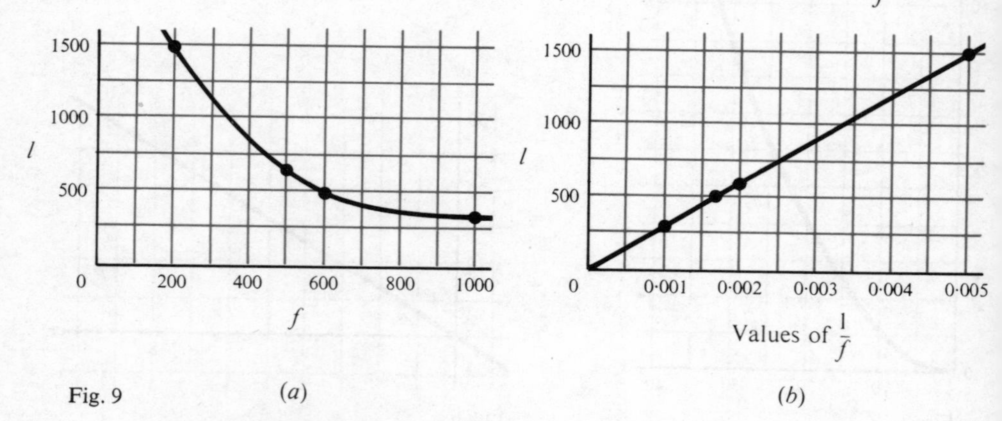

Fig. 9 (*a*) (*b*)

The graph in Figure 9(*a*) is called a *hyperbola*.
What is the gradient of the straight line in Figure 9(*b*)?
Do you agree that values of l and f are related by the rule $l = \frac{300\,000}{f}$?

Each value of l is increased by 100 to give the following table:

l_1: 400 600 700 1600

f: 1000 600 500 200

Check that l_1 and f satisfy the relation $l_1 = \frac{300\,000}{f} + 100$.

Summary

1 If the rule describing the relation between two sets of numbers has the form $y = \frac{a}{x}$, it is called a *reciprocal rule* and the numbers are said to be *inversely proportional*.

2 The graph of y against $\frac{1}{x}$ is a straight line through the origin with gradient a.

3 The graph of $y = \frac{a}{x}$ is a hyperbola.

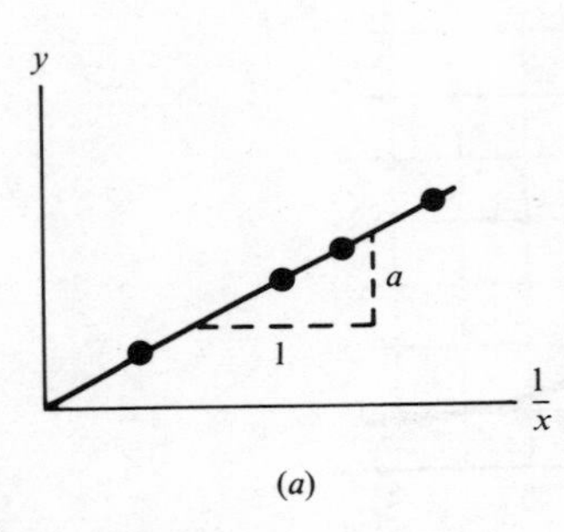

(*a*)

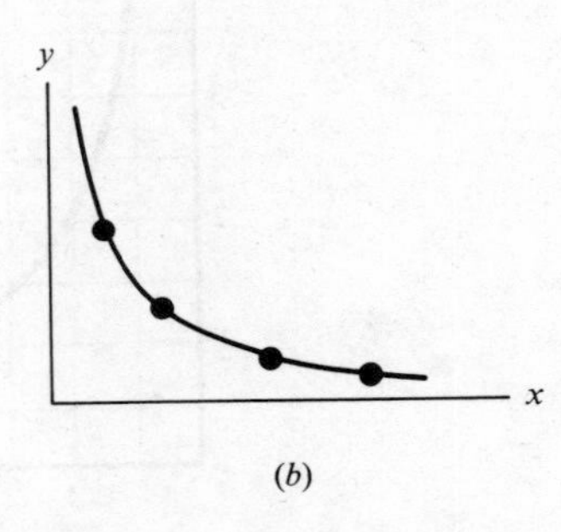

(*b*)

Fig. 10

Exercise C

1 In an experiment to investigate how the volume of a quantity of air varies under different pressures, the following table of results was obtained:

Pressure (N/cm²) p:	20	25	30	35	40	45
Volume (cm³) v:	158	125	105	90	79	70

Make a table showing the multipliers for p and v and the reciprocals of the multipliers for v. Calculate the corresponding values for $\frac{1}{v}$ and plot p against $\frac{1}{v}$ in a graph. Use your graph to find the rule relating p and v.

2 Figure 11 shows a curve which looks like part of a hyperbola. Make a table of six corresponding pairs of values of x and y. Calculate the sets of multipliers for x and y and also the reciprocals of the multipliers for x. Is the curve confirmed as a hyperbola? Find its equation by plotting a graph of values of y against $\frac{1}{x}$.

3 For photographic flash work, the table relates distance and aperture setting:

Distance, d:	1·00	1·40	2·00	2·75	4·00	5·50
Aperture, A:	22	16	11	8	5·6	4

Find an approximate rule relating d and A.

4 The following table of values is known to come from a reciprocal rule function.

x:	3	4	5	6	7
y:	105	87·5	63	52·5	45

One value of y is incorrect. Draw a suitable graph to find which it is and correct it. If $y = \frac{a}{x}$ what is the value of a?

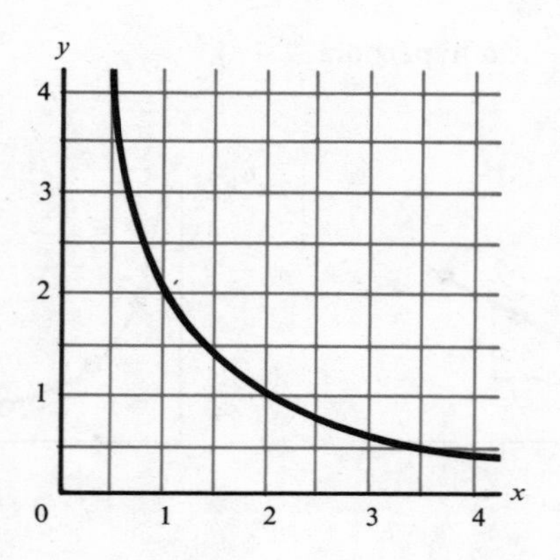

Fig. 11

4 Finding function rules in general

So far the ordered sets of data we have considered have been related by one of the following rules:

$$y = mx + c \text{ (linear)}$$

$$y = ax^2 \text{ (square)}$$

$$y = \frac{a}{x} \text{ (reciprocal)}$$

There are, of course, many other types of rule which relate experimental data. Finding the rule which applies in a particular case requires care both in computation and graph plotting. A calculating aid is almost a necessity.

Example 1

p:	0·32	0·68	1·28	2·12	3·20
t:	0·1	0·2	0·3	0·4	0·5

The graph of the data shown in Figure 12 suggests that the rule is not a linear one.

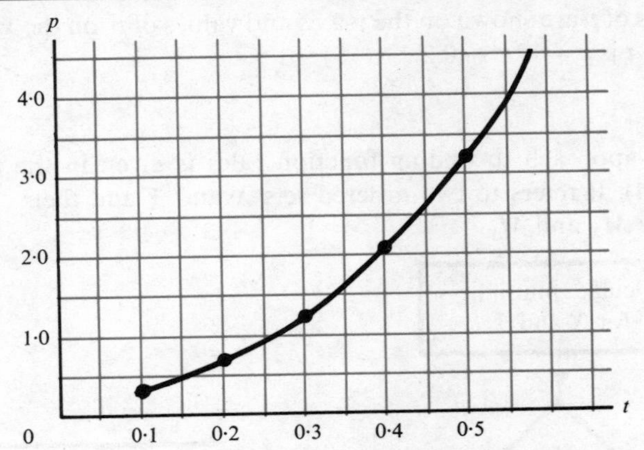

Fig. 12

Since plotting p against t does not produce a straight line, let us consider values of t^2 and $\frac{1}{t}$.

t^2:	0·01	0·04	0·09	0·16	0·25
$\frac{1}{t}$:	10·0	5·0	3·3	2·5	2·0

Figure 13 shows that a straight line graph is obtained when p is plotted against t^2. Do you think that plotting p against $\frac{1}{t}$ will produce a straight line? See if your prediction is correct by drawing a graph of p against $\frac{1}{t}$.

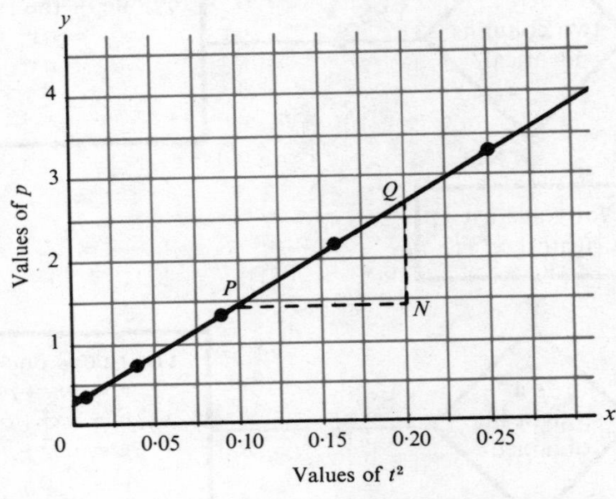

Fig. 13

Looking again at Figure 13 we see that the line crosses the y axis at $y = 0{\cdot}2$. By considering triangle PQN we obtain the value of the gradient as $\frac{2{\cdot}6 - 1{\cdot}4}{0{\cdot}20 - 0{\cdot}10}$ or 12. Do you agree that the equation of the graph in Figure 13 is

$$y = 12x + 0{\cdot}2?$$

Since values of p are shown on the y-axis and values of t^2 on the x-axis, the rule relating p and t is $p = 12t^2 + 0{\cdot}2$.

Summary

A systematic approach to finding function rules is given in the flow diagram (see Figure 14). It refers to two ordered sets X and Y and their corresponding multiplier sets M_X and M_Y.

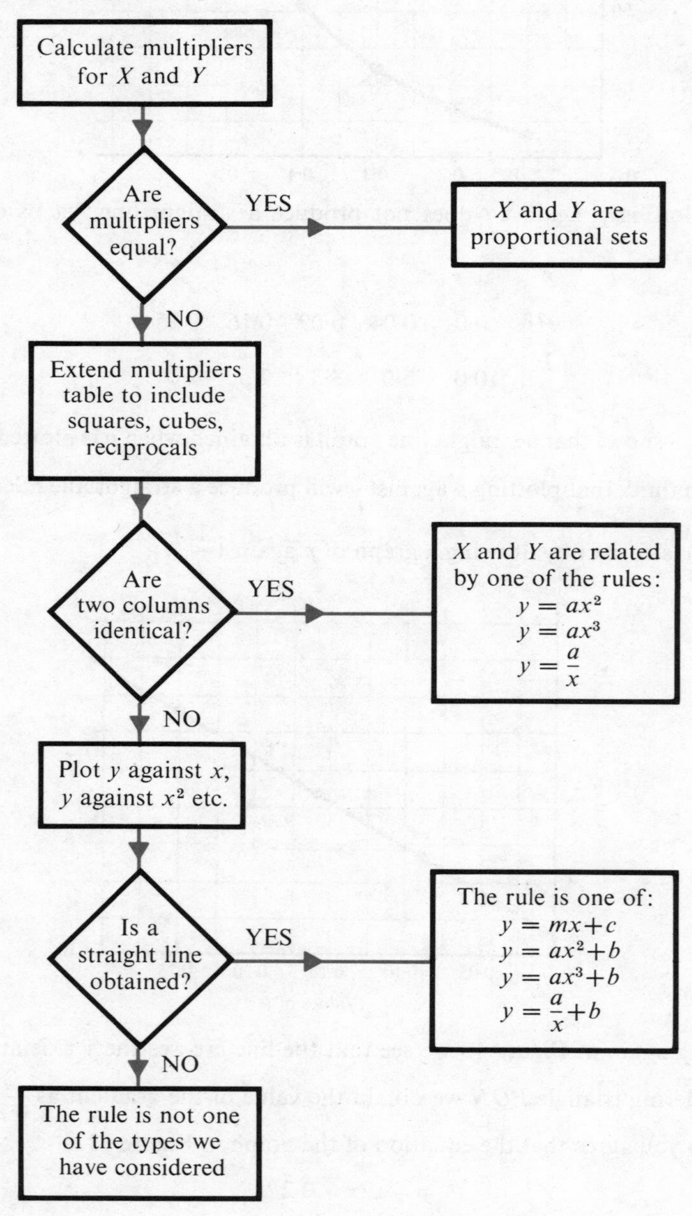

Fig. 14

Exercise D

1 The prices of fibre-glass paddling pools of different diameter are given in the table:

Diameter (m) d:	2	2·5	3	4	5
Cost (£) c:	12	13·8	16	21·6	28·8

The rule connecting c and d is of the form $c = ad^2 + b$ where a and b are constants. Draw a suitable graph to find the values of a and b.

2 Values of x and y are related by the rule $y = \frac{a}{x^2}$. Copy and complete the following table:

x	Multipliers for x	Reciprocals of squares of multipliers for x	y	Multipliers for y
2			5	
	?	?		?
?			1·25	
	?	?		?
10			?	

What is the value of a?

3 It is believed that a rule of the form $y = ax^2 + b$ connects values of x and y. The following results were obtained experimentally:

x:	0	1	2	3	4
y:	2·5	5·2	14·1	29·8	52·6

Find graphically, approximate values of a and b.

4 The following results are connected by a relation of the form $y = \frac{a}{x} + b$:

x:	1	2	4	7	10
y:	5·5	4·5	4	3·8	3·7

Plot values of $\frac{1}{x}$ against y and use your graph to find values of a and b.

5 Verify by a graph that the following figures satisfy approximately an equation of the form $y = ax^3 + b$, and find a and b as accurately as you can.

x:	4	5	6	7	8
y:	5·2	6·6	8·7	11·6	15·4

5 The growth function

The egg of a frog consists of a single cell. Once it has been fertilized, it divides into two. These two cells subdivide again giving four cells in all. If we assume that this process continues, the following sequence of numbers results:

$$1, 2, 4, 8, 16, 32 \ldots$$

If, further, we assume that the division of cells takes place at equal intervals of time, we obtain the following table:

Time	0	1	2	3	4	5
Number of cells	1	2	4	8	16	32

The function rule has a form which is quite different from any of the preceding examples in this chapter. Check that the rule is in fact

$$x \to 2^x.$$

If we call this function f, what is

$$f(6), f(10), f^{-1}(128), f^{-1}(2048)?$$

What value does the table suggest for 2^0?

The domain of the function f is the non-negative integers. It will prove worthwhile to extend it to include negative numbers. Copy and complete the following table:

$^-6$	$^-5$	$^-4$	$^-3$	$^-2$	$^-1$	0	1	2	3	4
					$\frac{1}{2}$	1	2	4	8	16

What are the values of 2^{-3}, 2^{-1}, 2^{-5}? What is the connection between 2^n and 2^{-n}?

The example we have considered is a growth function based on 2. What is the rule of the following table?

$^-4$	$^-3$	$^-2$	$^-1$	0	1	2	3	4
				1	5	25		625

Copy it and fill in the missing numbers.

Exercise E

1 If g is the growth function $g: x \to 4^x$, find:

(a) $g(2)$; (b) $g(3)$; (c) $g(5)$;
(d) $g^{-1}(4^7)$; (e) $g^{-1}(256)$.

2 Figure 15 is the graph of a growth function $h: x \to A^x$. Use the graph to find:

(a) $h(2)$; (b) $h(5)$; (c) $h^{-1}(2{\cdot}2)$; (d) $h^{-1}(4)$.

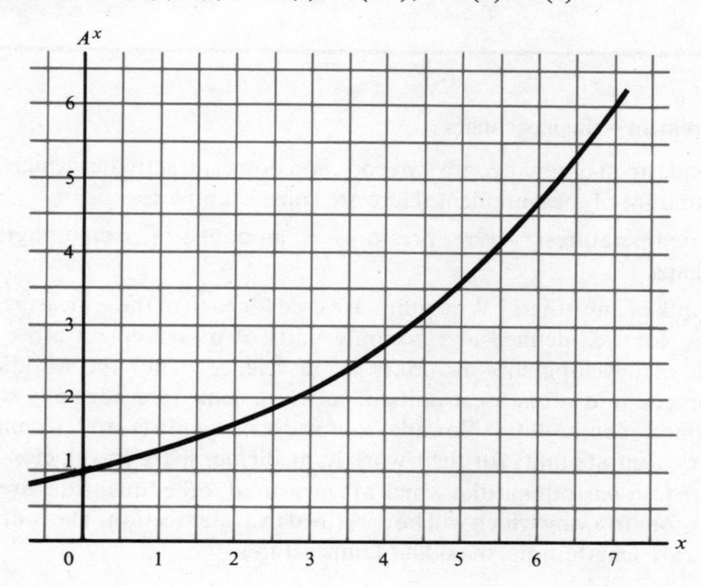

Fig. 15

What is the value of A?

3 The following table refers to the amounts of oxygen collected during a chemistry experiment:

Time (min) t:	1	2	3	4	5	6
Volume (ml) V:	20	24	29	35	42·5	51·5

Show this information in a graph.

For each interval of one minute, work out the fraction

$$\frac{\text{volume at end of interval}}{\text{volume at beginning of interval}}.$$

For example, the fraction for the first minute interval is $\frac{24}{20} = 1{\cdot}2$. Does the volume increase in a similar way to the frog cells in Section 5? Explain your answer carefully in your own words. Assuming the pattern of growth continues, estimate the volume at the end of the seventh minute.

4 f is the growth function $f: x \to 2^x$. Find the values of a, b, c, d, e, t in the following:

(a) $f(3) = f(2) \times f(a)$; (b) $f(5) = f(b) \times f(3)$;
(c) $f(2) = f(2) \times f(c)$; (d) $f(^-3) = f(^-2) \times f(d)$;
(e) $f(^-2) = f(e) \times f(1)$; (f) $f(t) = f(4) \times f(5)$.

If $f(p) = f(m) \times f(n)$, how do you think p is related to m and n?

Units and dimensions

(*a*) Measurement – the basic units

Cooking and car-maintenance are two common domestic activities which involve a certain amount of measurement. Here are some examples:

oven temperatures, tyre pressures, amounts of recipe ingredients, plug gaps.

Can you think of any others? What units are used for each of these measurements?

The need for well-defined and accurate units of measurement arose mainly as a result of developments in industry and science. Until the Middle Ages, fairly imprecise and often local units of measurement were normally sufficient for the typical community. Nowadays, however, scientists and technologists require very accurate units for their work in medicine, space travel, etc.

There are four *basic* quantities which are measured; other quantities are related to the basic four in a way which will be described in a later section. The four, shown in Table 1, are length, time, mass and temperature.

TABLE 1

Quantity	Standard metric unit	Comment
Length	Metre	
Time	Second	
Mass	Kilogramme	Mass is not the same as weight; the weight of an astronaut can vary but his mass in a sense remains constant.
Temperature	Degree Celsius	The word 'centigrade' is sometimes used in-instead of Celsius.

Investigation 1

Find out about the origins of the metric system and the methods of defining the basic units.

Investigation 2

Recently, more accurate methods of defining the metre and second have been found. What can you discover about these developments?

(*b*) Measurement – derived units

Apart from the basic four, many other quantities are measured – area, volume, speed for example. What others can you think of? The units for these quantities are *derived* from the basic units. Table 2 shows this in detail.

TABLE 2

Quantity	Unit	Symbol
Area	Square metre	m^2
Volume	Cubic metre	m^3
Speed	Metres per second	m/s or ms^{-1}

We can extend this idea and use the symbols M (mass), L (length), T (time) to show precisely how quantities such as area, volume, and speed are derived from the basic ones. For example, speed is a length divided by a time or $\frac{L}{T}$ for short.

We say that *speed has the dimensions* $\frac{L}{T}$. Do you agree with Table 3?

TABLE 3

Quantity	Unit	Symbol	Dimensions
Area	Square metre	m^2	L^2
Volume	Cubic metre	m^3	L^3
Speed	Metres per second	m/s or ms^{-1}	$\frac{L}{T}$ or LT^{-1}

Notice that the last two columns of Table 3 are closely related.

Investigation 3

Find out about the units and dimensions of acceleration, force, work, pressure.

Investigation 4

Find out all you can about the following electrical units:

volts, amperes, watts, ohms.

What do you know about the relationships between them?

(c) Checking the dimensions of a formula

Do you ever confuse $2\pi r$ and πr^2? Which is used to find the area of a circle? It is easy to decide using the idea of dimension. Since area has dimensions L^2, it must be $A = \pi r^2$ (π is a number and has no dimensions).

Here are two further examples of 'checking by dimensions'.

Example 1

A class has been asked to find a formula, with p as the subject, relating p, q, r, s, the lengths shown in Figure 1.
Two suggested answers are

(i) $p = \frac{rs}{q}$; (ii) $p = \frac{q}{rs}$.

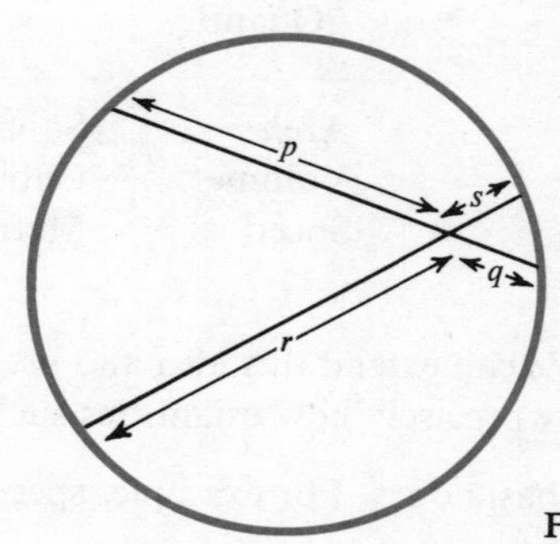

Fig. 1

$p = \frac{q}{rs}$ cannot be right because the quantities are all lengths and so $\frac{q}{rs}$ has dimensions $\frac{L}{L^2}$ or $\frac{1}{L}$ or L^{-1}. What can you say about the answer $p = \frac{rs}{q}$?

Example 2

The following two formulas are suggested for the area of the trapezium in Figure 2:

(i) $A = \frac{hab}{2}$; (ii) $A = \frac{h}{2}(a+b)$.

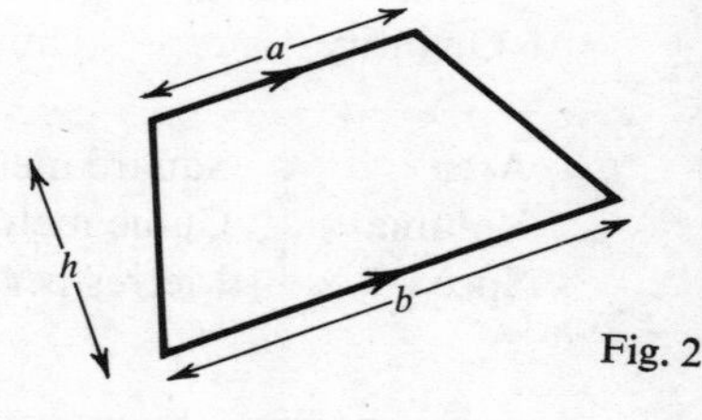

Fig. 2

(i) is obviously incorrect because $\frac{hab}{2}$ has dimensions $L \times L \times L$ or L^3 and so cannot be a measure of area.

(ii) could be correct: $a + b$ is a length and so is h. Hence $\frac{h}{2}(a+b)$ has dimensions L^2. Is $A = \frac{h}{2}(a+b)$ the correct formula for the area of a trapezium?

Exercise A

1 Look at the list of formulae for area and volume. Which ones are obviously wrong when you check the dimensions? Correct them and say which shapes they refer to.

(a) $A = \pi rl$; (b) $A = \pi r^2 h$; (c) $V = 2\pi r(r+h)$;
(d) $V = \frac{1}{3}\pi r^2 h$; (e) $A = 6a^2$.

2 The density of a substance is found by dividing its mass by its volume. Suggest a suitable unit for density. What are its dimensions?

3 Here are some answers given when a class was asked to find a formula for H in terms of the other lengths (see Figure 3).

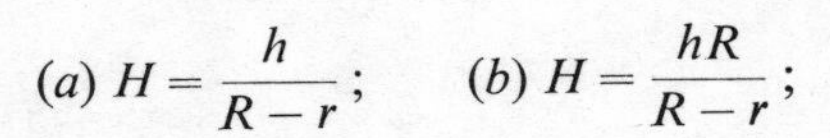

(a) $H = \dfrac{h}{R-r}$; (b) $H = \dfrac{hR}{R-r}$;

(c) $H = \dfrac{h-r}{R-r}$; (d) $H = \dfrac{h}{Rr}$.

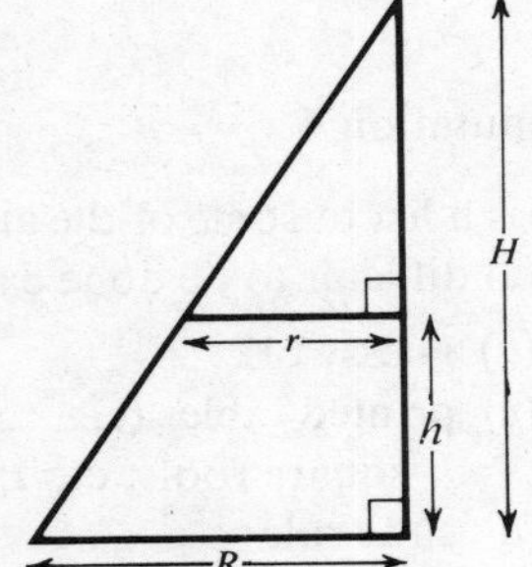

Fig. 3

Check the dimensions of each formula. Which could be correct?

4 The momentum of a moving object can be obtained by multiplying its mass by its velocity. State a suitable unit for momentum and find its dimensions.

5 The quantity 71 g/m^2 is printed on a box of typing paper. What information does it give about the paper? What are the dimensions of this quantity? Can you suggest a name for it?

6 An inflated rubber bathing ring has the measurements r cm and R cm shown in Figure 4. State the dimensions of each of the following quantities and say which could be a formula for the volume of the ring.

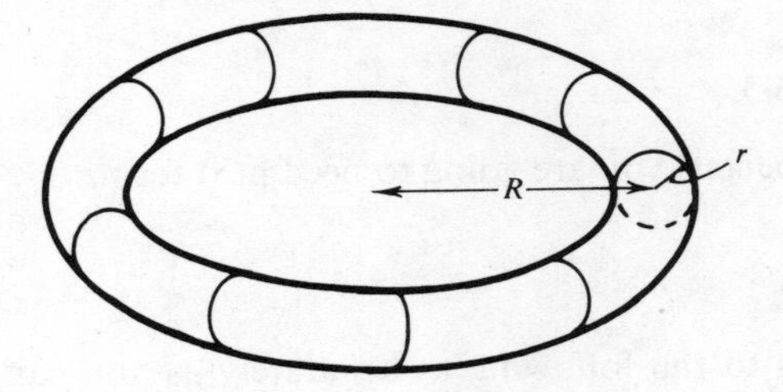

Fig. 4

(a) $\pi r R$; (b) $2\pi^2 r^2 R$; (c) $2\pi \dfrac{r}{R}$; (d) $\dfrac{\pi r}{R^2}$.

7 If A and B are areas and h is a length, what are the dimensions of the quantity

$$\frac{h}{3}\left(A + \sqrt{(AB)} + B\right)?$$

Can you suggest the shape associated with this formula?

8 The Ångström unit and the parsec are units of length used in atomic theory and astronomy respectively. Find out how each is defined.

Revision exercises

Computation 4

Here is a list of some of the aids which we employ to do computations when they are too difficult to be done exactly in our heads:

(*a*) guesswork; (*b*) paper and pen;
(*c*) printed tables (e.g. ready reckoner, square roots, etc.);
(*d*) slide rule; (*e*) logarithms;
(*f*) desk calculating machines; (*g*) electronic computers.

Say which aid or aids would be most appropriate if you were computing:

1 67×14p.

2 354×21.

3 The value of π to 1000 decimal places.

4 How long it would take you to drive from London to Edinburgh.

5 The density of copper, in a physics experiment.

6 7586^3.

7 The instant at which a moon rocket would reach its destination.

8 $\sqrt{496}$.

9 $2{\cdot}1 \times 3{\cdot}7 \times 5{\cdot}5$.

10 How many pencils you are going to need next term.

Computation 5

Find the answers to the following as accurately as you can. You are allowed complete freedom of method.

1 2345×3456. 2 $56{\cdot}7^2$. 3 $0{\cdot}065 \times 67$.
4 $0{\cdot}873^2$. 5 888×9898. 6 $66\,990 \div 154$.
7 $1\,917\,747 \div 333$. 8 $67 \div 23$. 9 $354 \div 9{\cdot}7$.
10 $\dfrac{231 \times 441}{21}$.

Exercise K

Calculate the following:

1 $\frac{64}{10} - \frac{32}{5}$.
2 $\frac{1}{5} + \frac{1}{15}$.
3 $\frac{1}{7} - \frac{4}{49}$.
4 $\frac{1}{13} + \frac{1}{2} + \frac{2}{3}$.
5 $(\frac{8}{9} \times \frac{3}{24}) + \frac{1}{9}$.
6 $3\frac{1}{2} \times 5\frac{1}{3}$.
7 $\frac{2}{7} \div \frac{3}{5}$.
8 $5\frac{1}{3} - 3\frac{5}{8}$.
9 $\dfrac{2\frac{1}{2} + 3\frac{3}{4}}{1\frac{1}{3} - \frac{7}{8}}$.

10 $\dfrac{\frac{3}{8} \times \frac{2}{15}}{\frac{5}{8} \times \frac{7}{15}}$.

Exercise L

State the letter(s) corresponding to the correct answer(s).

1 Figure 1 shows two cylinders both full of oil. The smaller one contains 7 kg of oil. What is the mass of the oil in the larger cylinder?

(*a*) $10\frac{1}{2}$ kg; (*b*) 21 kg; (*c*) 28 kg; (*d*) 42 kg.

The ratio of the areas of the curved surfaces of the cylinders is:

(*a*) 2:3; (*b*) 1:2; (*c*) 1:3; (*d*) 1:6.

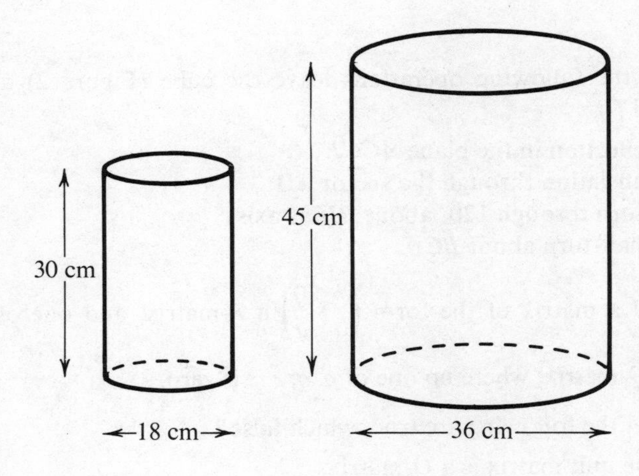

Fig. 1

2 $\sqrt{3} \approx 1{\cdot}73$. Which of the following square roots have the same significant figures as $\sqrt{3}$ (that is, differ only in the position of the decimal point)?

(*a*) $\sqrt{30}$; (*b*) $\sqrt{3000}$; (*c*) $\sqrt{0{\cdot}0003}$; (*d*) $\sqrt{300}$; (*e*) $\sqrt{0{\cdot}3}$.

3 The inverse of the matrix $\begin{pmatrix} 3 & 2 \\ 5 & 4 \end{pmatrix}$ is:

(*a*) $\begin{pmatrix} 2 & ^-1 \\ ^-2\frac{1}{2} & 1\frac{1}{2} \end{pmatrix}$; (*b*) $\begin{pmatrix} ^-3 & 5 \\ 2 & ^-4 \end{pmatrix}$; (*c*) $\begin{pmatrix} 4 & ^-2 \\ ^-5 & 3 \end{pmatrix}$; (*d*) none of these.

4 Figure 2 shows a cube. There is a line in the plane $EFGH$ which:

(*a*) meets AB;
(*b*) is parallel to AB;
(*c*) is perpendicular to AB;
(*d*) is none of these.

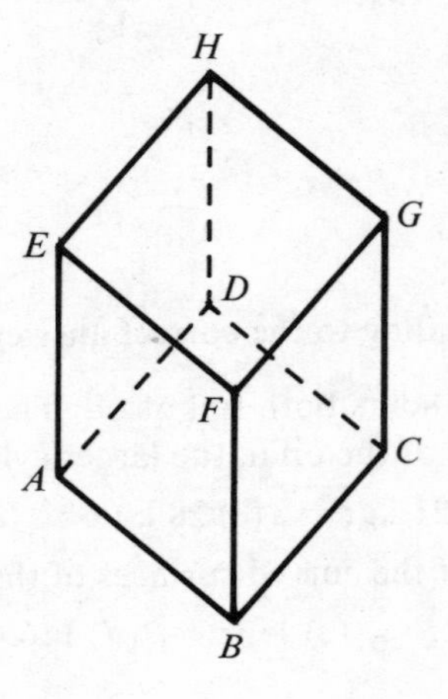

Fig. 2

5 Which of the following operations leave the cube (Figure 2) as a whole unchanged?

(*a*) Reflection in the plane $ACGE$;
(*b*) translation through the vector **AB**;
(*c*) a turn through 120° about BH as axis;
(*d*) a half-turn about BC.

6 Let us call a matrix of the form $\begin{pmatrix} p & 0 \\ 0 & q \end{pmatrix}$ a P-matrix, and one of the form $\begin{pmatrix} 0 & r \\ s & 0 \end{pmatrix}$ a Q-matrix, where no one of p, q, r, s is zero.

Which of the following are true, which false?

(*a*) The unit matrix is a Q-matrix.
(*b*) The square of a P-matrix is a Q-matrix.
(*c*) The product of a P-matrix and a Q-matrix can never be the zero matrix.
(*d*) The multiplicative inverse of a Q-matrix is a P-matrix.
(*e*) The additive inverse of a Q-matrix is a Q-matrix.

True: *a b c d e*
False: *a b c d e*

7 Let $\mathbf{P} = \begin{pmatrix} ^-1 & 0 \\ 0 & 1 \end{pmatrix}$ and $\mathbf{Q} = \begin{pmatrix} 0 & ^-1 \\ 1 & 0 \end{pmatrix}$. Which of the following are true, which false?

(*a*) $\mathbf{P}^2 = \mathbf{I}$, where $\mathbf{I}$ is the unit matrix.
(*b*) $\mathbf{PQ} = \mathbf{QP}$.
(*c*) $|\mathbf{P}| = |\mathbf{Q}|$.
(*d*) $\mathbf{Q}$ represents reflection in $x = y$.
(*e*) $\mathbf{Q}$ is its own multiplicative inverse.

True: *a b c d e*
False: *a b c d e*

8 You are given the following table showing values of x and y:

x:	5	15
y:	20	180

With which of the following statements are these data consistent?

(*a*) $y \propto x$;
(*b*) $y \propto x^2$;
(*c*) y is not proportional to any power of x;
(*d*) $y = 16x - k$, for some constant value of k.

Exercise M

State the letter(s) corresponding to the correct answer(s).

1 Of 16 students, 12 take economics, 8 take philosophy and all take at least one of these subjects. What is the probability that two students, chosen at random, take economics only?

(*a*) $\frac{1}{8}$; (*b*) $\frac{7}{30}$; (*c*) $\frac{1}{4}$; (*d*) $\frac{7}{16}$; (*e*) $\frac{1}{2}$.

2 If O stands for any odd number and E for any even number, then the matrix product $\begin{pmatrix} O & E \\ E & O \end{pmatrix}\begin{pmatrix} E & O \\ O & E \end{pmatrix}$ is:

(*a*) $\begin{pmatrix} E & E \\ E & E \end{pmatrix}$; (*b*) $\begin{pmatrix} O & O \\ E & E \end{pmatrix}$; (*c*) $\begin{pmatrix} E & E \\ O & O \end{pmatrix}$; (*d*) $\begin{pmatrix} O & E \\ E & O \end{pmatrix}$; (*e*) $\begin{pmatrix} E & O \\ O & E \end{pmatrix}$.

3 The value of $\sqrt{0{\cdot}009}$ is:

(*a*) 0·3; (*b*) 0·03; (*c*) 0·095 approx.; (*d*) none of these.

4 A price is increased by 20% to £96. The original price was:

(*a*) £76; (*b*) £116; (*c*) £80; (*d*) none of these.

5 Which of these is a fraction lying between $\frac{4}{5}$ and $\frac{5}{6}$?

(*a*) $\frac{9}{13}$; (*b*) $\frac{3}{4}$; (*c*) $\frac{6}{7}$; (*d*) none of these.

6 $3x + 5 > 11 - x$ implies:

(*a*) $x > 3$; (*b*) $x > 4$; (*c*) $x < 1\frac{1}{2}$; (*d*) none of these.

7 The gradient of the line $3x - 4y = 6$ is:

(a) $\frac{3}{4}$; (b) $\frac{4}{3}$; (c) 6; (d) none of these.

8 Two transformations have matrices

$$\mathbf{A} = \begin{pmatrix} ^-1 & 0 \\ 0 & 1 \end{pmatrix} \quad \text{and} \quad \mathbf{B} = \begin{pmatrix} 0 & ^-1 \\ 1 & 0 \end{pmatrix}.$$

Then:

(a) $\mathbf{A}^2 = \mathbf{I}$ (the identity matrix);
(b) $\mathbf{AB} = \mathbf{BA}$;
(c) $\mathbf{B}$ is the matrix for reflection in $x = y$;
(d) $\mathbf{B}$ is its own inverse.

Exercise N

1 Find a and b if $\begin{pmatrix} ^-2 & a \\ 4 & 1 \end{pmatrix}\begin{pmatrix} 3 \\ b \end{pmatrix} = \begin{pmatrix} 0 \\ 4 \end{pmatrix}$.

2 Write down any pair of positive integers x and y which satisfy *both* the inequalities:

$$x + 2y > 20; \qquad 2x - y > 30.$$

3 Solve the equations

$$2x + 3y = 2$$
$$^-3y = 4.$$

4 Find the equation of the straight line joining $A(0,3)$ and $B(2,7)$.
The line AB meets the line $x + 2y = 9$ at C. Calculate the coordinates of C.

5 $ABCD$ is a parallelogram. A is the point $(2,2)$ and B is the point $(3,4)$. The diagonals AC, BD meet at $(4,3)$. What are the coordinates of C and D?

6 Show on a sketch the set of points

$$\{(x,y): x(y - 2) = 0\}.$$

7 A certain kind of biscuit is made from a cylindrical disc of dough, whose radius is 3 cm and thickness $\frac{1}{2}$ cm.
Find how many biscuits you could make out of a lump of dough of volume 300 cm³.

8 Calculate the volume of the solid in Figure 3.

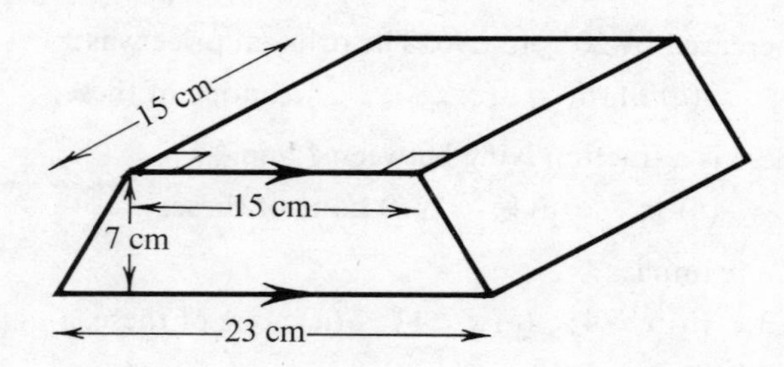

Fig. 3

Exercise O

1 If $4x + 5y > 8$ and $3x - 5y > 6$, deduce an inequality involving x but not y.

2 Find x and y if $\begin{pmatrix} 3 & 2 \\ -4 & 1 \end{pmatrix}\begin{pmatrix} x \\ y \end{pmatrix} = \begin{pmatrix} 6 \\ 14 \end{pmatrix}$.

3 Find x and y if $(x, y)\begin{pmatrix} 2 & 1 \\ -5 & 3 \end{pmatrix} = (20, 21)$.

4 Find x and y if $\begin{pmatrix} 1 \\ 2 \end{pmatrix} = \begin{pmatrix} 2 & -5 \\ -1 & 3 \end{pmatrix}\begin{pmatrix} x \\ y \end{pmatrix}$.

5 If $\mathbf{A} = \begin{pmatrix} 3 & 2 \\ 4 & 3 \end{pmatrix}$ and $\mathbf{B} = \begin{pmatrix} 7 & 6 \\ -2 & 5 \end{pmatrix}$, write down $\mathbf{A}^{-1}$ and hence find the 2×2 matrix $\mathbf{C}$ such that $\mathbf{AC} = \mathbf{B}$.

Find also the 2×2 matrix $\mathbf{D}$ such that $\mathbf{DA} = \mathbf{B}$.

6 Let $\mathbf{X} = \begin{pmatrix} 2 & 0 \\ 3 & 1 \end{pmatrix}$ and $\mathbf{Y} = \begin{pmatrix} 3 & 1 \\ 4 & 2 \end{pmatrix}$.

(*a*) If $\mathbf{a} = \begin{pmatrix} 2 \\ -5 \end{pmatrix}$ and $\mathbf{b} = \begin{pmatrix} 3\frac{1}{2} \\ -6\frac{1}{2} \end{pmatrix}$, show that $\mathbf{Yb} = \mathbf{Xa}$.

(*b*) Write down $\mathbf{Y}^{-1}$.

(*c*) Use your answers to (*a*) and (*b*) to help you find $\mathbf{P}$ such that $\mathbf{b} = \mathbf{Pa}$.

(*d*) Find $\mathbf{Q}$ such that $\mathbf{Xb} = \mathbf{Qa}$.

7 P is the point $(1, 0)$, Q is $(4, 6)$ and R is $(5, 2)$.

Triangle PQR is enlarged with centre of enlargement $(4, 0)$ and scale factor 3. What are the coordinates of P', Q', R'?

8 a and b are the binary numbers 10001·111 and 101·1. Find:

(i) $a + b$; (ii) $a - b$; (iii) $a \div b$ to 1 binary place.

Index

Index

Published by the Syndics of the Cambridge University Press
Bentley House, 200 Euston Road, London NW1 2DB
American Branch: 32 East 57th Street, New York, N.Y.10022

ISBNS: 0 521 20078 4 hard covers
0 521 08621 3 limp covers

First published 1973
Reprinted 1974 1975

Printed in Great Britain by William Clowes & Sons Ltd., London, Colchester and Beccles